TRANSDUCERS
and Their
ELEMENTS

Design and Application

TRANSDUCERS
and Their
ELEMENTS

Design and Application

Alexander D. Khazan

University of Massachusetts at Lowell

PTR Prentice Hall
Englewood Cliffs, NJ 07632

Library of Congress Cataloging-in-Publication Data

Khazan, Alexander D.
    Transducers and their elements : design and application /
Alexander D. Khazan.
        p.   cm.
    Includes bibliographical references and index
    ISBN 0-13-929480-5
    1. Transducers—Design and construction.   2. Detectors.
3. Measuring instruments.   I. Title.
TK7872. T6K47   1993
681'.2—dc20   93-22873
                                                              CIP

Editorial/production supervision: *The Wheetley Company, Inc.*
Buyers: *Mary Elizabeth McCartney and Alexis Heydt*
Cover design: *Karen Marsilio*

© 1994 by PTR Prentice Hall
Prentice-Hall, Inc.
A Paramount Communications Company
Englewood Cliffs, NJ 07632

The publisher offers discounts on this book when ordered
in bulk quantities. For more information, contact:

> Corporate Sales Department
> PTR Prentice Hall
> 113 Sylvan Avenue
> Englewood Cliffs, NJ 07632
>
> Phone: 201-592-2863
> Fax: 201-592-2249

Printed in the United States of America

10 9 8 7 6 5 4 3 2 1

ISBN 0-13-929480-5

ISBN 0-13-929480-5

90000

9 780139 294808

Prentice-Hall International (UK) Limited, *London*
Prentice-Hall of Australia Pty. Limited, *Sydney*
Prentice-Hall Canada Inc., *Toronto*
Prentice-Hall Hispanoamericana, S.A., *Mexico*
Prentice-Hall of India Private Limited, *New Delhi*
Prentice-Hall of Japan, Inc., *Tokyo*
Simon & Schuster Asia Pte. Ltd., *Singapore*
Editora Prentice-Hall do Brasil, Ltda., *Rio de Janeiro*

Contents

Preface

This book is an attempt to present the theory of transducers and their elements in an organized and elucidative manner. It is designed primarily as a text for courses at the graduate level and as a reference book for professionals who work with instrumentation in industry and in science. The book can also be useful for teaching one- or two-semester introductory or advanced courses at the undergraduate level. Those who deal with transducer design or application will find in the text many elements of a handbook with practical formulas, calculations, schemes, and so on. The author has summarized his long-term experience in research on transducers and in teaching the subject. It is hoped that the book will be helpful not only in the creation of new teaching programs but also in the improvement of existing programs on measurements, instruments, and transducers, particularly.

GOALS

The objective of this book is to satisfy the following goals:

1. Reviewing traditional as well as modern devices with an emphasis on their principles, methods of calculation, and application.
2. Supplying material for a graduate course.
3. Presenting material for advanced undergraduate courses.
4. Providing self-teaching and reference material for practicing engineers and scientists.

AUDIENCE

The comprehensive scope of the text and the specifics of the subject make the book helpful to students of electrical and mechanical engineering, industrial technology, physics, and chemistry. Engineers and scientists who are new to the field of transducers or who wish to improve or update their knowledge should also find this book useful.

This text can easily serve a variety of audiences through the selection of appropriate topics. Many topics are independent in content and can be easily combined.

PREREQUISITES

The physics and mathematics courses taught at the undergraduate level prepare the reader sufficiently for comprehension of the major part of the text. The material presented in this text can be used most readily if the reader has taken courses on circuit theory and electronics.

Some of the topics for graduates require knowledge in advanced mathematics and physics. These topics can be easily recognized in the text.

Flexibility in selecting the topics facilitates course planning for students with varying backgrounds.

CONTENTS

The book is divided into several parts. In the beginning of the book (Chapter 1), the reader becomes acquainted with some basics common to all transducers and their elements. This part contains introductory material essential for understanding the major topics of the book.

Chapters 2 to 6 constitute the core of the text and should form an integral part of the study. These chapters are divided into a number of sections on sensing and transduction elements. The chapters are arranged in a logical manner and cover the most common, widely used devices in instruments, excluding diminutive solid-state devices, which are described later. The calculations accompanying the descriptions of the elements are diversified in order to allow the reader to become familiar with the variety of problems inherent in transducer analysis.

Chapters 7 to 11 deal with solid-state sensors. Special attention is paid to microelectronic sensors and to the methods of their fabrication. The beginning of this part contains a brief review of some general aspects of the electrical and mechanical properties of semiconductors, since an understanding of these properties facilitates the comprehension of the material related to most solid-state sensors.

Two appendixes conclude the book. Appendix 1 presents transducer schemes organized in a table. In this appendix the emphasis is on the presentation of physical concepts and application rather than on the details of the construction and hardware.

Appendix 2 is devoted to tensors, which have become useful tools for the analysis of various solid-state sensors.

ORGANIZATION

In the core chapters, the emphasis on fundamental concepts and principles is intended to provide an understanding of structures designs rather than to expose the student to a wide variety of commercial components.

Sections on elements follow a traditional format, presenting first a physical description of the devices, then mathematical modeling and analysis, and, last, some features. In addition, whenever it is instructive, more complicated topics and mathematical models are developed, primarily for a graduate program.

Because the material is arranged in a modular structure, the instructor can easily form the curricula for courses of different content and complexity.

COURSE FORMATS

This book contains enough material for junior, senior, and graduate level courses. Through the suitable selection of topics and sequencing of their presentation, the book can be effectively used for a one- or two-semester course.

One of the features of this book is that it conveniently allows a simple structuring of different courses. The various topics are covered in such a way that a wide range of choices may be made in accordance with the interests of instructors.

Experience in teaching courses on transducers has shown that certain areas should be emphasized, others assigned as collateral reading, and still others deleted. The lectures on theory are easily combined with laboratory work and/or design projects.

NOTATIONS AND UNITS

The International System of Units (SI) is used throughout the book. However, some commonly used units for physical parameters are expressed in metric units in addition to SI. Powers of ten are, as a rule, multiples of ± 3.

REFERENCES

Books and articles published in several leading industrial countries have been used as references. A large part of the material introduced in the book constitutes the original topics on theory generated by the author.

ACKNOWLEDGMENTS

I am grateful to a number of my colleagues and workers in my field who have shared their knowledge with me, published the valuable material used in this book, and supported me throughout my career and during the writing of this book.

I thank the people of the publishing company for their valuable contribution to the work on this text.

I am thankful to Mrs. Sheila Kirschbaum for her assistance during the final stage of the work.

My appreciation is also extended to my wife Elena and my daughter Katherine, who have helped me greatly in the work on the manuscript.

Alexander D. Khazan
Amherst, New Hampshire

Introduction

An instrumentation transducer is a sensing device that provides a usable electrical output to a specific measurand. The term measurand stands for the quantity, property, or condition that is to be measured; for example, the pressure in a manifold, the roughness of a surface, or the number of parts on a conveyer.

The main function of the transducer is the conversion of the measurand into a signal that can be conveniently treated in a measurement or control system. In a measurement system, the value of some quantity is determined in terms of a standard unit. The output of a control system is forced to change or to stay in a predetermined manner as time progresses.

Transducing devices with pneumatic and hydraulic outputs are also termed transducers. However, electrical signal conversion techniques predominate in transducers because electrical methods are universal and provide a convenient interconnection between various components in measuring and control systems. Consequently, only transducers with electrical outputs are discussed in this book.

Traditionally, transducers are called by a variety of names: for instance, transmitters, in the process industry; sensors, in measurements of temperature, heat flux, and vacuums; detectors, in optical measurements; and load cells, in measurements of force. In some cases, they are also termed gages, pickups, or probes, or are given names reflecting the specific measurement: accelerometer, flowmeter, thermometer, and so on. Yet sensors or sensing devices should be more precisely defined. At present, they are usually regarded as devices that provide an instrumentation system with input signals. A transducer is a sensor and, generally, a subcategory of sensing devices or sensors.

Transducers are an integral part of our everyday life. We use them to monitor and measure everything from heartbeats to earthquakes. There is a demand for transducers in robotics, energy conservation, environmental control, automotive and consumer transport, alarms, agriculture, biomedicine, and health care, as well as in military, aircraft, and spacecraft equipment. Transducers are used in all scientific fields. If machines can be modeled after our muscles, and computers after our mind, transducers are the analogues of our senses.

New principles of transducers and their applications are constantly being discovered. Progress in the development of transducers is defined by achievements in electronics, physics, and chemistry. It is noteworthy that success in these sciences is defined in turn by advances in transducers because of their wide application in modern research. The rapid development of analog and digital electronics increases the sphere of transducer applications.

Transducers are part of a broad field of instruments, and they are among the most diversified electronic devices in the instrumentation industry. There are thousands of transducer types, and there are numerous places in the world where transducers are developed and produced. Within this moderate-length book, only a limited number of typical transducing devices can be described.

Transducers remain one of the developing fields of the high-tech industry. The level of growth of the transducer industry is significantly higher than the rate of economic growth forecast for most Western nations. There are several trends in contemporary transducer products. Mass-produced sensors are becoming smaller and less expensive because of accomplishments in precision electronic technology. Transducers compatible with digital signal conditioners are becoming more popular due to their high accuracy and flexibility in processing signals in digital systems. Many transducers are now loaded with amplifying and signal modifying elements that are placed inside the transducer's housing and sometimes are integrated with the sensing and transduction elements. This arrangement improves the transducers' characteristics and unifies their outputs. Further, optical sensors, employing fiber-optical systems, are finding wider applications in today's industry.

Transducer elements are the "bricks" of which any transducer is built. A sensing element is the part of a transducer that responds directly to the measurand. A transduction element is the electrical portion of a transducer from which the transducer output originates. These two elements are the most essential transducer components. A number of these devices will be described and discussed in this book.

chapter 1

Transducer Characteristics and Structures

For many systems, a transducer is a source of information. The operation of the transducer greatly defines the reliability of the information. In spite of a wide variety of different systems containing transducers, they can be divided into two big groups: measuring systems and control systems. These systems are important in our lives because the machines, processes, equipment, and systems of industrial, commercial, and military applications are associated with conditions that are continuously changing and so need to be monitored.

A transducer, like most measuring devices, is described by a number of characteristics and distinctive features. Among them are

1. the quantity to be measured by the transducer (measurand),
2. the operating principle describing the conversion of the measurand into an electrical signal,
3. design characteristics,
4. performance characteristics, and
5. reliability characteristics.

As a rule, the performance of a transducer element can be described by the same characteristics.

We will begin the discussion by considering measurands, which are quantities introduced at a transducer's input.

MEASURANDS

The International System of Units (SI) is used throughout the following chapters. Table 1.1 gives the base units of this system. The sections describing the elements' theory give various derived units.

When necessary, some special units for physical parameters are expressed in metric rather than SI units.

As a rule, powers of 10 are expressed in multiples of ± 3 in scientific notations.

TABLE 1.1

Base Units of the International System

Quantity	Name of unit	Unit symbol
length	meter	m
mass	kilogram	kg
time	second	s
electric current	ampere	A
temperature	kelvin	K
luminous intensity	candela	cd
amount of substance	mole	mol

The most common physical measurands and relevant units given in the SI system are listed on the following pages.

Linear and Angular Position, Motion, Length, and Thickness: Linear displacement is measured in meters (m) or in multiples and submultiples of the meter, which are the kilometer (km), centimeter (cm), millimeter (mm), and micrometer (μm, called also *micron*). An angular displacement is measured in radians (rad) or degrees (°). One radian is equal to 57.3 degrees. The minute of an arc, or *arc-minute,* is equal to 1/60 of 1°. The second of an arc, or *arc-second,* is equal to 1/3600 of 1°.

Linear and Angular Velocity: Linear velocity is measured in meters per second (m/s) or in multiples and submultiples: km/s and cm/s. A commonly used speed is kilometers per hour (km/h). Angular speed is measured in radians per second (rad/s). Submultiples are the milliradian per second (mrad/s) and microradian per second (μrad/s). Non-SI but frequently used units are the degree per second (°/s) and revolutions per minute (r/min.). One revolution is equal to 2π radians or to 360°.

Acceleration and Jerk: Linear acceleration is measured in meters per second squared (m/s^2). Angular acceleration is measured in radians per second squared (rad/s^2). Quite commonly, linear acceleration is expressed in *g*'s, where *g* is a symbol of acceleration caused by the force of the earth's gravity. $1g = 9.80665$ m/s^2. The symbol *g* should not be confused with the similar symbol used for the gram (italic script is utilized for denoting acceleration).

Attitude and Attitude Rate: Attitude is the position or orientation in motion or at rest of an aircraft, spacecraft, or other body. Attitude is determined by the relationship between axes of the body and some reference line, plane, fixed system, or reference axes.

Attitude is expressed in radians (rad) and in degrees (°), or in their submultiples: milliradians (mrad), minutes ('), and seconds (").

The attitude rate is measured in radians per second (rad/s) or in submultiples: milliradians per second (mrad/s).

Attitude-sensing instruments provide measurements with respect to a predetermined reference system. One typical system is three mutually orthogonal axes.

Stress and Strain: The deformation of solids is evaluated by measuring stress. Several mechanical parameters can be calculated using the values found for stress. Among these parameters are shear stress, Poisson's ratio, modulus of elasticity (Young's modulus), elastic limit, torsional deflection, and other parameters.

Strain is measured as a dimensionless ratio of the same length unit (m/m) which is sometimes called *strain.* It is also expressed in submultiples of strain, for instance *microstrains* ($\mu\epsilon$), $1\mu\epsilon = 1 \times 10^{-6}$m/m. The unit for the modulus of elasticity and stress is newtons per square meter (N/m^2).

Force and Mass (or weight):   Weight is the force of attraction of the body toward the earth. The mass of a body can be calculated by dividing weight by the acceleration due to gravity. The unit of force is the newton (N). In terms of basic SI units, N is equivalent to kg · m/s^2. The unit of mass is the kilogram (kg).

Torque: Torque is measured in newton-meters (N · m).

Pressure: Pressure is measured in units of force per unit of area, i.e., N/m^2. This unit is called the pascal (Pa). The pascal is a very small unit and the use of decimal multiples of the Pa is typical for practical measurements. Among these multiples are the hectopascal (hPa, 1hPa = 100Pa), the kilopascal (kPa, 1kPa = 1000Pa), and the megapascal (MPa, 1MPa = 1 × 10^{6}Pa).

Acoustic Pressure: Sound is measured in terms of the pressure (Pa) or power (W) which it develops. However, the most commonly used unit in practice for evaluating the sound pressure level is the decibel, which is 20 times the logarithm to the base 10 of the ratio of the effective values of the measuring and reference sound pressure levels. Unless a different reference pressure is specified, a pressure of 20μPa (2 × 10^{-4} μbar) is usually taken as the reference. Another unit used in acoustic engineering is the acoustic ohm, which is the ratio of the acoustic pressure to the flow rate (N · s/m^5).

Flow Rate: Flow rate can be expressed in terms of mass flow rate (kg/s) or volumetric flow rate (m^3/s).

Level of Liquid: The level of a liquid or other substance is measured as the height of its surface above a reference point. Quite often the measured height is used for calculating the volume or mass of a liquid if the dimensions of the tank and the density of the liquid are known. The direct unit for the measurement of level is the meter (m).

Humidity, Moisture, and Dew Point: Humidity is a measure of the water content in a gas. Moisture is a general term relating to water content in gases, liquids, and solids.

   Relative humidity is measured in percent (%RH). Absolute humidity is expressed in units of mass per unit volume (kg/m^3 and also g/m^3). Moisture is given in percent by volume or by weight.

Vacuum: It is assumed in practical engineering that a vacuum is a space in which the pressure is far below the normal atmospheric pressure, typically lower than 7kPa.

   Traditionally, a vacuum is measured in torrs (1torr = 1mmHg = 133.32Pa). The range of measurement is extended from 760torr (low vacuum) to 1 × 10^{-9} torr (ultra-high vacuum).

Density and Specific Gravity of a Substance: Density and specific gravity are the mass and weight of a given substance per unit volume, respectively. The corresponding units are kg/m^3 and N/m^3.

Viscosity of Fluid: Viscosity is a fluid's resistance to flow. Absolute (dynamic) viscosity is expressed in N · s/m^2. Kinematic viscosity is the absolute viscosity divided by the fluid's density. The units for kinematic viscosity are m^2/s. Relative viscosity is the ratio of absolute viscosity to the viscosity of a liquid, which is taken as a reference (quite commonly the liquid is water). Relative viscosity is dimensionless.

Temperature: The SI unit of temperature is the kelvin (K). Absolute zero is 0K. On the kelvin scale, the ice and steam points are 273.15K and 373.15K, respectively. On the Celsius scale, the ice and steam points are defined at 0°C and 100°C, respectively. The Celsius scale is widely used in measurements. Note that 1°C = 1K.

The following relationship can be used for the conversion of Fahrenheit (°F) and Celsius (°C) scales: °F = (9/5) × °C + 32.

Heat Radiation Intensity and Heat Flux: Radiation intensity is measured in units of power per unit of area, W/m^2 or W/cm^2.

A heated object radiates electromagnetic waves in the ultraviolet, visible, and infrared regions of the electromagnetic spectrum. The energy radiated by the object is proportional to its temperature. Typically, the outputs of the instruments are given in units of temperature: degrees Celsius (°C), degrees Fahrenheit (°F), or kelvin (K). The length of radiated waves is expressed in micrometers.

Heat flux is expressed in watts (W) or in watts per square meter (W/m^2). Quite often, measurements of heat flux are associated with the quantitative evaluation of the heat energy received from the sun. The measure of this energy is the solar constant which is the power per square meter received from the sun. One solar constant is equal to $1353W/m^2$.

Light Intensity: Light intensity is measured for the lengths of waves between 10nm and 1mm. The lengths of waves in the light spectrum lie between 10nm and 1mm. The rage for visible light is from 380 to 780nm, for ultraviolet (UV) radiation from 10 to 380nm, and for infrared (IR) radiation from 780 to 1×10^6nm (the portion between 780nm and $3\mu m$ is called near IR, and the portion between $3\mu m$ and $1000\mu m$ is called far IR).

Several units are used most often in optical measurements. Luminous intensity is measured in candelas (cd). The candela is one of the basic SI units. Luminous flux is expressed in lumens (lm). The unit of illuminance is the lux (lx). One lux is equal to one lumen per square meter. Brightness or luminance is expressed in candelas per square meter of light-emitting area (cd/m^2). Wavelengths of UV and visible light are usually given in nanometers (nm) and, for IR light, in micrometers (μm). Non-visual magnitudes are radiant flux (watts, W), radiant intensity (watts per steradian, W/sr, or watts per square meter, W/m^2), and radiance [watts per square meter per steradian, $W/(m^2 \cdot sr)$].

Voltage, Electrostatic Charge, Current, Electrical Power, Elapsed Time, and Frequency: The units corresponding to these quantities are the volt (V) for voltage, the coulomb (C) for electrostatic charge, the ampere (A) for current, the watt (W) for electrical power, the second (s) for elapsed time, and the hertz (Hz) for frequency.

Magnetic Field Intensity and Flux Density: The SI units for field intensity and flux density are the ampere per meter (A/m) and the tesla (T), respectively.

Radioactivity: The SI unit for radioactivity is the becquerel (Bq). It is the activity of a radionuclide decaying when its rate is one spontaneous transition per second.

Another quantity that is frequently used for measuring radiation is exposure dose. It is expressed in coulombs per kilogram (C/kg) and represents the exposure when nuclear radiation produces in air one coulomb of electrical charge per kilogram of dry air.

Chemical Properties and Chemical Composition: Almost all physical quantities are intermediate agents for chemical measurands; the relevant units are quite diversified.

There are a number of subdivisions of the measurands. For example, pressure can be absolute, gage, differential, barometric, atmospheric, static, or dynamic. The measurand can be associated with various-direction measurements, for example, one-, two-, and three-directional acceleration.

Many measurands are very specific. They are not shown in the given list of measurands (for instance, *turbidity* is one of these quantities; it is an optical property of liquid caused by the presence of suspended particles).

When saying "measurand" we have in mind the physical quantity, property, or condition that is measured by the transducer and not an intermediate quantity that

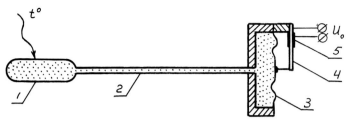

Figure 1.1 Gas-filled-bulb transducer for measuring temperature. 1 = gas-filled bulb, 2 = capillary tube, 3 = pressure-sensitive diaphragm, 4 = cantilever, 5 = strain gages.

is a product of the conversion inside the transducer. For example, the transducer in Figure 1.1 is intended for measuring temperature. In this transducer, the temperature-responsive gas fill in the bulb expands under a rise in temperature and increases the pressure in the sealed volume. Under the pressure, the diaphragm deflects and bends the cantilever. Strain in the strain gages causes a change in their resistances. Finally, the voltage developed in the strain-gage bridge is proportional to the measurand. As seen in the diagram, the conversion of the temperature induces several intermediate physical quantities: gas expansion, pressure increase, diaphragm deflection, change in the radius of curvature of the cantilever, strain in the gages, change in their resistance, and change in the voltage from the bridge. Only temperature, in this case, is regarded as a measurand. However, the signals at the inputs of the elements are also physical quantities and are actually intermediate measurands.

OPERATING PRINCIPLES

In order to describe the operating principles of a transducer, we should consider all of the sequences in the conversion, as we did in the preceding example. An electrical signal originates in the transduction element, which is introduced, in this example, by strain gages. It is obvious that the transduction element is the most important part of a transducer. This element defines the transducer's major functions. Many known physical phenomena are used in the design of transducers. It is difficult to describe all relevant physical relationships within this book. However, some of them are summarized in Table 1.2. More detailed descriptions are given in the chapters dealing with elements' principles. Reviewing Table 1.2 gives an overall picture of the conversion fundamentals.

TABLE 1.2

Principles of Transduction

No.	The change in the . . .	is caused by the change in the . . .
1	2	3
1	Resistance of a resistive element	a. Position of a wiper in a linear or angular potentiometer. b. Resistivity and dimensions of a strain-gage material that undergoes strain. c. Geometry of an elastic pipe carrying conductive liquid. d. Resistivity of a temperature-sensitive metallic or semiconductor material that experiences heating or cooling. e. Position of plates in a vacuum tube (change in transconductance). f. Voltage across or current in a varistor (an element having nonlinear volt-ampere characteristics). g. Resistivity of a metallic or semiconductor material sensitive to the strength of the magnetic field. h. Resistivity of a semiconductor material as a response to the incident light intensity. i. Concentration of chemicals in a solution.

(continued)

TABLE 1.2

Principles of Transduction *(continued)*

1	2	3
2	Self-inductance or mutual inductance of an inductive element	a. Length of the magnetic path. b. Cross-sectional area of the magnetic path. c. Permeability of the ferromagnetic material in the magnetic path due to the stress developed in the material; changes in its chemical composition or temperature. d. Inductive coupling between two or more coils, which is caused by the variation of the distance between the coils, their mutual orientation, inducement of eddy currents, and magnetic properties of the magnetic path under stressing or heating.
3	Capacitance of a capacitive element	a. Distance between two electrodes, which constitute a capacitor. b. Congruous area between two electrodes, which constitute a capacitor. c. Dielectric constant due to the displacement of a high-dielectric-constant-material spacer between the capacitor's electrodes; stress or heat applied to a stress- or temperature-sensitive spacer; variation of the chemical properties of a substance between the electrodes.
4	Electrical charges	Process of ionization of gases exposed to radiation from radioactive sources.
5	Voltage or current developed by a self-generating element	a. Relative position of a conductor and the lines of a magnetic field when they have a relative speed and the conductor crosses the lines. b. Temperature of two junctions when each of them contains two contacting conductors of different metals (Seebeck Effect in thermocouples). c. Electric charges developed in a crystal that undergoes heating (Pyroelectric Effect). d. Electric charges induced in a crystalline or ceramic material experiencing stress (Piezoelectric Effect). e. Radiant energy of light incident upon the p-n junction in a semiconductor (Photovoltaic Effect). f. Production of electrical energy during chemical action. g. Development of an opposite electrical charge in two dissimilar uncharged, contacting metals (Contact-Potential Effect). h. Motion of a liquid passing through a stationary capillary tube or porous body when the electric potential gradient is developed along the tube or pore (Electrokinetic Effect).
6	Voltage or current developed in an element fed from an ancillary source of energy	a. Transverse electric potential gradient in a current-carrying metal or semiconductor when a magnetic field is applied at right angles to the direction of current (Hall Effect). b. Electric potential gradient along a strip of metal carrying heat flow when the strip is exposed to a magnetic field perpendicular to its plane (Nernst Effect). c. Resistance and capacitance in a conductor, due to the field produced by an adjacent conductor (Proximity Effect). d. Velocity of discharged electrons liberated from a surface subjected to light providing a photoemissive current through the electrodes in a vacuum tube when a high voltage is applied between the electrodes (Photoemissive Effect).
7	Optical properties of a transparent material	Double refraction when the material is subjected to stress (Photoelastic Effect).
8	Emission of light	a. Heating of certain substances (e.g., diamond) below the red-hot temperature. b. Chemical action. c. Excitation by fast electron bombardment. d. Excitation by strong electric field. e. Excitation of phosphors by electromagnetic radiation of waves having various spectrums of lengths. f. Excitation of phosphors by radioactive particles.
9	Size or position of a body	a. Extension or contraction of the body as a function of temperature. b. Forces of acceleration. c. Pressure of heated or cooled gas confined in a sealed container. d. Pressure acting upon a diaphragm, bellows, Bourdon tube, or other pressure-sensitive element. e. Angular momentum (in a gyroscope).

TABLE 1.3

Transducer Design Characteristics (Adopted from [1])

Measurand	Electrical	Mechanical
Range	Excitation	Configuration
Overrange	Isolation	Dimensions
Recovery time	Grounding	Mountings
	Source impedance	Connections
	Load impedance	Case material
	Input impedance	Materials in contact
	Output impedance	with measured fluids
	Insulation resistance	Case sealing
	Breakdown voltage rating	Identification
	Gain instability	
	Output	
	End points	
	Ripple	
	Harmonic content	
	Noise	
	Loading error	

DESIGN CHARACTERISTICS

The design characteristics of a transducer include the ratings and descriptions of the major features. Among them are the characteristics pertaining to the measurand and also to electrical and mechanical design. Table 1.3 [1] gives the list of these characteristics.

PERFORMANCE CHARACTERISTICS

The performance characteristics of a transducer or an element can be divided into three groups: static, dynamic, and environmental (see Table 1.4 [1]).

Static characteristics are usually given for room conditions when a transducer is not subjected to acceleration, vibration, or shock and when a measurand is changed slowly.

Dynamic characteristics define the transducer's response to a time variation in the measurand. There are two types of environmental characteristics: nonoperating and operating, which pertain to the transducer's performance after or during exposure to specific external conditions, respectively.

RELIABILITY CHARACTERISTICS

When we qualify a transducer's reliability, we should take several aspects into account. First of all, the most essential characteristics that define reliability are operating life, cycling life, and storage life. Second, a number of the parameters that govern the extreme modes of operation or limits in certain conditions also relate to reliability, for example, burst-pressure rating and breakdown voltage rating. Third, some of the transducer's characteristics are important not only in terms of its operational reliability but also in terms of the safety of the operators and the systems in which the transducer is utilized. These characteristics must be specified.

In addition to providing brief descriptions of the given basic transducer characteristics, we will review some of them with more detail and with graphic interpretations of the most common cases. We will start off with a transducer whose theoretical input-output characteristics can ideally be represented by a straight line (Fig. 1.2). In practice, the performance of a real transducer is associated with zero

TABLE 1.4

Transducer Performance Characteristics (Adopted from [1])

Static	*Dynamic*	*Environmental*
1	2	3
Resolution	Frequency response	Operating environmental effects
Threshold	Transient response:	operating temperature range
Creep	response time	thermal zero shift
Hysteresis	rise time	thermal sensitivity shift
Friction error	time constant	—or—
Repeatability	Natural frequency	temperature error
Linearity	Damping	—or—
(+reference line)	Damping ratio	temperature error band
Sensitivity	Overshoot	temperature gradient error
Zero-measured output	Ringing frequency	acceleration error
Sensitivity shift		—or—
Zero shift		acceleration error band
Conformance		attitude error
(+reference curve)		vibration error
—or—		—or—
Static error band		vibration error band
(+reference line or curve)		ambient-pressure error
Reference lines:		—or—
theoretical slope		ambient-pressure error band
terminal line		mounting error
end-point line		Nonoperating environmental effects
best straight line		Type-limited environmental effects
least-squares line		conduction error
Reference curves:		strain error
theoretical curve		transverse sensitivity
mean-output curve		reference-pressure error

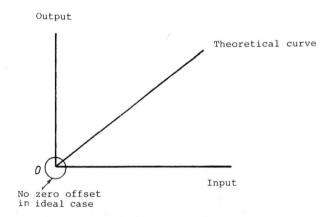

Figure 1.2 Ideal transducer characteristics.

drift, a change in sensitivity, nonlinearity, nonrepeatability, and hysteresis. The effect of zero error (Fig. 1.3) may be caused by imperfections in the initial calibration, threshold sensitivity, or dead zone; the influence of temperature; or the aging process (in Figure 1.3 and in the following figures, the deviations of different characteristics are graphically exaggerated against the real magnitudes which do not usually exceed several percent).

The deviation of the observed output from the correct value is constant over the entire range; in other words, the slope of the characteristics is constant. With a change in the sensitivity, the transducer's output (Fig. 1.4) deviates from the correct value by a factor that is proportional to the measurand.

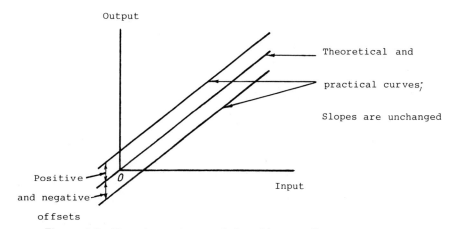

Figure 1.3 Transducer characteristics with zero offset.

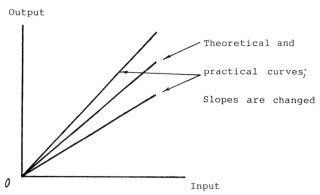

Figure 1.4 Transducer characteristics with change in sensitivity.

Under certain conditions the slope of the characteristic is changed. Some factors affecting this error are variations in temperature, excitation voltage, and degradation of materials. In our example, we assume that the transducer is nominally linear. The deviations from linearity (Fig. 1.5) can manifest themselves in various shapes. The nonlinearity can be specified in terms of the maximum departure from the

1. straight line passing through zero and full-scale output (Fig. 1.6);
2. tangent at the origin (Fig. 1.7); or

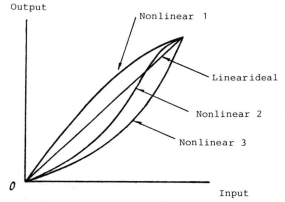

Figure 1.5 Nonlinear characteristics of transducer.

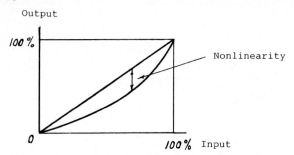

Figure 1.6 Nonlinearity with reference to zero-to-full-scale-output line.

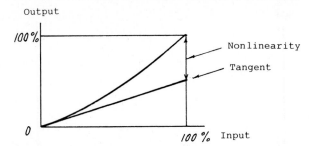

Figure 1.7 Nonlinearity with reference to tangent at origin.

3. best straight line (Fig. 1.8), or least-squares line, or another line which must be specified.

Some transducers do not naturally possess the straight-line characteristics described above. Nonlinearity for such transducers (nonconformity) is defined in terms of departure from the true curve (Fig. 1.9). A transducer's output may depend not only on the applied input but also on the formerly applied input; and a different output is obtained depending on whether the input is increasing or decreasing (Fig. 1.10). This behavior is evaluated by hysteresis, which is the maximum difference in outputs, corresponding to one measurand.

The nonideal behavior of a transducer is evaluated by errors, which are the differences between the practical and theoretical or specified outputs expressed in terms of percents of full-scale output (%FSO) or in terms of measurands.

The accuracy is a ratio of error to the full-scale output. It is usually expressed in terms of %FSO or in units of the measurand.

So far we have discussed the imperfections in statics. The dynamic errors occur when the output of a transducer depends on time functions of the input—for instance, the rate of its change or frequency. It is a common practice to determine the dynamic characteristics by applying to the input a step-changed or sinusoidally varied measurand with a simultaneous recording at the output. Depending on the transducer's energy-dissipating characteristic (damping ratio), three common types of outputs (Fig. 1.11) can be produced in response to a step change in the measurand. In one case, the output oscillates before it comes to the final output value (underdamped system). In the other case, it changes slowly without oscillation (overdamped system). Finally, a critically damped system has a response that is between these two.

The most relevant time characteristics are illustrated in Figure 1.12. Note that the response time and rise time can be arbitrarily specified—for example, 90, 95, 98 percent (for response time); 2 to 90 percent, 5 to 95 percent, 10 to 98 percent (for rise time). The transducer's frequency response is determined by calculating or measuring

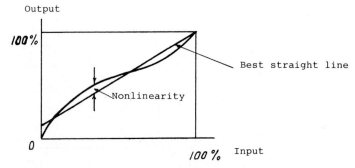

Figure 1.8 Nonlinearity with reference to best straight line.

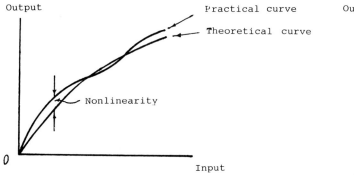

Figure 1.9 Nonconformity with reference to theoretical curve, which is naturally nonlinear.

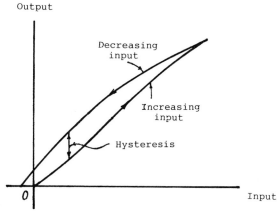

Figure 1.10 Transducer hysteresis.

the changes in the output amplitudes and output-to-input phases when a measurand varies sinusoidally within a specified range of frequencies (see representative curves in Figure 1.13), but its amplitude is kept constant.

Frequently, the amplitude change is evaluated by taking the ratio of the measured output amplitude to the amplitude at the reference frequency, which is chosen within the usable frequency range. This ratio is expressed in decibels and the phase difference is expressed in angular degrees.

When the individual specifications of the transducer's characteristics are not needed, and the maximum deviation in the output is essential, the term *error band* becomes a convenient definition. It represents the band of maximum deviations of the output from the specified reference line or curve (Fig. 1.14) and is expressed in percent of full-scale output (it should not be confused with *instrument bandwidth,* which is a measure of the useful frequency range shown in Figure 1.13). A stepped error band (Fig. 1.15) can be used in those cases where one portion of the transducer characteristic has more limited deviations than the others.

The reader will find more detailed descriptions of certain transducer characteristics while working with the chapters on element theory.

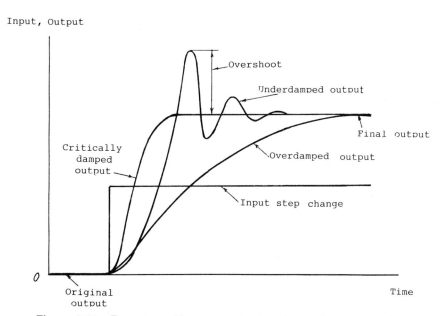

Figure 1.11 Response of transducer to step change in measurand.

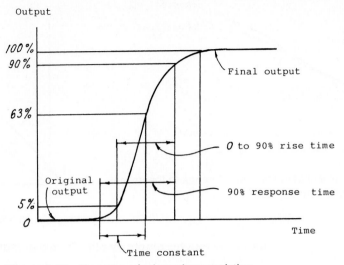

Figure 1.12   Transducer's time characteristics.

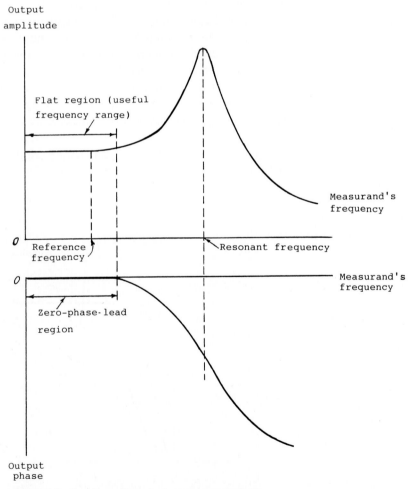

Figure 1.13   Transducer's frequency response.

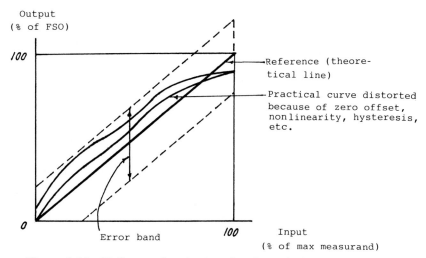

Figure 1.14   Static error band referred to theoretical line.

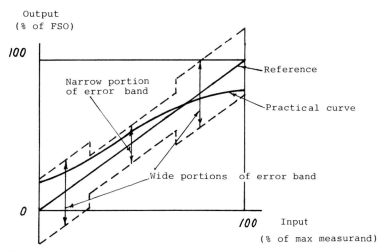

Figure 1.15   Stepped error band referred to theoretical line.

TRANSDUCER STRUCTURE

Depending on its construction, the transducer can be simple, consisting of one transduction element, or complex, with different cells providing conversion of signals, delivering power to parts, filtering, or having other functions. Inclusion in the transducer housing of either element is dictated by many factors such as convenience of signal conditioning, reduction of noise level, and improvement of dynamic response. Quite often, the transducer is a complex system that not only converts a physical quantity into an electrical signal but also provides signal conditioning or compensation for temperature, for example. Strictly speaking, all of the functional assemblies inside the transducer housing are the transducer's elements, but it should be noted that only two of them are most commonly present in the construction—namely, transduction and sensing elements.

When a transducer is designed, many different features of its structure are taken into account in order to create a device consistent with specifications. There are a variety of engineering approaches in choosing the principle of operation and in determining the synthesis of the elements in the design. The structure of a transducer greatly defines its operation. It is known from practice that a simple-structure transducer is less expensive and more reliable than one having a complex structure. However, loading the transducer with additional components leads to better performance (higher accuracy, lower response time and other improvements).

In spite of the diversity of transducer structures, they can be divided into several basic schemes, each with a distinctive connection of sensing and transduction elements: series, differential, ratio metering, and compensating (servo). In the transducer with a series connection of elements, the input of the first element is a measurand, and the output of each foregoing element is an input for the following element. For example, the structure of the transducer in Figure 1.1 is represented by four blocks (Fig. 1.16), of which block 1 is the first sensitive element (gas-filled bulb) performing the temperature-to-pressure conversion, block 2 is a second sensitive element (diaphragm) developing the force in response to pressure, block 3 is a third sensitive element (cantilever) converting the force into the deformation of the cantilever, and, finally, block 4 is a transduction element (strain gage) converting the deformation into the resistance change. Assuming a linear conversion, the transfer characteristic of the transducer is calculated as follows:

$$G_1 = y_1/x, \qquad G_2 = y_2/y_1, \qquad G_3 = y_3/y_2,$$
$$G_4 = y_4/y_3, \qquad G = G_1G_2G_3G_4 = y_4/x \qquad (1.1)$$

where $\qquad x$ = measurand,
y_1 to y_4 = input and output signals at blocks,
G_1 to G_4 = sensitivities of sensing and transduction elements, and
G = transfer characteristic of transducer.

If the conversion is associated with errors Δy_i's which are additive to the input signals of each block, the resulting error at the output Δy is calculated as shown below:

$$y_1 = y_1' + \Delta y_1, \qquad (1.2)$$

$$y_2 = y_2' + \Delta y_1 G_2 + \Delta y_2, \qquad (1.3)$$

$$y_3 = y_3' + \Delta y_1 G_2 G_3 + \Delta y_2 G_3 + \Delta y_3, \qquad (1.4)$$

$$y_4 = y_4' + \Delta y_1 G_2 G_3 G_4 + \Delta y_2 G_3 G_4 + \Delta y_3 G_4 + \Delta y_4, \qquad (1.5)$$

$$\Delta y = y_4 - y_4' = \Delta y_1 G_2 G_3 G_4 + \Delta y_2 G_3 G_4 + \Delta y_3 G_4 + \Delta y_4 \qquad (1.6)$$

where y_1' to y_4' = free-of-error outputs.

As seen from equation 1.6, for the series connection of transducer elements, the resulting error at the output is a sum of the errors of the elements referred to the output. In the design of this simplest-type transducer, there are no special measures for the depression of errors.

A differential transducer (Fig. 1.17) contains two channels of elements connected in series. This connection is similar to those shown in Figure 1.16. The outputs of channels y_1 and y_2 are applied to the input of a cell providing subtraction of signals. If sensitivities G of each channel are the same, and the operation of the

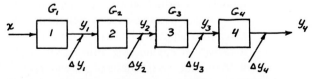

Figure 1.16 Block diagram of transducer with series connection of elements.

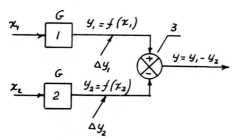

Figure 1.17 Block diagram of differential transducer.

transducer can be described by a linear function with constant bias y_0, the outputs can be given as

$$y_1 = Gx_1 + y_0 \quad \text{and} \quad y_2 = Gx_2 + y_0, \quad \text{and} \qquad (1.7)$$

the output of the transducer will be

$$y = y_1 - y_2 = G(x_1 - x_2). \qquad (1.8)$$

There are two modes of operation of a differential transducer: (1) $x_1 = $ var and $x_2 = $ const and, (2) $x_1 = -x_2 = $ var. For the first mode, when x_1 changes from x_1' to x_1'' and, respectively, y is changed from y' to y'', the sensitivity is determined as

$$\frac{y'' - y'}{x_1'' - x_1'} = G. \qquad (1.9)$$

For the second mode, this ratio is $2G$; therefore, the sensitivity of the two-channel system is twice that of the one-channel system.

If errors Δy_1 and Δy_2 are added to the outputs, they become

$$y_1 = y_1' + \Delta y_1 \quad \text{and} \quad y_2 = y_2' + \Delta y_2 \qquad (1.10)$$

where y_1' and y_2' are the outputs that are free of errors.

The magnitudes of Δy_1 and Δy_2 are equal or very close to each other since, in the design of differential transducers, the two channels are usually exposed to the same factors that stimulate the emergence of the errors. The difference of equations 1.10,

$$y_1 - y_2 = y_1' - y_2', \qquad (1.11)$$

shows that the differential connection of the channels gives a substantial decrease or elimination of errors. The improvement of the linearity of the transfer characteristic is another advantage in using the differential scheme. Let us assume that the channels are modeled by nonlinear functions

$$y_1 = f(a + x), \quad y_2 = f(a - x), \qquad (1.12)$$

and x is a small perturbation around point a. The first four terms of the Taylor expansion for y_1 and y_2 are

$$y_1 = f(a) + \frac{x}{1!}f'(a) + \frac{x^2}{2!}f''(a) + \frac{x^3}{3!}f'''(a) + \ldots \text{ and} \qquad (1.13)$$

$$y_2 = f(a) - \frac{x}{1!}f'(a) + \frac{x^2}{2!}f''(a) - \frac{x^3}{3!}f'''(a) + \ldots . \qquad (1.14)$$

The difference $y_1 - y_2$ is

$$y_1 - y_2 = 2\frac{x}{1!}f'(a) + 2\frac{x^3}{3!}f'''(a) + \ldots . \qquad (1.15)$$

The term containing x squared, the most powerful contributor of nonlinearity, is canceled. It is obvious that the deviation of $y_1 - y_2$ from a straight line is smaller as compared with y_1 or y_2 for the same variations in x_1 or x_2.

Similarly to the differential transducer, a ratio-metering transducer (Fig. 1.18) includes two channels 1 and 2 containing in series connected elements and one cell 3 forming output y proportional to the ratio of the two input signals x_1 and x_2:

$$y = F(x_1/x_2). \qquad (1.16)$$

If the output signals y_1 and y_2 of each channel are linear functions of inputs and if the uniform sensitivity of each channel G is affected by the same error ΔG, the output will not respond to the error:

$$y_1 = (G + \Delta G)x_1, \qquad y_2 = (G + \Delta G)x_2 \qquad \text{and} \qquad y_1/y_2 = x_1/x_2. \qquad (1.17)$$

The reduction or elimination of errors generated by the change in sensitivity is the major advantage in the application of this system.

A compensating (servo) transducer (Fig. 1.19) is built of three major blocks: (1) error detector 1, which provides the difference (Δx) of the signal from measurand x and the signal from the feedback network; (2) feedforward elements 2; and, (3) feedback elements. In this transducer, the measurand is applied to one of the inputs of an error detector, whereas the other input is fed from the output of the feedback elements creating the signal x_c, having the physical nature of the measurand x. The difference between x and x_c is treated in the feedback loop in such a manner that the input signal is constantly compensated and the signal from the error detector is near to zero. We can illustrate the basic features of this transducer if we consider the operation of a simple scheme shown in Figure 1.19 and assume that the conversions in blocks 1,2, and 3 are linear. It is clear from the scheme that the relationships between the signals are

$$y = G\,\Delta x, \qquad x_c = Hy, \qquad \Delta x = x - x_c \qquad (1.18)$$

where G and H are sensitivities of blocks 2 and 3, respectively.

From equations 1.18,

$$x_c = GH\Delta x = GH(x - x_c) \qquad \text{or} \qquad (GH + 1)x_c = GHx. \qquad (1.19)$$

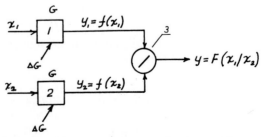

Figure 1.18 Block diagram of ratio-metering transducer.

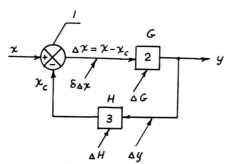

Figure 1.19 Block diagram of compensating (servo) transducer.

Since GH is usually much greater than unity, $x \approx x_c$, and

$$y = \frac{G}{1 + GH} x \tag{1.20}$$

Now we can determine how the error Δy at the output of this scheme is affected by the errors ΔG and ΔH in G and H. If $x = \text{const}$, then

$$\Delta y = \left(\frac{\partial y}{\partial G} \Delta G + \frac{\partial y}{\partial H} \Delta H \right) x. \tag{1.21}$$

This expression is a modified formula for the full differential of y with substitutions of increments Δy, ΔG, and ΔH for differentials dy, dG, and dH. After taking derivatives, we will find that

$$\Delta y = \left[\frac{1}{(1 + GH)^2} \Delta G - \frac{G^2}{(1 + GH)^2} \Delta H \right] x. \tag{1.22}$$

Unlike in a simple system with a series connection of elements, in the compensating circuit, the influence of the variation of the sensitivity of the feedforward elements is greatly reduced [factor $1/(1 + GH)^2$ in equation 1.22].

The dependence on the variation of the feedback circuit sensitivity can be significantly diminished by simple measures for stabilization of this circuit. A similar calculation developed for the additive errors $\delta \Delta x$ and Δy (see Fig. 1.19) shows an identical depression of these errors due to the function of the feedback loop. The use of the feedback loop in the transducer structure makes it possible to have an instrument not only with increased accuracy but also with a faster response to the measurand, a wider range of measurands, higher output power, and smaller dependence on the load conditions. The functions of a servo element are considered in *Elements of Servo Transducers*.

In presenting the structure of transducers, we have discussed only those elements that directly treat the signals carrying the information. However, some contemporary transducers contain a number of auxiliary elements that play an important role in the operation of the transducer. Among these elements are electrical dc feeding circuits (stabilizers, filters, overload switches and others); ac excitation circuits; amplifiers, filters of signals, mixers of signals, and detectors; analog summing, subtracting, differentiating, and integrating circuits; acceleration, pressure, humidity, and temperature sensors for the circuits of compensation; temperature-measuring or -controlling electronics; analog-to-digital converters; acoustic blocking filters, and other components.

In some designs, a microprocessor is placed inside the transducer housing to modify (e.g., linearize), memorize, or average the output signal; to provide compensation for the change of physical parameters; or to check automatically the characteristic points of the calibration curve. With advances in silicon technology, several silicon chips combined function as an integrated sensor-instrument system placed inside the transducer housing (Fig. 1.20).

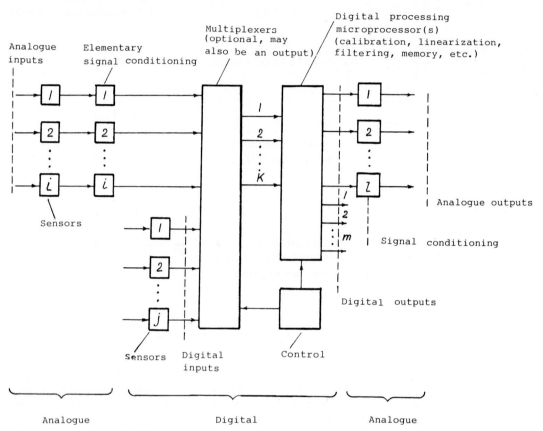

Figure 1.20   Block diagram of integrated sensor/instrument system.

chapter 2

Contact, Resistive, and Electrode Elements

Beginning in this chapter and through Chapter 6, we will be describing basic sensing and transduction elements, excluding tiny solid-state and optical devices, which are considered in Chapters 8 through 11. Diversity is the most notable feature of these elements, whose concepts reflect the variety of measurands and physical principles of conversion on which the elements are based.

The emphasis in the presentation of the elements is on physical concepts and analysis of performances rather than on details of construction and hardware. The latter can be found in special references and in handbooks on transducers [e.g., 1,2].

A number of advances have been made in the classification and organized representation of transducers and elements [3–6]. A specific interest is in classification by a three-dimensional diagram which combines input and output energy-conversion effects with the most common measurands. Considering this approach, we found it expedient to introduce elements in groups related to specific physical concepts of conversion. While reading about the elements, one can easily imagine how the described constructions can be used in a transducer or how they can be changed and/or combined to obtain more complex modifications as shown in Appendix 1 of this book.

The absolute majority of the devices introduced in this chapter are based on a conversion of a measurand into a change of resistance at the output of the element. Depending on the conversion mechanism, an electrode element develops a change of resistance or impedance at the output or produces electrical charges.

CONTACT ELEMENTS

A contact element converts displacement into an electrical signal by opening or closing electrical contacts.

The simplest element (Fig. 2.1) is composed of a moving rod 1 that closes the insulated contacts 2 and 3 while it moves up. Figure 2.2 illustrates constructions with several contacts. A system with one or more levers (Fig. 2.3) can be used in order to increase the element's sensitivity to displacement.

The contact couples are shaped like a cone (Fig. 2.4), disk, or hemisphere.

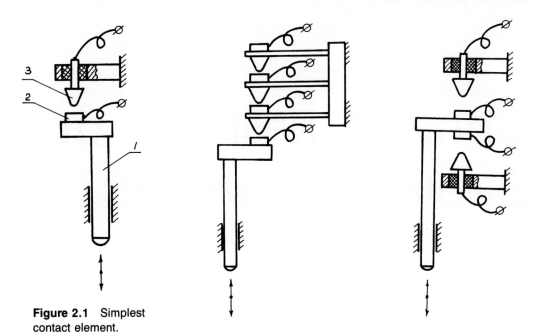

Figure 2.1 Simplest
contact element.

Figure 2.2 Contact elements with several contacts.

The most commonly used materials for the contacts are tungsten, silver, platinum, rhenium, rhodium, and alloys of platinum with iridium and palladium; gold plating of contacts is often used to improve reliability.

The condition of the contact surface is critical for the performance of the element. During the operation, the contacts are exposed to multiple collisions and they should be firm enough to withstand a mechanical destruction and a change in geometry of the contacting surfaces.

Table 2.1 illustrates [7] a wear of the contacts made of three different materials depending on the number of closing cycles of the contacts. For the soft materials (for instance, silver), this wear is substantially larger. The wear is most intense during the first stage of work. The threshold of response of the contact element is defined by the minimum distance between the contacts, which is in many instances not lower than 1 micrometer. A decrease of this distance causes the breakdown of gas or ignites the arc.

The maximum voltage and current [8] for the silver contacts should not exceed 13V and 0.4A; for the platinum and tungsten contacts it should not exceed 17V and 0.9A, respectively. Spark is a source of electrical erosion of the contacts. A striking voltage is smaller than the breakdown voltage and is defined by the gas pressure at the area of the contacts and by the distance between the electrodes. Figure 2.5 reveals the function

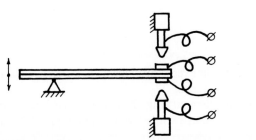

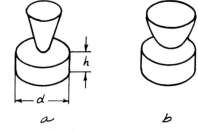

Figure 2.3 Contact element with lever for increasing sensitivity to displacement.

Figure 2.4 Contact couples:
a = cone and disk, b = hemisphere and disk.

TABLE 2.1

Wear of Contacts

Materials of the contacts	Number of cycles		
	25×10^3	400×10^3	1000×10^3
	Wear of the contacts in micrometers		
tungsten	1–5	1–5	1–5
silver	2–11	6–12	28
rhenium	0–2	0–2	0–4

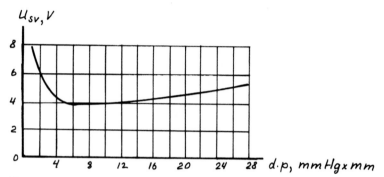

Figure 2.5 Striking voltage as function of product: distance between electrodes *d* and gas pressure *p*.

$$U_{sv} = f(d \cdot P) \qquad (2.1)$$

where U_{sv} = striking voltage, V;

    d = distance between electrodes, mm; and

    P = pressure of gas, mmHg.

The erosion can occur if the free path of electrons is larger than the distance between the contacts. If the voltage applied to the electrodes is 20V and the gap is 1 micrometer, the electrical field gradient will be 200 000V/cm. The electrons are extracted from a cathode accelerated by the field toward the anode. Due to the electronic bombardment, the anode loosens. Ions from the anode migrate to the cathode and form a hill on its surface and a cavity on the surface of the anode. This process disturbs the normal operation of the contacts. The erosion of platinum and platinum alloys is higher than for the hard materials such as tungsten and molybdenum. A metal bridge can be built of the anode material if the contacts are opened slowly.

Oxide and sulphide films on the surface of the contacts create corrosion. The contacts of platinum do not have this problem; however, it is typical for contacts made of tungsten.

For good electroconductivity the oxide film must be punctured when the contacts are closed. As a rule, the resistance of the film is much higher than that of the metal. There is a certain transition during the opening and closing of the contacts. In other words, the contacts are not opened and closed instantly. Figure 2.6 illustrates the four basic phases of closing and opening the contacts:

Phase 1—stable, open condition;

Phase 2—unstable closing;

Phase 3—stable, closed condition; and

Phase 4—unstable opening.

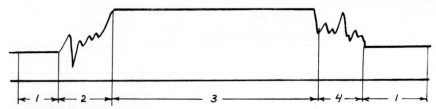

Figure 2.6 Closing and opening of contacts. 1, 2, 3, and 4 are phases of commutation.

The mechanism of the contact element or of the transducer associated with the element has to provide forces across the contacts of 1 to 10 cN in order to puncture the film when the contact is closed (usually it is 10cN for tungsten and 3cN for other metals). It was found experimentally [7] that after a one-year storage the forces needed for puncturing the film are 0.1–0.2cN for silver, 0.2–7.1cN for tungsten, and 0.2–1.4cN for rhenium.

A junction resistance R_j varies from a fraction of an ohm to several ohms. It depends primarily on the shape of the contacts, materials, surface roughness, and force across the contacts. The following empirical formula can be used to compute R_j in ohms:

$$R_j = \frac{a}{(0.1P_c)^b} \qquad (2.2)$$

where a is a factor depending on the conditions of the contact surfaces and materials—it varies from 0.06×10^{-3} for silver to 0.28×10^{-3} for other materials; and

b is a factor defined by the shape of the contacts—for instance, if point contacts (cone-to-plane or hemisphere-to-plane) are used, $b = 0.5$ and P_c is a force in N across the contacts.

The junction resistance is increased proportionally to the number of cycles of work [7]. For example, for tungsten contacts this change ΔR_j in ohms is described by the formula

$$\Delta R_j = 1.8 \times 10^{-2}N^{0.6} \qquad (2.3)$$

where N is the number of cycles.

The voltage and current for the contacts are not taken larger than 5–10V and 100mA, respectively.

The current is calculated by the equation of thermal balance:

$$I^2R_j = \mu S_c \Delta\theta \qquad (2.4)$$

where $\quad I$ = current through contacts, A;

$\quad R_j$ = junction resistance, Ω;

$\quad \mu$ = heat transfer coefficient, W/(m$^2 \cdot$ °C);

$\quad S_c$ = entire cooling surface of contacts, m^2; and

$\quad \Delta\theta$ = accepted difference between the contact and ambient temperature, °C ($\Delta\theta \approx 70$°C).

Some practical dimensions for the disk contact are given in Table 2.2. In order to reduce contact wear when the direct current flows through them, different materials can be used for the cathode and anode. For instance, the anode could be made of platinum or silver and the cathode of tungsten or molybdenum.

The error of devices with the contact element is between 0.1 and 0.5 micrometers for the tungsten contacts and between 0.1 and 1.5 micrometers for the silver and rhenium ones.

TABLE 2.2

Dimensions of Disc Contacts

Dimensions in mm	Current I in A		
	below 1	1–5	5–10
d	1.5–3	3–5	5–8
h	1	1.5	2

It is a good practice to exercise the element by cycling. The error for the element after one million cycles is much smaller than for one without exercising.

The stability of performance can be improved if during the operation the current flows through the contacts only when they are closed (no transition period).

JUNCTION RESISTANCE ELEMENTS

A junction resistance element converts force or displacement into a change of resistance of the conducting parts' junctions.

The typical element (Fig. 2.7) includes moving and fixed electroconductive armatures 1 and 2. A column of electroconductive washers or disks 3 between the armatures is exposed to force F or displacement δ, which develops a compressing stress at the surface of the conductors. With the change of stresses, the path of the current is changed as well. This leads to a change of the entire resistance of the column measured between leads 4 and 5. The differential connection of the column is shown in Figure 2.8, where the upper and lower armatures 1 and 2 are fixed, but armature 3 in the middle of the column transfers force F and displacement δ to the structure. In this construction, one of the elements is under an increased compression, while the other has a decreased compression. When $F = 0$ and $\delta = 0$, the elements in the column have the initial compression stress offset. Due to the offset, the change of the direction of the force or displacement does not rupture the current path through the column.

The disks and washers are made of electrode carbon, graphite, ceramics with graphite filling, or plastics with colloid particles of a semiconductor material. The specific resistance of the materials varies from 30 to $100\Omega \cdot mm^2/m$.

The resistance of the element as a function of the force or displacement is substantially nonlinear (2 to 30%) and is described by the following two functions:

$$R_j = R_0 + \frac{a}{F} \tag{2.5}$$

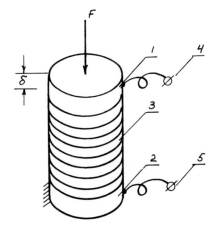

Figure 2.7 Junction resistance element.

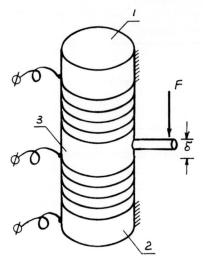

Figure 2.8 Differential junction resistance element.

$$R_j = R_0 + \frac{b}{\delta} \tag{2.6}$$

where R_j = resistance of the column, Ω;

   R_0 = constant component of R_j, Ω;

   F = force, N;

   δ = displacement, m; and

   a and b are factors expressed in $\Omega \cdot$ N and $\Omega \cdot$ m, respectively.

Formulas 2.5 and 2.6 are empirical, and factors a and b are valid for a certain range of variables F and δ:

$$F_{min} \leq F \leq F_{max}, \qquad \delta_{min} \leq \delta \leq \delta_{max}. \tag{2.7}$$

The thickness of the conducting circles is between 0.5 and 3.5 mm.

Two functions, $R_j = f(F)$ and $R_j = f(\delta)$, are given in Figure 2.9. They are obtained [8] experimentally for disks with a 10mm diameter, 1mm thickness, and current of 2A. The disks have the following composition: 80% porcelain clay, 12% coke, and 8% graphite.

The element allows the development of a high power which is limited by heating the parts. The equation for calculating the electrical and constructional characteristics of the element is

$$I^2 R_j = \mu S_c \Delta\theta \tag{2.8}$$

where I = current through the circles, A;

   R_j = column resistance, Ω;

   μ = heat transfer coefficient, W/(m$^2 \cdot$ °C),

      μ is 12 to 15 W/(m$^2 \cdot$ °C);

   S_c = the side surface of the column in m^2; and

   $\Delta\theta$ = the allowed difference between the column and the ambient

      temperatures, °C ($\Delta\theta$ is 140 to 180°C).

The maximum compressing stress is not taken larger than 6.5×10^6 Pa. The ranges of force and displacement are 1N to 10MN and 0.05×10^{-3} to 0.2×10^{-3} m, respectively. The hysteresis of the characteristics is between 1 and 2% after aging. The temperature coefficient of the characteristics is 0.1 to 0.5%/°C.

The construction of a transducer with a junction resistance is durable. It drives recorders without amplification due to a high-power output provided by the element.

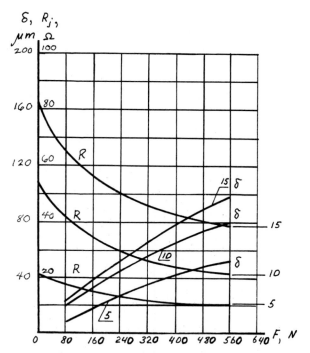

Figure 2.9 Resistance of electroconductive column R_j as function of axial force and displacement. F = force, R_j = resistance, δ = deformation, 5, 10, and 15 denotes number of conducting disks.

However, the accuracy of the device is usually low (approximately 10%); therefore, devices with this element should be used primarily in cases where the indications of forces and displacements (not precision measurements) are required.

Another version of the junction resistance element is shown in Scheme 26a.* Carbon (anthracite) electroconductive powder is slightly compressed in a cylindrical, nonelectroconductive housing. A carbon electrode and a diaphragm are in contact with the powder.

The electrode and diaphragm are electroconductive and can be made of metal as well. Leads bring a current to the structure. The operation of the element is based on the ability of the carbon grains to change the junction resistance due to the applied force. The diaphragm is deformed when pressure P is applied to its surface, and the grains undergo a higher contact stress. The resistance between them is changed, creating an electrical signal proportional to the applied pressure. This element is chiefly used to convert alternating pressure in microphones. The characteristics of the element are defined by the spring factors of the diaphragm, powder, and geometry of the contacting parts. One of the electrical connections of the element is shown in Figure 2.10. Element 1 is excited by acoustic pressure P. The element is fed by direct current from battery 2. With a variation of pressure, current I has an alternating component that induces an output voltage U_{OUT} in transformer 3.

The output-input characteristic (Fig. 2.11) is substantially nonlinear due to the specific properties of the powder. The characteristic is composed of three parts. Part A corresponds to the excitation threshold [0.02 to 0.03 bar (-40db) or 2×10^3 to 3×10^3Pa], where the element is practically insensitive to variations of pressure. Part B is a linear or operational zone. Part C is saturation. The element provides a large output signal. However, its dynamic range is small, and its nonlinear distortion is high. The element is not used for measuring static pressures because the resistance

---

*The word *Scheme* followed by a number denotes a scheme in the table of Appendix 1.

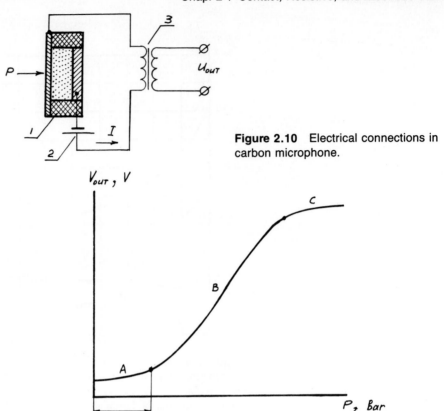

Figure 2.10 Electrical connections in carbon microphone.

Figure 2.11 Transfer characteristic of carbon microphone.

changes with time. The electrical current develops microarcs, grains are heated, and contacts between the grains are formed due to sintering. When these contacts are broken and the osculation between the diaphragm and grains is disturbed, electrical noise or crackling is created in the electrical circuit.

The junction resistance element is simple, inexpensive, and durable. These qualities have made it very popular in mass production.

POTENTIOMETRIC ELEMENTS

A potentiometric element is used to convert a linear or angular displacement into a change of resistance. The element (Fig. 2.12) includes resistor 1 with leads 2 and 3 and a sliding contact 4 (wiper), which can move along the resistor, changing the resistance between the wiper and one of the leads.

The electroconductive part of the element is made by winding a thin wire over a mandrel (supporting structure). It also can be made of electroconductive plastic, of a mixture of metal and ceramics, or of a thin film of metal or semiconductor on a glass or ceramic substrate.

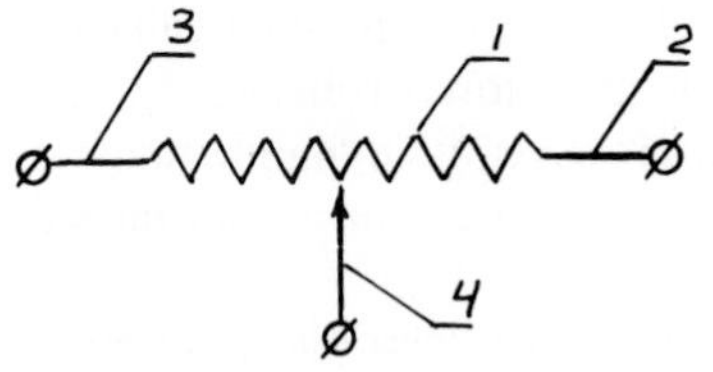

Figure 2.12 Potentiometric element.

Figure 2.13 shows different constructions of the element. They are all furnished with a wound resistor, which can be replaced in a particular construction by plastic or film ones.

The element for the conversion of translatory displacement (Fig. 2.13a) has shaft 1, which is restricted to move along the x-axis. Wire 2 is wound around a flat card or round mandrel 3. The wire is insulated, but the path along which wiper 4 moves is cleaned of insulation.

In a rotary element (Fig. 2.13b), shaft 1, carrying wiper 2, is free to rotate around the x-axis. A wire is wound around the bent flat mandrel 3, which can also be shaped like a toroid.

In a multiturn element (Fig. 2.13c), wiper 1 moves along a helical supporting structure with winding 2. In this case, the angle α to be measured can be much larger than 360°. The winding can be provided with connection 1 (tap) at the middle of the winding (Fig. 2.13d) for a convenient connection to an electrical bridge or differential circuit.

Several taps are used (Fig. 2.13e) in order to change the characteristic of the element by connecting additional resistors 1, 2, and 3. The change of the characteristic is also achieved by using a shaped mandrel (Fig. 2.13f and 2.13g).

An element (Fig. 2.13h) for providing a sine or cosine function of angle α of shaft 1, carrying wiper 2, is a simple structure with the potentiometer having one tap.

The potentiometer is most commonly connected (Fig. 2.14) as a voltage divider with resistor R_L introducing a load element.

Resistance R_1 of the potentiometer between points A and B will be

$$R_1 = \rho \frac{x}{A} \tag{2.9}$$

where ρ = wire resistivity, $\Omega \cdot$ m;
     x = length of wire between points A and B, m; and
     A = cross-sectional area of wire, m^2.

Assuming that the winding is uniform and the wire does not have any change in diameter and resistivity, the output voltage U_{OUT} from the element is written

$$U_{OUT} = U \frac{x}{l} \cdot \cfrac{1}{1 + \dfrac{x}{l} \cdot \dfrac{R}{R_L} - \dfrac{x^2}{l^2} \cdot \dfrac{R}{R_L}} \tag{2.10}$$

where U = voltage of excitation, V;
     l = overall length of wire, m; and
     R = total resistance of potentiometer, corresponding to length l, Ω.

The last expression is easily obtained by assuming that

$$R_1 + R_2 = R,$$
$$R_1 = ax, \quad \text{and}$$
$$R_2 = a(l - x) \tag{2.11}$$

where a = constant coefficient in Ω/m.

If $R_L \gg R$, the output is linear:

$$U'_{OUT} \approx \frac{U}{l} x \quad \text{(see Figure 2.15).} \tag{2.12}$$

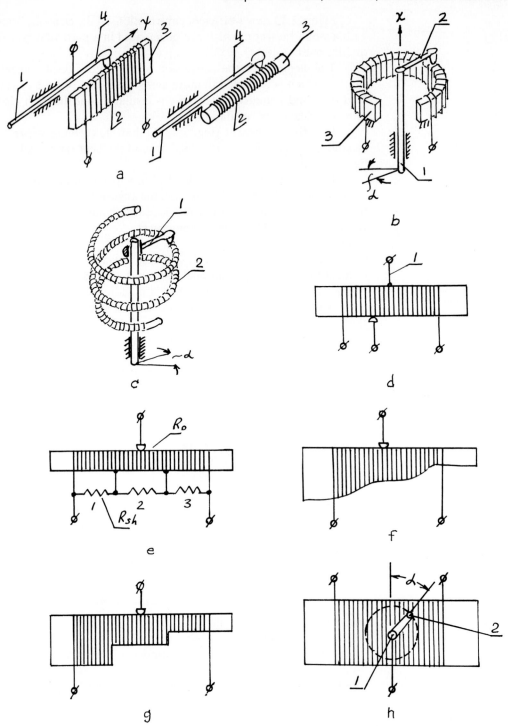

Figure 2.13 Constructions of potentiometric elements. a = linear displacement element, b = rotary displacement element, c = multiturn element, d = element with tap in middle of winding, e = element with several taps for nonlinear correction of characteristics, f = shaped mandrel element for obtaining nonlinear characteristics, g = step-shaped-mandrel element for obtaining characteristic with different slopes at subintervals, h = element for providing sine or cosine functions of angle α.

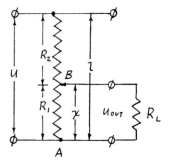

Figure 2.14 Electrical circuit of potentiometric element loaded with resistor R_L.

If R_L is commensurable with R, the output becomes appreciably nonlinear (curves 1 to 3 in Figure 2.15). The smaller the R_L, the larger the nonlinearity γ, which can be evaluated with respect to the chord:

$$U'_{OUT} = \frac{U}{l}x$$

$$\gamma = \frac{U'_{OUT} - U_{OUT}}{U'_{OUT}} = \frac{1}{1 + \dfrac{l^2}{x(l - x)} \cdot \dfrac{R_L}{R}} \qquad (2.13)$$

By taking the derivative of γ and equating it to zero, the maximum for γ is found:

$$\gamma_{max} = \frac{1}{1 + 4\dfrac{R_L}{R}}. \qquad (2.14)$$

This maximum will be at $x = l/2$. It is clear from equation 2.14 that for $R_L \gg R \; \gamma_{max} \approx 0$.

The nature of the curve $U_{OUT} = f(x)$, with an additional resistor R'_L, connected to the potentiometer (see Figure 2.16a), is shown in Figure 2.16b. Curves 1, 2, and 3 reveal the output voltages when $R_L \gg R$ and $R'_L \gg R$; when R_L is comparable with R but $R'_L \gg R$; and when R_L and R'_L are both comparable with R, respectively. It is noteworthy that the sensitivity of the output voltage to the change of x can be increased at a certain area of curve $U_{OUT} = f(x)$ by a simple addition of R'_L.

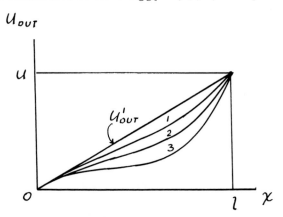

Figure 2.15 Output voltage of potentiometer versus wiper's position.

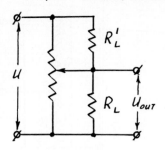

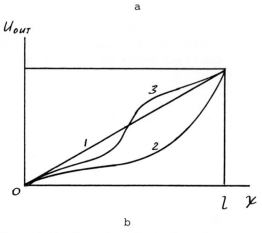

a

b

Figure 2.16 Operation of potentiometric element with two loading resistances. a = circuit, b = nonlinear characteristics.

    Two more connections (Fig. 2.17) of the element have a limited use: simple reostat (Fig. 2.17a) and current divider (Fig. 2.17b). In the first connection, the output current I_{OUT} is inherently a nonlinear function of R_1:

$$I_{OUT} = \frac{U}{R_1 + R_L} = \frac{U}{R} \cdot \frac{1}{\dfrac{x}{l} + \dfrac{R_L}{R}} \tag{2.15}$$

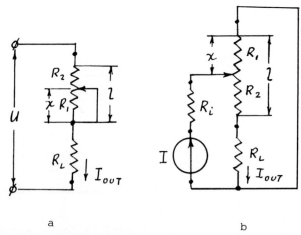

a
                                                b

Figure 2.17 Connection of potentiometric element as rheostat (a) and as current divider (b).

In the second circuit, the internal resistance R_i of the current source is much larger than the equivalent resistance of the element [$R_i \gg R_1 R_2 / (R_1 + R_2)$]. The load resistance R_L is taken to be smaller than R. The output current I_{OUT} is calculated as for the simple current divider:

$$I_{OUT} = I \frac{R_1}{R + R_L} = I \frac{\dfrac{x}{l}}{1 + \dfrac{R_L}{R}}. \qquad (2.16)$$

If $R_L \ll R$,

$$I_{OUT} = I \frac{x}{l}. \qquad (2.17)$$

A potentiometric element can be easily adopted for providing a prescribed function of the output voltage versus displacement of the wiper. The nature of this function is defined by the shape of the mandrel or by its variable height $h(x)$ if it is made of a flat plate (Fig. 2.18) of uniform thickness b.

For the given $U_{OUT} = f(x)$, function $h(x)$ can be found by making two assumptions:

1. the winding is uniform along the mandrel, and
2. the resistance of the load is much greater than the overall resistance of the element.

The output voltage is

$$U_{OUT} = U \frac{R_x}{R} = U \frac{l_x}{l_w} \qquad (2.18)$$

where R_x = resistance between points A and B, Ω;
   R = total resistance of the winding, Ω;
   l_x = length of wire between points A and B, m; and
   l_w = overall length of wire corresponding to resistance R, m.

R and R_x are simply calculated:

$$R_x = \rho \frac{l_x}{A} \quad \text{and} \quad R = \rho \frac{l_w}{A} \qquad (2.19)$$

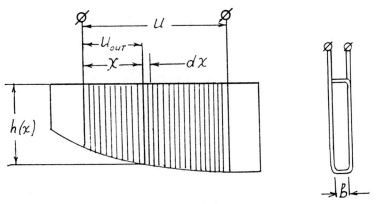

Figure 2.18   Potentiometric element for providing prescribed function.

where ρ = resistivity of wire, $\Omega \cdot m$, and
      A = cross-sectional area of wire, m^2.

The length of one turn at coordinate x is approximately $2[h(x) + b]$. If the total number of turns is N and the length of the potentiometer is l, N/l represents the number of turns per unit of length. This value multiplied by the length of one turn gives the length of wire per unit of length of the potentiometer:

$$\frac{N}{l}\,[2h(x) + 2b]. \qquad (2.20)$$

Thus, the differential change of the wire's length dl_x when the wiper is displaced by dx, is written

$$dl_x = \frac{2N}{l}\,[h(x) + b]dx. \qquad (2.21)$$

The differential of l_x is found from equation 2.18:

$$dl_x = \frac{l_w}{U}\,dU_{OUT}. \qquad (2.22)$$

The value of $h(x)$ is calculated from equations 2.21 and 2.22:

$$h(x) = \frac{l_w l}{2UN} \cdot \frac{dU_{OUT}}{dx} - b. \qquad (2.23)$$

Substituting l_w taken from equation 2.19, we find that

$$h(x) = m\,\frac{dU_{OUT}}{dx} - b \qquad (2.24)$$

where m is a constant defined by the construction of the potentiometer:

$$m = \frac{AlR}{2NU\rho}. \qquad (2.25)$$

A derivative of U_{OUT} with respect to x must be taken in order to find the thickness of the mandrel and to shape it properly. Note that a slope of the mandrel's profile $\Delta h/\Delta x$ should not be larger than 0.35, otherwise winding becomes complicated.

If a potentiometer with sections (Fig. 2.13e) is used, the shunting resistance R_{SH} for obtaining the given resistance of section R is calculated

$$R_{SH} = \frac{R_0 R}{R_0 - R} \qquad (2.26)$$

where R_0 = resistance of the section without shunting, Ω.

The required characteristic 1 (Fig. 2.19) is divided into sections. Chords 2, 3, and 4 approximate curve 1 and help to determine the shunting resistances.

During fabrication, the wire is wound with a certain tension, which creates stress σ_w. This stress, along with the stress due to the temperature expansion of the

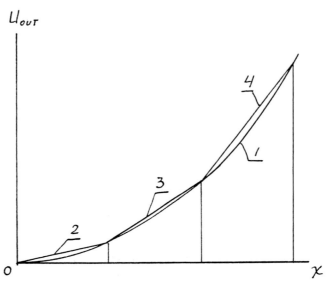

Figure 2.19 Approximation of nonlinear performance of potentiometer with chords.

wire and mandrel σ_t, must not exceed limit σ_m which is defined by the properties of the wire:

$$0 < (\sigma_w + \sigma_t) < \sigma_m \tag{2.27}$$

$$\sigma_w = \frac{F_w}{A} \tag{2.28}$$

where F_w = force during winding, N; and
A = wire's cross-sectional area, m^2.

So,

$$\sigma_t = E_w(\alpha_m - \alpha_w)(t_a - t_w) \tag{2.29}$$

where E_w = modulus of elasticity of wire material, N/m^2;
α_m and α_w = thermal coefficients of expansion for the materials of the mandrel and wire, respectively, 1/°C;
t_a = ambient temperature, °C; and
t_w = temperature at which wire was wound, °C.

The dimensions of the potentiometer are determined by the power that is dissipated in it. A current through the resistive conductor should not exceed the level that is restricted by the operational conditions and heat balance:

$$\frac{I^2 R}{\mu A_c} \leq (t_{m\,max} - t_{a\,max}) \tag{2.30}$$

where I = maximum current allowed due to heating, A;
R = resistance of the element, Ω;
μ = heat transfer coefficient, W/(m^2 · °C) [μ has a range of 12 to 15W/(m^2 · °C)];
$t_{m\,max}$ = maximum temperature allowed for the materials or for the specific conditions, °C (e.g., drift of the potentiometer's resistance due to the temperature change);
$t_{a\,max}$ = maximum temperature of ambient, °C; and
A_c = cooling surface, m^2.

Potentiometers with resistive conductors made of plastic, metal film, or metal ceramics have practically an indefinite resolution. However, in the wire-type potentiometer, the resistance is changed in steps, not gradually.

Figure 2.20a illustrates three consequent positions of wiper 1 when it moves along the x-axis and contacts round metal turns 2 with insulation 3. They are shown in sections. The width of the wiper is slightly larger than two diameters of wire as it is in the real construction (between 2 and 3 diameters of wire). Figure 2.20b illustrates on a large scale a step change of the output. When the wiper moves along the x-axis, it shorts the turns, creating "jumps" at the output. If the length of winding l carries N turns, the output will have $2N$ steps. The resolution of the element will be defined by these steps or by the geometry and size of the contact area of the wiper and wire. In position A, sections a and b are shorted. The output voltage U'_{OUT} for the unloaded element with uniform winding is defined as follows:

$$U'_{OUT} = U \frac{n}{N - 2} \tag{2.31}$$

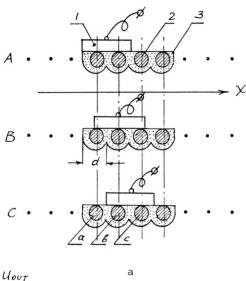

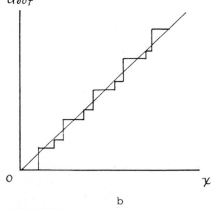

Figure 2.20 Positions of wiper and output of element. a = consecutive positions of wiper when it is in contact with wire conductors, b = discrete output voltage as function of gradual motion of wiper.

where N = total number of turns;

$\quad n$ = number of turns to the left (before turn a); and

$\quad U$ = voltage applied to the element, V.

In position B (displacement by $d/2$) sections a, b, and c are shorted, and voltage U''_{OUT} at the output will be

$$U''_{OUT} = U\frac{n}{N-3}. \tag{2.32}$$

The output voltage U'''_{OUT} for position C (additional displacement by $d/2$) is

$$U'''_{OUT} = U\frac{n+1}{N-2}. \tag{2.33}$$

The approximate differences between these voltages, which define the steps at the output voltage, are found as follows:

$$U''_{OUT} - U'_{OUT} = U\frac{n}{N^2}$$

$$U'''_{OUT} - U''_{OUT} = U\left(\frac{1}{N} + \frac{n}{N^2}\right). \tag{2.34}$$

They do not have the same magnitude, so the steps vary in height. It is obvious that the resolution cannot be higher than the diameter of the wire.

The sensitivity of the element's resistance to temperature change must be taken into account. The resistance as a function of temperature is usually calculated from the linear function as follows:

$$R = R_0(1 + \alpha_t \Delta t) \tag{2.35}$$

where $\quad R$ = resistance of the element, Ω;

$\quad R_0$ = original resistance, Ω;

$\quad \alpha_t$ = thermal coefficient of resistance, $1/°C$; and

$\quad \Delta t$ = change of temperature, $°C$.

The strain gage effect due to thermal expansion and stress can also create a change in resistance. Therefore, the thermal coefficients of the wire and mandrel should not differ significantly. A voltage between the wiper and wire can be generated due to the thermoelectric effect. The combination of materials must be correct, especially if the element is excited with direct current. Dc drifts are reduced by using an alternating current. This current also provides more stable signal conditioning. However, the frequency of the excitation voltage should not be very high (radio frequency); otherwise the influence of distributed capacitance and inductance can influence the accuracy of the device. An equivalent circuit of the element for an alternating current is shown in Figure 2.21a. The total impedance of the element and phase angle can be written as follows:

$$Z = \frac{(R + j\omega L)\dfrac{1}{j\omega C}}{R + j\omega L + \dfrac{1}{j\omega C}} = \frac{R}{(1 - \omega^2 LC)^2 + \omega^2 C^2 R^2}$$

$$+ j\omega \frac{L - CR^2 - \omega^2 L^2 C}{(1 - \omega^2 LC)^2 + \omega^2 C^2 R^2} = r + jx = |Z|e^{j\varphi}. \tag{2.36}$$

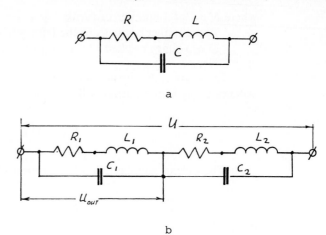

Figure 2.21   Equivalent electrical circuits of element for ac excitation.

Since the reactance is much smaller than ohmic resistance, $\tan \varphi \approx \varphi$ and

$$\varphi = \frac{\omega[L(1 - \omega^2 LC) - CR^2]}{R} \tag{2.37}$$

where Z = element's impedance, Ω;
        R = ohmic resistance of wire, Ω;
        L = equivalent inductance, H;
        C = equivalent capacitance, F;
        r = equivalent active resistance of the element, Ω;
        x = equivalent reactance, Ω;
    $|Z|$ = magnitude of Z, Ω;
        φ = phase angle, rad; and
        ω = current's angular frequency, rad/s.

For frequencies below 200Hz,

$$\frac{\omega L}{R} \leq 1, \qquad \omega CR \leq 1 \qquad \text{and} \qquad \omega^2 LC \leq 1. \tag{2.38}$$

Therefore,

$$\varphi \approx \frac{\omega(L - CR^2)}{R} = \omega \tau \tag{2.39}$$

where τ = time constant, s.

So,

$$\tau = \frac{L}{R} - CR. \tag{2.40}$$

Figure 2.21b illustrates an equivalent circuit when the wiper is in an intermediate position A. The following ratios should be provided in the design in order to reduce errors due to frequency:

$$\frac{R_1}{R_2} = \frac{L_1}{L_2} = \frac{C_1}{C_2}. \tag{2.41}$$

A skin effect is defined as the tendency of the alternating current to flow near the surface of a conductor. Being thus restricted to a small part of the total sectional area, the ac produces the effect of increasing the resistance. The effective resistance of the element can be calculated from the following formulas:

$$R = R_0\left(1 + \frac{\epsilon^4}{3}\right), \text{ for } \epsilon < 1, \tag{2.42}$$

$$R = R_0\left(\frac{1}{4} + \epsilon + \frac{3}{64\epsilon}\right), \text{ for } \epsilon > 1, \text{ and} \tag{2.43}$$

$$\epsilon = \frac{a}{2}\sqrt{\pi f \mu \gamma} \tag{2.44}$$

where R = effective resistance, Ω;
$\quad R_0$ = resistance to direct current, Ω;
$\quad a$ = radius of wire, m;
$\quad f$ = frequency, Hz;
$\quad \gamma$ = conductivity of material, $1/(\Omega \cdot m)$; and
$\quad \mu$ = magnetic permeability, H/m—it can be taken $4 \times \pi \times 10^{-7}$ H/m for nonmagnetic materials; for magnetic materials it has a wide range of variation (for annealed iron: from 0.5×10^{-3} to 8×10^{-3} H/m).

The materials that are used for the resistive layer must possess the following qualities:

1. the resistivity must be high in order to have a high resistance and voltage at the output;
2. the thermal coefficient of resistance must be low in order to diminish the influence of temperature on the element's resistance and on the performance of the electrical circuit;
3. the reluctance to friction wear must be high in order to have a long life of winding;
4. they should resist oxidation or the building up of chemicals on the surface, either of which can disturb the contact with the wiper, create electromotive force, or develop electrical noise;
5. they should not change the resistance with applied voltage (this happens with elements made of carbon); and
6. their characteristics should not change with time.

Noble metals and special alloys are used for the high-accuracy elements. Resin with electroconductive powder is a typical material for inexpensive devices. Using plastic allows an easy trimming of the conductive layer to obtain necessary characteristics during fabrication. Plastic elements have a long life. In some instances, a combination of wire wound with plastic is used to gain advantages of both types. The following materials are most commonly used for the wire: constantan, nickeline, nichrome, manganine, advance, tungsten, platinum, gold and silver in alloys with iridium, indium, and palladium. Precious metals with additives are harder and better withstand abrasive wear. Table 2.3 illustrates some of these materials and their properties.

The wiper is made of precious metals, which are softer than the wire material. The wiper contains two or more fine bristles of different lengths or one flat plate split into two or more reeds of different lengths. They slide together over the conductive path. The difference in length protects the sliding part from bouncing or mechanical resonance when the wiper moves quickly or when the element is exposed to vibra-

TABLE 2.3

Materials for Wire of Potentiometric Elements

Material	Composition (figures denote percentage)	Resistivity at 20°C, $\Omega \cdot mm^2/m$	Thermal coefficient of resistance 10^{-6} 1/K
Constantan	Cu 60, Ni 40	0.48	5
Nichrome	Ni 80, Cr 20	1.30	100
Manganine	Cu 84, Mn 12, Ni 4	0.43	10
Nickeline	Cu 62, Ni 18, Zn 20	0.42	20
Advance	Cu 54, Ni 44, Fe 0.45	0.49	15
Platinum	Pt 100	0.10	2450
Tungsten	W 100	0.06	4600
Platinum-Iridium alloy	Pt 90, Ir 10	0.23	1500

tions. The contacting element can also be shaped like a cylinder or rectangular prism. The material of the wiper should match the material of the wire. Silver or copper is used with graphite; an alloy of nickel and silver is used for constantan, manganine, or nichrome wire. Besides these materials, rhodium, copper with palladium, gold and silver, osmium and iridium, ruthenium with platinum, platinum and iridium, and platinum and beryllium are also utilized [1] for the wiper.

The mandrel is made of plastic and is cast in a precision mold for the low-cost elements. It can be also machined of micarta, ebonite, resin-dipped fabric laminate, and plexiglass. Some of these materials should be coated with varnish to avoid hydroscopicity and growth of fungus. The low temperature conductivity and high heat capacity of nonmetal supporting structures limit the maximum current density through the wire to 5A/mm^2. More precise and more thermal-conductive is the mandrel machined of aluminum alloys, which are anodized and/or varnished for electrical insulation. They allow a current density of 10A/mm^2 and greater. The wire for winding is also insulated with varnish. The contact path is formed by removing the insulation and polishing the path. A good quality of this path reduces electrical noise, which greatly depends on the stability of the contact resistance and on the electromotive force caused by friction.

The pressing force at the contact is between 3×10^{-3} and 5×10^{-3}N for the bristle slider and 50×10^{-3} to 100×10^{-3}N for the reed-type. The starting and running torque defined by the forces of friction in the rotational element can be between 2×10^{-6} and 200×10^{-6}N $\cdot$ m, but for conventional devices it is 100 times greater. The friction force in the translational element is 0.1 to 0.5N. The moment of inertia or mass of the moving parts is important for the dynamic characteristic of the sensor. Their ranges are from 0.1×10^{-9} to 10×10^{-9}kg $\cdot$ m^2 and 3×10^{-3} to 100×10^{-3}kg. Typical numbers describing the accuracy of the element are as follows: the error due to the discrete character of the element is 0.02 to 0.03% of full scale, nonlinearity 0.05 to 0.10%, electrical instability 0.03 to 0.1%, and temperature error about 0.1%/10°C. The life of the element varies from 2×10^6 cycles for the wire-wound type to 50×10^6 cycles for the plastic made. The major shortcomings of the wire-wound potentiometric element are the low resolution, relatively low resistance of the device, and limited usage with alternating current (capacitance and inductance effects). However, the high output signal, accuracy, stability, and predictable low temperature drifts are important advantages of this device.

STRAIN-GAGE ELEMENTS

A strain-gage element employs the effect of the change in the resistance of an electrical conductor as a response to a measured deformation.

The conductor can be a liquid, a plastic with electroconductive filling, a metal, or a semiconductor.

Liquid Strain Gages

A liquid strain gage has one of the simplest constructions. It is composed (Fig. 2.22) of an elastic (rubber) pipe 1 with two end electrodes 2 and 3, connected to leads 4 and 5. The pipe is filled with a liquid conductor 6 and is deformed by the force F, applied to the ends of the pipe. The deformation causes a change in the length of the pipe l and its diameter d. However, volume V, filled with incompressible liquid, remains constant.

The resistance of the conducting column is

$$R = \rho \frac{l}{A} \tag{2.45}$$

where R = resistance, Ω;
l = length of column, m;
ρ = resistivity of liquid, $\Omega \cdot$ m; and
A = cross-sectional area, m^2.

By multiplying the numerator and denominator by l, the resistance R can be introduced as a function of one variable, l (V and ρ are constant):

$$R = \rho \frac{l^2}{Al} = \rho \frac{l^2}{V}. \tag{2.46}$$

For a small variation in the length Δl,

$$\frac{\Delta R}{R} = 2 \frac{\Delta l}{l} \tag{2.47}$$

where ΔR and Δl are changes in R and l, respectively.

This expression is readily obtained by differentiating equation 2.46 and assigning $\Delta R \approx dR$ and $\Delta l \approx dl$.

Factor 2 in equation 2.47 is the gage factor. It shows that the relative change in the resistance $\Delta R/R$ is twice as large as the strain, $\Delta l/l$. This element can be used for a large deformation (max $\Delta l/l = 1.5$) and operates linearly up to $\Delta l/l \approx 0.5$. Mercury and electrolytes are the most common materials employed for the column.

The alternating current feeds the element to reduce the effects of polarization and chemical decomposition. Some polarization can create a capacitance effect and phase shifts (2 to 5°) between the voltage and current.

One of the electrolytes [7] has the following composition: 1 liter of water, 150g of $CuSO_4$, 50g of H_2SO_4, and 50g of ethyl alcohol. The element with this electrolyte

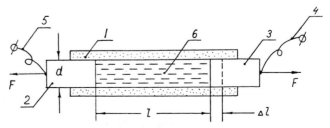

Figure 2.22 Liquid strain gage.

and platinum electrodes (d = 1.2mm, l = 30mm) has a resistance of 5.6 kΩ. The error is no larger than ±2%FS for $\Delta l/l$ = 0.5. An element with a mercury conductor (d = 0.135mm, l = 15mm) has $\Delta l/l$ = 0.2 and a resistance of 5Ω.

Alternating deformations with frequencies up to 2kHz can be measured with this element. Disadvantages of the electrolytic transducer are a high temperature coefficient of resistance (0.016 to 0.024 1/°C) and instability in elastic properties of the rubber pipe and in the electrodes' conditions. However, it is a convenient device because of its simplicity and ability to provide a signal proportional to the average deformation of a relatively long length.

Plastic Strain Gages

These elements exhibit a high gage factor reaching 300. They are shaped as ribbons or threads [9] and fabricated from the following compositions:

1. graphite, fine quartz powder, and resin;
2. graphite, chalk, and shellac (or resin); or
3. coal (or soot) and bakelite lacquer.

The elements of 40 to 60mm in length consist of threads 1 to 2mm in diameter or ribbons 4 to 6mm wide and 1mm thick.

The elements' constructions are simple and inexpensive, but they are not widely used because of mechanical instability, resistance drift with time, high hysteresis (2 to 3%), and dispersion of parameters (up to ± 20%) during manufacture.

Metal Strain Gages

A model of an *unbonded*-metal-wire element (Fig. 2.23) is represented by wire 1 that is prestrained between two electrically insulated posts 2 and 3. The wire is rigidly affixed to posts that are restricted to move only along the x-axis. Under force F the wire's length changes. If l is the effective length of the wire exposed to the deformation, its resistance is

$$R = \rho \frac{l}{\left(\dfrac{\pi D^2}{4}\right)} \qquad (2.48)$$

where R = resistance, Ω;
  ρ = resistivity of the wire's material, Ω · m;
  l = effective length, m; and
  D = diameter of wire, m.

By taking the logarithm and differential of equation 2.48 and substituting d's with Δ's, the relative change of resistance is found:

$$\frac{\Delta R}{R} = \frac{\Delta l}{l} - 2\frac{\Delta D}{D} + \frac{\Delta \rho}{\rho} \qquad (2.49)$$

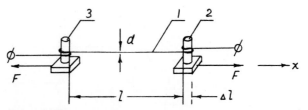

Figure 2.23 Unbonded strain gage.

By definition, Poisson's ratio ν is the ratio of the transverse contracting strain to the elongation strain when a rod is stretched by the forces that are applied at its ends and that are parallel to the rod's axis:

$$\nu = -\frac{\Delta D/D}{\Delta l/l} \tag{2.50}$$

where $\Delta l/l$ and $\Delta D/D$ are elongation and transverse strains, respectively.

They are dimensionless but for convenience are often denoted as m/m, mm/mm, or *microstrain* [μ strain = (m/m) $\times$ 10^{-6}].

For all materials, $0 \leq \nu \leq 0.5$, but for metals, $0.24 \leq \nu \leq 0.4$. After dividing both sides of equation 2.49 by $\Delta l/l$ and substituting ν from equation 2.50, the expression for the gage factor K is obtained:

$$K = \frac{\Delta R/R}{\Delta l/l} = 1 + 2\nu + \frac{\Delta \rho/\rho}{\Delta l/l}. \tag{2.51}$$

The gage factor shows how sensitive the element is to strain. In formula 2.51,

$$1 = \text{sensitivity due to a change in the length,}$$

$$2\nu = \text{sensitivity due to a change in the cross-sectional area, and}$$

$$\frac{\Delta \rho/\rho}{\Delta l/l} = \text{sensitivity to the change in resistivity under strain.}$$

The latter ratio is defined by the properties of the materials. For most metals used in strain gages, the K is between 2 and 5.

According to Hooke's law, the force F that is needed to obtain the strain $\epsilon = \Delta l/l$, is written

$$F = \sigma A = E\epsilon A \tag{2.52}$$

where F = force, N;
σ = stress, N/m^2;
A = area, m^2; and
E = Young's modulus, N/m^2.

The change in the resistance that is due to the applied force for a single wire is

$$\frac{\Delta R}{R} = F\frac{K}{EA}. \tag{2.53}$$

In many electrical circuits associated with strain gages, the output signal is proportional to the ratio $\Delta R/R$ and/or to ΔR.

The unbonded element is usually made of wire 1, which is wound around movable pins 2 (Fig. 2.24) in order to increase the total length of the wire and the change in the resistance ΔR. It is also useful for diminishing the influence of variation in resistances of leads and joints on the reading. A scheme with a differential arrangement of elements (Fig. 2.25) is more common. Two strain gages 1 and 2 form a differential mechanical and electrical structure. Pin supports 3 and 4 are stationary, whereas the center part 5 moves horizontally. The two sets of resistance wire are preloaded and never become slack. When the force F is applied to the center part, as shown in Figure 2.25, the right set of wire undergoes additional tension, while the tension in the left set is decreased. In some constructions, four

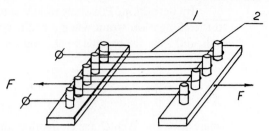

Figure 2.24 Unbonded strain gage with wound wire and movable pins.

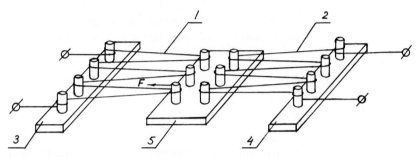

Figure 2.25 Unbonded strain gages in differential scheme.

coils are wound with the initial tension. They form two or four active arms of a Wheatstone bridge, thus providing an instrument with both a large output signal and compensation for thermal and mechanical drifts. The construction with unbonded elements is relatively expensive because precise fabrication and adjustment of parts are necessary for the normal operation of the device. However, this gage is very sensitive to force and displacement because they are directly converted to the deformation of the wire without a loss of mechanical energy in the intermediate members. Copper-nickel, chrome-nickel, and nickel-iron alloys are employed for a wire of 8 to 30 μm in diameter. The wire can sustain stress up to $2 \times 10^6 N/m^2$. The resistance of one grid can range from 100Ω to 1000Ω. Depending on the material, the gage factor ranges from 2 to 4. A typical bridge circuit with the element can develop 20 to 50mV at the output for the maximum deformation of the wire.

Originally, a *bonded* strain gage was designed and used as a zigzag grid of fine wire 1 (Fig. 2.26) bonded to a thin paper or plastic backing 2. The wire is soldered or welded to electrical leads 3 and 4. A protective film 5 covers the surface of the element. A foil element (Fig. 2.27) has a similar construction. A thin foil pattern 1 is used instead of wire for a strain-sensitive conductor. A big area 2 at the end of the

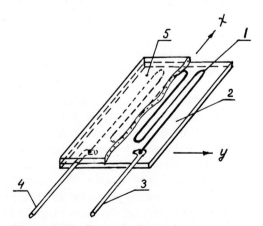

Figure 2.26 Bonded wire strain gage.

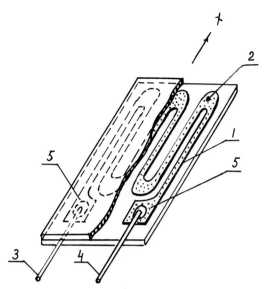

Figure 2.27 Bonded film strain gage.

loop creates better adhesiveness and decreases the sensitivity to transversal stresses. Leads 3 and 4 are connected by soldering tabs 5 to the grid. The tabs are integral with the grid pattern.

There are two basic applications of bonded gages: for measuring strains or stresses on the elements of construction and for measuring deformations in the spring elements of transducers. In both cases, the gages are cemented to a member whose deformation is transmitted through the cement and backing to the wire or foil. The deformation causes the change in the resistance, which is a measure of the deformation. The strain of the member is detected directly by measuring the strain of the resistive grid, and the stress is readily calculated from Hooke's law:

$$\sigma_m = E_m \epsilon \qquad (2.54)$$

where σ_m = stress in the member, N/m^2;

E_m = modulus of elasticity (Young's modulus) of the member's material, N/m^2; and

ϵ = strain evaluated by measuring the resistance of the gage, m/m.

The strain gage is sensitive not only to longitudinal but also to lateral strains because of the end loops. Therefore,

$$\frac{\Delta R}{R} = K_x \epsilon_x + K_y \epsilon_y \qquad (2.55)$$

where ϵ_x and ϵ_y = strains in the direction of x and y axes, respectively; and

K_x and K_y = axial and transverse gage factors, respectively.

Due to the end loops, the magnitude of the gage factor K is somewhat less than the sensitivity of material K_0. With an increase in the gage length l, the difference between K and K_0 decreases. That difference becomes negligibly small for commonly used strain gages with the increase in the area at the end of the loop and with the gage length l = 15mm or greater. The factor K_y is the same order of magnitude as the other factors. For strain gages with a short gage length ($l < 5$mm), the values K_x and K_y are comparable in magnitude and the transverse gage factor must be taken into account in the evaluation of stresses. There are a wide variety of patterns that can be combined in various numbers on the same backing for taking different readings. The single-element gages shown in Figures 2.26 and 2.27 are intended to measure the stress along the x-axis. A gage in Figure 2.28a measures the deformation along the x- and y-axes. Three measurements along the x-, y-, and z-axes are possible with the elements shown in Figures 2.28b and 2.28c.

Gage combinations in Figure 2.28d, called *rosettes,* contain three gages that form a "sandwich." They are intended to measure stresses in small areas and along three axes. The grids are superimposed over each other. The pattern in Figure 2.28e is affixed to the diaphragm of the pressure transducer. The center part of this pattern responds to tangent strains of the diaphragm and the edge part reacts to radial strain.

Metal-wire gages are superseded by the bonded metal-foil elements in many applications because of the following characteristics:

1. the foil element has a larger resistive area—it provides better grid bonding and transmission of deformations;
2. due to the larger surface, the heat transfer from the element is better and the current feeding the element can be greater, which provides a higher output signal—it more quickly follows the temperature variation of the member to which it is attached;
3. the foil element has much less sensitivity to transversal stress;
4. the size of the element can be very small, down to 0.2mm;
5. its stability is better during a prolonged loading and influence of temperature; and
6. the photoetching technique for the fabrication of the grid pattern is inexpensive and insures a low cost for the device.

It should be noted that the wire gage has a higher resistance of the grid per unit area.

It was mentioned before that a common practice in the analysis of the state of stress is determining not only the magnitude but also the direction of principal stresses. In such cases, three strain gages (Fig. 2.28c) at angles of 45° are mounted simultaneously. From the three measured strains, it is a simple matter to determine [10] the principal strains and the angle defining the principal axes (i.e., the axes corresponding to such a system of axes in which the shear strains are zero). The following formulas illustrate the basic calculation of the principal stresses:

$$\epsilon_{,} = \epsilon_x \cos^2 \varphi + \epsilon_y \sin^2 \varphi,$$

$$\epsilon_{,,} = \epsilon_x \cos^2(\varphi + 45°) + \epsilon_y \sin^2(\varphi + 45°),$$

$$\epsilon_{,,,} = \epsilon_x \cos^2(\varphi + 90°) + \epsilon_y \sin^2(\varphi + 90°). \tag{2.56}$$

After simple transformations, we obtain

$$\tan 2\varphi = \frac{\epsilon_{,} - 2\epsilon_{,,} + \epsilon_{,,,}}{\epsilon_{,} - \epsilon_{,,,}},$$

$$\epsilon_x = \frac{\epsilon_{,} + \epsilon_{,,,}}{2} + \frac{1}{2}\sqrt{(\epsilon_{,} - \epsilon_{,,,})^2 + (\epsilon_{,} - 2\epsilon_{,,} + \epsilon_{,,,})^2},$$

$$\epsilon_y = \frac{\epsilon_{,} + \epsilon_{,,,}}{2} - \frac{1}{2}\sqrt{(\epsilon_{,} - \epsilon_{,,,})^2 + (\epsilon_{,} - 2\epsilon_{,,} + \epsilon_{,,,})^2} \tag{2.57}$$

where $\epsilon_{,}$, $\epsilon_{,,}$, and $\epsilon_{,,,}$ = strains along axes I, II, III, respectively, m/m;
ϵ_x, and ϵ_y = principal strains along axes x and y, respectively, m/m; and
φ = angle defining the principal axes, degrees.

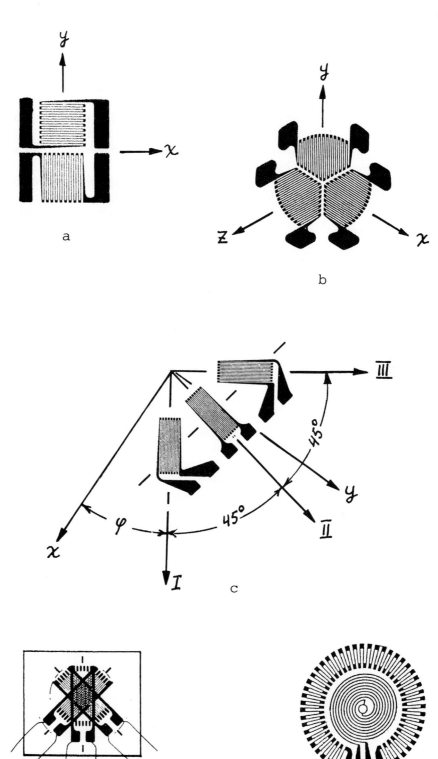

Figure 2.28 Different grid patterns of strain gages. a = two-axis, 90° pattern; b = three-axis, 120° pattern; c = three-axis, 45° pattern; d = stacked patterns; e = pattern for pressure-sensitive diaphragm. (a), (b), (c) courtesy of BLH Electronic, Inc., (d) courtesy of Measurements Group, Inc.

Advances in strain-gage technology have developed a method of measuring internal stresses in very thin specimens of hardened material. In this method [11, 12], a thin-film bismuth or semiconductor strain gages are deposited or affixed to one of the surfaces of the specimens. The other surface is etched by removing thin layers of the material. During etching, the internal balance of stresses is disturbed, and the specimen undergoes deformation that is sensed by the gages. The distribution and magnitude of stresses are then determined from the known values of strains. This method can have a number of applications. The basic calculations related to the method are given below.

A flat specimen with balanced internal stresses becomes bent (Fig. 2.29), when the outer layers are taken off. If the specimen rests on two supports, the relation between stress and bent deflection are given by the following formula [13]:

$$\sigma_a = \frac{E}{b^2}\left[\frac{(\delta - a)^2}{3} \cdot \frac{\Delta f_a}{\Delta a} \cdot \frac{1}{1 + \dfrac{\Delta a}{\delta - a}} - (\delta + \Delta a - 2a)f_{a - \Delta a}\right.$$

$$\left. - \sum_0^{a - \Delta a} \frac{(\delta - x)}{3} \cdot \frac{\Delta f}{\Delta x} \cdot \frac{1}{1 + \dfrac{\Delta x}{\delta - x}} - \sum_0^{a} x \cdot \Delta f\right]\left(1 + \frac{\nu}{1 - \nu} \cdot \frac{t - 2\delta}{t}\right) \qquad (2.58)$$

where σ_a = stress in the layer with coordinate a (Fig. 2.29), N/m²;
E = modulus of elasticity, N/m²;
b = half of support span, m;
δ = original thickness of specimen, m;
Δf_a = deflection increase due to removing the layer with thickness of Δa, m;
Δa = thickness of removed layer during one cycle (e.g., one cycle of etching), m;
a = coordinate of layer (Fig. 2.29), for which the stress is calculated, m;
x = coordinates of layers removed before the layer with coordinate a, m;
Δx = thicknesses of layers with coordinate x, m;
$f_{a - \Delta a}$ = deflection due to removing the layer with thickness $a - \Delta a$, m;
ν = Poisson's ratio, dimensionless; and
t = specimen's width (in the direction perpendicular to the plane of drawing in Figure 2.29), m.

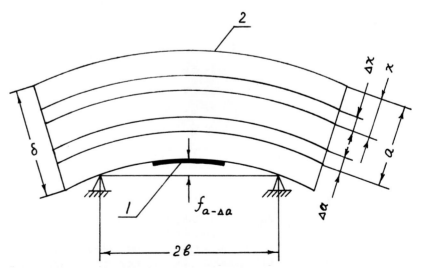

Figure 2.29 Deformation of specimen after removal of surface layers.

Using formula 2.58 as an initial expression, the relationship between the internal stresses and surface strain is calculated by simple substitutions.

When the layer with thickness x is removed, the radius of curvature ρ_x as a function of stress at the surface ϵ_x is expressed

$$\rho_x = \frac{\delta - x}{2\epsilon_x} \quad \text{and} \quad \rho_x = \frac{b^2}{2f}. \tag{2.59}$$

Therefore, the bending deflection f corresponding to ρ_x is

$$f = b^2 \frac{\epsilon_x}{\delta - x}. \tag{2.60}$$

Differential deflection df can be expressed as

$$df = \frac{\partial f}{\partial \epsilon_x} d\epsilon_x + \frac{\partial f}{\partial x} dx. \tag{2.61}$$

Taking the differential of equation 2.60 and replacing df, $d\epsilon_x$, and dx by increments Δf, $\Delta \epsilon_x$, and Δx, respectively, we obtain

$$\Delta f = b^2 \left[\frac{\Delta \epsilon_x}{\delta - x} + \frac{\epsilon_x \Delta x}{(\delta - x)^2} \right]. \tag{2.62}$$

Using equations 2.60 and 2.62 we will find that

$$f_{a - \Delta a} = b^2 \frac{\epsilon_a}{\delta - a} \tag{2.63}$$

$$\Delta f_a = b^2 \left[\frac{\Delta \epsilon_a}{\delta - a} + \frac{\epsilon_a \Delta a}{(\delta - a)^2} \right] \tag{2.64}$$

where $\Delta \epsilon_a$ and ϵ_a = strains at the surface when layers of thicknesses Δa and a are removed.

Now, by replacing in the formula 2.58 the terms containing deflections with the terms containing strains, we will obtain the function for calculations:

$$\sigma_a = E \left\{ \frac{1}{3} \left[(\delta - a) \frac{\Delta \epsilon_a}{\Delta a} + \epsilon_a \right] \frac{1}{1 + \dfrac{\Delta a}{\delta - a}} - \frac{\delta + \Delta a - 2a}{\delta - a} \epsilon_a \right.$$

$$\left. - \sum_0^{a - \Delta a} \left[(\delta - x) \frac{\Delta \epsilon_x}{\Delta x} + \epsilon_x \right] \left(\frac{x}{\delta - x} + \frac{1}{3} \cdot \frac{1}{1 + \dfrac{\Delta x}{\delta - x}} \right) \frac{\Delta x}{\delta - x} \right\}$$

$$\times \left(1 + \frac{\nu}{1 - \nu} \cdot \frac{t - 2\delta}{t} \right). \tag{2.65}$$

If only surface stress σ_{as} is calculated, formula 2.65 can be reduced to the following:

$$\sigma_{as} = \frac{E\delta}{3} \left(1 - \frac{\Delta a}{\delta} \right)^3 \frac{\Delta \epsilon}{\Delta a} \left(1 + \frac{\nu}{1 - \nu} \cdot \frac{t - 2\delta}{t} \right). \tag{2.66}$$

For a rapid analysis of internal or surface stresses and/or graphical presentation of the stresses, a computer program can be written for calculations with the preceding formulas.

Signals from the strain gage can be fed to a signal processor and microprocessor for simultaneous analysis. If the rate of layer removing (e.g., etching) is known, the inputs for thicknesses in formula 2.65 can be introduced as functions of time. During the experiments, the side of the specimen with the strain gages is insulated by an etchant-resistive layer (e.g., beeswax). Hydrofluoric acid is widely used as a main component for etching many vitreous materials.

The described method determines stresses in microstructures in which the understanding of the state of internal, residual stress is frequently important.

The maximum current through the strain-gage element is calculated from the equation of thermal balance:

$$I_m^2 R = \mu \, S_c \Delta\theta \tag{2.67}$$

where I_m = maximum current, A;
$\quad R$ = resistance of strain gage, Ω;
$\quad \mu$ = heat transfer coefficient, W/(m$^2 \cdot$ °C);
$\quad S_c$ = cooling surface of the gage, m^2; and
$\quad \Delta\theta$ = temperature of overheat, °C.

For air, μ is 10W/(m$^2 \cdot$ °C), but for the element attached to a metal base with cement, μ is between 2000 and 3000W/(m$^2 \cdot$ °C). The current can be increased if the element is excited by pulses and if its time constant is small enough to cool the element between pulses. The material of the element is sensitive to temperature. Its resistance is approximated by a linear function:

$$R_t = R_o(1 + \alpha_r \, \Delta\vartheta) \tag{2.68}$$

where R_t = element's resistance, Ω;
$\quad \Delta\vartheta$ = deviation of temperature from the original, K;
$\quad R_o$ = original resistance (for $\Delta\vartheta = 0$), Ω; and
$\quad \alpha_r$ = thermal coefficient of resistance, 1/K.

A material with a low magnitude of α_r should be chosen for an accurate device. There are a number of compensation techniques that are used to diminish the change in resistance caused by thermal drift. The compensation is arranged more easily for elements with a small α_r.

The difference between the coefficients of expansion of the element and its base is also a source of change in the resistance caused by temperature. If element 1 (Fig. 2.30) of length l_o is attached to base 2, its deformation under temperature will

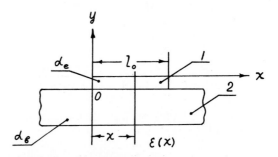

Figure 2.30 Illustration to formulas describing deformation of element on base having different coefficients of expansion.

be a function of the coefficients of expansion of the element and base. Assuming that the base is much thicker than the element and their moduli of elasticity are of the same order of magnitude, the following expressions can be written to describe the deformation of the structure:

$$\Delta l_e = l_o \alpha_e \Delta \vartheta, \tag{2.69}$$

$$\Delta l_b = l_o \alpha_b \Delta \vartheta, \tag{2.70}$$

$$\Delta l_e - \Delta l_b = l_o \Delta \vartheta \, (\alpha_e - \alpha_b), \tag{2.71}$$

$$\alpha = \frac{\Delta l_e - \Delta l_b}{l_o \Delta \vartheta} = \alpha_e - \alpha_b \tag{2.72}$$

where Δl_e and Δl_b = deformations of the unbonded element and base, respectively, when their temperature change is $\Delta \vartheta$; and
α = apparent temperature coefficient, which is an additional strain per 1K.

The apparent thermal coefficient of resistance, α_a, is found by multiplying α by the gage factor K:

$$\alpha_a = K\alpha \tag{2.73}$$

Hence, the overall thermal coefficient of resistance is

$$\alpha = \alpha_r + K \, (\alpha_e - \alpha_b). \tag{2.74}$$

The apparent deformation ϵ_a due to the action of heat is found from equations 2.51 and 2.63:

$$\epsilon_a = \frac{\Delta l}{l} = \left[\frac{\alpha_r}{K} + (\alpha_e - \alpha_b) \right] \Delta \vartheta. \tag{2.75}$$

It is obvious that it is advantageous to have a small difference between α_e and α_b.

A good compensation is possible if

$$\frac{\alpha_r}{K} + (\alpha_e - \alpha_b) = 0 \tag{2.76}$$

which corresponds to a special selection of materials.

The thermal coefficient of the gage factor is relatively small for metal elements. For instance, it is $-30.00 \times 10^{-6} \mathrm{K}^{-1}$ for constantan.

Another important characteristic of the element is determined by the properties of the cement. It must truly transmit the deformation from the base to the element. Plastic cements have a creep, which increases with temperature and is revealed when the element is exposed to a prolonged deformation. A quantitative evaluation of the creep can be made using the following formula [14]:

$$\beta = \left(\frac{R_1 - R_{01}}{R_{01}} - \frac{R_2 - R_{02}}{R_{02}} \right) \bigg/ \frac{R_1 - R_{01}}{R_{01}} \tag{2.77}$$

where
β = creep, dimensionless; and
R_{01}, R_1, R_2, R_{02} = element's resistances before loading, after loading, after a prolonged load, after a prolonged operation, but without the load, respectively, Ω.

The β for plastic cements with a one-hour load is 1 percent or less.

Johnson or white noise caused by the random motion of electrons in the element's conductors is the natural limit for signal conditioning and must be much smaller than the signal from the element.

The root mean square voltage V generated across the element is introduced as

$$V = \sqrt{4KTR\Delta f} \qquad (2.78)$$

where V = ac rms voltage, V;
   K = Boltzmann's constant—1.38×10^{-23} J/K;
   T = absolute temperature, K;
   R = resistance of the element, Ω; and
   Δf = frequency range of signal conditioning, Hz.

The typical order of magnitude for V is 1×10^{-9} V compared with the output signal of 1×10^{-5} V, which corresponds to a deformation of 1×10^{-6} m/m.

The element provides signals that are proportional to the average deformation ϵ_{av} along its length (see Fig. 2.29):

$$\epsilon_{av} = \frac{1}{l_o} \int_o^{l_o} \epsilon_x dx \qquad (2.79)$$

where l_o = element's base length, m; and
   ϵ_x = strain at the point with the x coordinate.

If a vibratory deformation or a rapidly varying transient is to be measured, the length of the element's base must be much smaller than the wave length of the sound in the material of the base. Figure 2.31 illustrates this requirement.

A bar of length l vibrates under the excitation with force $F(x)$ and has a number of compressions C and decompressions D. If the length l'_o of the element's A base is comparable with the wave length λ, a false reading is possible because of the difference in the effects from the positive and negative strains. This is true if the amplitude of deformation is measured. Measurements with element B ($l''_o \ll \lambda$) are preferable.

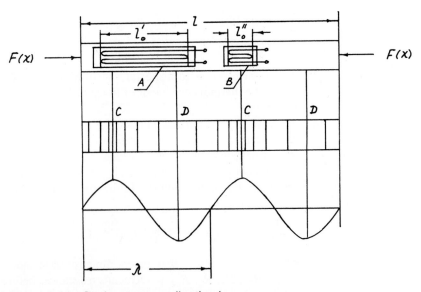

Figure 2.31   Strain gages on vibrating bar.

The wave length of sound is calculated from the following formula:

$$\lambda = \frac{v}{f} \qquad (2.80)$$

where λ = wave length of sound, m;
v = velocity of sound, m/s; and
f = frequency of sound, 1/s.

Table 2.4 illustrates the velocity of sound for different materials. For the materials that are not shown in the table, the velocity is calculated

$$v = \sqrt{\frac{E}{m}} \qquad (2.81)$$

where E = modulus of elasticity, N/m^2; and
m = density of material, kg/m^3.

The application of wire and foil elements has disadvantages because of the creep and "zero shift" associated with using cements. Thin-film technology eliminates a relatively thick layer of glue by directly depositing the strain-sensitive film on the deformed member. If this member is made of metal, an intermediate layer of a thin-film insulation is required, but it is much thinner than any cement and can be made of a material with high elastic properties.

Two basic methods are used to form a strain-gage film on the surface of the base: one, evaporation and deposition of a thin-metal film and, two, sputter deposition. During evaporation, the base is placed in a vacuum chamber and the insulating material is vaporized and condensed on the surface in order to form an insulating dielectric coating. If the base is not made of metal, this stage is omitted. The next stage is the evaporation of the metal (vacuum of 1×10^{-9}mmHg) through templates to form a desired gage pattern on the top of the base or on the insulating coating. In addition to this, relatively thick tabs are deposited at the ends of strain-sensitive patterns for soldering the electrical leads. Often, an additional insulating film is deposited on the top to protect the sensitive metal pattern. Regular evaporation can be combined with ion implantation for better adhesion of the metal with the base. In sputter technology, the first stage is the same: insulation of the base surface in a vacuum chamber by depositing the dielectric film. After this step, the metal is sputtered on the top of the insulator. The next steps consist of applying the photosensitive coating, masking, and microimaging. At the end of this entire process, a specific shape of the pattern forms on the base.

TABLE 2.4

Velocity of Sound for Some Materials

Material	Velocity in m/s
Aluminum	5110
Beryllium copper	3730
Brass	3640
Phosphor bronze	3520
Covar	4080
Constantan	4030
Invar	4010
Ni-Span C	4190
Fused Silica	5760
Tool steel	5130
Stainless steel	4970
Tungsten	4300

Next, a chemical etching or removal of unmasked metal by sputter etching occurs in the vacuum chamber. The described process is convenient for a simultaneous forming of a two- or four-active-arm electrical bridge on the surface of the sensing element. The arms are uniform in their characteristics, thus simplifying the compensation and trimming technique. The patterns can be made very small and thin. They can be integrated with a thin spring element and do not significantly change the elastic properties. The resistance of the elements can be 1000 Ω or greater. The magnitude of the gage factor is similar to that of the foil gages. The time and temperature stability are improved by the elimination of glue.

For hostile environments (temperature up to 1200°C), ceramic bonding is employed. In this case [1] a thin metal-wire gage is applied to the base by means of flame-sprayed ceramics that form both the insulating layer and the bonding cement.

Similar materials are used for different types of gages. The materials must be strong enough to withstand tensile and compressive strains and have a high gage factor, low thermal coefficient of resistance, and high resistivity. Their combination with electrical leads should not generate a junction thermoelectric potential.

Good characteristics are found in copper-nickel alloys such as constantan, nickel-chromium alloys, iron-chromium-aluminum alloys (for very high and low temperatures), and platinum-tungsten alloys. Some foils and films are made of gold-silver and titanium-aluminum alloys. Constantan is most commonly used. Table 2.5 shows the characteristics of some metals used for gages.

Table 2.6 illustrates the maximum operating temperatures for several materials employed in strain gages. These materials can have a low gage factor or resistivity but must withstand a relatively high temperature. Special alloys are fabricated for operation at 1000°C and above. The thermal coefficient of the gage factor for metal gages is usually small within the typical range of strains. The thickness of wires and foils for the gages varies from 4 to 30 μm. The thickness of the deposited film can be less than 1μm.

A number of cements used for bonding gages work at about room temperature. As a rule, cements are plastics, which become soft with a rise in temperature and do not properly transmit the deformation from the base. The cement's yielding can restrict the strain resolution to 1×10^{-7} m/m. The plastic backing can be made of nitrocellulose paper for moderate temperatures, phenolic-impregnated paper for high temperatures, and glass-fiber reinforcement and polyamide resins for a wide range of temperatures. For extremely high temperatures, the element is capsuled and can be welded to the base. The base's thickness is 0.03mm or smaller. Sometimes, to insure good bonding, an extra-thin (15μm) layer of a special enamel or ceramics is deposited on the surface of the specimen; after this procedure, the sensitive grid is cemented

TABLE 2.5

Materials for Metal Resistors of Strain Gages

Material	Composition, %	Gage factor	Resistivity, $\Omega \cdot mm^2/m$	Thermal coefficient of resistance, 10^{-6}1/K	Expansion coefficient, 10^{-6}1/K
Constantan	Cu 60, Ni 40	2.0	0.48	5	12.5
Nichrome	Ni 80, Cr 20	2.0	1.3	100	18
Manganine	Cu 84, 12Mn, 4Ni	2.2	0.43	10	17
Nickel	Ni 100	−12	0.11	6000	12
Chromel	Ni 65, Fe 25, Cr 10	2.5	0.90	300	15
Platinum	Pt 100	5.1	0.10	2450	8.9
Elinvar	Fe 55, Ni 36, Cr 8, Mn 0.5	3.8	0.84	300	9
Platinum-Iridium alloy	Pt 80, Ir 20	6.0	0.36	1700	8.9
Platinum-Rhodium alloy	Pt 90, Rh 10	4.8	0.23	1500	8.9
Bismuth	Bi 100	22	1.19	300	13.4

TABLE 2.6

Maximum Temperature at Which the Materials Used for Strain Gages Can Operate

Material	Temperature, °C
Constantan	400
Chromel	800
Nichrome	1000
Platinum	1300
Platinum-Iridium alloy	1300

under pressure, thus providing a thin cement film (about 1μm). Waterproofing is important to protect the cements against the absorption of moisture, which leads to an increase in their volume and resistance and to changes in mechanical and electrical characteristics of the element. For dynamic measurements, the low yielding of cements is not as critical as for static measurements. For instance, elements having a limit of 600°C for static measurements are good for a temperature of 800°C in dynamics.

Gluing the gages is a delicate operation and requires a certain expertise. Cements need a setting time and a thermal treatment for proper curing.

Most commonly, gages are used to measure the longitudinal deformation within 2.5×10^{-3}m/m. For the "average" element with a gage factor between 2 and 3, the change in resistance is 5×10^{-3} to $7 \times 10^{-3}\Omega/\Omega$. However, there are special designs that measure strain in a wider range [1] reaching as high as 100×10^{-3} or 200×10^{-3}m/m. Constantan, which is annealed in a vacuum and mounted on high-elongation backings, is employed for measuring the post-yield strains when a specimen is loaded above the yield point.

Fatigue-life gages contain a specially treated foil, which can cumulate deformation under a cyclic loading. The change in the gages' resistance is irreversible and is proportional to the number of cycles and to the magnitude of deformation during each cycle.

Manganine wound around a spool forms an element sensitive to hydrostatical (three-dimensional) pressure. Immersing the pressure-sensitive spool in gases and liquids measures relatively high pressures in the media.

The metal elements are manufactured with resistances of 5 to 1000Ω. The vibration strength of the elements depends on the maximum deformation during one cycle.

A correlation (Table 2.7) is found [15] between the deformation and the number of cycles without deterioration of characteristics of the elements.

Depending on their sizes, metal strain gages can be used for measurements in dynamics with frequencies between 0 and 100kHz and to respond to shock waves with components higher than 500kHz. They can work at accelerations exceeding 1.6×10^6m/s^2, ambient pressures greater than 1×10^9Pa, and magnetic field densities greater than 2T.

A compensated element provides an accuracy of the sensor of 0.2 to 0.5%FS. Precision devices with special measures can reach an accuracy of 0.02%FS.

In spite of the small output signal (small change in resistance) and sensitivity to temperature, the metal-made strain gages have a number of positive features: small size and weight, small inertia, simplicity of mounting, and low cost.

Semiconductor strain gages will be discussed in Chapter 9.

TABLE 2.7

Deformation per Cycle Versus Number of Cycles

Maximum deformation per cycle, m/m	5×10^{-4}	1×10^{-3}	2×10^{-3}	3×10^{-3}
Number of cycles	$\geq 10^9$	$\geq 10^8$	$\geq 10^7$	$\geq 10^5$

ELECTRODE ELEMENTS

An electrode element converts certain chemical quantities into an electrical signal or parameter. It can also form a signal proportional to mechanical, thermal, electrical, and other nonchemical quantities where chemical transformations are an intermediate stage.

Generally, the element is composed of a container carrying an electro-conductive fluid (usually liquid) with two or more electrodes immersed in it. The electromotive force, current, and impedance are usually measured. They give information about the composition of the fluid, concentration of chemicals, displacement, velocity, and so on.

The fluid should be electroconductive (an electrolyte). The electrodes are employed to pick up the electrical signal. They also react to the substances in the electrolyte, creating a specific electrochemical conversion. The electrical characteristics of the element are defined by the composition and concentration of the electrolyte, the materials of the electrodes, the type and magnitude of the voltage applied to the electrodes, or the current flowing through the electrolyte. They also depend on factors such as the type of chemical reaction developed, temperature, and speed of migration of the liquid. Most commonly, the elements are used to evaluate the composition and solution concentration (alkalinity or acidity).

Electrical Potentials and Conductivity of Solutions

The following is a brief description of some phenomena that relate to the operation of electrode elements.

The electroconductive substance of the element can be gas, liquid, or solid. Generally, an aqueous solution is used. The dissociation of water is very small due to strong bonds between the atoms:

$$H_2O \leftrightarrows H^+ + OH^-. \tag{2.82}$$

Having a high dielectric constant, water is very active in forming ions of acids, alkalies, and salts. Electrolytic dissociation is the separation of the molecule into two or more fragments or ions that carry electrical charges. Positive ions are called cations because they are attracted to the cathode ($-$) in a cell with electrodes. Negative ions are called anions because they are attracted to the anode ($+$) in a cell with electrodes. Dissociation occurs in a melted salt (without a solvent) due to broken bonds in molecules when a substance is heated and liquified. The melted salt reveals properties of a solution.

Ions move in the solution due to action of the electrical current, mechanical motion of the liquid's particles, or different concentrations of ions in the different parts of the electrolyte and members associated with the electrolyte. When ions move, the solution becomes electroconductive. The electrical conductivity γ is defined by the type of ions in the solution, chemical concentration, and mobility of ions. Ohm's law for this case is written

$$J = \gamma E \tag{2.83}$$

where J = current density, A/m^2;
    γ = conductivity, S/m (siemens per meter); and
    E = electrical field intensity along the path of ions' migration, V/m.

The concentration is the weight of a substance per unit of volume: for instance, milligrams per liter (mg/L), grams per liter (g/L), and kilograms per liter (kg/L).

More commonly, concentration is expressed as molarity, which is the number of gram-molecular weights of substance in one liter of solution. In this instance, the numbers showing the concentration precede capital M (1M, 0.01M, 5.6×10^{-6}M, etc.). The normal solution is an aqueous solution containing one equivalent weight of the active reagent in grams in one liter of the solution. Equivalent weight is the number of parts by weight of an element or compound that will combine with or replace, directly or indirectly, 1.008 parts by weight of hydrogen, 8.00 parts of oxygen, or the equivalent weight of any other element or compound.

A normal solution is an aqueous solution containing one equivalent weight of the active reagent in grams in one liter of solution.

The conductivity is a function of concentration [16]:

$$\gamma = \lambda f c = \lambda a, \tag{2.84}$$

$$a = c \cdot f \tag{2.85}$$

where λ = equivalent conductivity;
$\quad f$ = coefficient of activity;
$\quad c$ = concentration of the solution; and
$\quad a$ = chemical activity.

Chemical properties depend on chemical activity, and a is often a parameter to be measured. $f = 1$ for very low concentrations, and it becomes smaller for the higher concentrations.

The conductivity greatly depends on the temperature. For small concentrations (lower than 0.05M)

$$\gamma_t = \gamma_0[\, 1 + (t - t_0)\beta] \tag{2.86}$$

where $\quad \gamma_t$ = conductivity at temperature t, S/m;
$\quad \gamma_0$ = conductivity at temperature t_0, S/m;
t and t_0 = temperatures, K; and
$\quad \beta$ = temperature coefficient of conductivity, 1/K.

For solutions with small concentrations and a temperature of 18°C, β is assumed to be 0.016, 0.019, and 0.024K^{-1} for acids, alkalies, and salts, respectively. With a temperature increase over 40°C or 50°C, γ_t (t) becomes nonlinear with a decrease of β. With the increase of concentration, function $\gamma = f(c)$ becomes nonlinear with a slope reversal (Fig. 2.32 and Fig. 2.33).

If a metal electrode is immersed in a solution with a small concentration of positive ions, some atoms of the electrode will dissolve in the solution, turn into positive ions, and give the liquid a positive charge. Since the electrode receives an excess of electrons, it is charged negatively. In the solutions with a high concentration of positive ions, the ions are deposited on the surface of the electrode, giving it a positive charge. The electrode potentials are very informative. Usually two or more electrodes are placed into the solution for measuring the difference between two potentials. In this case, one of the electrodes is a reference, as hydrogen is with respect to the solution with a normal activity of hydrogen ions. The electrode is made of a platinum sponge with hydrogen absorbed on the surface. This sponge is immersed into a solution containing positive hydrogen ions. Gaseous hydrogen is then pumped into the solution. Platinum does not react, but does conduct electrical current and absorb hydrogen on the surface:

$$H_2 = 2H^+ + 2e. \tag{2.87}$$

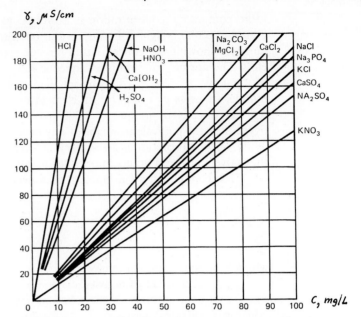

Figure 2.32 Conductivity as function of concentration for solutions with low concentrations. Conductivity is expressed as microsiemens (submultiple of siemens). (Courtesy of Siemens A.G.)

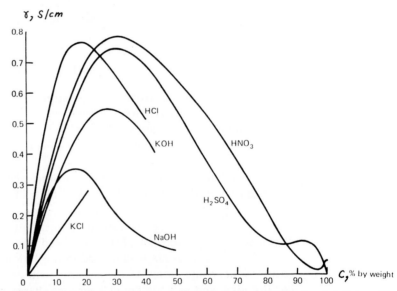

Figure 2.33 Conductivity as function of concentration for solutions with high concentrations. (Courtesy of Siemens A.G.)

An electrode potential of some substances relative to the hydrogen electrode is calculated from the Nernst equation:

$$E = E_0 + \frac{RT}{nF} \ln fc \tag{2.88}$$

where E = electrode potential, V;

E_0 = (a) normal potential of the electrode at 18°C and at the normal concentration of solution or

(b) standard potential that the electrode has in the solution when the electrode-ion activity equals unity, V;

R = 8.317 = universal gas constant, J/K;

T = absolute temperature, K;

n = ion valency;

F = Faraday's constant (or faraday) = 96487.0 ± 1.6, C/g-eq. (the electric charge required to liberate 1 gram-equivalent of a substance by electrolysis);

c = ion concentration, mol/L; and

f = coefficient of ion activity.

Equation 2.88 can be modified for a temperature of 18°C:

$$E = E_0 + \frac{0.058}{n} \log fc. \qquad (2.89)$$

E_0 varies within ± 3V: for instance, for Zn, $E_0 = -0.76$V, and for Cu, $E_0 = +0.34$V.

An element for measuring the electrode potential can be composed of two half-cells with two separated electrodes (their construction will be described below). The difference of potentials E_{12} is determined from the Nernst equation:

$$E_{12} = E_1 - E_2 = E_{01} - E_{02} + \frac{RT}{nF} \ln \frac{f_1 c_1}{f_2 c_2} \qquad (2.90)$$

where indexes 1 and 2 are related to half-cells 1 and 2, respectively. Explanation of the terms in equation 2.90 is the same as in formula 2.88. The difference in potentials is defined by the electrodes' materials and by the ratio of activities. If the electrodes of the same material are in contact with similar substances of different concentrations,

$$E_{01} = E_{02} \quad \text{and} \quad E_{12} = \frac{RT}{nF} \ln \frac{f_1 c_1}{f_2 c_2}. \qquad (2.91)$$

Some difference in potentials can also be developed at the interface of two solutions as a result of different speeds of diffusion of ions of the solutions (a diffusion difference in potentials). The order of magnitude of these potentials is between units and dozens mV.

If two solutions are used and the potential must be reduced, the intermediate concentrated solution is used with similar activities of anions and cations: for instance, KCl and KNO_3. Semipermeable membranes made of porous glass are employed at the interfacing of the solutions. Some of the potentials can be created if there is a selective passage or reluctance for the particular ions. Nonmetal electrodes can also have a charge if ions deposited on their surfaces.

The passage of electrical current through a solution causes the cations to migrate to the cathode. Here they combine with electrons, which are supplied by a current source, and form hydrogen or atoms of metals. The anions, being attracted to the anode, lose their electrons and form atoms. The electrons then flow to the current source and to the cathode. According to Faraday's law, the same amount of electricity (faraday) for different substances is required to convert one gram-equivalent of a substance. The current through the electrodes depends on the type of current source (dc or ac), ion activity, electrode potentials, and rate of diffusion of ions from the solution to the layer at the electrode's surface. Polarization is a significant factor in the current flow through the electrolyte. Due to electrolysis, the concentration of the ions at the electrodes changes. Their potential also changes. A voltage applied to the electrodes is divided between the voltage drop on the resistance of the solution and the potentials of polarization of the electrodes. Ohm's law for this case is introduced by the following formula:

$$I = [U - (E_a - E_c)]/R \qquad (2.92)$$

where I = current through solution, A;
      U = voltage applied to the electrodes, V;
      E_a = anode potential, V;
      E_c = cathode potential, V; and
      R = resistance of a column of solution between two electrodes, Ω.

Voltages of polarization that form E_a and E_c depend on the current and on certain properties of the solution and electrodes. Figure 2.34 illustrates the change of E_a and E_c as a function of current.

    If two metal electrodes are immersed in a salt solution, both will gain the normal potential E_0, and the difference in the potentials of the two electrodes will be zero. With an increase of current I, the potentials of the anode and cathode (E_a and E_c) are slightly increased and decreased, respectively. A smooth change of the potentials for the small current ($0 < I < I_m$) is defined by the diffusion of the ions; but the rate of diffusion is limited. The final number of charge carriers approaches the electrode, and the current's rise is stopped at I_m. When the potential is increased above I_m, the current reaches another plateau, which will correspond to the different ions (for the solution of several salts) or to the dissociation of water where the charges are brought to the cathode by hydrogen ions. Therefore, each step in the curve corresponds to the specific type of cation. The maximum current I_m (in A) is defined as follows:

$$I_m = ASc/\Delta \qquad (2.93)$$

where A = constant, which depends on the valency, Faraday's constant, and coefficient of diffusion, A · L/mol · m;
     S = area of the electrodes, m^2;
     c = concentration of ions, mol/L; and
     Δ = thickness of the diffusion layer, m.

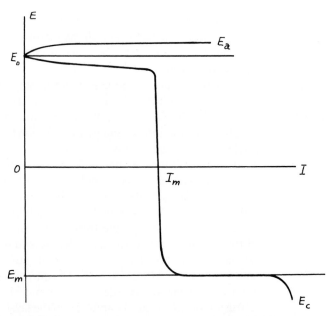

Figure 2.34 Change of anode and cathode potentials as function of current through electrodes.

Any solution containing ions is conductive for ac, which is used to measure electroconductivity and to reduce the effects of polarization. For the alternating current, the polarization is also alternating. In other words, the concentration of ions at the electrodes will be composed of a certain bias and a sinusoidal part. If the frequency of ac is not too high, the effects of polarization will follow the current with a certain phase shift. During the positive half-cycle, the anode, being at the positive potential, releases cations, and their concentration at the electrode is increased. At the end of the positive half-cycle, the ion concentration and voltage of polarization reach the maximum. A reverse process will occur at the negative half-cycle when the former anode becomes a cathode and absorbs cations from the solution, thereby decreasing its potential to the minimum at the end of the negative half-cycle. As a result, the voltage of polarization lags the current by 90°, but the voltage drop on the column of the solution between the electrodes coincides with the current. To avoid the effects of polarization, the electrodes in some devices are not immersed in liquid, but have an inductive or capacitive coupling with it.

Electrokinetic or zeta potentials are associated with phenomena related to the movement of charged particles through a continuous medium or with the movement of a continuous medium over a charged field. When a polar fluid passes through a permeable refractory-ceramic or fritted glass member, an electrical potential is generated between two chambers. If the members are furnished with electrodes, an electrical potential proportional to the speed of particles can be picked up. This is called an electrokinetic or zeta potential.

The effect is reversible: the liquid will move through a semipermeable partition if it is subjected to the electrical field.

In water solutions, the activity of hydrogen ions depends on the alkalinity or acidity of the solution. The activity of the H^+ ions becomes greater with the increase of acidity, and the activity of the negative ions decreases. The reverse takes place for the increased alkalinity.

For water or water solutions, K_{H_2O}, the product of activities of H^+ (a_{H^+}) and OH^- (a_{OH^-}) ions is constant at a constant temperature:

$$K_{H_2O} = (a_{H^+}) \times (a_{OH^-}) = \text{const.} \qquad (2.94)$$

At 25°C, $K_{H_2O} = 1.0 \times 10^{-14}$ (g-ion/L)^2. In pure water and neutral solutions,

$$a_{H^+} = a_{OH^-} = 1.0 \times 10^{-7} \text{ g-ion/L.} \qquad (2.95)$$

In acid solutions, $a_{H^+} > a_{OH^-}$, and in alkaline solutions, $a_{H^+} < a_{OH^-}$.

Since the range of a_{H^+} and a_{OH^-} variation is wide, it is practical to describe the properties of solutions by a negative logarithm of activities, which is denoted by pH:

$$\text{pH} = -\log a_{H^+}. \qquad (2.96)$$

The logarithmic relationship between pH and a_{H^+} means that the change for one unit of pH corresponds to one order of magnitude of a_{H^+} change. The pH changes from 0 for strong acids to 14 for strong alkali. Number 7 corresponds to a neutral solution. An increase in numbers from 0 to 7 corresponds to a decrease of acidity. An increase from 7 to 14 corresponds to an increase of alkalinity. Some solvents have much higher numbers for pH. For example, ammonia's pH reaches 33. The most common method of determining the pH is to measure the electrode potentials of different electrodes.

Some other phenomena and relationships are employed in electrode elements; however, the main ones have been described above, and the rest are often a combination or modification of those considered here.

Elements for the Measurement of Conductivity

An element for measuring conductivity is made of electrodes with direct and indirect contact with a solution. In the first version (Fig. 2.35), two flat plates 1 and 2 with leads 3 and 4 insulated from solution 5 are attached to the opposite sides of container 6. Due to the definite geometry and dimensions of the electroconductive column, resistance R or conductance $G = 1/R$ is determined as follows:

$$R = \frac{1}{\gamma} \frac{l}{hb} = \frac{1}{\gamma} K_e = \frac{1}{\lambda a} K_e \qquad (2.97)$$

where b, h, and l are dimensions in cm as in Figure 2.35.

$K_e = l/hb$ is a constant of the element. The unit for K_e is 1/cm (it is common to express γ in S/cm and submultiples of it such as μS/cm and K_e in 1/cm). γ, λ, and a are as in formula 2.84.

The constructions of electrodes and containers with more complicated shapes (Fig. 2.36) are used in practice more frequently. For these cases, factor K_e is found experimentally or calculated in a more complicated way than in formula 2.97. The electrodes are introduced by two parallel cylinders (Fig. 2.36a), concentric members (Fig. 2.36b), annular rings (Fig. 2.36c), two arches (Fig. 2.36d), and so on.

In order to reduce polarization, the four-electrode cell (Fig. 2.37) is also used. In this cell, the current-carrying electrodes 1 and 2 are separated from electrodes 3 and 4 for picking up the voltage drop along the separating volume. This voltage can be a measure of the resistance between electrodes 3 and 4. To eliminate polarization, the input of the voltage-measuring circuit should not draw a large current. This circuit is also immune to the contamination of current electrodes. The typical electrode materials are nickel, graphite, stainless steel, ferrous-nickel alloys, platinum gold, noble-metal plated substrates, and other materials inert to the solution. The electrodes are isolated [1] by polyvinylchloride, epoxies, silicone rubbers, and high-temperature plastics.

Using an ac current is also an effective measure against polarization. As mentioned earlier, when an element with two electrodes is fed by an alternating current, the voltage of polarization lags behind the current by 90° and the voltage drop on the separating volume has no phase shift with the current. Therefore, the equivalent circuit and phasor diagram (Fig. 2.38) for ac contain two voltage drops U_{C1} and U_{C2} across capacitors C_1 and C_2. The voltages and capacitors introduce effects of polarization. They are in a vectorial sum with voltage U_R on the column of electrolyte. The sum of the voltages U is applied to the cell. This voltage produces current I. Strictly speaking, C_1 and C_2 are not constant since U_{C1} and U_{C2} are not proportional to current I. The order of magnitude of U_{C1} and U_{C2} is between a fraction of one volt and several volts. C_1 and C_2 are relatively large and depend on the size of the electrodes (fraction of a farad per square meter). A dielectric constant of the solution contributes a small capacitance parallel to the leads in the equivalent circuit.

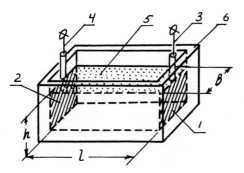

Figure 2.35 Electrode element with flat plates for measuring electroconductivity.

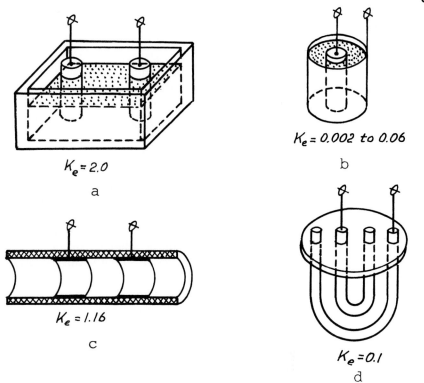

Figure 2.36 Different configurations of conductance element. Corresponding electrode constants K_e [1] are as shown.

Electrode elements are used for measurements in a wide range of conductivities: from 1×10^{-8}S/cm (for the water solutions) to 1S/cm (for the melted salts).

Another approach, which eliminates problems associated with polarization and contamination of electrodes, is a noncontact interfacing between an electrolyte and members sensitive to the conductivity. Figure 2.39 illustrates one of the devices employing this concept. A transformer carrying primary winding 1 has a secondary winding 2 introduced by a pipe with a hollow shunt 3. The pipe and bridge are made

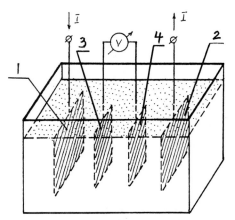

Figure 2.37 Four-electrode cell for measuring electrical conductance and reducing effect of polarization.

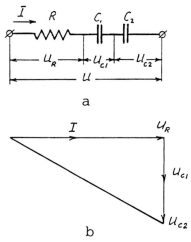

Figure 2.38 Ac equivalent circuit (a) and phasor diagram (b) for element with two electrodes and solution.

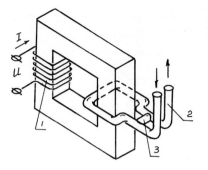

Figure 2.39 Element with transformer for noncontact measurement of conductivity.

of nonconductive material and carry a solution to be tested. The liquid in the loop and bridge forms one short turn that can be shown in the equivalent circuit as the resistance of liquid in the closed path R_s (Fig. 2.40) parallel to capacitance C_s due to the dielectric constant of the solution. An equivalent circuit in Figure 2.40a shows the terms referred to the primary. In this circuit the terms are defined as the following:

U = voltage applied to primary, V;
I = current through primary, A;
ω = angular frequency of voltage U, rad/s;
R_1 = resistance of primary, Ω;
L_1 = leakage inductance of primary, H;
L_2' = leakage inductance of secondary, referred to primary, H;
R_c = equivalent resistance of losses in the core, Ω;
L_c = magnetizing inductance, H;
R_s' = resistance of one turn of liquid referred to primary, Ω;
x_s' = capacitive reactance of one turn referred to primary, Ω; and
a = turns ratio; that is the ratio of the number of turns of primary (W_1) to secondary (W_2).

It can be assumed for many cases that $L_1 \gg L_2'$, $L_c \gg L_1$, $R_c \gg \omega L_1$, $R_s \ll 1/\omega C_s$, and an equivalent circuit can be reduced to that shown in Figure 2.40b. If $R_s' \gg R_1$ and $R_s' \gg \omega L_1$, further reduction results in the circuit in Figure 2.40c.

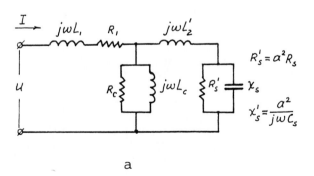

a

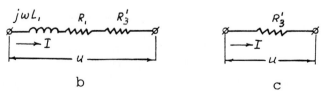

b   c

Figure 2.40 Equivalent circuits for element in Figure 2.39. a = full circuit, b and c = reduced circuits.

Since $W_2 = 1$,

$$R'_s = W_1^2 R_s. \tag{2.98}$$

Assuming that the coupling coefficient is equal to unity (no loss of energy),

$$I = \frac{U}{\sqrt{(\omega L_1)^2 + (R_1 + W_1^2 R_s)^2}}. \tag{2.99}$$

With the assumptions for the circuit in Figure 2.40c,

$$I = \frac{U}{W_1^2 R_s}. \tag{2.100}$$

Representing the resistance of the liquid path as

$$R_s = \frac{l}{\gamma A}, \tag{2.101}$$

the conductivity is found as follows:

$$\gamma = \frac{I}{U} \cdot \frac{l}{A} W_1^2 \tag{2.102}$$

where I, U, and W_1 were as previously defined;
$\quad$ l = length of the current path in the liquid turn; and
$\quad$ A = cross-sectional area of the path.

If γ is expressed in S/cm, l and A must be expressed in cm and cm^2, respectively.
$\quad$ U, l, A, and W_1 are constructional parameters; therefore, γ is defined by current I.
$\quad$ Three more versions of this element are shown in Figure 2.41. In one version, a coil is placed on the test tube. In another version, a coil is placed on a manifold. The coil can also be sealed and immersed in the solution.
$\quad$ Capacitive coupling is also used for noncontact elements. The electrodes of a capacitor are separated from the solution by the wall of a test tube or manifold (Fig. 2.42a and Fig. 2.42b) or they can have an insulating coating and be immersed in liquid (Fig. 2.42c). An equivalent circuit of the element is shown in Figure 2.43.

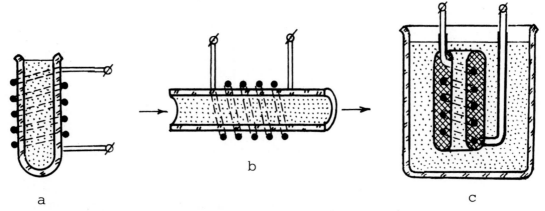

a $\qquad$ b $\qquad$ c

Figure 2.41 Noncontact inductive elements for measuring conductivity. a = coil on test tube, b = coil on manifold, c = coil immersed in solution.

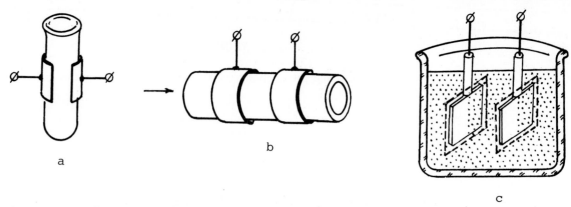

Figure 2.42 Noncontact capacitive elements for measuring conductivity. a = electrodes on test tube, b = electrodes on manifold, c = insulated electrodes immersed in solution.

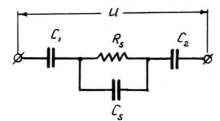

Figure 2.43 Equivalent circuit for element with capacitance coupling.

R_s is the active resistance of the solution, C_s is the capacitance that is defined by the dielectric properties of the solution, and C_1 and C_2 are capacitances due to the dielectric properties of the insulating walls or coatings. If C_1 and C_2 are known, R_s and C_s can be determined by measuring the magnitude and phase of current I. For the case where $R_s \gg 1/j\omega(C_1 + C_2)$ and $1/j\omega C_s \gg R_s$, R_s is readily found:

$$R_s = \frac{U}{I}. \tag{2.103}$$

The following ranges of frequencies are usually employed for measuring conductivity: for direct-contact elements—50Hz to 5kHz; for inductive elements—40 to 100kHz; and for capacitive elements—10 to 30MHz. Measurements are made on solutions with conductivities of 1×10^{-2} to 100S/cm.

In some constructions, the electrodes are connected in series with a temperature-sensitive resistor of copper or nickel (Fig. 2.44a). A semiconductor thermistor with

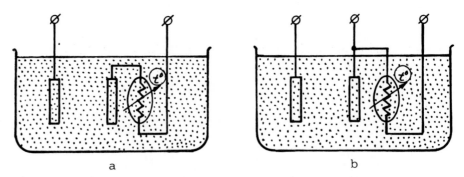

Figure 2.44 Element with temperature-sensitive resistor (TSR) for thermal compensation. a = TSR in series with electrodes, b = connection of electrodes and TSR for incorporating with electrical bridge.

a negative thermal coefficient of resistance is also used. It is connected to the adjacent arm (Fig. 2.44b) of the electrical bridge. The resistors are exposed to the same temperature as a solution and are used for thermal compensation.

Noncontact elements are often used for measuring conductivity at increased temperatures of solutions (up to 100°C).

Displacement-sensitive Elements with Electrodes

The ability of electrolytes to conduct electrical current and be as mobile as any liquid is utilized in the construction of displacement-sensitive elements. The element is usually composed (Fig. 2.45) of a nonelectroconductive container 1, which carries two or more electrodes 2 and 3, and electrolyte 4 with a gas bubble 5, located just at the area of electrodes. When the container deviates from the horizontal position, the bubble is displaced along the electrodes. The area of conductance between the electrodes is changed, making the resistance between the two electrodes a function of angular displacement. A similar effect can be achieved by changing the distance between two electrodes (Fig. 2.46). An element sensitive to two-directional deflection from the horizontal plane is shown in Figure 2.47. At the horizontal position, bubble 1 is uniformly projected on electrodes 2, 3, 4, and 5 and the resistances between the electrodes are equal. When the element is tilted, the balance is disturbed and the differences in the resistances become the measures of the angles of the deflection.

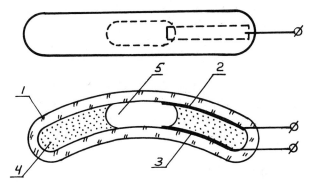

Figure 2.45 Displacement-sensitive element with electrolyte and bubble.

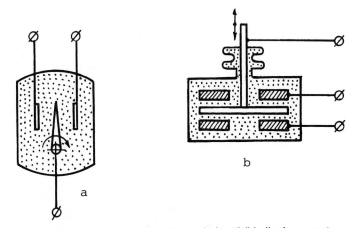

Figure 2.46 Angular (a) and translational (b) displacement-sensitive elements with electrolyte and variable-distance electrodes.

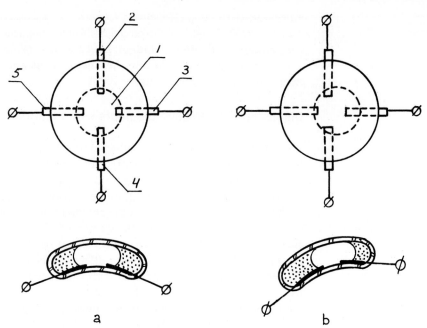

Figure 2.47 Element sensitive to deflection in two directions. a = horizontal position of element, b = tilted position of element.

Resistive Elements for Hygrometry

The operation of an element for hygrometry is based on a change of resistance between two electrodes in response to the moisture absorption of chemicals serving as spacers between the electrodes. In one construction (Scheme 128), a pipe of nonconductive materials carries strips of metal electrodes. In the internal surface of the pipe, electrodes form two insulated arrays with leads. The strips and space between them are coated with phosphorus pentoxide (P_2O_5), which has a high resistance when it is dry and a low resistance when it absorbs moisture. The absorption is accompanied by the following process:

$$P_2O_5 + H_2O \rightarrow 2HPO_3. \tag{2.104}$$

Forming the phosphoric acid defines electroconductivity. When the current flows through the element, electrolysis produces the following three components:

$$2HPO_3 \rightarrow H_2 + 0.5O_2 + P_2O_5. \tag{2.105}$$

The current of electrolysis is proportional to the absolute humidity of gas. This current is defined as follows:

$$I = FnQq/M \tag{2.106}$$

where I = current, A;
F = Faraday's constant, C/g;
n = basicity;
Q = gas flow, m^3/s;
q = absolute humidity, g/m^3; and
M = molecular weight of water.

Measurements using the element can be provided in a wide range of humidities (10^{-4} to 1% of moisture content by volume); however, the accuracy of the method is not very high, 5 to 10%.

Memistors

A memistor is a variable resistor that is used as a memory cell, integrator, or element of electrical circuits for controlling or generating slow electrical signals at infralow frequencies. The typical structure of a memistor is illustrated in Figure 2.48. It is composed of a resistive substrate 1 with two leads 2 and 3 in electrolyte 4 with the ions of metal of a controlling anode 5. A sealed ampule 6 is made of nonconducting material.

When dc voltage is applied between the anode and one of the leads, the metal of the anode is dissolved in the electrolyte and the same amount of metal is deposited on the substrate, reducing the resistance between leads 2 and 3. If the polarity of the dc voltage is reversed, the opposite process takes place: the metal is removed from the substrate and deposited on electrode 5, increasing the resistance of the substrate. In both cases, the amount of the deposited metal and the resistance change is proportional to the electricity through the controlling electrodes.

The time constant of the device is high because the deposition of metal is a relatively slow process.

There are devices that are built of solid electrolyte.

According to Ohm's law and Faraday's law of electrolysis,

$$G = G_0 + \frac{A\gamma Q}{ldnF} \tag{2.107}$$

$$Q = \int idt \tag{2.108}$$

where G = conductance of the substrate between the electrodes, S;
G_0 = initial conductance (before the action of the current causing the metal deposition), S;
A = atomic weight of deposited metal;
γ = conductivity of deposited metal, S/cm;
Q = charge of electricity, C;
l = length of the substrate along the current flow, cm;
d = density of deposited metal, g/cm^3;
n = valency of the metal; and
F = Faraday's constant, C/g.

It can be found from formulas 2.107 and 2.108 that conductivity G is a measure of the charge or current and time. In practical designs, G_0 is between 0.01 and 0.1S; the relative change of G, G_{max}/G_{min} is from 10 to 100. The dc controlling current is from 50μA to 20ma. The change of the conductance from the maximum to the minimum at the maximum controlling current is 5 to 150S (in the devices with a solid electrolyte this time is much greater). The devices can operate at a wide range of temperatures (for instance, from -200 to $+100°C$), which primarily depends on the type of electrolyte.

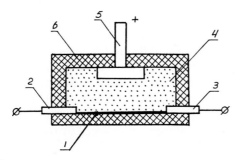

Figure 2.48 Construction of memistor.

PH-sensitive Elements with Electrodes

The pH of solutions is evaluated from the measurements of the difference in potentials of two electrodes, one of which is a measuring electrode and the other a reference electrode or standard hydrogen electrode. They form two half-cells which are employed in combination. The half-cells 1 and 2 (Fig. 2.49) are immersed in fluid 3, usually along with temperature sensor 4. In the Nernst equations, 2.88 and 2.89, the emf developed by the cells is written

$$E = E_0 + b\text{pH} \tag{2.109}$$

where E_0 = emf when pH = 0, V;

b = constant, which is defined by the type of the electrodes and temperature of the solution, V/pH.

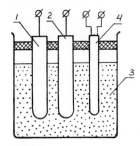

Figure 2.49 Two-half-cell system for measuring pH.

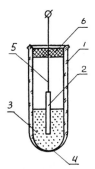

Figure 2.50 Glass-membrane electrode for measuring pH.

There are several constructions of measuring electrodes, but the most common is a glass electrode (Fig. 2.50).

The electrode assembly includes ampule 1, made of high-resistance glass. At the bottom of the ampule, electrode 2 of AgCl is dipped in a constant-pH solution 3 (0.1M HCl). pH-responsive, special formula glass (SiO_2 72%, CaO 8%, Na_2O 20% [17]) membrane 4 is formed at the lower end of the ampule. The assembly also contains an electrical lead 5 and a sealing cap 6. The outside surface of the membrane is in contact with the solution that is being tested. The difference in potentials is developed at the interface of the glass-solution. Its magnitude depends on the activity of hydrogen ions. Alkaline ions of glass, such as sodium, migrate from the membrane to the solution. Their places are occupied by hydrogen ions, which are more movable. The surface becomes saturated with hydrogen ions. A thin glass membrane (about 0.05mm) is a conductor with high resistance. In order to pick up the potential from the internal surface, the ampule is filled with a solution of known pH. The electrode is usually made of a platinum wire covered with silver chloride or silver bromide. The emf developed by the electrode has three components. The first component is proportional to the pH to be measured. It is at the interface of the solution and glass membrane. The second component is at the glass membrane and a constant-pH solution. The third component is at a constant-pH solution and an electrode. The last two components are constant; therefore, the potential at the electrode is proportional to the pH in question. The glass electrode can be used for solutions with different solvents, but it requires special calibration. Keeping the same structure of the electrode, the shape of the sensitive diaphragm can be changed. This shape (spherical, cylindrical, or flat) depends on factors such as the object of measurement and the position of the element during measurement. Besides liquid, the object can be pulp, skin, a sheet of leather, and so on.

Two basic types of reference electrodes are used to work with the measuring electrode: the silver-silver chloride electrode and the calomel electrode. A generalized structure of a reference electrode is illustrated in Figure 2.51.

The electrode assembly is composed of a glass ampule 1. A reference-metal-ion solution is enveloped in a small container 3 with a semipermeable membrane. The container carries a reference-metal-wire electrode 4. The container is dipped into salt solution 5, which interfaces through the porous plug 6 with the measured fluid outside the ampule. Plug 7 seals the ampule, and lead 8 conducts an electrical signal.

Unlike the measuring electrode, all three potentials developed at the interfaces remain constant. The interfaces are: the measured solution and the salt solution, the salt solution and the reference-metal-ion solution, and the reference-metal-ion solution and the reference-metal-wire solution. In the silver-chloride electrode, the reference electrode is made of silver or platinum coated with AgCl. KCl is usually used as a salt solution (sometimes it is replaced by the acid HCl). The silver-chloride

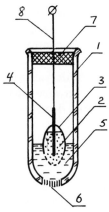

Figure 2.51 Generalized structure of reference electrode for measuring pH.

electrode can work at temperatures reaching 250°C and under vibration. The electrode is simple and its potentials are repeatable. The calomel or mercurous-chloride (Hg_2Cl_2) electrode is made of mercury. It is placed into a hard-to-dissolve solution of Hg_2Cl_2, connected with a saturated solution of potassium chloride (KCl). At the interfacing of KCl, diffusion potentials are small. The system acts as a chemical bridge in contact with the measured fluid.

The junction between KCl and the fluid is provided through fine capillaries or a porous partition.

PH elements are used for measuring the activities of the ions of many elements. The electrodes are made of various materials, including copper, gold, tin, and ion-exchange resins.

There exist electrodes sensitive to a particular type of ion (for example, anions, cations, monovalent, and divalent). Their selectivity depends on their composition and ion-exchange properties, which are extremely high for ion-exchange resins. Low resistivity is another valuable characteristic of the resins. Due to the high resistance of the glass membranes, the equivalent internal resistance of electrode elements is also high (10^6 to $10^9 \Omega$). The current must be small to reduce polarization. Under these circumstances the input resistance of the voltmeter associated with the electrodes must also be high.

Calculations with the Nernst equation show that the voltage developed by the measuring electrode is about 60mV/pH at room temperature. For the glass-electrode element, the temperature coefficient of the characteristic is about 0.2 (mV/°C)/pH.

The time constant of glass-electrode devices is usually defined by the thickness of the glass membrane. If the thickness is about 0.05mm, the potential proportional to the pH immediately follows the change of pH. The typical range for the time constant is 1 to 5 seconds.

All errors are primarily defined by the influence of temperature and diffusion potentials.

An accuracy of ± 0.02 to ± 0.01pH at temperatures of 0 to 100°C is common.

The methods of compensation of temperature drifts are similar to those for the conductance elements—the temperature-sensitive resistor is placed into the solution. It is connected directly to the electrodes, or it is a component of the feedback circuit for the signal conditioning. The temperature characteristics are obtained experimentally for the given types of electrodes, and they represent a family of lines (Fig. 2.52) passing through one characteristic equipotential point (E_{ep}, pH_{ep}). In the graph, three lines, $E(t_1)$, $E(t_2)$, $E(t_3)$, for three temperatures, t_1, t_2, and t_3, give voltages E developed by the element as functions of the pH. The slopes $a_1 = \tan \alpha_1$, $a_2 = \tan \alpha_2$, . . . , $a_n = \tan \alpha_n$ can be determined from the graph:

$$a_1 = \frac{E_1}{10 - pH_{ep}}, \qquad a_2 = \frac{E_2}{10 - pH_{ep}}, \qquad \cdots, \qquad a_n = \frac{E_3}{10 - pH_{ep}} \qquad (2.110)$$

From the last set of expressions, a can be approximated as a linear function of temperature t:

$$a = bt \qquad (2.111)$$

where b is a factor, (mV/pH)/°C.

Potentials in the (E', pH) coordinate system are expressed as follows:

$$E' = bt pH' \qquad (2.112)$$

$$pH = pH' + pH_{ep} \qquad or \qquad pH' = pH - pH_{ep} \qquad (2.113)$$

$$E = E_{ep} + E' \qquad (2.114)$$

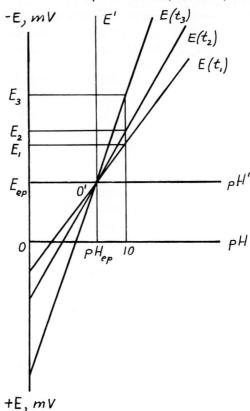

Figure 2.52 Electromotive force developed by pH-sensitive element as function of temperature.

From the last equation, remembering that E's are negative and are usually expressed in mV:

$$E = -[E_{ep} + bt(\text{pH} - \text{pH}_{ep})].\qquad(2.115)$$

Using this formula with the known values of E_{ep}, pH_{ep}, and b, algorithms can be developed for temperature compensation with analog or digital circuits.

In some designs, the measuring and reference electrodes along with temperature sensor are combined in one assembly.

During measurements, the potentials of electrodes are usually evaluated with respect to the so-called hydrogen electrode. It is a spongy platinum 1, saturated with hydrogen (Fig. 2.53). The electrode is placed into solution 2 with the activity of hydrogen ions equal to unity (1g-ion/L). Conditionally, the potential of this electrode is equal to zero. This standard reference half-cell is connected to another half-cell 3 by means of a chemical bridge 4. The bridge is filled with KCl and has two semipermeable plugs 5 and 6. A difference in potentials is usually developed at the interface of the two solutions. This occurs due to the differences in ion activities and the rate of diffusion. In order to reduce this effect, a saturated solution of potassium chloride is employed in the bridge. This substance contains anions and cations with similar activities.

The set of two half-cells shown in Figure 2.53 is chiefly used in laboratory practice, since supplying hydrogen is not always convenient and economical. The potentials of the reference electrodes are measured and given with respect to the hydrogen electrode. For instance, for AgCl and Hg_2Cl_2 this potential E_0 in volts is given as a function of temperature [17].

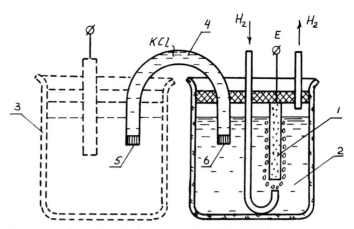

Figure 2.53 Half-cells for measuring pH with hydrogen electrode.

For the AgCl electrode,

$$E_0 = 0.22239 - 645.52 \times 10^{-6}\,(T - 298) - 3.284 \times 10^{-6}\,(T - 298)^2$$
$$+\ 9.948 \times 10^{-9}\,(T - 298)^3. \tag{2.116}$$

For the calomel electrode with KCl of 0.1 M-concentration,

$$E_0 = 0.335 - 0.00007\,(T - 298); \tag{2.117}$$

with KCl of 1 M-concentration,

$$E_0 = 0.2810 - 0.00024\,(T - 298);\ \text{and} \tag{2.118}$$

with saturated solution of KCl,

$$E_0 = 0.2420 - 0.00076\,(T - 298), \tag{2.119}$$

where E_0 = electrode potential, V, and
$\quad\ T$ = absolute temperature, K.

Using the Nernst equation and knowing the reference potentials, one can readily calculate the potentials for pH. In formulas, the natural logarithms should be converted to the common logarithm ($\ln M = 2.30259 \log_{10} M$) in order to calculate $E = f(\text{pH})$, which is reduced to

$$E = E_0 + 0.058\ \text{pH} \tag{2.120}$$

ORP or Redox Elements

ORP and Redox are abbreviations for "oxidation-reduction potential" and "reduction-oxidation." The electrodes for measuring ORP have the same structure and construction as regular pH electrodes. The measuring electrode is made of a noble metal, platinum, gold, or silver, or of a combination of these materials when one of them is used as a substrate and another as a thin coating on its surface. The reference electrode is actually the same as for pH-measurements. ORP is calculated on the basis of the logarithm of the ratio of activities of the oxidized and reduced states of ions (see the Nernst equation).

In the scale of ORP, positive numbers correspond to the content of oxidizing agents in the solution; zero, to the absence of the agents; and negative, to the reducing agents. The construction of the measuring electrode contains a wire, foil, and thick-film deposited structures.

Oxygen-sensitive Elements

Oxygen-sensitive elements are used to determine the amount of oxygen in a mixture of gases, such as air. The element (Fig. 2.54) is made of zirconium oxide ceramics and shaped like pipe 1, which carries platinum electrodes 2 and 3 on the external and internal surfaces. The inside of the chamber is filled with a reference gas, whereas the external surface is exposed to the gas mixture to be tested. The atmospheric air can be taken as the reference; its oxygen is 20.9%. The wall of the element becomes permeable for oxygen ions at a temperature above 400°C. The material between the electrodes behaves as a solid electrolyte.

At temperatures between 400 and 900°C, the surface that is in contact with oxygen of a higher concentration becomes an anode, the other one a cathode. Molecules of oxygen absorb electrons at the anode and lose them at the cathode. The difference in potentials between the two electrodes is found from the following Nernst equation:

$$E = 0.0215T \ln \frac{O_{2r}}{O_{2m}} \qquad (2.121)$$

where E = emf in question, V;
T = absolute temperature, K;
O_{2r} = concentration of oxygen in the reference gas, %; and
O_{2m} = concentration of oxygen in the measured gas, %.

Figure 2.55 illustrates $E = f(O_{2m})$ for different temperatures. The device must be combined with a temperature sensor for an adequate temperature compensation. A special heater must be incorporated in the design unless the temperature of the medium is naturally high. The accuracy of measurements is about 0.1% and the time constant is lower than 5s.

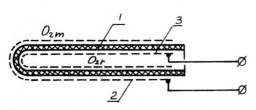

Figure 2.54 Zirconium-oxide ceramic element sensitive to oxygen.

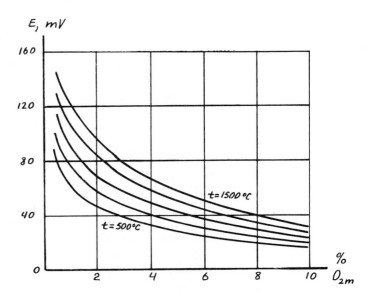

Figure 2.55 Emf developed by zirconium-oxide element as function of oxygen content at different temperatures.

Polarographic Cells

The processes of polarization with the evaluation of the sizes of the plateaus described earlier are employed in a polarographic cell to analyze the ion content in solutions.

The cell (Fig. 2.56) consists of container 1 with solution 2 to be tested, capillary 3 with mercury 4 dropping from the capillary (cathode), and a mercury reservoir 5 feeding the capillary. A mercury pool 6 at the bottom is an anode. During measurements, the mercury drops are continuously formed at the capillary and drop to the bottom. It takes from 1 to 6 seconds for the drops to form and detach. The size of each drop is approximately 1mm (the diameter of the capillary is about 0.1mm; the length is 150mm). Due to the significant difference in the dimensions of the cathode and anode, the current density at the capillary is much greater than that at the pool. The potential of the anode can be assumed constant, whereas all polarization processes are concentrated at the cathode, which is constantly renewed in order to protect it from contamination by the secondary species.

The reaction that occurs at the cathode is the reduction of ions and the formation of atoms at the specific potential of the cathode. For instance, if the solution contains metal ions M_1^{2+}, M_2^{2+}, ... M_i^{2+}, the typical reaction at the cathode is

$$M_i^{2+} + 2e \rightarrow M_i \tag{2.122}$$

$$M_i + x\,\text{Hg} \rightarrow \text{amalgam}. \tag{2.123}$$

The polarogram (Fig. 2.57) will reflect the presence of $M_1, M_2, \ldots, M_i$ by the corresponding plateaus at the specific voltages $V_1, V_2, \ldots, V_i$. Concentrations of substances are characterized by currents $I_1, I_2, \ldots, I_i$. The higher the step of the current $(I_i - I_{i-1})$, the greater the amount of atoms in the solution.

In order to test the unknown spectrum of atoms in the solution, the voltage should be increased gradually between 0 and 3V. The characteristic potentials for different ions are known. For instance, in a neutral medium they are 0.45, 0.60, 1.05, and 1.86V for Pb^{2+}, Cd^{2+}, Zn^{2+}, and K^+, respectively.

The mercury pool is substituted in some designs by a calomel reference electrode combined with a chemical bridge having a salt filling. The mercury electrode is primarily used for testing cations. For anions and melted salt, the platinum, gold, nickel, carbon, and carbide electrodes are also used.

A higher sensitivity of the cell can be reached if the dc voltage across the electrodes is modulated by sinosoidal or trapesoidal alternating pulses.

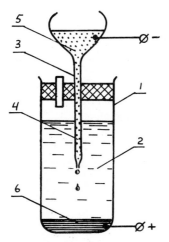

Figure 2.56 Polarographic mercury cell.

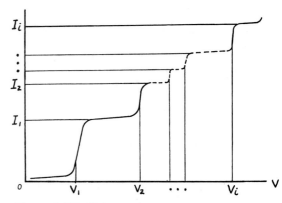

Figure 2.57 Polarogram.

The polarographic cell allows the measurement of concentrations from 1×10^{-4}M to 1×10^{-3}M. The threshold of response is between 1×10^{-7}M and 1×10^{-9}M.

Coulomb-meter Elements

Coulomb-meter elements are used to register the quantity of electricity by electrolytic means and are categorized as integrating devices. The elements' input is a charge of electricity, and their output is the amount of substance created due to electrolysis. The change of the size of the electrodes, their resistance, and the optical transparency of the solution can also be output values. The substances created during the electrolysis are solids, liquids, or gases. The memistor, which was described previously, has a similar structure.

An element [16] that provides the direct conversion of a quantity of electricity into displacement is shown in Figure 2.58.

Two columns of mercury 1 and 2 are separated by drop 3 of a liquid solution of salt HgI_2 and enveloped in a glass capillary 4 (its diameter is approximately 0.3mm). Anode 5 and cathode 6 are thin wires in contact with the columns. The capillary is furnished with scale 7. When a current flows through the element, the mercury is dissolved at the anode and forms a salt:

$$Hg + 4I^- \rightarrow HgI_4^{2-} + 2e. \tag{2.124}$$

At the same time it is reduced from the solution at the cathode:

$$HgI_4 + 2e \rightarrow Hg + 4I \tag{2.125}$$

During electrolysis, the overall volume of the substances in the capillary remains the same. However, the position of the drop along the capillary is changed by length δ. The displacement of the drop can be read by means of a scale and it is proportional to the amount of mercury transferred from the left to the right column.

According to Faraday's law, the quantity of electricity and mass of substance extracted from the solution are directly proportional to each other:

$$Q = \int_{t_1}^{t_2} i \, dt = MnF/A \tag{2.126}$$

where Q = charge of electricity, C;
       t_1 and t_2 = beginning and ending of time counting, s;
              i = current, A;
              t = time, s;
              M = mass of substance, g;
              n = valency of metal;
              F = Faraday's constant, C/g; and
              A = atomic weight of metal.

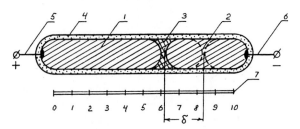

Figure 2.58   Mercury-capillary integrating element.

The mass of the substance that was transferred during the electrolysis is

$$M = dS\delta \qquad (2.127)$$

where d = density of metal, g/cm^3;
$\quad S$ = cross-sectional area of the capillary, cm^2; and
$\quad \delta$ = change of the length of the metal column or the displacement of the drop, cm.

From formulas 2.126 and 2.127,

$$\delta = Q\frac{A}{FndS} \qquad \text{or} \qquad (2.128)$$

$$\delta = I \cdot \Delta t\,\frac{A}{FndS} \qquad (2.129)$$

where I = constant dc current, A, and
$\quad \Delta t = t_2 - t_1$ = period of integration, s.

It is obvious that the displacement of the drop can be informative for evaluating the charge of electricity, the average current over a certain period of time, or the time for the constant or known variation of the current.

An element similar to the mercury cell is shown in Figure 2.59. It is composed of a copper anode 1 and cathode 2. Capillary 3 and anode reservoir 4 are filled with blue vitriol, which is dissociated, forming Cu^{+2} and SO_4^{-2} ions. Scale 5 is used for reading. Due to oxidation at the anode and reduction at the cathode, the anode material is dissolved, and copper is deposited on the cathode. The increase of cathode δ is proportional to the charge of electricity or current and time, as was explained for the mercury cell.

The relationships between δ, Q, I, and Δt are exactly the same as in formulas 2.126–2.129.

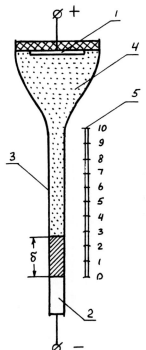

Figure 2.59 Copper-capillary integrating element.

The reading can be visual, optoelectronic, capacitive, and so on (with substitution for the scale). A zero adjustment can be provided by a simple change of the direction of the current. Depending on the construction of the cell, the ranges for current I and time Δt can be 0.01 to 1mA and 5 to 10×10^3 hours, respectively. To count time, the current should be stabilized. The accuracy of the measurement that can be reached using the cells is about 5 percent. It is determined by the quality of temperature compensation, methods of reading the growth of the columns, and accuracy in the fabrication of different parts of the element.

The element can be used as a component of devices for integrating voltages and currents with a high time constant, measuring the time of operation of electrical devices, generating infralow frequencies, delaying in time relays and in delay lines, and so on.

One more device of this group is shown in Figure 2.60. The element consists of a porous diaphragm 1 with grid platinum electrodes 2 and 3 at both sides of the diaphragm. The pores in the diaphragm and pipes 4 and 5 are filled with a sulphuric acid (H_2SO_4) solution. The pipes and compartment for the diaphragm form a sealed envelope. Scale 6 is attached to one of the pipes. It can be substituted by an electronic pickup sensitive to the displacement of the column of liquid 7 in the pipe. When current I flows through the diaphragm's electrolyte, hydrogen is liberated at the cathode and absorbed at the anode, creating excessive pressure at the cavity next to the cathode. The column of liquid will be displaced as shown in Figure 2.60. The displacement is a measure of the charge of electricity.

This device is fairly sensitive to electrical input and can operate with an error of about 2 percent.

A current-to-frequency converter is built [18] on the basis of the gaseous cell. It has a similar structure (Fig. 2.61), which includes a porous diaphragm 1 carrying an electrolyte, two electrodes 2 and 3, and a needle electrode 4 with a drop of electrolyte 5. The structure is sealed in housing 6. An increased pressure at the left chamber due to current I pushes drop 5 out of the gap to the right. This opens the path for the gas, equalizing the pressures in both chambers. The drop is pulled back in the gap due to the capillary effect, and the cycle is repeated once again. When the drop moves, it periodically breaks the junction between electrodes 3 and 4. The rate

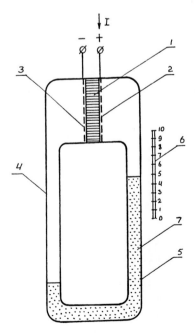

Figure 2.60 Gaseous integrating element.

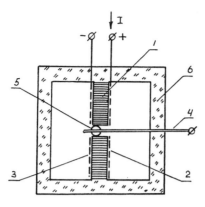

Figure 2.61 Gaseous converter of current to frequency.

of building of the pressure and frequency of travel of the drop are proportional to the current. Thus the current can modulate the frequency, which can be regarded as the output of the element. The upper frequency is limited by electrochemical and hydrodynamical processes.

Electrokinetic or Zeta Potential Elements

As was mentioned in the beginning of this chapter, the potentials developed across a capillary or porous body can be used as a measure of speed of an electrolyte passing through them. Indirectly, these potentials can be proportional to the force, pressure, acceleration, and other mechanical quantities. The elements for the conversion of these values into the electrical signal are similar (Fig. 2.62). A porous partition 1 is the core of these devices. The partition carries two electrodes 2 and 3, and it is surrounded with a polar liquid 4. When a certain mechanical agent induces the liquid to flow through the partition, the difference in potentials is created, and it is directly proportional to the speed of the flow. The agents for the examples given in Figure 2.62 are flow F (a), dynamic difference in pressure $P_1(t) - P_2(t)$ (b), acceleration of housing (c), and vibratory displacement of the detecting head 5, which transfers vibrations through spring 6 (d).

The operation of the cells in Figure 2.62 is convertible, that is the voltage applied to the electrodes causes the liquid to flow through the pores of the partition (electroosmosis).

The average velocity of the flow is proportional to the voltage: [17]

$$V = \frac{Q}{S} = \frac{\xi \epsilon I}{4\pi \eta x l} = \frac{\xi \epsilon E}{4\pi \eta}$$

(2.130)

where V = average velocity, m/s;
Q = flow rate, m^3/s;
S = cross-sectional area of the flow path (sum of cross-sectional areas of pores), m^2;
ξ = electrokinetic (zeta potential), V;
ϵ = dielectric constant of liquid, F/m;
I = current through partition, A;
l = thickness of partition, m;
π = 3.1416;
η = dynamic viscosity of liquid, N · s/m^2;
x = conductance of partition between two electrodes, S; and
E = field intensity of the electrical field along partition, V/m.

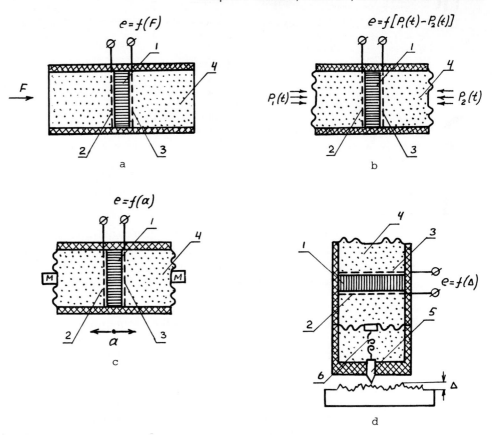

Figure 2.62 Schematic diagrams of electrokinetic elements. a = flowmeter, b = dynamic differential-pressure meter, c = accelerometer, d = vibrating displacement-sensing head; e = voltage developed across partition, $P_1(t)$ and $P_2(t)$ = alternating pressures, a = acceleration, Δ = amplitude of roughness, M = seismic mass.

If the liquid flows through the partition due to the difference in pressures $\Delta P = P_1(t) - P_2(t)$, the voltage across the electrodes is written as follows:

$$U = \frac{\xi\epsilon \cdot \Delta P}{\eta\gamma} \tag{2.131}$$

where U = voltage induced at the electrodes, V;
   ΔP = difference in pressure, Pa; and
   γ = conductivity of liquid, S/m.

The liquids that are used for the elements are acetone, nitrobenzene, and acetonitrile. Porous partitions are made of fused quartz, glass, porcelain, polyethylene, fluorine plastic, and other materials. The electrodes at the partition are made of platinum, silver, aluminum alloys, and graphite. Diaphragms can be made of metal and nonmetal materials.

The elements are applicable when low-frequency signals have to be measured. The frequency range of the signals is from 0.01 to 10 000Hz. The output signal's range is from microvolts to fractions of a volt. The accuracy of the method is several percent.

The element cannot be used for measuring steady components of measurands such as pressure and displacement. It is temperature-sensitive and needs a special sensor and network for temperature compensation. However, it is a simple device with a high sensitivity to low-frequency signals, and it provides high output signals.

chapter 3

Capacitive and Inductive Elements

In this chapter we will consider elements that are components of transducers belonging to the group of the most traditional instruments. Simplicity and reliability are attractive characteristics, which make the elements desirable for application in instruments. In spite of the differences in construction, these elements have a quite common usage, as we shall see in the following discussion.

CAPACITIVE ELEMENTS

A capacitive element converts a change in the position of the electroconductive plates forming a capacitor, and a change of the properties of a dielectric between the plates into an electrical signal.

An illustrative scheme of the element is shown in Figure 3.1. It is composed of two parallel plates 1 and 2 and a dielectric separator 3. Leads 4 and 5 from each conducting plate are connected to the circuit. Ignoring leakage and fringing, the capacitance of such an element is written as follows:

$$C = \frac{\epsilon A}{\delta} \tag{3.1}$$

where $\epsilon = \epsilon_0 \epsilon_r$ is permittivity of dielectric, F/m, or $s^4 A^2 / kg \cdot m^3$;

ϵ_0 = permittivity of empty space (the same units);

ϵ_r = dielectric constant of the media between the plates, dimensionless (also called relative permittivity or relative dielectric constant);

A = congruous area, m^2; and

δ = plate separation, m.

Alternation of any of the three physical parameters causes a change of the capacitance. For instance, the plate can be displaced along the x- and y- axes, providing a signal proportional to the displacement. If ϵ is affected by temperature or moisture, the magnitude of C responds to these variables.

The displacement of plates parallel to their planes gives the change of the overlapping area (A). It is usually used to measure big displacements ($>$ 1mm).

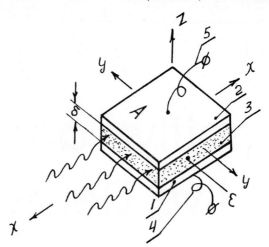

Figure 3.1   Model of capacitive element.

A shift in longitudinal position with a variation of δ is typical for measurements of small displacements (< 1mm).

An element having a structure of the common r-f variable-tuning capacitor (Fig. 3.2) used in radios is successfully used for the measurement of angular displacement α. In this capacitor, a set of metal plates 1 at the rotor can be rotated inside the assembly of spaced plates of stator 2. The area between two sets is varied and proportional to angle α.

With mechanical elements linked to the moving plate or with a special substance placed between the plates, the element can be adapted for measuring displacements, sizes, proximity to a target, velocities, forces, accelerations, vibrations, sound intensity, pressure, and levels of liquid; density, composition, and continuity of materials; moisture content in a substance, strain in materials, and temperature. It also serves as an actuator in force-balance transducers.

The element provides a convenient output to be fed into a frequency converter, and an electronic counter for producing digital signal conditioning. It is also widely used in microelectronic sensors due to its simple structure and ease of microfabrication. For the typical element, the range of capacitance change is small. However, contemporary techniques allow a measurement of capacitance with an accuracy of better than 1×10^{-18}F. This allows a high degree of accuracy in the evaluation of the measurand.

Structures of Capacitive Elements

Table 3.1 illustrates different structures of the capacitive element along with basic relationships for the capacitance as a function of physical variables and parameters.

The values of capacitances are calculated without regard to fringing and leakage effects, which can affect the field between the plates. Their influence should be taken into account when the separation between the plates is comparable with the sizes of the plates.

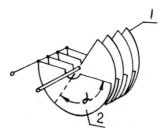

Figure 3.2   Angular-displacement capacitive element.

TABLE 3.1

Structures of Capacitive Elements

No.	Variable	Structure	Basic relationship	No.	Variable	Structure	Basic relationship
1	2	3	4	1	2	3	4
1	Gap δ		$C = \varepsilon A \dfrac{1}{\delta \pm \Delta\delta}$	9	Position of a shaped dielectric		$C = f[\delta,\, A,\, \varepsilon_0,\, \varepsilon,\, \delta_1,\, x]$
2	Gaps δ for a differential scheme		$C_1 = \varepsilon A \dfrac{1}{\delta \pm \Delta\delta}$ $C_2 = \varepsilon A \dfrac{1}{\delta \pm \Delta\delta}$	10	Position of a liquid dielectric		$C = \dfrac{W}{\delta}[H\varepsilon_0 + h(\varepsilon - \varepsilon_0)]$
3	Area A		$C = \dfrac{\varepsilon}{\delta}(A \pm \Delta A)$ $\Delta A = f(\Delta)$	11	Dielectric constant is a function of force, strain, temperature, and moisture content (F, λ, t°, and φ, respectively)		$C = \dfrac{A}{\delta}\varepsilon(F,\, \lambda,\, t^\circ,\, \varphi)$
4	Areas A for a differential scheme		$C_1 = \dfrac{\varepsilon}{\delta}(A \pm \Delta A)$ $C_2 = \dfrac{\varepsilon}{\delta}(A \mp \Delta A)$ $\Delta A = f(\Delta)$	12	Thickness of a dielectric film		$C = \dfrac{A}{\dfrac{\delta - \delta_f}{\varepsilon_0} + \dfrac{\delta_f}{\varepsilon_f}}$
5	Area A for an angular displacement		$C = \dfrac{\varepsilon}{\delta}(A \pm \Delta A)$ $\Delta A = f(\Delta)$	13	Thicknesses and dielectric constants in a multi-dielectric structure		$C = \dfrac{A}{\dfrac{\delta_1}{\varepsilon_1} + \dfrac{\delta_2}{\varepsilon_2} + \dfrac{\delta_3}{\varepsilon_3} + \ldots + \dfrac{\delta_n}{\varepsilon_n}}$
6	Areas A for a differential angular displacement		$C_1 = \dfrac{\varepsilon}{\delta}(A \pm \Delta A)$ $C_2 = \dfrac{\varepsilon}{\delta}(A \mp \Delta A)$ $\Delta A = f(\Delta)$	14	Thicknesses and a dielectric constant in a multiplate structure. n is the number of plates		$C = \dfrac{A\varepsilon}{\delta}(n - 1)$
7	Presence or absence of a dielectric		$C' = \dfrac{\varepsilon_0}{\delta}A$ $C'' = \dfrac{\varepsilon}{\delta}A$	15	Angular displacement, w is plate's width		$C = \dfrac{\varepsilon w}{\alpha}\ln\dfrac{R}{r}$
8	Position of a dielectric		$C = \dfrac{w}{\delta}[L\varepsilon_0 + l(\varepsilon - \varepsilon_0)]$				

The differential structures allow the reduction of the nonlinearity of the transfer characteristic and electrostatic force on the moving part. It is also helpful in compensation for the temperature drifts.

Some elementary cells are outlined in Table 3.1.

In practical designs, the elements frequently have more complicated shapes and are combined with parts of a transducer or measuring system (diaphragm, spring, wheel, liquid, human body, and so on). The most typical structures will be analyzed in the following paragraphs.

Equivalent Circuits and General Characteristics

When voltage is applied to the capacitor's plates, the plates become charged due to the concentration of the electrons at one plate and repulsion from the other. The charge is proportional to the applied voltage and capacitance:

$$Q = V \cdot C \qquad (3.2)$$

where Q = quantity of charge, C;
C = capacitance of the capacitor, F; and
V = applied voltage, V.

It is desirable to have a high voltage or charge to obtain a significant signal proportional to the measurand. The maximum voltage is usually restricted by the dielectric strength of a plate separator, which is defined as the breakdown potential. The values for this potential are given in Table 3.2.

Many elements have air between the plates. It is commonly accepted that the dielectric constant for dry air ϵ_{air} is unity and the permittivity for free space is:

$$\epsilon_0 = 8.85415 \times 10^{-12} F/m. \qquad (3.3)$$

However, ϵ_{air} is a function of pressure, temperature, and humidity. For precision calculation, the following formulas should be used:

$$\epsilon_{air} = 1 + \frac{P}{T}\left[28 + \frac{\varphi P_H}{P}\left(\frac{135}{T} - 0.0039\right)\right] \qquad (3.4)$$

TABLE 3.2

Characteristics of Some Insulators

Material	Dielectric constant	Dielectric strength kV/mm
Vacuum	1.00000	∞
Air	1.00054	0.8
Water	78	—
Paper	3.5	14
Ruby mica	5.4	160
Amber	2.7	90
Porcelain	6.5	4
Fused quartz	3.8	8
Pyrex glass	4.5	13
Bakelite	4.8	12
Polyethylene	2.3	50
Polystyrene	2.6	25
Teflon	2.1	60
Neoprene	6.9	12
Pyranol oil	4.5	12
Titanium oxide	100	6

$$\log P_H = 7.45 \frac{T - 273}{T - 38.3} + 2.78 \tag{3.5}$$

where
P = air pressure, Pa;
T = absolute temperature, K;
φ = relative humidity, %; and
P_H = saturated vapor pressure (Pa) for a given temperature.

For dry air, normal atmospheric pressure, and temperature range $-60°C$ to $+60°C$, the temperature coefficient for ϵ_0 is -2×10^{-6} 1/°C. It increases and turns to positive values with an increase in humidity ($+10 \times 10^{-6}$ 1/°C for $\varphi = 100\%$).

When the space between the plates is filled with a dielectric, the value of the dielectric permittivity ϵ should be substituted in the formulas for calculating capacitances.

The dielectrics can be polar and nonpolar. The nonpolar elements or compounds, unlike the polar ones, have no permanent electric dipole moment.

The polar substances are characterized by a high dielectric constant ($\epsilon_r > 12$), high conductivity G, ($G \approx 10^{-3}$S/m), and a high dielectric loss, which depends on the frequency of excitation voltage. Water, acetone, ethyl, and methyl alcohols are categorized as polar substances.

The permittivity of nonpolar material is low ($\epsilon_r < 3$), as are the conductivity and dielectric losses (G is between 50×10^{-12} and 0.5×10^{-15}S/m). Mineral, organic, and petroleum oils (kerosene and gasoline) have ϵ_r between 1.25 and 1.5. For condensed gases (H_2, O_2, and N_2), $1.25 \le \epsilon_r \le 1.5$. These materials are good electrical insulators. The intermediate group of materials (weak-polar) have $3 \le \epsilon_r \le 6$ and G between 0.1×10^{-12} and 10.0×10^{-9}S/m. Some numbers for ϵ_r are given in Table 3.2 [19].

The dielectric constants of many gases differ insignificantly from the permittivity of free space. For instance, air, hydrogen, oxygen, and nitrogen have an ϵ_r of 1.00054, 1.00027, 1.00065, and 1.00068, respectively.

An ideal dielectric has a zero conductivity. This means that the electrical circuit of the element with this dielectric is a pure capacitance, which does not absorb any real power. The reactive power is

$$Q_r = U^2 \omega C, \tag{3.6}$$

and impedance is

$$Z_c = \frac{1}{j\omega C} = -\frac{j}{\omega C} \tag{3.7}$$

where Q_r = reactive power, var;
ω = angular frequency, rad/s; and
C = capacitance, F.

A real dielectric between the capacitor plates contributes to losses, which can be evaluated by the power factor. The power represents the fraction of the input volt-amperes that are dissipated in the capacitor:

$$P = UI \cos \theta \tag{3.8}$$

where
P = dissipated power, W;
U = voltage applied to the capacitor, V;
I = current through the capacitor, A;
$\cos \theta$ = power factor; and
θ = power factor angle, degree.

As a rule, the power factor becomes higher with a rise in temperature and absorption of moisture. The losses are also evaluated in terms of a phase angle δ, by which the current flowing into the capacitor fails to be 90° out of phase with the voltage:

$$\cos \theta = \sin \delta. \tag{3.9}$$

The tangent of the phase angle is termed the dissipation factor, and its reciprocal is the capacitor's quality factor Q_f, which is a ratio of the capacitor's reactance to the equivalent series resistance. Typically, δ is small and

$$\cos \theta = \sin \delta \approx \tan \delta = \frac{1}{Q_f} \approx \delta \tag{3.10}$$

where δ is expressed in radians.

The actual element carrying the real dielectric can be replaced (Fig. 3.3) with a perfect capacitor C_s or C_p with the series or parallel resistances:

$$R_s = \frac{\cos \theta}{\omega C_s} \quad \text{and} \quad R_p = \frac{1}{\omega C_p \cos \theta}. \tag{3.11}$$

Eliminating $\cos \theta$ from formula 3.11 gives

$$R_s = \frac{1}{R_p \omega^2 C^2}. \tag{3.12}$$

The equivalent resistances are independent of the voltage applied to the element, but vary inversely with capacitance and frequency. For some dielectrics, $\cos \theta$ is a function of frequency. For the materials used in the element, $\cos \theta$ varies from 0.0001 to 0.0200.

In addition to formulas 3.8–3.12, the relationships for the dissipated powers in parallel (P_p) and series (P_s) equivalent circuits, along with formulas for converting C_s and R_s to C_p and R_p, are given below.

$$P_p = U^2 \omega C_p \tan \delta, \quad \tan \delta = \frac{1}{\omega C_p R_p} \tag{3.13}$$

$$P_s = U^2 \omega C_s \frac{\tan \delta}{1 + \tan^2 \delta}, \quad \tan \delta = \omega C_s R_s \tag{3.14}$$

$$C_s = C_p (1 + \tan^2 \delta), \quad R_s = \frac{R_p}{1 + \dfrac{1}{\tan^2 \delta}} \tag{3.15}$$

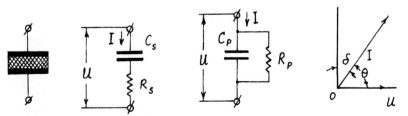

Figure 3.3 Equivalent circuits of element with dielectric, and element's phasor diagram.

The medium between the capacitance plates can be warmed by the absorption of power, which can affect the accuracy of measurements. In terms of signal conditioning, U and ω should be high but with a definite limit.

Single Displacement Elements

Transfer characteristics for a single-capacitance-displacement element are listed in Table 3.1. The capacitance is a linear function of the common area and a nonlinear (hyperbolic) function of the gap. However, if the reactance rather than capacitance is regarded as an output of the element, it has a linear response to δ and a nonlinear response to A and ϵ since the reactance is the reciprocal of the capacitance.

For the single capacitor, the incremental change of the capacitance

$$C = \frac{\epsilon A}{\delta} \qquad (3.16)$$

can be found by taking a differential:

$$dC = \frac{\partial C}{\partial \epsilon}\, d\epsilon + \frac{\partial C}{\partial A}\, dA + \frac{\partial C}{\partial \delta}\, d\delta. \qquad (3.17)$$

Presentation of this differential with finite increments gives:

$$\Delta C = \frac{A_1}{\delta_1}\, \Delta\epsilon + \frac{\epsilon_1}{\delta_1}\, \Delta A - \frac{A_1\epsilon_1}{(\delta_1 + \Delta\delta)^2}\, \Delta\delta. \qquad (3.18)$$

The relative change of C:

$$\frac{\Delta C}{C_1} = \frac{\Delta\epsilon}{\epsilon_1} + \frac{\Delta A}{A_1} - \frac{1}{\left(1 + \dfrac{\Delta\delta}{\delta_1}\right)^2}\, \frac{\Delta\delta}{\delta_1}. \qquad (3.19)$$

The relative changes of capacitance due to the relative changes of variables are:

$$S_\epsilon = \frac{\Delta C/C_1}{\Delta\epsilon/\epsilon_1} = 1, \qquad S_A = \frac{\Delta C/C_1}{\Delta A/A_1} = 1, \qquad \text{and}$$

$$\frac{\Delta C/C_1}{\Delta\delta/\delta_1} = -\frac{1}{\left(1 + \dfrac{\Delta\delta}{\delta_1}\right)^2} \qquad (3.20)$$

where C_1, ϵ_1, A_1, and δ_1 are initial values of C, ϵ, A, and δ, respectively, ΔC, $\Delta\epsilon$, ΔA, and $\Delta\delta$ are incremental changes of C, ϵ, A, and δ, respectively.

In all these and the following formulas, ϵ must be replaced by ϵ_0, if the medium between the plates is air or any other substance with $\epsilon_r = 1$.

By introducing the capacitive reactance x as

$$x = \frac{1}{\omega C} = \frac{\delta}{\omega\epsilon A} \qquad (3.21)$$

and assuming that the frequency of excitation remains constant, the x-change Δx and relative sensitivities $S_{\epsilon x}$, S_{Ax}, and $S_{\delta x}$ can be written as follows:

$$\Delta x = \frac{\partial x}{\partial \epsilon} \Delta \epsilon + \frac{\partial x}{\partial A} \Delta A + \frac{\partial x}{\partial \delta} \Delta \delta, \tag{3.22}$$

$$S_{\epsilon x} = \frac{\Delta x/x_1}{\Delta \epsilon/\epsilon_1} = -\frac{1}{\left(1 - \dfrac{\Delta \epsilon}{\epsilon_1}\right)^2}, \qquad S_{Ax} = \frac{\Delta x/x_1}{\Delta A/A_1} = -\frac{1}{\left(1 - \dfrac{\Delta A}{A_1}\right)^2},$$

$$S_{\delta x} = \frac{\Delta x/x_1}{\Delta \delta/\delta_1} = 1 \tag{3.23}$$

where x_1 is an initial value of x.

When the signal-capacitance element operates at a nonlinear mode [for instance, $C = f(\delta)$], its nonlinearity can be significant. By expanding the expression for the capacitance in series, the "weight" of the nonlinear terms can be evaluated as follows:

$$C = \frac{\epsilon A}{\delta + \Delta \delta} \approx \frac{\epsilon A}{\delta}\left[1 - \frac{\Delta \delta}{\delta} + \left(\frac{\Delta \delta}{\delta}\right)^2 - \left(\frac{\Delta \delta}{\delta}\right)^3 + \dots\right]. \tag{3.24}$$

A pair of capacitors arranged in a differential connection provides a much better linear output.

Differential Displacement Elements

The output signal from the dual element can be the difference between two capacitances or their reactances. As an example, structure No. 2 from Table 3.1 will be analyzed (Fig. 3.4).

If the middle electrode is displaced to the right by $\Delta \delta$, the two capacitances and reactances are

$$C_1 = \frac{\epsilon A}{\delta + \Delta \delta}, \qquad X_1 = \frac{\delta + \Delta \delta}{\omega \epsilon A} \qquad \text{and} \tag{3.25}$$

$$C_2 = \frac{\epsilon A}{\delta - \Delta \delta}, \qquad X_2 = \frac{\delta - \Delta \delta}{\omega \epsilon A}. \tag{3.26}$$

Assuming $\Delta \delta/\delta \ll 1$, the differences will be

$$C_2 - C_1 \approx 2\frac{\epsilon A}{\delta}\left[\frac{\Delta \delta}{\delta} - \left(\frac{\Delta \delta}{\delta}\right)^3\right], \tag{3.27}$$

$$X_2 - X_1 = -\frac{2}{\omega \epsilon A}\Delta \delta. \tag{3.28}$$

To calculate the differences for the single-capacitor element, we assign

$$C_1 = \frac{\epsilon A}{\delta}, \qquad \text{hence} \tag{3.29}$$

$$C_2 - C_1 = \frac{\epsilon A}{\delta}\left[\frac{\Delta \delta}{\delta} + \left(\frac{\Delta \delta}{\delta}\right)^2\right] \qquad \text{and} \tag{3.30}$$

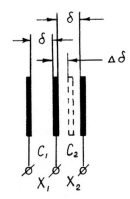

Figure 3.4
Differential-
displacement
element.

$$X_2 - X_1 = -\frac{1}{\omega \epsilon A} \Delta \delta. \tag{3.31}$$

It is obvious from formulas 3.27–3.31 that the output signal from the element is doubled for the differential type and its transfer characteristic $C_2 - C_1 = f(\Delta \delta)$ is more linear since the nonlinear cubic term $(\Delta \delta / \delta)^3$ in equation 3.27 is much smaller than the quadratic term $(\Delta \delta / \delta)^2$ in equation 3.30.

Any other differential structure can be analyzed in a similar way, which leads to the same conclusion. The improvement over a single capacitance is obvious. Like any differential construction, this structure provides better compensation for temperature.

Angle Displacement Elements

An angular motion of semicircular conductive plates (Fig. 3.2 and Fig. 3.5) between another set of arc-shaped electrodes provides a greater absolute change of capacitance per one degree of rotation than does a single-plate element. This is desirable for any capacitive element since a small change of capacitance versus a change of a measurand is one of the drawbacks of the element.

When moving plates 1 are displaced against fixed plates 2, the capacitance is proportional to the angle of displacement because it determines the congruous area 3, which is $A_m (180° - \alpha)/180°$. The area corresponding to $\alpha = 0°$ is denoted by A_m. Therefore, capacitance C as a function of angle α is

$$C = \frac{\epsilon A_m (n - 1)}{\delta} \cdot \frac{(180 - \alpha)}{180} \tag{3.32}$$

where δ = adjacent plates' separation, m, and
 n = total number of moving and stationary plates.

Sensitivity S_α $(F/1°)$ is expressed as

$$S_\alpha = \frac{\partial C}{\partial \alpha} = -\frac{\epsilon A_m (n - 1)}{180 \delta}. \tag{3.33}$$

Function $C = f(\alpha)$ is linear in this instance, but if required, it can be made nonlinear by a change of the plates' profile.

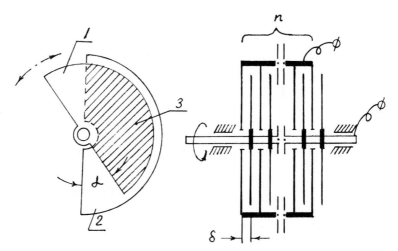

Figure 3.5   Model of angular-displacement element.

Film-thickness Elements

It is convenient to measure the thickness of a thin dielectric film 1 (Fig. 3.6) by placing or pulling it between two plates 2 and 3 of the capacitor. In this case, the film is not subjected to any force during measurements, and a small displacement of the film toward one of the plates does not affect the accuracy of measurement.

Capacitance C, which is measured at the terminals, can be introduced by two capacitances in a series connection. One of them, C_1, is a capacitance of the film having thickness δ_f, permittivity ϵ_f, and area A equal to the area of the plate. The other one is a capacitor formed by the plates that have spacing $\delta - \delta_f$. A calculation of the transfer characteristic and sensitivity S_f for this element is performed as follows:

$$C = \frac{C_1 C_2}{C_1 + C_2}, \tag{3.34}$$

$$C_1 = \frac{\epsilon_f A}{\delta_f}, \qquad C_2 = \frac{\epsilon_0 A}{\delta - \delta_f}, \tag{3.35}$$

$$C = \frac{A}{\dfrac{\delta - \delta_f}{\epsilon_0} + \dfrac{\delta_f}{\epsilon_f}}, \tag{3.36}$$

$$S_f = \frac{\partial C}{\partial \delta_f} = \frac{A\left(\dfrac{1}{\epsilon_0} - \dfrac{1}{\epsilon_f}\right)}{\left(\dfrac{\delta - \delta_f}{\epsilon_0} + \dfrac{\delta_f}{\epsilon_f}\right)^2}, \tag{3.37}$$

If $\epsilon_f \gg \epsilon_0$,

$$S_f \approx \frac{A\epsilon_0}{(\delta - \delta_f)^2}. \tag{3.38}$$

Large Displacements and Level Elements

The motion of a dielectric between the capacitor plates can be used for measuring both large displacements and levels of liquids or compounds. If the media between the electrodes is electroconductive, the electrodes are coated with an insulating layer. Figure 3.7 shows a model of a movable along-the-x-axis dielectric 1 that is placed between two electrodes 2 and 3.

If the gap between the electrodes, length, and width are δ, l, and w, respectively, and if the rectangular-bar dielectric with permittivity ϵ has dimensions $l \times w \times \delta$ and moves a distance x, capacitance C between the electrodes is calculated as follows:

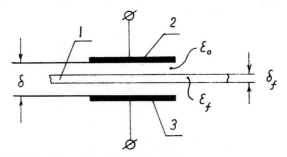

Figure 3.6 Film-thickness element.

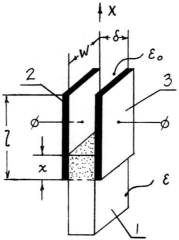

Figure 3.7 Dielectric displacement element.

$$C = \frac{\epsilon_0 l w}{\delta} + \frac{(\epsilon - \epsilon_0)w}{\delta} x \qquad (3.39)$$

and sensitivity S_x is

$$S_x = \frac{(\epsilon - \epsilon_0)w}{\delta}. \qquad (3.40)$$

The same structure is used to measure the liquid level between two parallel electrodes. The permittivity of liquid is usually greater than that of the air or gas above the liquid. If we imagine that the volume above the dashed line is occupied by liquid, the reading of the capacitance value between the electrodes is directly proportional to the height of the column.

An element with two concentric electrodes 1 and 2 (Fig. 3.8) is superior to the parallel-plate arrangement. It has a more rigid configuration and provides better

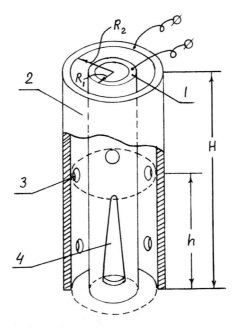

Figure 3.8 Liquid-level element with two concentric electrodes.

electrical protection against parasitic external capacitances. The walls of the electrodes can have perforations 3 for quick penetration of liquid in the space between the electrodes when they are immersed in a tank. Sometimes special contour holes 4 are fabricated in the electrodes, or they have a special shape to provide a nonlinear correction of the characteristic capacitance versus level, and to have a linear signal from the transducer.

To derive a transfer function for the element, we calculate the elementary reactance dx_c (Fig. 3.9), formed by capacitor C' with two circular imaginary electrodes at radius r, separation dr from each other, and length l. Capacitance C' is

$$C' = \frac{\epsilon_0 A}{dr} = \frac{\epsilon_0 2\pi r l}{dr}. \tag{3.41}$$

For angular frequency ω:

$$dx_c = \frac{1}{\omega C'} = \frac{1}{\omega \epsilon_0 2\pi r l} \cdot \frac{dr}{r} \quad \text{and} \tag{3.42}$$

$$x_c = \frac{1}{\omega \epsilon_0 2\pi r l} \int_{R_1}^{R_2} \frac{dr}{r} = \frac{ln\left(\dfrac{R_2}{R_1}\right)}{\omega \epsilon_0 2\pi l} = \frac{1}{\omega C} \tag{3.43}$$

where a capacitance between the electrodes with radii R_1 and R_2 is

$$C = \frac{\epsilon_0 2\pi l}{ln\left(\dfrac{R_2}{R_1}\right)}. \tag{3.44}$$

This relation can be simplified if the separation

$$\delta = R_2 - R_1 \ll R_2 + R_1. \tag{3.45}$$

Using the first term of expansion for $ln(R_2/R_1)$, we can write

$$ln\frac{R_2}{R_1} \approx 2\frac{R_2 - R_1}{R_2 + R_1} = 2\frac{\delta}{R_2 + R_1} \quad \text{and} \tag{3.46}$$

$$C = \frac{\epsilon_0 \pi l (R_1 + R_2)}{\delta}. \tag{3.47}$$

Now the transfer function for the element can be derived similarly to that for a parallel-plate element;

$$C = \frac{\epsilon_0 2\pi H}{ln\left(\dfrac{R_2}{R_1}\right)} + \frac{(\epsilon - \epsilon_0)2\pi}{ln\left(\dfrac{R_2}{R_1}\right)} h \tag{3.48}$$

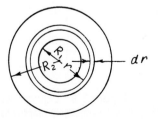

Figure 3.9 Model for derivation of concentric-electrode element transfer function.

or, for a small gap

$$C = \frac{\epsilon_0 \pi (R_1 + R_2)}{\delta} \left[H + \left(\frac{\epsilon}{\epsilon_0} - 1 \right) h \right].$$ (3.49)

Sensitivity S_h of C to a change of h is

$$S_h = \frac{\partial C}{\partial h} = \frac{(\epsilon - \epsilon_0) 2\pi}{ln \left(\dfrac{R_2}{R_1} \right)} \approx \frac{\pi (R_1 + R_2)(\epsilon - \epsilon_0)}{\delta}.$$ (3.50)

There are a number of variations in the construction of the capacitive level measuring technique. The upper surface of the liquid can act as a movable plate with respect to the fixed part on the container. If the tank is made of metal, it can be used as one electrode.

The capacitive cells can be employed not only for an analog measurement. A discrete or point-level sensing of a liquid's presence or absence at a particular height of the tank is also a common technique.

For precision measurements, an additional element is submerged in liquid to compensate for changes in the fluid characteristics.

A height of liquid above another liquid can be determined by the element if the liquids have different dielectric constants.

It is easy to modify derived formulas for the calculation of a moving rotor element for measurements of large displacements (Fig. 3.10). In this design, a sliding conductive electrode 1 is restricted to movement along the x-axis. It is mounted inside a stationary hollow electroconductive cylinder 2, forming a capacitor with a linear variation of capacitance versus displacement x. Assuming that the media between the electrodes is air, the capacitance is defined as follows:

$$C = \frac{\epsilon_0 2\pi (l - x)}{ln \left(\dfrac{D_2}{D_1} \right)}, \quad \text{and sensitivity is}$$ (3.51)

$$S_x = \frac{\partial C}{\partial x} = - \frac{\epsilon_0 2\pi}{ln \left(\dfrac{D_2}{D_1} \right)}.$$ (3.52)

Displacement-to-phase Converters

These devices have functions similar to those of the electromagnetic synchro for motion sensing. They provide an output voltage of constant amplitude, but with a

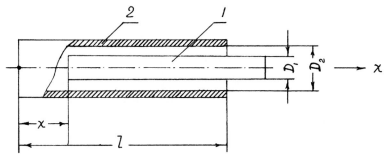

Figure 3.10 Moving-rotor element.

phase angle shift directly proportional to the translational or angular displacement. Unlike the electromagnetic device, the capacitive converter does not contain windings, a ferromagnetic core, and so on. Its moving parts are simple and light, which decreases a time constant of the measuring or control system. Figure 3.11 shows a model of the element for a translational motion to help elucidate the concept. A similar arrangement is used in angular displacement measurements. The scheme in Figure 3.11 can be regarded as a representation of the involute of a circle. A stationary plate 1 includes sinusoidally shaped electroconductive films 2, 3, 4, and 5 and strip 6. Slider 7 is also electroconductive and is separated from the plate by a thin air gap. The slider and films form capacitances. When the slider moves along the x-axis, the capacitances change sinusoidally. The slider has also a capacitive coupling with strip 6. The phase shift between the sinusoidal waves on the plate is 90°. The films are fed by ac voltages U', U'', U''', and U'^V, having phase angle shifts between each other:

$$U' = U_m \cos \omega t;$$

$$U'' = U_m \cos(\omega t - 90°) = -U_m \sin \omega t;$$

$$U''' = U_m \cos(\omega t - 180°) = -U_m \cos \omega t;$$

$$U'^V = U_m \cos(\omega t - 270°) = -U_m \sin \omega t \qquad (3.53)$$

where U_m = peak value of a phase voltage, V, and
ω = angular frequency, rad/s.

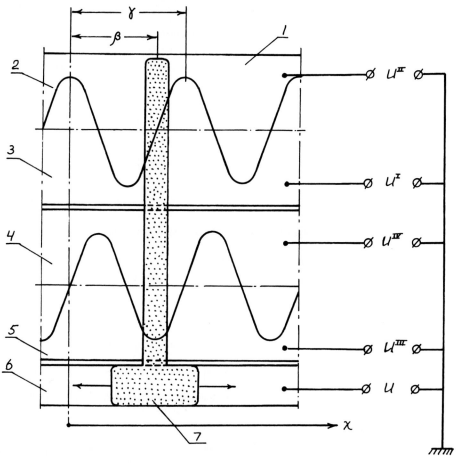

Figure 3.11 Model of displacement-to-phase converter.

A measuring circuit is built in such a way that the output voltage U is proportional to the sum of the voltage drops between the slider and films. In turn, they are proportional to the sinusoidal change of capacitances between the slider and strips.

This change is a function of a linear or angular displacement β:

$$U_1 = KU' \cos \beta;$$

$$U_2 = KU'' \cos(\beta - 90°);$$

$$U_3 = KU''' \cos(\beta - 180°);$$

$$U_4 = KU'^V \cos(\beta - 270°) \tag{3.54}$$

where U_1 to U_4 are phase voltages, V, and

K is a constant that is defined by constructional parameters and the electrical circuit.

It should be noted that the capacitance between the lower part of the slider and strip 6 is used to avoid a sliding contact on moving parts. This capacitance is in series with the set of measuring capacitances.

Substituting for U' through U'^V in equations 3.54 and taking the sum of voltages yields

$$U = 2U_mK[\cos \beta \cos \omega t + \sin \beta \sin \omega t] = 2U_m K \cos (\omega t + \beta). \tag{3.55}$$

It is obvious from this equation that the phase of the output voltage is proportional to β, whereas the amplitude of this voltage remains constant. A period of wave γ corresponds to a 360° change in phase.

Depending on the requirements of the instrument, the number of phases can be increased (6, 8, and so on). It is desirable to feed each phase with a stable voltage amplitude; for instance, by splitting the phase from the common single-phase source.

Pressure-sensitive Capacitive Elements

There are two basic structures of pressure-sensitive capacitive elements. In one of them, a flexible diaphragm 1 (Fig. 3.12), exposed to pressure P, transmits (by stem 2) deflection h (proportional to P) to a movable electrode 3, which forms a variable capacitor with a second, stationary electrode 4. It is obvious that the change of the capacitance is proportional to pressure P. This construction allows a good separation of the media of the measurand from the area of two electrodes, which is sensitive to mechanical and electrical interferences.

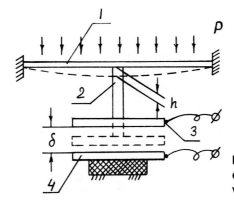

Figure 3.12 Pressure-sensitive capacitive element with separation of diaphragm and variable capacitance.

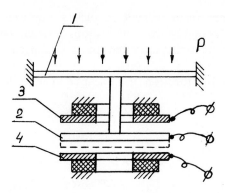

Figure 3.13 Differential pressure-to-capacitance element with separation of diaphragm and capacitor.

In the other construction (Fig. 3.13) diaphragm 1, sensitive to pressure, drives electrode 2 between two stationary electrodes 3 and 4, providing a differential mode of operation. There are a variety of transducers with different intermediate cells transmitting pressure into displacement. A calculation of these devices is divided into two parts: first the stroke of the diaphragm under pressure is evaluated; then the change of capacitance is calculated.

In a direct pressure-converting element (Fig. 3.14), the electroconductive diaphragm, affected by pressure, moves toward a stationary electrode 2, positioned alongside the diaphragm. The diaphragm and electrode form a capacitor sensitive to pressure. This is the second basic structure.

Quite popular is a differential structure (Fig. 3.15) where diaphragm 1 moves under pressure between two fixed plates 2 and 3, increasing one capacitance and decreasing the other as a response to the change of pressure.

These elements can be furnished either with a flexible diaphragm [20] made of material with a small modulus of elasticity or with a stiff diaphragm with a high modulus of elasticity. The first type is more sensitive and is used for measurements of low pressure: down to a fraction of mm H_2O. It can be assumed that this round diaphragm 1 becomes spherical (Fig. 3.16) when pressure P is applied to one of its sides. The diaphragm is fixed in holder 2, and the capacitance change is provided due to the change of the gap between the diaphragm and the stationary electroconductive plate 3.

Deflection y of any point of the diaphragm of radius r is written as follows [20]:

$$y = \frac{2\sigma}{P}\left[\sqrt{1 - \left(\frac{rP}{2\sigma}\right)^2} - \sqrt{1 - \left(\frac{aP}{2\sigma}\right)^2}\right], \tag{3.56}$$

where y = diaphragm's deflection at any point of radius r, m, and
$\quad\sigma$ = radial tension, N/m.

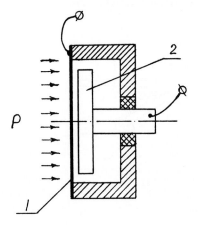

Figure 3.14 Single-capacitor pressure-sensitive element.

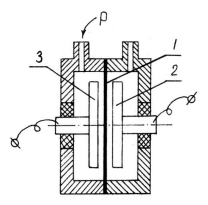

Figure 3.15 Differential-capacitance pressure-sensitive element.

Expansion in series of expression 3.56 gives

$$y = \frac{P}{4\sigma}\left[(a^2 - r^2) + \frac{1}{16}\left(\frac{P}{\sigma}\right)^2 (a^4 - r^4) + \frac{1}{512}\left(\frac{P}{\sigma}\right)^4 (a^6 - r^6) + \ldots\right]. \quad (3.57)$$

For a small diaphragm deflection ($h/a \ll 1$), a linear approximation of equation 3.57 is acceptable:

$$y = \frac{P}{4\sigma} (a^2 - r^2). \quad (3.58)$$

The value of capacitance ΔC for the incremental ring of width dr, that is located on the spherical surface when the ring is deflected a distance y, is given as

$$\Delta C = \frac{\epsilon 2\pi r dr}{\delta - y}. \quad (3.59)$$

For $y/\delta \ll 1$, $1/(\delta - y)$ is approximated:

$$\frac{1}{\delta - y} = \frac{1}{\delta}\left(1 + \frac{y}{\delta}\right). \quad (3.60)$$

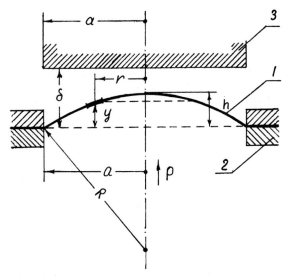

Figure 3.16 Deformation of flexible pressure-sensitive diaphragm.

The full capacitance $C + \Delta C$ after deflection is the integral:

$$C + \Delta C = \frac{2\pi\epsilon}{\delta}\int_0^a\left(1 + \frac{y}{\delta}\right)r\,dr. \tag{3.61}$$

Substitution of y from equation 3.58 gives

$$C + \Delta C = \frac{\pi\epsilon}{\delta}\,a^2 + \frac{\pi\epsilon}{\delta}\cdot\frac{P}{2\delta\sigma}\int_0^a(a^2 - r^2)r\,dr \tag{3.62}$$

where ϵ = permittivity of media between the electrodes, F/m; and
$(\pi\epsilon/\delta)a^2$ = capacitance before deflection ($h = 0$), F.

Therefore,

$$\Delta C = \frac{\pi\epsilon P}{2\delta^2\sigma}\int_0^a(a^2 - r^2)r\,dr = \frac{\pi\epsilon Pa^4}{8\delta^2\sigma}, \tag{3.63}$$

and the relative change of the capacitance is

$$\frac{\Delta C}{C} = \frac{a^2 P}{8\sigma\delta^2}. \tag{3.64}$$

The sensitivity is

$$\frac{\partial(\Delta C)}{\partial P} = \frac{\pi\epsilon a^4}{8\delta^2\sigma}. \tag{3.65}$$

The stiff diaphragm (Fig. 3.17) is usually used without initial tension (radial stress). Its deformation is described as follows:

$$y = \frac{3}{16}\,P\,\frac{1 - \mu^2}{Et^3}\,(a^2 - r^2)^2 \tag{3.66}$$

where y, P, a, and r are as in equation 3.56 and
μ = Poisson's ratio, dimensionless;
E = modulus of elasticity, N/m^2; and
t = diaphragm's thickness, m.

The capacitance corresponding to the deflection of the diaphragm under pressure P is found in the same way as for the flexible diaphragm:

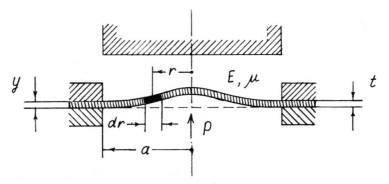

Figure 3.17   Deformation of stiff diaphragm.

$$C + \Delta C = \frac{2\pi\epsilon}{\delta}\int_0^a\left(1 + \frac{y}{\delta}\right)r\,dr. \tag{3.67}$$

It is assumed as before that $y/\delta \ll 1$.

Substituting y from formula 3.66 and integrating, the capacitance change is calculated as follows:

$$\Delta C = \frac{\pi}{16} \cdot \frac{\epsilon(1 - \mu^2)a^6}{E\delta^2 t^3}P. \tag{3.68}$$

The relative change of the capacitance and sensitivity is

$$\frac{\Delta C}{C} = \frac{1}{16} \cdot \frac{(1 - \mu^2)a^4}{E\delta t^3}P, \tag{3.69}$$

$$\frac{\partial(\Delta C)}{\partial P} \cdot \frac{1}{C} = \frac{1}{16} \cdot \frac{(1 - \mu^2)a^4}{E\delta t^3}. \tag{3.70}$$

All these formulas are derived for statics; in other words, for a slow process of pressure change when dynamic properties of the diaphragm do not affect its response to the pressure to be measured. The relation between the frequency of the alternating pressure and resonance frequency of the diaphragm is important for the design of the element. This frequency [21] for the flexible diaphragm is

$$f_0 = k\frac{0.38}{a}\sqrt{\frac{\sigma}{\rho}} \tag{3.71}$$

where f_0 = resonance frequency, 1/s;
ρ = density of diaphragm's material, kg/m^3; and
k = coefficient for a specific mode of vibration.

Coefficient k is equal to unity for the fundamental mode, and its values are given in Table 3.3 for the higher modes. In the table, the number of nodal diameters and circles are termed m and n, respectively.

σ should not exceed a yield stress $\sigma_{\max}$. For instance, $\sigma_{\max}$ for aluminum is $0.7 \times 10^6 \text{N/m}^2$, and for duralumin it is $1.5 \times 10^6 \text{N/m}^2$. It follows from equation 3.71 that the resonance frequency can be increased by increasing the tension of the diaphragm.

The equivalent mass M_{eq} of the diaphragm is a fraction of its real mass M in the range of frequencies between zero and resonance frequency.

$$M_{eq} = 0.3M \tag{3.72}$$

The equivalent stiffness S_{eq} of the diaphragm is calculated as

$$S_{eq} = (2\pi f_0)^2 M_{eq}. \tag{3.73}$$

TABLE 3.3

Coefficient k in Formula 3.71

m	0	1	2	0	1	2	0	1	2
n	1	1	1	2	2	2	3	3	3
k	1.0	1.6	2.1	2.3	2.9	3.5	3.6	4.2	4.8

A flexible circular diaphragm with a rigid center of radius b has an equivalent stiffness:

$$S_{eq} = 3.14 \frac{E}{1 - \mu^2} \cdot \frac{\alpha + 1}{(\alpha - 1)^3} b\beta^3 \tag{3.74}$$

where b is in meters, $\alpha = a/b$, $\beta = t/b$, and t is the thickness of the flexible part of the diaphragm.

An equivalent mass of such a diaphragm is given by

$$M_{eq} = M_b + 0.3M_a \tag{3.75}$$

where M_b and M_a are the real masses of the rigid and flexible parts of the diaphragm, respectively.

The resonance frequency is calculated as

$$f_0 = \frac{1}{2\pi} \sqrt{\frac{S}{M_{eq}}} \tag{3.76}$$

where S = diaphragm's relevant stiffness, kg/m^2.

For the rigid plane diaphragm, the values of f_0, S_{eq}, and M_{eq} are written as follows:

$$f_0 = 0.47 \frac{t}{a^2} \sqrt{\frac{E}{\rho(1 - \mu^2)}}, \tag{3.77}$$

$$S_{eq} = 4.19 \frac{E}{1 - \mu^2} \cdot \frac{t^3}{a^2}, \tag{3.78}$$

$$M_{eq} = 0.151M. \tag{3.79}$$

These formulas are given for the fundamental mode.

A layer of air between two plates creates additional stiffness S_a for the vibratory structure.

It is calculated as follows:

$$S_a = \frac{\rho c^2 A}{\delta_a} \tag{3.80}$$

where A = area of the diaphragm, m^2,
  c = air speed, m/s, and
  δ_a = air's density, kg/m^3.

This formula is correct for $\delta \ll \lambda_{min}$, where λ_{min} is the minimum length of acoustic waves exciting the diaphragm. The characteristic of output versus input pressure frequency can be controlled. For a specified response of the output signal to the change of the frequency of the input signal, the volumes adjacent to the diaphragm should provide a definite acoustic load on the diaphragm. This response is obtained by forming labyrinths above and behind the diaphragm, making holes in the diaphragm's cover, and so on. These measures are frequently used to make a frequency characteristic more flat. Some more relations for the calculation of systems with diaphragms can be found in Chapters 4, 5, and 9.

Among the widely used condenser microphones are the acoustic-pressure transducers containing the diaphragm-capacitive element. Sound waves set the diaphragm in vibration, producing capacitance variations that are converted into audiofrequency signals. The element requires a stable dc polarization voltage applied through a high resistance in order to form an ac output signal. An electrical equivalent circuit (Fig. 3.18) of the microphone contains a dc voltage (U_0) source, variable-gap (δ) acoustic pressure sensitive capacitor C, and resistance of load R_L. Due to the gap variation δ_1, an ac output voltage U_1 proportional to the current change i_1 is developed across R_L.

In order to find $U_1 = f(\delta_1)$, we will take the derivative with respect to time of both sides of the voltage-balance equation for the circuit in Figure 3.18:

$$U_0 = iR_L + \frac{1}{C}\int idt, \tag{3.81}$$

$$CU_0 = CiR_L + \int idt, \tag{3.82}$$

$$U_0\frac{dC}{dt} = R_LC_0\frac{di}{dt} + iR_L\frac{dC}{dt} + i. \tag{3.83}$$

Using operator (D) notation, the last relationship is rewritten as follows:

$$U_0C_1D = R_Li_1C_0D + i_0R_LC_1D + i_1 \tag{3.84}$$

where C_0 and i_0 are a capacitance and current corresponding to steady conditions. Current i_1 extracted from equation 3.84 is

$$i_1 = \frac{DC_1(U_0 - i_0R_L)}{R_LC_0D + 1}. \tag{3.85}$$

For the steady conditions,

$$U_0 \gg i_0R_L. \tag{3.86}$$

The output ac voltage U_1 is

$$U_1 = i_1R_L. \tag{3.87}$$

The combination of equations 3.85–3.87 gives

$$U_1 = \frac{U_0C_1R_LD}{C_0R_LD + 1}. \tag{3.88}$$

For the capacitor,

$$dC = -\frac{\epsilon A}{\delta_0^2}d\delta = -\frac{C_0}{\delta_0}d\delta \tag{3.89}$$

where δ_0 is δ for steady conditions.

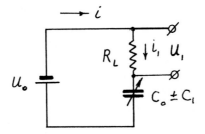

Figure 3.18 Equivalent circuit of capacitive microphone.

Replacing differentials by increments and dropping the minuses give

$$C_1 = \frac{C_0}{\delta_0} \delta_1 \qquad (3.90)$$

A time constant τ of the circuit is

$$\tau = C_0 R_L \qquad (3.91)$$

Finally, the transfer characteristic is calculated by substituting C_1 and τ in the equation (3.88)

$$U_1 = \delta_1 \frac{U_0}{\delta_0} \cdot \frac{\tau D}{\tau D + 1} \qquad (3.92)$$

Replacing D by $j\omega$,

$$U_1 = \delta_1 \frac{U_0}{\delta_0} \frac{\tau j\omega}{\tau j\omega + 1} \qquad (3.93)$$

where ω is an angular frequency of δ_1.

If $C_1/C \ll 1$, which means $\delta_1/\delta_0 \ll 1$ and $\tau = C_0 R_L \gg 1/\omega$, $\qquad (3.94)$

$$U_1 \approx \delta_1 \frac{U_0}{\delta_0}. \qquad (3.95)$$

This means that the output signal is proportional to the displacement and practically insensitive to the frequency of input (down to very low frequencies).

For $\tau \ll 1/\omega$,

$$U_1(j\omega) \approx j\delta_1 \frac{U_0}{\delta_0} \tau\omega. \qquad (3.96)$$

The output is directly proportional to the frequency; in other words, to the speed of change of the input signal:

$$U_1(t) \approx \delta_1 \frac{U_0}{\delta_0} \tau \frac{d\delta_1}{dt} \qquad (3.97)$$

τ change can be provided by connecting additional resistances and/or capacitances parallel to or in series with C_0 and R_L.

The phase angle between U_1 and δ_1 is calculated as follows:

$$\varphi = -\arctan \frac{1}{\omega C_0 R_L}. \qquad (3.98)$$

An equivalent electrical circuit of the microphone output includes capacitances and resistances of a preamplifier input and capacitance of a cable. They shunt the output, increasing the equivalent capacitance and decreasing its resistance. The additional capacitance gives attenuation of the output signal at high frequencies. This is one of the reasons why the preamplifier is usually mounted close to the microphone, reducing the length of the connecting leads and their capacitance.

Many contemporary microphones and vibratory sensors contain an electret as a source of polarization. An electret is a plastic or ceramic solid dielectric (for instance, teflon or $CaTiO_3$) possessing persistent electric polarization by virtue of a long time constant for decay of a charge instability. The electret is aged to stabilize its characteristics. An element with an electret (Fig. 3.19) contains a metal diaphragm 1, vibrating against electret 2, which is metallized at two surfaces 3 and 4.

The surface charge density σ_{ind}, induced on the metallized surface, is calculated as

$$\sigma_{ind} = \frac{\sigma}{\dfrac{\epsilon_r \delta}{\delta_e} + 1} \tag{3.99}$$

where σ = density of surface charge, C/m^2;
$\quad \delta$ = gap, m; and
$\quad \delta_e$ = electret's thickness, m.

Charge Q of the surface is

$$Q = A\sigma_{ind}. \tag{3.100}$$

The gap varies:

$$\delta = \delta_0 + \delta(t) \tag{3.101}$$

where $\delta(t)$ = time variation of gap, m.

Therefore, current I, which can be developed in the circuit, is evaluated as follows:

$$I = \frac{dQ}{dt} = A\frac{d\sigma_{ind}}{dt} = \frac{A\sigma\epsilon_r}{\delta_e\left(\dfrac{\epsilon_r\delta_0}{\delta_e} + 1\right)^2} \cdot \frac{d\delta}{dt}. \tag{3.102}$$

The element with an electret does not need a polarization voltage source. This is the most important advantage. The electret's charge is a function of temperature, which should not exceed the limit at which the electret loses its specific properties. Nonstable charge versus time can be a point of concern in the application of the device.

There are many original designs of pressure-sensitive capacitors. In one of them, for high-pressure measurements (Fig. 3.20), a thick diaphragm 1 incorporates a movable electrode 2 fastened to its center. The electrode is insulated from the

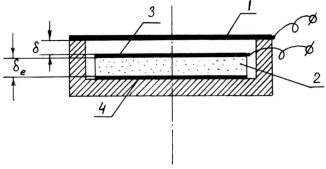

Figure 3.19 Capacitive element with electret.

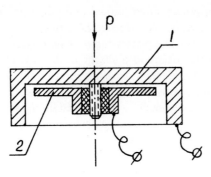

Figure 3.20 Pressure-sensitive structure with movable electrode fastened to center of diaphragm.

diaphragm. As the pressure increases, the capacitance is decreased. This structure has low temperature drifts due to the reduced influence of temperature stress gradients on the deformation under pressure.

Other Elements

Moisture content in bulk material can be determined by measuring the capacitance of two parallel plates. This capacitance is a function of the dielectric constant, which is in turn a function of the moisture content (η in %).

The dielectric constant of material ϵ_m is calculated as follows:

$$\epsilon_m = \epsilon_d \left(1 + \frac{\epsilon_w - \epsilon_d}{a\epsilon_w + b\epsilon_d} \cdot \frac{\gamma}{\gamma_w} \cdot \frac{\eta}{100} \right). \tag{3.103}$$

Taking into account formula 3.1, we can write the sensitivity of the element, S_η:

$$S_\eta = \frac{\partial C}{\partial \eta} = \frac{\epsilon_0 \epsilon_d A}{100\delta} \left(\frac{\epsilon_w - \epsilon_d}{a\epsilon_w + b\epsilon_d} \cdot \frac{\gamma}{\gamma_w} \right) \tag{3.104}$$

where ϵ_m = dielectric constant of material;
$\quad\epsilon_d$ = dielectric constant of dry material;
$\quad\epsilon_w$ = dielectric constant of water ($\epsilon_w = 80$);
$\quad\gamma$ = density of material, kg/m³;
$\quad\gamma_w$ = density of water, kg/m³; and
$\quad a$ and b are constant coefficients, which vary as follows:

$$0.01 \leq a \leq 0.03$$

$$0.5 \leq b \leq 1.5. \tag{3.105}$$

One of the humidity-sensitive elements is built by anodizing a small aluminum strip and forming aluminum oxide on the surface. The strip is one of the electrodes; the other electrode is gold deposited on oxide. The oxide is porous and absorbs moisture from the surrounding medium. The gold coating is thin enough to allow molecules of water to penetrate inside the layer and to reach an equilibrium state corresponding to the vapor pressure of the moisture in the gas. Moisturizing of the layer is accompanied by a change in the electrical leakage of the dielectric, combined with a change in the dielectrical constant, providing a signal correlating with humidity. A finely ground, activated alumina-type desiccant placed between plates is also used as a hygroscopic dielectric with variation of the dielectric constant versus humidity.

A proximity- or distance-sensing element contains one capacitor plate, and the target surface constitutes the other. A dimensional change can be detected without mechanical contact with the object. For instance, in biomedical applications, the element can be a microphone pickup of respiratory and cardiac sounds. If the target surface is electroconductive, it is grounded and behaves as one of the electrodes. If it is not grounded, an electroconductive light plate (foil) is fixed on the target, grounded, and, moving with the target, it changes the capacitance. The proximity element (Fig. 3.21) usually includes a guarding electrode 1, in addition to sensing 2 and object electrodes 3, to protect the sensing electrode against the stray-capacity effect from the surrounding elements of construction. This arrangement also helps to orient the lines of the electrical field between the electrodes perpendicular to their surfaces and parallel to each other.

In a *capacitive tachometer element,* two techniques can be used to produce a capacitance variation as a function of shaft position. The relative position of the plates can be varied, or the dielectric constant is changed. Figure 3.22 illustrates the concepts.

A *capacitive acceleration element* contains one fixed plate and a diaphragm or cantilever with a seismic mass. The spring elements are deflected under acceleration. They act as moving electrodes that provide a change of capacitance proportional to the acceleration.

The capacitance can be an index of strain.

A *solid-state strain-capacitive element* contains a piezoceramic thin bar (for instance, of barium titanate), which is cemented on a deformed part. The deformation causes the dielectric constant of the material to change, providing a proportional variation of the capacitance.

Another working principle is shown in Figure 3.23. Two initially bent flexible strips 1 and 2 carry two electrodes 3 and 4, insulated from each other. The strips are attached to the holders 5 and 6, which are fixed on the deformating surface. During deformation, the gap between the electrodes is changed, alternating the capacitance. The element can be made with low stiffness and can allow operation at a wide range of temperatures.

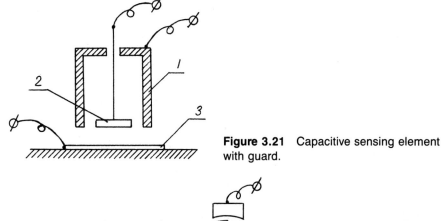

Figure 3.21 Capacitive sensing element with guard.

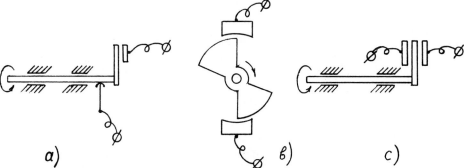

Figure 3.22 Capacitive tachometer elements. a = variation of relative positions of plates, b = split stator construction, c = variation of dielectric constant (rotor's part is made of dielectric).

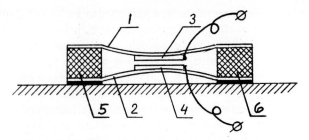

Figure 3.23   Capacitive strain-sensitive element.

A *capacitive element sensitive to temperature* has a solid-state structure of multilayer glass/ceramic-metallized electrode. The change of capacitance as a function of temperature allows the measurement of temperatures in the range from 0.01K to 400K. The element has a fast response to temperature.

Elements sensitive to the density, composition, or continuity of materials have a similar working principle. A change of the characteristics of the material between the plates causes a change in the dielectric constant, which is in turn proportional to the specific characteristics to be measured. The electrodes are made of two flat plates or concentric cylinders, which are submerged in the stationary or moving medium.

A *dynamic capacitor* is used for the modulation of current or voltage in electrical circuits. Frequently, this capacitor converts a weak dc signal into alternating current or voltage for a subsequent treatment. Two high-quality (gold-plated) electrodes are placed in an electrically and magnetically shielded container with inert gas. One of the electrodes is fixed; the other one is forced by an electromagnetic system to vibrate, creating a sinusoidal alternation of the gap between the electrodes:

$$\delta = \delta_0 + \delta_a \sin \omega t \tag{3.106}$$

where δ_a = peak value of a sinusoidal change of gap, m, and
  δ_0 = gap for steady conditions, m.

For an electrical circuit of capacitor C (Fig. 3.24), connected in series with resistor R, the performance of the device is described with the assumption that $\delta_a \ll \delta_0$:

$$C = \frac{\epsilon A}{\delta} = \frac{\epsilon A}{\delta_0} \cdot \frac{1}{1 + \alpha \sin \omega t} \approx C_0(1 - \alpha \sin \omega t) = C_0 - C_a \sin \omega t \tag{3.107}$$

where

$$\alpha = \delta_a/\delta_0, \quad C_0 = \frac{\epsilon A}{\delta_0}, \quad C_a = \frac{\delta_a}{\delta_0} C_0. \tag{3.108}$$

If $\omega RC \ll 1$, $U_c \approx U$,

$$q \approx (C_0 - C_a \sin \omega t) U \tag{3.109}$$

where q is an electrical charge, C.

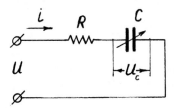

Figure 3.24   Electrical circuit of dynamic capacitor.

The peak value of current i is defined as

$$i = \frac{dq}{dt} = U\omega C_a.$$ (3.110)

For ω = const and C_a = const, the ac current in the circuit is directly proportional to the dc voltage.

For another case when $\omega RC \gg 1$, $q \approx C_0 U$ = const:

$$U_c = \frac{q}{C} = \frac{C_0 U}{C_0 + C_a \sin \omega t} \approx U\left(1 - \frac{C_a}{C_0} \sin \omega t\right).$$ (3.111)

The ac component of voltage U_c is directly proportional to the dc voltage for the constant values of C_a, C_0, and ω.

The list of capacitive elements is not limited to the given examples. There are additional industrial and scientific applications where the elements supply information about chemical reactions, the curing of plastic material, and so on.

Electrostatic Forces and Nonlinear Dynamics

In most instances, an electrostatic force exerted between capacitance plates is very small and can be completely negligible. However, sometimes it plays an important role in the performance of a device.

Examples of such devices include capacitive actuators in microelectronic sensors, driving elements in vibratory transducers, and so on. It is important to evaluate the force of attraction between different parts, including a large-diameter diaphragm highly sensitive to pressure.

The force developed between the capacitor's plates is a partial derivative of the energy stored in the capacitor with respect to the displacement of the electrodes along the lines of the field. This force always tends to reduce the gap between the plates:

$$F = -\frac{\partial W}{\partial y}, \qquad W = \frac{1}{2} U^2 C = \frac{1}{2} \frac{q^2}{C}, \qquad U = \frac{q}{C}, \qquad C = \frac{\epsilon A}{y}$$ (3.112)

where F = electrostatic force, N;
   W = energy stored in the capacitance, J; and
   y = variable separation for the parallel-plate capacitor, m.

Using formulas from the last string, the force is calculated for the parallel-plate capacitor:

$$F = \frac{1}{2} U^2 \frac{dC}{dy} = \frac{1}{2} U^2 \frac{\epsilon A}{y^2} = -\frac{1}{2} \frac{q^2}{\epsilon A}.$$ (3.113)

The negative signs indicate that the energy in the gap is reduced with the motion of the plates and that the forces attract the plates to each other. These signs are frequently neglected in calculations when the magnitude of the forces is in question.

If an ac voltage $U = U_m \sin \omega t$ is applied to the plates, the force is calculated by substituting U in formula 3.113 and assigning $y = \delta$:

$$F = \frac{1}{2} \frac{\epsilon A}{\delta^2} U_m^2 \sin^2 \omega t = \frac{1}{4} \frac{\epsilon A}{\delta^2} U_m^2 - \frac{1}{4} \frac{\epsilon A}{\delta^2} U_m^2 \cos 2\omega t.$$ (3.114)

This force has two components. One of them ($\epsilon A U_m^2/4\delta^2$) is an average force. The other one is an alternating, double-frequency force.

The equations describing statics or dynamics of the system with the element must include the electrostatic force if it is comparable with external forces and/or the system's internal forces of inertia, friction, or elasticity. Quite frequently, this analysis leads to the solution of nonlinear differential equations. The following is an illustration of a typical case.

Let us assume that we have a parallel-plate element (Fig. 3.25) that is composed of a fixed plate 1 and movable plate 2 with spring 3. Its motion is translational along the x-axis only and is characterized by lump parameters of mass m, friction coefficient b, and spring factor k. An external force on system F is applied to plate 2. The electrical circuit of the system includes the capacitor and resistor r in series with the capacitor. The voltage at the terminals is V.

Our goal is to write electromechanical equations describing behavior of the system and to reduce them to a form convenient for solution.

A mechanical force F_M is in balance with force F and electrostatic force F_E:

$$F + F_E = F_M. \tag{3.115}$$

F_E is calculated by using the relationship given in formula 3.112:

$$W = \frac{1}{2}\frac{q^2}{C} = \frac{1}{2}\frac{q^2 y}{\epsilon A}, \qquad y = \delta - x, \tag{3.116}$$

$$F_E = -\frac{\partial W}{\partial y} = -\frac{1}{2}q^2\frac{1}{\epsilon A}. \tag{3.117}$$

The mechanical force F_M, opposing F, contains three components: force of inertia, $m(d^2x/dt^2)$, force due to friction, $b(dx/dt)$, and spring-restraining force, kx:

$$F_M = m\frac{d^2x}{dt^2} + b\frac{dx}{dt} + kx. \tag{3.118}$$

Equation 3.115 can be rewritten as

$$F = D^2mx + Dbx + kx + \frac{1}{2}q^2\frac{1}{\epsilon A}. \tag{3.119}$$

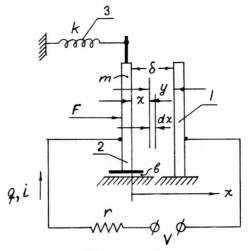

Figure 3.25 Model of nonlinear, electromechanical, capacitive system.

The equating of voltages in the electrical circuit gives

$$Ri + \frac{1}{C} \int i \, dt = V. \tag{3.120}$$

Since

$$i = \frac{dq}{dt}, \qquad q = \int i \, dt \qquad \text{and} \qquad C = \frac{\epsilon A}{y},$$

the preceding relationship is rewritten as follows:

$$R \frac{dq}{dt} + q \frac{1}{C} = V = DRq + \frac{y}{\epsilon A} q. \tag{3.121}$$

Equations 3.119 and 3.121 describe the mechanical and electrical behavior of the system, whose transfer characteristic can be determined if the functional relation between the mechanical and electrical quantities is found. This leads to the solution of linearized, nonlinear, differential equations 3.119 and 3.121 whose explicit analytical solutions cannot be obtained.

The nonlinear nature of the equations becomes obvious from their appearance and from substitutions of x and q from one equation to another. The linearization is possible since the plate displacement is small and the changes of electrical quantities are also small. In addition, a stable equilibrium point exists. Because of this, the problem can be solved in terms of small perturbations about the variables. The simplest technique of introducing the small variations is taking differentials of functions. This is accomplished in the following steps using equations 3.119 and 3.121:

The differential for F is given:

$$dF = (mD^2 + bD + k)dx + \frac{q_0}{\epsilon A} \, dq \tag{3.122}$$

where q_0 = charge for a stable equilibrium point, C.

The differential for V is written:

$$dV = DR \cdot dq + \frac{q_0}{\epsilon A} dy + \frac{y_0}{\epsilon A} dq = \left(RD + \frac{\delta - x_0}{\epsilon A} \right) dq - \frac{q_0}{\epsilon A} dx \tag{3.123}$$

where $dy = -dx$, and

x_0 = displacement of plate for a stable equilibrium point, m.

Replacing differentials dF, dx, dV, and dq with incremental changes ΔF, Δx, ΔV, and Δq, the system of two simultaneous, linear differential equations is obtained:

$$\begin{cases} (mD^2 + bD + k) \, \Delta x + \dfrac{q_0}{\epsilon A} \, \Delta q = \Delta F \\[2mm] -\dfrac{q_0}{\epsilon A} \, \Delta x + \left(RD + \dfrac{\delta - x_0}{\epsilon A} \right) \Delta q = \Delta V \end{cases} \tag{3.124}$$

These equations in matrix form are

$$
\begin{bmatrix} (mD^2 + bD + k) & \dfrac{q_0}{\epsilon A} \\[2ex] -\dfrac{q_0}{\epsilon A} & \left(RD + \dfrac{\delta - x_0}{\epsilon A} \right) \end{bmatrix} \begin{bmatrix} \Delta x \\[2ex] \Delta q \end{bmatrix} = \begin{bmatrix} \Delta F \\[2ex] \Delta V \end{bmatrix} \tag{3.125}
$$

Now they can be easily solved.

One of the radical methods of reducing the electrostatic forces is based on the use of a differential structure in force-inducing elements. In this case, the element's parts should be exposed to two oppositely oriented forces whose resultant is small or zero.

Attenuation of Sensitivity in a Microelectronic Capacitive Element

Reducing the sizes of microelectronic capacitive elements leads to a decrease of a nominal capacitance. Increasing the frequency of the current feeding the capacitance is a reasonable way to decrease the effective impedance of the element. There is a limit in the increase of the frequency, which is defined by the specific structure of the element, its shape, and the orientation of its parts. The analysis [22] clarifies the problem for a two-plate capacitor, but the theory, with a minor modification, is applicable to more complex structures.

The microcapacitors can be fabricated by using thin-film deposition, diffusion, and microassembly technology. High capacitance per unit of area and high resistance of the electrodes are the major differences between the macro and microcapacitors. Losses at the electrodes are increased, and the currents are not distributed uniformly along the length of the electrode.

The calculation of a nominal capacitance and transfer characteristics differs from that of the regular capacitor. The analysis has been provided for two models of a capacitor when current flows to the electrodes through leads attached to plates at different sides or at the same side and when the current experiences attenuation within the capacitor as in a transmission line. A characteristic parameter P_0 for the qualification of the capacitor value is a function of several quantities:

$$
P_0 = \sqrt{\frac{(r_1 + r_2)\omega C_0}{2}} \cdot l, \tag{3.126}
$$

where r_1 and r_2 = resistances of electrodes per unit of length, Ω/m;

  ω = angular frequency of current, rad/s;

  C_0 = capacitance of dielectric per unit of length, F/m; and

  l = length of electrodes, m.

If P_0 is small ($P_0 < 0.2$; the electrodes' resistance, their length, and the frequency of excitation are small), the capacitance is practically insensitive to the change of frequency ω.

When P_0 is large ($P_0 > 5$), the capacitance functions irregularly:

1. The value of the capacitance is not a function of the length of the electrodes. This means that the nominal capacitance can be insensitive to the longitudinal displacement of the electrodes.
2. The capacitance is inversely proportional to $\sqrt{\omega r_1}$, and $\sqrt{\omega r_2}$.
3. The resistance of electrodes, rather than leak resistance, affects the capacitance and its quality factor.
4. The element's geometry and the method of the electrodes' connection define the effective capacitances and resistances. A wide and short element

is preferable. Square electrodes are expedient for usage. However, if the electrodes are long, it is desirable to attach feeding leads at the electrodes' middle. The evaluation of parameter P_0, is extremely desirable for structures with a solid dielectric whose conductance cannot be neglected.

Temperature Instabilities

Any capacitive element responds to a temperature variation in the dielectric constant of the dielectric material placed between the electrodes, although the primary source of temperature errors is thermal deformation of the parts of the element's construction. It is a good practice to compensate for temperature drifts just at the place where they are induced. In a capacitive element, the temperature deformations quite frequently produce a change of capacitance comparable with that from the measurand. There are a number of techniques of compensation in the signal-processing stages. We will discuss here only the example that is related directly to the performance of the element and that illustrates how a proper choice of materials can significantly improve temperature stability.

   The representative device is a round, pressure-sensitive element (Fig. 3.26) containing a metal casing 1 with an attached thin, electroconductive, pressure (P)-sensitive diaphragm 4, which forms a variable capacitance with a stationary electrode 3. This electrode is supported by posts 2, and is electrically insulated from casing 1. The pertinent dimensions are shown in the figure.

   Due to a possible difference in the thermal coefficients of expansion of parts 1 and 2 (α_1 and α_2, respectively), gap δ can be changed with a temperature change of Δt.

   For the initial temperature,

$$\delta = l_1 - l_2. \tag{3.127}$$

For the changed temperature, a new gap δ_t is

$$\delta_t = l_1(1 + \alpha_1\Delta t) - l_2(1 + \alpha_2\Delta t). \tag{3.128}$$

   The error γ in the reading of the initial capacitance C_0 due to the temperature change Δt is

$$\gamma = \frac{C_t - C_0}{C_0} = \frac{\delta_0 - \delta_t}{\delta_t} = \frac{l_2\alpha_2 - l_1\alpha_1}{\delta_0 + (l_1\alpha_1 - l_2\alpha_2)\Delta t}. \tag{3.129}$$

where C_t = capacitance due to Δt change in temperature, F.

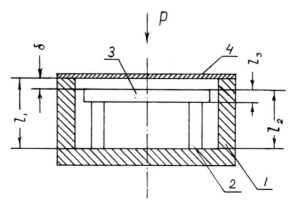

Figure 3.26   Pressure-sensitive, temperature-stabilized element.

In this calculation, a deformation of electrode 3 is neglected; therefore, the congruous area of the electrodes remains the same. It is also assumed that $l_2 \gg l_3$. If the numerator in formula 3.129 is zero, the error γ is also zero. Therefore, the condition for compensation is

$$\frac{\alpha_1}{\alpha_2} = \frac{l_2}{l_1} \quad \text{or} \quad l_2 = \frac{\delta}{\dfrac{\alpha_2}{\alpha_1} - 1}. \tag{3.130}$$

This condition is satisfied for the given gap if $\alpha_2 > \alpha_1$.

In the preceding calculation, the temperature of the parts was assumed to be equalized. In practice, the temperature gradients during the operation can significantly affect the performance of the device.

For instance, if the gas contacting the diaphragm changes its temperature rapidly while the temperature of the casing remains the same, the stress balance in the diaphragm is disturbed. This leads to a change of the transfer characteristic of the diaphragm (see formula 3.58) and induces instabilities. One of them (buckling) is shown schematically in Figure 3.27. A plain (dashed line) diaphragm 1 is buckled due to elongation of the material under temperature. Deformation of casing 2 is delayed due to a slow heat exchange. The diaphragm deformation can be evaluated by calculating the center displacement h, assuming that the curved surface is spherical:

$$h = \sqrt{(l^2 - 4a^2)\frac{3}{16}} \tag{3.131}$$

where l = length of arc, m, and
        a = diaphragm radius, m.

The length of the arc is the diameter of the diaphragm increased due to the temperature elongation:

$$l = 2a(1 + \alpha\Delta t) \tag{3.132}$$

where α = thermal coefficient of expansion, 1/K, and
        Δt = temperature, K.

Substitution of l in formula 3.131 gives

$$h = 1.22a\sqrt{\alpha\Delta t}. \tag{3.133}$$

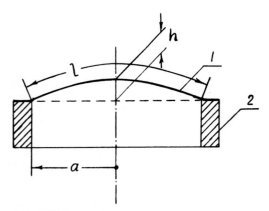

Figure 3.27 Temperature deformation of diaphragm.

Even if the diaphragm is made of a low-α material, the magnitude of h is high. For instance, for Ni-Span-C with $\alpha = 8.0 \times 10^{-6}$, $\Delta t = 5K$, and $2a = 50$mm, the center displacement h is 0.20mm. It is very high for the capacitive element.

Thermal deformations of electrical parts associated with the element (leads, wires, and so on) also contribute to temperature error. Therefore, they must be designed with care.

Features

The stable operation of any type of the capacitive element is possible if the materials of the parts are mechanically durable, resistive to corrosion, and have low temperature coefficients of physical parameters (nickel alloys, titanium, stainless steel). The members supporting the electrodes should be rigid to avoid deformation due to mounting or temperature effects. The elements must be electrostatically and magnetically shielded. The device is not inherently sensitive to the magnetic field. However, an induced electromotive force in the conductors due to the alternating magnetic field can create a spurious signal.

Ceramics, ruby, and fused silica, rather than plastic, should be used for insulators. The surface resistance of ceramics depends on humidity and contamination. The surfaces of the insulators should be coated by moisture-repelling materials, and the gaps between the electrodes should be protected from moisture, dust, fumes, and so on. The working area of the electrode should be fabricated with special measures to prevent corrosion (for example, plating by gold and rhodium). A combination of metals and alloys that are in contact with each other provides protection from electrolytic corrosion. In a high-quality capacitive transduction element, the quality of the electrode surfaces facing each other must be high. They should be smooth without any micropeaks, which concentrate charges and develop a silent discharge, creating a noise in the electrical signal.

The major advantages of capacitive elements are high sensitivity, simplicity of construction, high frequency response, small size, and small mechanical loading effect. In some applications, no contact with the process is required. In addition, the element can work at high temperatures, is insensitive to permanent magnetic fields, and has good linearity, repeatability, stability, and resolution. It is easy to incorporate the element into an experimental system. Due to a reactive power consumption and a very small fraction of the real power, the self-heating of the element is negligible. All these features, along with a low production cost, make this element valuable for many applications.

The capacitance of the contemporary elements is 20 to 1000pF, output impedance 1kΩ to 10MΩ, and excitation voltage 20 to 200V dc or ac. The frequency (for ac) lies between several kHz and several tens of MHz. The gap between the electrodes varies for different constructions from several micrometers to several millimeters. It is possible to measure the displacement of the electrodes below 1×10^{-6}mm at the frequency of the displacement up to 1MHz. The accuracy of measurements may be nearly 0.25 percent to 0.05 percent. The compensation for temperature is usually achieved by using a heater with a system of control and by stabilizing the temperature of the parts. The compensation for a humidity change can be provided by mounting silica gel dehumidifiers in the element's casing.

In measuring devices employing a capacitive element, the signal-conditioning electronics is integrated with the element or placed close to it. Long or loose leads from the element cause variation in the effective capacitance to be measured. The interconnecting cable should be shielded; its capacitance should be low and remain fixed in its value. The cable's movement can generate an unwanted signal.

The equivalent capacitance C_{EQ} of the element with the cable is calculated as

$$C_{EQ} = C[1 + \omega^2 L_{CA}(C_{CA} + C)] + C_{CA} \qquad (3.134)$$

where C = element's capacitance, F;
         C_{CA} = cable's capacitance, F;
         L_{CA} = cable's inductance, H; and
          ω = angular frequency of excitation, rad/s.

A capacitance parallel to the element decreases the sensitivity of transduction.

Assuming $L_{CA} = 0$ in the last expression, the sensitivity $\Delta C_{EF}/C_{EF}$ is found as follows:

$$\frac{\Delta C_{EF}}{C_{EF}} = \frac{\Delta C}{C + C_{CA}} = \frac{\Delta C}{C} \cdot \frac{1}{1 + \dfrac{C_{CA}}{C}}. \qquad (3.135)$$

With a C_{CA} increase, the sensitivity is decreased.

Some limitations are typical for the capacitive element; the major ones are presented here. The nominal capacitance and its change as a response to the measurand are small, and the impedance of output is high. The device has a high sensitivity to the variation of temperature. Instability in the positioning of connecting elements causes erratic signals. Accuracy is affected by wetting the parts, capillary effects, some build-up on electrodes, and an unpredictable change of the permittivity of the medium (for example, forming gas fractions, bubbles, fumes on the surface of liquid, and so on). The construction and assembly should be precise. Signal conditioning is relatively more complex than for other elements, and a parasitic potential can be induced in the ungrounded plate. All of these drawbacks cause difficulties, but they can be largely overcome by a rational design.

INDUCTIVE ELEMENTS

In an inductive element, inductance is alternated by varying the characteristics of the magnetic circuit. These characteristics can be the length or the area of the air gap, permeability of the ferromagnetic portion of the circuit, and connection or location of the winding's turns.

A representative example of the element structure is shown in Figure 3.28. It contains a pivoted armature 1, which can be turned around point O, increasing or

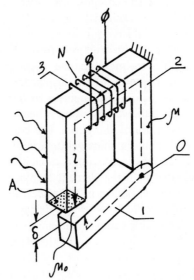

Figure 3.28 Model of inductive element.

decreasing gap δ with a fixed core 2. The core of length l carries winding 3 having N turns. The cross-sectional area of the gap is A. It is assumed to be the same for the core. The permeabilities of air and of the core's material are μ_0 and μ, respectively. Reluctance R for this circuit is expressed as follows:

$$R = \frac{1}{G} = \frac{\delta}{\mu_0 A} + \frac{l}{\mu A}. \qquad (3.136)$$

where units for R, δ, l, μ, and A are 1/H, m, m, H/m, and m^2, respectively. G stands for permeance [H].

The inductance of the coil L in henries is

$$L = \frac{N^2}{R} = N^2 G = \frac{N^2}{\dfrac{\delta}{\mu_0 A} + \dfrac{l}{\mu A}}. \qquad (3.137)$$

As follows from this formula, a deflection of the armature, giving a change of gap δ, affects the value of the inductance. By a small modification, this structure becomes sensitive to a change of area A. A transducer containing this element can convert a translational or angular motion into an electrical signal. With small additions, it forms signals proportional to force, velocity, acceleration, pressure, flow rate, and so on.

An element with a variation of μ, when the other parameters are kept constant, allows the measurement of force, hydrostatic pressure, temperature, magnetic field intensity, and other quantities. A number of contemporary materials made of metal alloys and oxides of ferromagnetics have a permeability sensitive to the listed physical values. The inductive element is widely used in a reverse mode for generating force or displacement proportional to the current in its coil. Similarly to the capacitive element, the variable inductance is an appropriate component for interfacing with digital circuits.

Structures of Inductive Elements

Various compositions of inductive elements are illustrated in Table 3.4. Single-element devices are simple and inexpensive, but their metrological characteristics are inferior to those for a differential-type device. Only the circuits providing a direct conversion are listed in the table. Some relationships are given, assuming that $\mu \gg \mu_0$ and neglecting the fringing and leakage effects. A wide variety of transducers employing the inductive element are built with the addition of different sensing elements to make the transducer sensitive to physical values that are primarily mechanical in nature. Some typical elements will be discussed in the following paragraphs in greater detail.

Equivalent Circuits and General Characteristics

The permeability of air and various nonmagnetic materials can be assumed equal to the permeability of free space:

$$\mu_0 = 4\pi \times 10^{-7} \text{H/m}. \qquad (3.138)$$

The permeability of ferromagnetics varies depending on the magnetic field intensity and frequency of core excitation. This permeability μ is frequently expressed with relative permeability μ_r:

$$\mu = \mu_0 \mu_r. \qquad (3.139)$$

TABLE 3.4

Structures of Inductive Elements

No.	Variable	Structure	Basic relationship	No.	Variable	Structure	Basic relationship
1	2	3	4	1	2	3	4
1	Gap δ in a single closed-magnetic-path circuit	$\delta \pm \Delta\delta$	$L = \dfrac{N^2}{R}$, $\quad R = \dfrac{2(\delta \pm \Delta\delta)}{\mu_0 A}$	9	Displacement Δx of halves of two coils	$x \pm \Delta x$; L_1, L_2	$L = L_1 + L_2 \pm 2L_{12}$, $\quad L_{12} = f(x \pm \Delta x)$
2	Gaps δ_1 and δ_2 in a differential closed-magnetic-path circuit	$\delta_1 \pm \Delta\delta$; $\delta_2 \mp \Delta\delta$; N_1, N_2	$L_1 = \dfrac{N_1^2}{R_1}$, $\quad L_2 = \dfrac{N_2^2}{R_2}$, $\quad R_1 = \dfrac{2(\delta_1 \pm \Delta\delta)}{\mu_0 A_1}$, $\quad R_2 = \dfrac{2(\delta_2 \mp \Delta\delta)}{\mu_0 A_2}$	10	Displacement x of a sliding contact along the length of the inductor	l; x; N	$L = \dfrac{(N\frac{x}{l})^2}{R}$
3	Area A in a single closed-magnetic-path circuit	δ_1, δ_2; $A \pm \Delta A$; N	$L = \dfrac{N^2}{R}$, $\quad R = \dfrac{\delta_1 + \delta_2}{\mu_0(A \pm \Delta A)}$	11	Displacement Δx of a countered core	$\pm \Delta x$; x; y	$L = f[\Delta x, y(x)]$
4	Areas A_1 and A_2 in a differential closed-magnetic-path circuit	δ_1; $A_2 \mp \Delta A_2$; N_1, N_2; $A_1 \pm \Delta A_1$; δ_2	$L_1 = \dfrac{N_1^2}{R_1}$, $\quad L_2 = \dfrac{N_2^2}{R_2}$, $\quad R_1 = \dfrac{\delta_1 + \delta_2}{\mu_0(A_1 \pm \Delta A_1)}$, $\quad R_2 = \dfrac{\delta_1 + \delta_2}{\mu_0(A_2 \mp \Delta A_2)}$	12	Displacement x of a short circuited turn	x	$L = f(x)$ See text
5	Area A in a single, closed-path, angular-displacement circuit	$A \pm \Delta A(\varphi)$; φ; δ_1; δ_2; N; or	$L = \dfrac{N^2}{R}$, $\quad R = \dfrac{\delta_1 + \delta_2}{\mu_0[A \pm \Delta A(\varphi)]}$	13	Thickness d of an electro-conductive rod (wire)	d	$L = f(d)$ See text
6	Areas A_1 and A_2 in a differential, closed-path angular-displacement circuit	A; φ; δ; δ_1, δ_2; A_1, A_2	$L_1 = \dfrac{N_1^2}{R_1}$, $\quad L_2 = \dfrac{N_2^2}{R_2}$, $\quad R_1 = \dfrac{1}{\mu_0}\left[\dfrac{\delta}{A} + \dfrac{\delta_1}{A_1 \pm \Delta A_1(\varphi)}\right]$, $\quad R_2 = \dfrac{1}{\mu_0}\left[\dfrac{\delta}{A} + \dfrac{\delta_2}{A_2 \mp \Delta A_2(\varphi)}\right]$	14	Proximity Δx of an electro-conductive part to a coil inducing eddy currents in the part	Δx	$L = f(x)$ See text
7	Core's displacement Δx in a single, open-path circuit	$\pm \Delta x$; x	$L = f(\Delta x)$ See text	15	Permeability μ, which is a function of force, pressure, strain, temperature, etc. (F, P, λ, and t^0, respectively)	F, P, λ, t^0; A; l; R_0; N	$L = \dfrac{N^2}{R}$, $\quad R = R_0 + \dfrac{l}{A\mu(F, P, \lambda, t^0, H)}$
8	Armature's displacement Δx in a single, open-path circuit	$\pm \Delta x$; x	$L = f(\Delta x)$ See text	16	Position of a discrete-output-modulating member	D; x; L_1; L_2	L; L_1; L_2; $L_1 > L_2$

There are three major types of materials used in the elements in which μ is not a function of a measurand. They are laminated silicon transformer steels, ferrites, and permalloys. The configuration of the magnetization curves for these materials greatly depends on composition and technology.

Figure 3.29 shows the shapes of these curves and their ranges. The element usually operates at the linear part of the curves to reduce nonlinear distortion in signals.

For low frequencies of excitation, μ is calculated simply as a ratio (Fig. 3.30):

$$\mu = \frac{B_a}{H_a}. \tag{3.140}$$

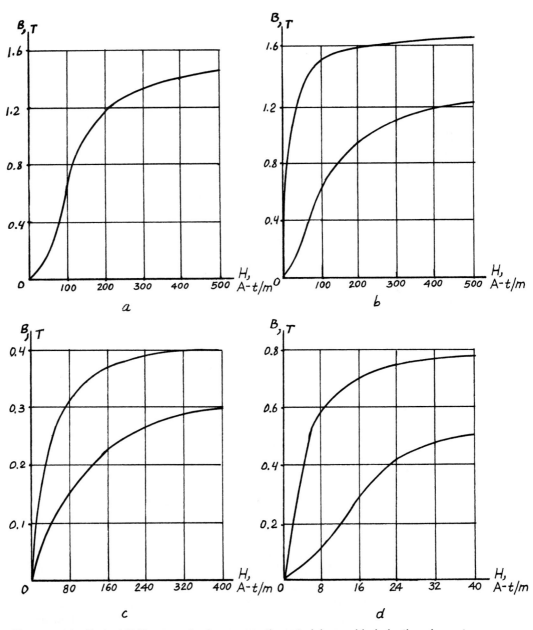

Figure 3.29 Typical B-H curves for ferromagnetic materials used in inductive element. a = soft iron, b = transformer steels, c = ferrites, d = permalloys.

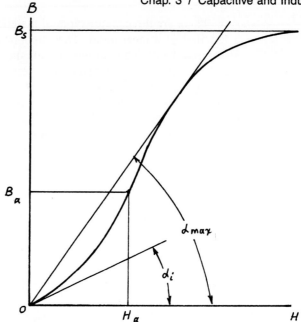

Figure 3.30 Magnetization curve's characteristic parameters.

However, at higher frequencies the effective value μ'_{ef} should be substituted in the formulas for calculating reluctances, especially if the core is made of electro-conductive material. Due to ac excitation, the alternating magnetic field induces eddy currents in the core. These currents produce a field that opposes the main field developed by the system and prevents the field from penetrating inside the core. The field decreases from the surface exponentially toward the interior, and the skin depth β denotes the distance at which the field is decreased to $1/e$ of its value at the surface. This effect markedly reduces the effective permeability of the material. μ'_{ef} can be expressed [23] as

$$\mu'_{ef} = \mu \frac{H_{ef}}{H_a} = \frac{1}{t} \sqrt{\frac{\mu}{\pi f \sigma}} \qquad \text{for } \beta < t \qquad (3.141)$$

where H_a = applied field, A/m;

H_{ef} = effective field, A/m;

t = thickness of core, m;

f = frequency of field, Hz; and

σ = electrical conductivity of material, S/m.

Coefficient β is calculated using the following formula:

$$\beta = \frac{1}{\sqrt{\pi f \mu \sigma}} \qquad (3.142)$$

If $\beta > t$, $\mu'_{ef} \approx \mu$. $\qquad (3.143)$

Almost all inductive elements have a gap in the magnetic system. Magnetic charges are concentrated on the faces of the gaps. The presence of the magnetic charges brings a negative component into the field inside the core, demagnetizing the material. This effect is reflected in the reduction of μ_r to $\bar{\mu}_c$. From analysis and experiment [24], $\bar{\mu}_c$ is calculated as follows:

$$\bar{\mu}_c = \frac{\mu_r}{1 + \dfrac{M}{4\pi}(\mu_r - 1)} \tag{3.144}$$

$$\frac{M}{4\pi} = \left[1 + 0.211\left(\frac{l_w}{l_c}\right)^{-1.116}\right] e^{(6.855 - 8.074\lambda^{0.1353})} \tag{3.145}$$

$$\lambda = \frac{l_c}{d_c} = \frac{l_c}{2}\sqrt{\frac{\pi}{A_c}} \tag{3.146}$$

where μ_r = relative permeability of core's material;
$\quad\quad l_w$ = length of winding, m;
$\quad\quad l_c$ = length of core, m;
$\quad\quad \lambda$ = relative length of core;
$\quad\quad d_c$ = diameter of core, m (for a round shape); and
$\quad\quad A_c$ = cross-sectional area of core (for a rectangular shape), m^2.

Table 3.5 lists the typical values [23] and ranges for essential parameters that are used to calculate the magnetic systems.

In the Table,

μ_{ir} = relative initial permeability;

$\mu_{\text{max } r}$ = maximum relative permeability; and

B_s = saturation flux density.

μ_{ir} and $\mu_{\text{max } r}$ are calculated by evaluating angles α_i and α_{max} for the lines beginning at the origin and tangent to the *B-H* curve (Fig. 3.30):

$$\mu_{ir} = \tan \alpha_i, \quad\quad \mu_{\text{max } r} = \tan \alpha_{\text{max}}. \tag{3.147}$$

An equivalent electrical circuit of a single-coil element is shown in Figure 3.31. The following notations are used for the elements of the circuit:

R_0, *L,* and *C* = resistance, inductance, and intrinsic capacitance [Ω, *H,* and *F*], respectively.

R_c and R_d = resistances introducing active losses in the core and insulation [Ω], respectively.

TABLE 3.5

Essential Characteristics of Some of the Magnetic Materials (Source: [23])

Material (composition[1])	μ_{ir}	$\mu_{\text{max } r}$	B_s [T]	σ [S/m]
Commercial iron (0.2imp.[2])	250	9 000	2.15	10×10^6
Purified iron (0.05imp.)	10 000	200 000	2.15	10×10^6
Silicon-iron (4 Si)	1 500	7 000	1.95	1.7×10^6
Silicon-iron (3 Si)	7 500	55 000	2.00	2.0×10^6
Mu metal (5 Cu, 2 Cr, 77 Ni)	20 000	100 000	0.65	1.6×10^6
78 Permalloy (78.5 Ni)	8 000	100 000	1.08	6.3×10^6
Supermalloy (79 Ni, 5 Mo)	100 000	1,000 000	0.79	1.7×10^6
Permendur (50 Cs)	800	5 000	2.45	14.3×10^6
Mn-Zn ferrite	1 500	2 500	0.34	0.05×10^{-6}
Ni-Zn ferrite	2 500	5 000	0.32	1.0×10^{-9}

[1] Percent by weight; remainder is Fe.
[2] imp. = impurities.

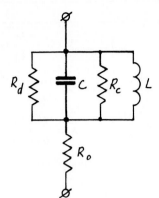

Figure 3.31 Equivalent electrical circuit of single-coil element.

The impedance Z_{eq} between the terminals is

$$Z_{eq} = R_{eq} + \omega L_{eq} \tag{3.148}$$

where ω = angular frequency of current in the circuit, rad/s.

The equivalent values of resistance and inductance are

$$R_{eq} = R_0 + r\frac{\omega^2 L^2}{r^2(1 - \omega^2 LC)^2 + \omega^2 L^2} \tag{3.149}$$

$$L_{eq} = L\frac{r^2(1 - \omega^2 LC)}{r^2(1 - \omega^2 LC)^2 + \omega^2 L^2} \tag{3.150}$$

In these formulas,

$$r = \frac{R_c R_d}{R_c + R_d}. \tag{3.151}$$

R_0 is calculated as a function of the defined diameter of the winding's wire d (bare), the dimensions of the spool (Fig. 3.32), and the number of turns N, which depends on two empirical factors: $\alpha = 1.25$ and $\beta = 0.003$ to 0.0016 mm^2. These factors should be taken into account for different diameters of enamel-insulated wire (the thinner the diameter, the greater β):

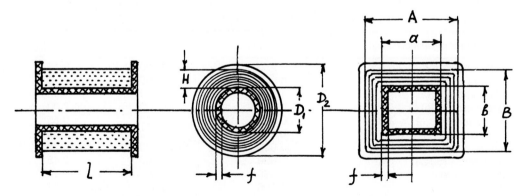

Figure 3.32 Round and rectangular coils with relevant dimensions of spools.

$$N = \frac{lH}{\alpha d^2 + \beta}, \tag{3.152}$$

$$R_0 = \frac{4\rho l_{av} N}{\pi d^2}. \tag{3.153}$$

Copper's resistivity ρ varies from 0.0175 to 0.0182 $\Omega \cdot mm^2/m$. It depends on impurities (Al and Mn) and on the conditions of the drawing during fabrication. The average length of one turn l_{av} in meters is calculated for the round spool as follows:

$$l_{av} = 0.5\pi(D_1 + D_2). \tag{3.154}$$

For the rectangular spool,

$$l_{av} = 2(a + b) + \pi \frac{B - b}{2}. \tag{3.155}$$

All dimensions but l_{av} in formulas 3.152–3.155 should be taken in millimeters to obtain R_0 in Ω. l_{av} is expressed in meters when it is substituted in formula 3.153.

The intrinsic capacitance, C, is calculated by evaluating capacitance C_1, formed by the first layer of the wire and core. Another component, C_2, is the capacitance of the layers as it is seen from the terminals. For the round spool,

$$C_1 = \frac{\pi 8\epsilon_1 D_1 l}{8f - \pi d} \quad \text{and} \quad C_2 = \frac{4\epsilon_1 \pi(D_1 + D_2)ld}{(4f - \pi r)(D_2 - D_1)}. \tag{3.156}$$

For the rectangular spool,

$$C_1 = \frac{16\epsilon_2 l(a + b)}{8f - \pi d} \quad \text{and} \quad C_2 = \frac{8\epsilon_2 ld(4a + 0.88b + \pi B)}{(8f - \pi d)(B - b - 2d)} \tag{3.157}$$

where ϵ_1 and ϵ_2 are permittivities for the material of spool and wire insulation, respectively.

An experimental evaluation of C_2 is useful for obtaining precise values and variations due to different factors, such as temperature. By connecting a variable capacitance C_v parallel to the coil, the resonance frequencies (ω_1 and ω_2) for the two pre-set values of C_v (C_{v1} and C_{v2}) are found. The capacitance is calculated by solving two simultaneous equations:

$$\omega_1 = \frac{1}{\sqrt{L(C + C_{v1})}} \quad \text{and} \tag{3.158}$$

$$\omega_2 = \frac{1}{\sqrt{L(C + C_{v2})}}. \tag{3.159}$$

From these equations,

$$C = \frac{\omega_2^2 C_{v1} - \omega_1^2 C_{v2}}{\omega_1^2 - \omega_2^2}. \tag{3.160}$$

It is assumed in this measurement that $R_0 \approx 0$ and $R_d = R_c \approx \infty$ in the equivalent circuit.

For all calculations,

$$R_c = \frac{U^2}{P_c} \tag{3.161}$$

where U = voltage applied to the coil, V, and
P_c = power loss in the core, W.

Power loss in the core has two components [25].

One of them is eddy current losses, P_e; the other is hysteresis losses P_h:

$$P_e = k_e f^2 t^2 B_m^2 V = k_e f^2 t^2 B_m^2 \frac{M}{\gamma}, \tag{3.162}$$

$$P_h = k_h f B_m^n \frac{M}{\gamma} \tag{3.163}$$

where k_e = constant depending on the conductivity of the material, S/m;
$\quad f$ = frequency of excitation, 1/s;
$\quad t$ = thickness of material, m;
$\quad B_m$ = maximum flux density in the core, T;
$\quad V$ = material's volume, m^3;
$\quad M$ = mass of core, kg;
$\quad \gamma$ = density of the core material, kg/m^3;
$\quad k_h$ = constant depending on properties of the material
$\qquad$ (It is proportional to the reaction of the material to the alignment of the magnetic particles in the material.), $A^2 s^2/\text{kg} \cdot \text{m}$; and
$\quad n$ = Steinmetz exponent, which varies from 1.5 to 2.5.

The unit for k_h is obtained assuming that $n = 2$.

For the given values M, δ, and t, the total core loss is written as follows:

$$P_c = K_h f B_m^n + K_e f^2 B_m^2 \tag{3.164}$$

where

$$K_h = k_h \frac{M}{\gamma} \quad \text{and} \quad K_e = k_e t^2 \frac{M}{\gamma}. \tag{3.165}$$

Even at the audio-frequency excitation, R_c can be comparable to the reactive part of the impedance (ωL) and can significantly affect the operation of the element. In some cases, knowledge of P_h, P_e, K_h, K_e, and n is desirable to qualify the contribution of each constituent part of the losses (for example, a search for temperature instability). The best way to evaluate P_h, P_e, and R_c is by experimenting. The element is connected to a variable voltage and frequency source. The voltage and frequency are then changed; B is measured along with voltage U, current I, and phase shift angle θ between them. Since R_0 is known, P_c can be calculated for several combinations of f and B. The following substitution of P_c, f, and B in formula 3.164 allows a system of equations to be written and solved for K_e, K_h, and n. Thus, P_e and P_h are also determined.

Loss in insulation P_d is calculated similarly to that for a capacitive element. For the voltage U applied to the coil,

$$R_d = \frac{U^2}{P_d}. \tag{3.166}$$

where R_d = resistance of insulation, Ω.

For many cases, R_d can be assumed to be infinitely large and ignored in calculations.

A quality factor Q_f is readily calculated by taking the ratio of imaginary and real parts of the formula 3.148 for the impedance Z_{eq}:

$$Q_f = \frac{\omega L_{eq}}{R_{eq}} = \frac{\omega L}{R_0}\left[(1 - \omega^2 CL) + \frac{\omega^2 L^2}{(1 - \omega^2 CL)r}\left(\frac{1}{r} + \frac{1}{R_0}\right)\right]^{-1}. \tag{3.167}$$

The element usually operates at frequencies of excitation ω much lower than the resonant frequency ω_0 for an L-C circuit:

$$\omega \ll \omega_0 = \sqrt{\frac{1}{LC}} \quad \text{or} \quad \omega^2 CL \ll 1 \tag{3.168}$$

Taking into account this condition and that $r \gg R_0$, equation 3.167 for Q_f is reduced:

$$Q_f = \left(\frac{1}{Q_\omega} + \frac{1}{Q_c} + \frac{1}{Q_d}\right)^{-1} \tag{3.169}$$

where $Q_\omega = \omega L/R_0 = Q$ of coil;
$\quad\;\; Q_c = R_c/\omega L = Q$ of core; and
$\quad\;\; Q_d = R_d/\omega L = Q$ of insulation.

Since

$$R_d \approx \infty \quad \text{and} \quad \frac{1}{Q_d} \approx 0$$

$$Q_f \approx \left(\frac{1}{Q_\omega} + \frac{1}{Q_c}\right)^{-1}. \tag{3.170}$$

Single Displacement Elements

Calculating reluctances and their variations due to the changes of measurands is a common operation in the analysis of elements. It is important to understand that the reluctance is calculated as a resistance to the propagation of the flux along its path. Usually, this path is divided into sections. Each section has a uniform cross-sectional area A_i and permeability μ_i. Thus, the total reluctance R_Σ to the flux along its path is expressed as a sum of the components connected in series:

$$R_\Sigma = \sum_{i=1}^{m} \frac{l_i}{A_i \mu_i} \tag{3.171}$$

where m = number of sections.

For a parallel connection,

$$G_\Sigma = \frac{1}{R_\Sigma} = \sum_{i=1}^{m} \frac{1}{\dfrac{l_i}{A_i \mu_i}} \tag{3.172}$$

where G_Σ = total permeance, H.

If a member has length l, variable $A(x)$, and $\mu(x)$ along the flux propagation (x-axis), its reluctance is calculated as the integral:

$$R = \int_l \frac{dx}{A(x)\mu(x)}.$$ (3.173)

It is useful to draw an electrical equivalent circuit for calculating reluctances by presenting the reluctances as electrical resistances. The lengths l_i in formulas 3.172 and 3.173 are calculated as mean lengths.

As an example, we will consider a variable-gap-displacement element (Fig. 3.33) in which cylindrical plunger 1 moves vertically in the O-shaped stationary core 2. Winding 3 carrying current i creates a magnetomotive force M, which stimulates flux ϕ to flow through the circuit. The flux is divided into two parallel components: ϕ_1 and ϕ_2. Displacement of the plunger changes the overall reluctance of the circuit and, consequently, the inductance of the system. This inductance is a measure of displacement. The lower part of the plunger is a truncated cone (for demonstration purposes). The permeability of the plunger's and core's material is μ. For this instance, we assume that μ is constant and does not depend on the field intensity. This condition notably simplifies the problem, which can be solved with linear equations. A nonlinear case will be discussed in the following paragraphs. The gap

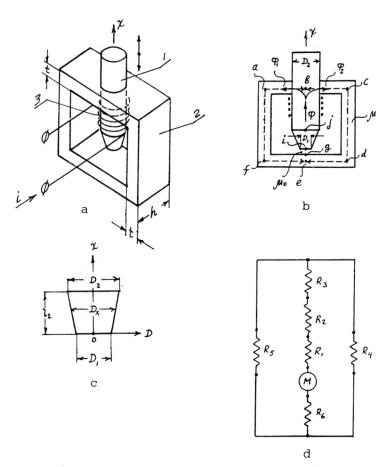

Figure 3.33 Variable-gap-displacement inductive element.
a = construction, b = pertinent dimensions, c = drawing for cone reluctance calculation, d = equivalent circuit.

between parts 1 and 2 at the cylindrical joint is small and its reluctance is neglected. Fringing and leakage fluxes in the system are also neglected. All pertinent dimensions are shown in the figure. In order to find the reluctance, we should calculate its magnitude as it is seen from the coil. It is easy to recognize that the equivalent circuit is composed of several series and parallel terms as shown in the figure. The meaning of each term is explained in Table 3.6.

The reluctance of the conical part is calculated by expressing the variable diameter D_x and area $A(x)$ (Fig. 3.33c) as functions of length x:

$$D_x = D_1 + x \frac{D_2 - D_1}{l_2}.$$ (3.174)

Hence,

$$R_2 = \int_0^{l_2} \frac{dx}{A(x)\mu} = \int_0^{l_2} \frac{dx}{\mu\left(D_1 + x \dfrac{D_2 - D_1}{l_2}\right)} = \frac{4l_2}{\mu\pi D_1 D_2}.$$ (3.175)

Thus, the total reluctance R is

$$R = \frac{R_S R_P}{R_S + R_P}$$ (3.176)

where

$$R_S = R_1 + R_2 + R_3 + R_6 \quad \text{and}$$ (3.177)

$$R_P = \frac{R_4 R_5}{R_4 + R_5}.$$ (3.178)

TABLE 3.6

Reluctances of Parts of Magnetic Circuit

Part of circuit between the labeled points	Length of the part	Area of the part	Reluctance
g-i	l_1	$\dfrac{\pi D_1^2}{4}$	$R_1 = \dfrac{4l_1}{\mu_0 \pi D_1^2}$
i-j	l_2	$A(x)$	$R_2 = \dfrac{4l_2}{\mu\pi D_1 D_2}$ see calculation
j-b	l_3	$\dfrac{\pi D_2^2}{4}$	$R_3 = \dfrac{4l_3}{\mu\pi D_2^2}$
b-c-d-e	l_4	th	$R_4 = \dfrac{l_4}{\mu th}$
b-a-f-e	l_5	th	$R_5 = \dfrac{l_5}{\mu th}$
e-g	$l_6 = \dfrac{t}{2}$	$\dfrac{\pi D_1^2}{4}$	$R_6 = \dfrac{4l_6}{\mu\pi D_1^2}$

Since $\mu \gg \mu_0$, the reluctance of the gap is much larger than the reluctances of the ferromagnetic parts.

Therefore,

$$R \approx R_1 = \frac{4l_1}{\mu_0 \pi D_1^2}. \tag{3.179}$$

The preceding approximation is commonly used for a rough evaluation of element performance.

It is seen from formulas 3.137 and 3.179 that the element's output (inductance L or reactance ωL) is inversely proportional to the measurand. In a similar manner, transfer characteristics of an element with the gap area variation can be calculated. The area is usually directly proportional to a change of one of the dimensions in the construction. Therefore, the output can be a linear function of the displacement. For instance, if armature 1 (Fig. 3.34) of a magnetic system is restricted to moving only along the x-axis and core 2 remains fixed, the only variation in reluctance is due to the change of the area at pole A, which is $h(h - x)$. The area at pole B does not change; therefore, the inductance can be practically a linear function of x.

Reluctances for typical parts [24] of magnetic systems are shown in Table 3.7. By combining them in one structure, a particular system can be easily calculated. The directions of flux ϕ are indicated by arrows. For calculating the reluctances of gaps having the shapes introduced in Table 3.7, μ_0 has to be substituted for μ.

Solutions of the nonlinear equations give more precise results. Using a manual or computer iteration technique is desirable.

The simplified calculation can be composed of several steps.

1. First, a magnetic circuit is chosen; then it is reduced to a series-connected source of magnetomotive force, with reluctance of magnetic parts and reluctance of a variable air-gap part.
2. The circuit is divided into sections with uniform cross-sectional areas and permeabilities of the material.
3. The gap flux density B_δ is assigned and taken between 0.1T and 1.0T. For the chosen B_δ, the common flux ϕ is calculated as a product:

$$\phi = B_\delta A_\delta \tag{3.180}$$

where A_δ is an area of the air part of the circuit.

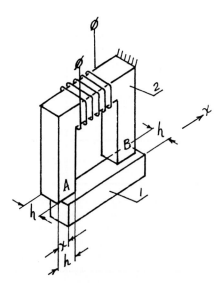

Figure 3.34 Variable-area-displacement inductive element.

TABLE 3.7

Reluctances of Various Shapes

Shape	Reluctance	Shape	Reluctance
	$\dfrac{l}{\mu th}$		$\dfrac{4l}{\mu\pi(D_2^2 - D_1^2)}$
	$\dfrac{4l}{\mu\pi D^2}$		$\dfrac{4l}{\mu\pi D_1 D_2}$
	$\dfrac{l\cdot \ln\frac{t_2}{t_1}}{\mu h(t_2 - t_1)}$		$\dfrac{\ln\frac{D_2}{D_1}}{2\pi\mu h}$
	$\dfrac{2T - 2t + 0.5l}{\mu th}$		$\dfrac{4l}{\pi\mu(D_1 D_2 - D_3^2)}$

4. Flux densities B_i in the different parts of the circuit are then calculated as follows:

$$B_i = \frac{\phi}{A_i} \qquad (3.181)$$

where A_i is an area of section i.

5. For the found B_i, a corresponding value of permeability μ_i can be determined from the magnetization curve:

$$\mu_i = \frac{B_i}{H_i} \qquad (3.182)$$

where H_i is the field intensity found from the magnetization curve for a given section i.

6. Reluctances R_i of the sections are calculated as follows:

$$R_i = \frac{l_i}{\mu_i A_i} \qquad (3.183)$$

where l_i is a known length of section i.

7. The overall reluctance R is a sum of the sections' reluctances and reluctance of the gap:

$$R = \sum_{i=1}^{n} \frac{l_i}{\mu_i A_i} + \frac{\delta}{\mu_0 A_\delta} \qquad (3.184)$$

where δ = length of gap, m.

8. Inductance L of the element with a coil having N turns is

$$L = \frac{N^2}{R}. \qquad (3.185)$$

9. If rms and frequency ω of the voltage U feeding the element are defined, the current I through the winding is

$$I = \frac{U}{\sqrt{r^2 + (\omega L)^2}} \qquad (3.186)$$

where r = coil's resistance, Ω.

At this stage, only one point of the transfer characteristic is determined. The further analysis can include the calculation of the following functions:

$$L = f(\delta), \qquad L = f(A_\delta), \qquad I = f(l_\delta), \qquad \text{or} \qquad I = f(A_\delta) \qquad (3.187)$$

when the other parameters of the circuit remain constant. For instance, if we have chosen characteristic $I = f(\delta)$, the current I, calculated from formula 3.186, is an output corresponding to input l_δ.

In order to obtain all points of the characteristic, we should vary B's and l_δ's, calculate ϕ's, and find the current that must satisfy conditions in formulas 3.180–3.186 along with Ohm's law for this magnetic circuit:

$$IN = \phi R. \qquad (3.188)$$

It should be remembered that this equation is nonlinear, since the flux and reluctance are functions of current.

For many practical purposes, approximation $\mu \gg \mu_0$ is acceptable, and for-mula 3.137 is simplified for analysis to

$$L = \frac{N^2 \mu_0 A}{\delta}. \qquad (3.189)$$

The inductance and reactance are linear functions of a variable area A and hyperbolic functions of the gap.

For an evaluation of sensitivity characteristics, we will take a differential of in-duction, given in formula 3.189, and replace the differentials with finite increments:

$$\Delta L = \frac{N^2 \mu_0}{\delta_1} \Delta A - \frac{N^2 \mu_0 A}{(\delta_1 + \Delta \delta)^2} \Delta \delta. \qquad (3.190)$$

The relative change of L is

$$\frac{\Delta L}{L} = \frac{\Delta A}{A_1} - \frac{1}{\left(1 + \dfrac{\Delta \delta}{\delta_1}\right)^2} \frac{\Delta \delta}{\delta_1}. \qquad (3.191)$$

The relative sensitivities due to the relative changes of the area and gap are

$$S_A = \frac{\Delta L / L_1}{\Delta A / A_1} = 1 \qquad \text{and} \qquad S_\delta = \frac{\Delta L / L_1}{\Delta \delta / \delta_1} = -\frac{1}{\left(1 + \dfrac{\Delta \delta}{\delta_1}\right)^2} \qquad (3.192)$$

where L_1, A_1, and δ_1 are initial values of L, A, and δ, respectively; and ΔL, ΔA, and $\Delta \delta$ are incremental changes of L, A, and δ, respectively.

Ignoring the ohmic and capacitive components in the equivalent circuit, the relative sensitivities for the inductive reactance $X = \omega L$ are

$$S_{AX} = \frac{\Delta X / X_1}{\Delta A / A_1} = 1 \quad \text{and} \quad S_{\delta X} = \frac{\Delta X / X_1}{\Delta \delta / \delta_1} = -\frac{1}{\left(1 + \dfrac{\Delta \delta}{\delta_1}\right)^2} \quad (3.193)$$

where X_1 is the initial value of X.

These sensitivities have the same magnitudes as S_A and S_δ. The output signal from the element can be substantially linear if the area of the gap is proportional to the measurand, but the signal is not a linear function of the length of the gap change. The degree of nonlinearity is shown in the given expansion:

$$L = \frac{N^2 \mu_0 A}{\delta + \Delta \delta} \approx \frac{N^2 \mu_0 A}{\delta} \left[1 - \frac{\Delta \delta}{\delta} + \left(\frac{\Delta \delta}{\delta}\right)^2 - \left(\frac{\Delta \delta}{\delta}\right)^3 + \ldots \right]. \quad (3.194)$$

The major factor of nonlinearity is the quadratic term $(\Delta \delta / \delta)^2$. An improvement of the characteristic is achieved in the differential element.

Differential Displacement Elements

The structure of these elements (see Numbers 2, 4, and 6 of Table 3.4) is composed of two identical single elements that are arranged in such a way that an increase or decrease of the length or area of the gap in one element is accompanied by opposite changes in the other. The output signal, which is equal or proportional to the difference ΔL between two inductances, is [24]

$$\Delta L = L_2 - L_1 = \frac{N^2}{\dfrac{\delta - \Delta \delta}{\mu_0 A} + \displaystyle\sum_1^{n-1} \frac{l_n}{\mu_0 A_n} + \sum_1^k \frac{l_k}{\mu_k A_k}}$$

$$- \frac{N^2}{\dfrac{\delta + \Delta \delta}{\mu_0 A} + \displaystyle\sum_1^{n-1} \frac{l_n}{\mu_0 A_n} + \sum_1^k \frac{l_k}{\mu_k A_k}} \approx \frac{N^2 \mu_0 A}{\delta - \Delta \delta} - \frac{N^2 \mu_0 A}{\delta + \Delta \delta}$$

$$\approx \frac{2 N^2 \mu_0 A}{\delta} \left[\frac{\Delta \delta}{\delta} - \left(\frac{\Delta \delta}{\delta}\right)^3 \right] \quad (3.195)$$

where n and k = number of air and ferromagnetic sections, respectively;

l_n and l_k = lengths of air and ferromagnetic parts, respectively, m;

A = area at variable gap, m^2; and

A_n and A_k = areas of air and ferromagnetic parts, respectively, m^2.

This equation is reduced with an assumption that the structure is symmetrical. Each half of the element has one alternating gap, an equal number of coil turns, equal areas of the gap, and uniform geometry of ferromagnetic parts (the expansion in the brackets is taken up to the fifth power term).

It is easy to see after comparing equation 3.194 with 3.195 that the sensitivity for the differential element is twice as high as for the single type. The linearity of the transfer characteristic is improved significantly. A depression of adverse influences such as temperature and electromagnetic fields is typical for the element.

Solenoid Elements

Solenoid elements are introduced in two versions (see Numbers 7 and 8 of Table 3.4) with a displacing core in a single-open-path circuit and with a displacing armature in the same circuit.

The most noteworthy feature of this element is a high linearity of transfer characteristics. The calculations [24, 26] for this system are more complicated than for the circuit with a closed-flux path through the magnetic parts. Calculations with a computer are very helpful.

The element consists of a ferromagnetic core 1 (Fig. 3.35) and coil 2. The core is restricted to moving along the longitudinal axis of the coil. Due to insertion of the core in the coil, its inductance L is increased by ΔL:

$$\Delta L = \frac{\pi r^2}{8} \mu_0 (\bar{\mu}_c - 1) \frac{N^2 R}{P_c P_w} \left[\sqrt{(m + P_1)^2 + 1} \right.$$

$$\left. + \sqrt{(m - P_1)^2 + 1} - \sqrt{(m + P_2)^2 + 1} - \sqrt{(m - P_2)^2 + 1} \right] \qquad (3.196)$$

$$\text{where } m = \frac{l}{R}; \quad P_1 = \frac{P_c + P_w}{R}; \quad \text{and} \quad P_2 = \frac{P_c - P_w}{R}. \qquad (3.197)$$

The inductance of the coil is

$$L = L_0 + J \left[\sqrt{(m + P_1)^2 + 1} + \sqrt{(m - P_1)^2 + 1} - \sqrt{(m + P_2)^2 + 1} \right.$$

$$\left. - \sqrt{(m - P_2)^2 + 1} \right] \qquad (3.198)$$

where L = inductance of coil, H;
   r = radius of core, m;
   μ_0 = permeability of air ($4\pi \times 10^{-7}$), H/m;
   $\bar{\mu}_c$ = effective relative permeability (reduced due to demagnetizing effect in the core; see formula 3.144);
   N = number of turns;
   R = average radius of coil, m;
   P_c = half the core length, m;
   P_w = half the winding length, m;
   l = distance between the centers of the coil and core, m;
   l_c = length of core ($l_c = 2P_c$), m;
   l_w = length of winding ($l_w = 2P_w$), m; and
   L_0 = inductance of the coil without core, H.

Parameter J in formula 3.198 is written:

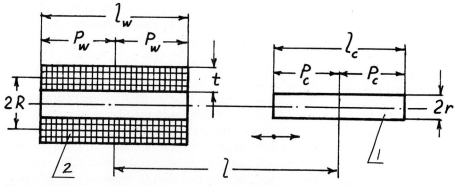

Figure 3.35 Schematic diagram of solenoid element.

$$J = \frac{\pi r^2}{8} \mu_0(\bar{\mu}_c - 1) \frac{N^2 R}{P_c P_w}. \tag{3.199}$$

L_0 is calculated with any formulas for a multiple-layer coil of solenoidal form. The following is one such formula:

$$L_0 = \frac{2\pi^2 R N^2 (1 - 0.667\rho + 0.333\rho^3)10^{-7}}{0.45 + 0.375\rho + \alpha} \tag{3.200}$$

$$\text{where } \rho = \frac{t}{2R} \quad \text{and} \quad \alpha = \frac{P_w}{R}. \tag{3.201}$$

The calculation of permeability for a nonlinear B-H curve is performed in several steps.

1. An average field intensity in the core is calculated using the following formula:

$$H_{av} = \frac{IN}{8P_c P_w} \left\{ \sqrt{[l_c + (P_c + P_w)]^2 + R^2} + \sqrt{[l - (P_c + P_w)]^2 + R^2} \right.$$

$$\left. - \sqrt{[l + (P_c - P_w)]^2 + R^2} - \sqrt{[l - (P_c - P_w)]^2 + R^2} \right\} \tag{3.202}$$

where H_{av} = average field intensity in the core, A-t/m, and
I = current in the coil, A.

2. If μ is calculated manually, the next step is to find from the B-H graph the value of inductance B_{av} corresponding to H_{av} and to take the ratio $\mu = B_{av}/H_{av}$. For computer programming, the function $B = f(H)$ should be presented as an algebraic expression. The accuracy that can be achieved with a polynomial of order four in an approximation of the B-H curve is sufficient for practical calculations. For this approximation, a set of linear simultaneous equations is written. In these equations (see the following expressions) B's and H's are taken from the tables or from graphs for B-H curves (Fig. 3.36):

$$B = a_1 H + a_2 H^2 + \ldots a_n H^n. \tag{3.203}$$

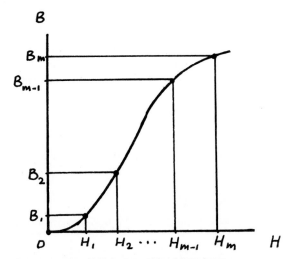

Figure 3.36 B-H curve with points for approximation.

In a matrix form, the equations are

$$
\begin{bmatrix}
H_1 & H_1^2 & \cdots & H_1^n \\
H_2 & H_2^2 & \cdots & H_2^n \\
\cdots & \cdots & \cdots & \cdots \\
H_m & H_m^2 & \cdots & H_m^n
\end{bmatrix}
\begin{bmatrix}
a_1 \\
a_2 \\
\vdots \\
a_n
\end{bmatrix}
=
\begin{bmatrix}
B_1 \\
B_2 \\
\vdots \\
B_m
\end{bmatrix}
\qquad (3.204)
$$

where a_1 to a_n are unknown polynomial coefficients, and B_i ($i = 1$ to n) and H_j ($j = 1$ to m) are the coordinates of the curve for this approximation. As an illustration, the data for calculations at low-level fields ($0 < H < 50$ A-t/m) and at higher-level fields ($50 < H < 300$ A-t/m) are given in Table 3.8 for one of the low-carbon steels.

The approximation for $0 < H < 50$ A-t/m range is

$$
\begin{aligned}
B = {}& 3.267 \times 10^{-4} H + 1.067 \times 10^{-5} H^2 - 1.667 \times 10^{-7} H^3 \\
& + 3.333 \times 10^{-9} H^4.
\end{aligned}
\qquad (3.205)
$$

For the range, $50 < H < 300$ A-t/m:

$$
\begin{aligned}
B = {}& -1.2328 \times 10^{-2} H + 3.5315 \times 10^{-4} H^2 - 1.9478 \times 10^{-6} H^3 \\
& + 3.1880 \times 10^{-9} H^4.
\end{aligned}
\qquad (3.206)
$$

Programming the computer to calculate μ and μ_r is easy with these approximations.

3. The next step is calculating $\bar{\mu}_c$ with formula 3.144. If the core material is electroconductive, the value of $\bar{\mu}_c$ should be corrected (formula 3.141). The calculation of the transfer characteristic $L = f(l)$ is conducted by assigning a different combination of the parameters of the coil and core and varying the distance l. It is useful to evaluate at the same time the sensitivity, which is calculated as a product of two partial derivatives:

$$
\begin{aligned}
S_l = \frac{\partial L}{\partial l} = \frac{\partial L}{\partial m} \cdot \frac{\partial m}{\partial l} = \frac{J}{R} \Bigg(& \frac{1}{\sqrt{1 + 1/(m + P_1)^2}} + \frac{1}{\sqrt{1 + 1/(m - P_1)^2}} \\
& - \frac{1}{\sqrt{1 + 1/(m + P_2)^2}} - \frac{1}{\sqrt{1 + 1/(m - P_2)^2}} \Bigg).
\end{aligned}
\qquad (3.207)
$$

The sensitivity is maximum at

$$
m = \frac{P_1 + P_2}{2} = \frac{P_c}{R} \qquad \text{and} \qquad l = P_c. \qquad (3.208)
$$

By substituting these values in equation 3.207, the maximum sensitivity can be calculated. The final selections of the parameters should be based on a compromise between the desirable linearity and sensitivity.

Coil-displacement Elements

A total inductance L of two closely located coils 1 and 2 with self-inductances L_1 and L_2 (Fig. 3.37) depends on the distance x between the coils. The mutual inductance L_{12} is a function of the distance, since a coupling coefficient k depends on the distance:

TABLE 3.8

H as a Function of B for Low-carbon Steel

B	[T]	0	10	20	40	50	100	200	300
H	[A-t/m]	0	0.0042	0.010	0.028	0.043	0.67	1.18	1.32

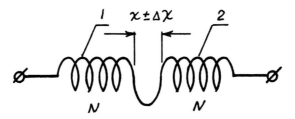

Figure 3.37 Coil-displacement element.

$$L = L_1 + L_2 \pm L_{12} \tag{3.209}$$

$$L_{12} = k\sqrt{L_1 L_2}. \tag{3.210}$$

If the coils have a series-aiding connection (plus at L_{12}), a decrease in the coil separation leads to the increase in the overall inductance. And the reverse, a series-opposing connection (minus at L_{12}), gives a decrease of the inductance with a decrease of the distance between the coils. This simple concept is utilized for building a displacement element. As a rule, the element contains two identical coaxial solenoids with the same number of turns N. For this system, the coupling coefficient can be taken from the chart in Figure 3.38. The chart gives coefficient k versus a specified spacing between the coils. The spacing is measured in terms of the coils' diameters. The mutual inductance is calculated from formula 3.211 [27, 28].

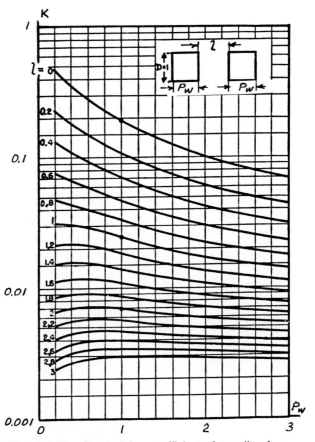

Figure 3.38 Chart giving coefficient of coupling for specified spacing between two coaxial solenoids (adopted from [28]).

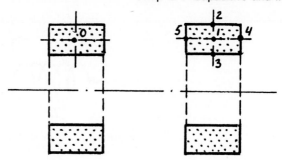

Figure 3.39 Location of points for calculating mutual inductance of two identical coaxial coils (adopted from [28]).

$$L_{12} = L_{21} = \frac{N^2}{3}(L_{02} + L_{03} + L_{04} + L_{05} - L_{01}) \qquad (3.211)$$

where N is the number of turns per coil and the separate terms are the mutual inductances between the coaxial circles, corresponding to the subscript, as indicated in Figure 3.39. Inductance L_{11} between the parallel coaxial circles is given by

$$L_{11} = 2.54 \times 10^{-6}B\sqrt{Aa} \qquad (3.212)$$

where A = radius of the large circle, m (Fig. 3.40);
     a = radius of the small circle, m; and
     B = factor depending on r_1/r_2 as given in Figure 3.40.

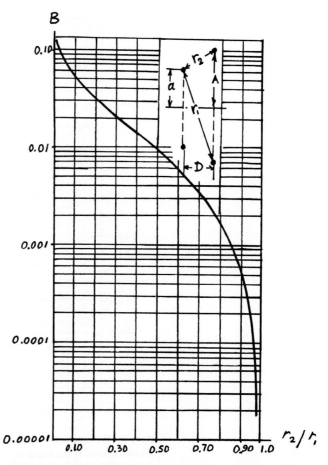

Figure 3.40 Factor B for use in equation 3.212, applied to two parallel coaxial circles (adopted from [28]).

The quantities r_1 and r_2 are the longest and the shortest distances, respectively, between points on the circumferences of the two circles.

This ratio is

$$\frac{r_2}{r_1} = \sqrt{\frac{(1 - a/A)^2 + (D/A)^2}{(1 + a/A)^2 + (D/A)^2}}. \tag{3.213}$$

This element is fed with a radio-frequency voltage and has to be well-shielded. Its simple construction is very appropriate for models used in experiments.

Long-displacement Elements

Most frequently, an inductive element is employed for small displacement measurements. Only a few devices allow the measurement of a large displacement, and the typical element (Fig. 3.41) to be described is one of them. The construction includes a magnetic fork-type core 3 with coil 1 and shorted turn 2 around one of the fork's bars. The magnetic flux ϕ_x linking the turn is at maximum (ϕ_{max}) when the turn is closer to the coil, and it is at minimum when it is at the end of the fork. With an assumption that ϕ_x is a linear function of distance x, the value of the flux is calculated as follows:

$$\phi_x = \phi_{min} + \frac{x}{x_{max}} (\phi_{max} - \phi_{min}). \tag{3.214}$$

Since the shorted turn is a one-turn winding with self-inductance L_2 and resistance r_2, the coupling coefficient k for the two coils will be a function of ϕ_x. By definition, k is the portion of the flux due to coil 1 that links turn 2:

$$k = \frac{\phi_x}{\phi_{max}}. \tag{3.215}$$

The mutual inductance L_{12} for the coils is

$$L_{12} = k\sqrt{L_1 L_2} \tag{3.216}$$

where L_1 = self-inductance of coil 1, H.

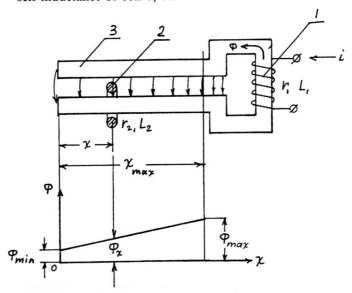

Figure 3.41 Schematic diagram of long-displacement element with moving shorted turn.

The impedance Z, "looking in" to the terminals of coil 1, is

$$Z = R_{eq} + j\omega L_{eq} = \left(r_1 + \frac{\omega^2 L_{12}^2 r_2}{r_2^2 + \omega^2 L_2^2}\right) + j\omega\left(L_1 - \frac{\omega^2 L_{12}^2 L_2}{r_2^2 + \omega^2 L_2^2}\right) \quad (3.217)$$

where R_{eq} = equivalent resistance, Ω;

$\quad\quad\; L_{eq}$ = equivalent inductance, H;

$\quad\quad\; r_1$ = resistance of coil 1,Ω; and

$\quad\quad\; \omega$ = angular frequency of current, rad/s.

Therefore,

$$L_{eq} = L_1 - \frac{\omega^2 L_{12}^2 L_2}{r_2^2 + \omega^2 L_2^2}. \quad (3.218)$$

Since $L_{12} = f(k)$, $k = f(\phi_x)$, and $\phi_x = f(x)$, inductance L_{eq} is a function of displacement.

ϕ_x is gradually changed from ϕ_{min} to ϕ_{max} with an increase in x. This means that the displacement of the turn can gradually change the inductance, which is "seen" from the terminals. The sensitivity is found by taking the derivative of the transfer function $L_{eq} = f(x)$ with respect to x.

Another version of the element (Fig. 3.42) has a magnetic system that includes a moving ferromagnetic armature 1 inside long bars 2 and 3, which constitute together with coil 4 a magnetic system with a variable reluctance. If the permeability of the material of the magnetic parts is not substantially higher than that for air, the reluctance of the magnetic path is controlled by distance x, which is the length of one of the components of the magnetic path. The calculation of the coil's induction for this case does not differ from that for the single displacement element and is given by the following formulas:

$$L = \frac{N^2}{R}, \quad (3.219)$$

$$R = \frac{2g}{\mu_0 A} + \frac{l_c}{\mu A} + \frac{l_a}{\mu A} + \frac{2x}{\mu A} \quad (3.220)$$

where $\quad\quad\quad\quad L$ = inductance, H;

$\quad\quad\quad\quad\quad\; N$ = number of coil's turns;

$\quad\quad\quad\quad\quad\; R$ = reluctance, 1/H;

g, l_c, and l_a = lengths of the magnetic path in air gap, part of core, and in the armature, respectively, m;

μ and μ_0 = permeability of core and air, respectively, H/m; and

$\quad\quad\quad\quad\quad\; A$ = cross-sectional area of the flux-conducting parts (it is assumed to be uniform), m^2.

The sensitivity of the inductance to the change of distance x can be markedly reduced if the air gap and permeability of magnetic parts are large.

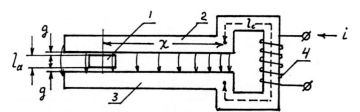

Figure 3.42 Schematic diagram of long-displacement element with moving ferromagnetic armature.

Proximity Elements

Proximity to a massive ferromagnetic member 1 (Fig. 3.43) is easily measured with a closed-loop magnetic-system inductor 2, in which member 1 plays the role of a displaced armature. If the permeabilities of metal parts μ_1 and μ_2 are much greater than that for air, the inductance is calculated using formula 3.189:

$$L = \frac{N^2 \mu_0 A}{2\delta}.$$
(3.221)

This device is applicable for measuring small distances and has all features of the element with a variable gap. An element with an open magnetic circuit (Fig. 3.44) is more suitable for measuring a larger remoteness of an object. A core of this device is shaped as a straight bar 1 with coil 2 attached at the end of the bar, facing target 3.

The inductance is calculated with formula 3.198, which can be modified for specific shapes and dimensions of the magnetic circuit. The relative change of inductance δL [24] when gap g is changed from 0 to ∞ characterizes the sensitivity of the element and can be calculated in the beginning of the design to make preliminary selection of the main parameters. This sensitivity is written as follows:

$$\delta L = \frac{(\bar{\mu}_{c0} - 1)\left[\sqrt{(m_0 + P_1)^2 + 1} + \sqrt{(m_0 - P_1)^2 + 1} -\right.}{(\bar{\mu}_{c\infty} - 1)\left[\sqrt{(m_\infty + P_1)^2 + 1} + \sqrt{(m_\infty - P_1)^2 + 1} -\right.} \rightarrow$$

$$\rightarrow \frac{\left.- \sqrt{(m_0 + P_2)^2 + 1} - \sqrt{(m_0 - P_2)^2 + 1}\right]}{\left.- \sqrt{(m_\infty + P_2)^2 + 1} - \sqrt{(m_\infty - P_2)^2 + 1}\right]} - 1$$
(3.222)

where $\bar{\mu}_{c0}$ and $\bar{\mu}_{c\infty}$ are calculated with formula 3.144 for $g = 0$ and $g = \infty$, respectively.

The values of m_0 and m_∞, defined in formulas 3.197, are found from the following substitutions:

$$m_0 = \frac{0.5 l_w}{R} \quad \text{and} \quad m_\infty = \frac{l_0}{R}$$
(3.223)

(for $g = 0$ and $g = \infty$, respectively).

For calculating the ratio $M/4\pi$ (formula 3.145), parameter λ is determined for $g = 0$ and $g = \infty$, respectively:

$$\lambda_0 = l_c \sqrt{\frac{\pi}{A_c}} \quad \text{and} \quad \lambda_\infty = \frac{l_0}{2} \sqrt{\frac{\pi}{A_c}}.$$
(3.224)

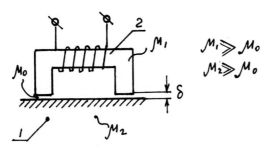

Figure 3.43 Proximity element with closed-loop magnetic system.

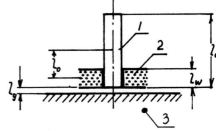

Figure 3.44 Proximity element with open-loop magnetic system.

The element radiates an electromagnetic field and is sensitive to the field from outside since the magnetic circuit is open. An adequate shielding is important for reliable operation.

If an electroconductive object is located close to the coil generating an alternating field, the eddy currents in the object change the parameters of the coil. Its inductance and quality factor are reduced because the field induced by the eddy currents tends to reduce the field created by the coil. This is displayed in the reduction of the inductance. The loss of the circulation power of the eddy currents decreases the quality factor. The change of these two parameters depends on the geometry and conductivity of the object and on the distance between the coil and the object.

Figure 3.45 shows one of the configurations of the proximity elements. A flat spiral coil 1 is located alongside target 2 at distance l to be measured. The coil's impedance Z is calculated [29] as follows:

$$Z = R_0 + \frac{1}{\mu_0 \sigma \beta} \left(-\frac{\partial L_{12}}{\partial l} \right) + j\omega \left[(L_0 - L_{12}) + \frac{1}{\mu_0 \sigma \beta \omega} \left(-\frac{\partial L_{12}}{\partial l} \right) \right] \quad (3.225)$$

where R_0 = coil's resistance, Ω;
$\quad\quad \sigma$ = electrical conductivity of material, S/m;
$\quad\quad \beta$ = skin depth, m (formula 3.142);
$\quad\quad L_{12}$ = mutual inductance with a mirror-image presentation of the coil located at the distance of $2l$ from the surface of the target (Fig. 3.45), H;
$\quad\quad l$ = distance from the coil to the target's surface, m;
$\quad\quad \omega$ = angular frequency of the excitation current, 1/s; and
$\quad\quad L_0$ = inductance of the coil in free space (without interaction with any object), H.

L_0 in henries is calculated with a standard formula:

$$L_0 = 2\mu_0 D_{av} N^{5/3} ln\left(\frac{4D_{av}}{t}\right) \quad\quad (3.226)$$

where

$$D_{av} = \frac{D_1 + D_2}{2} \quad \text{and} \quad t = \frac{D_2 - D_1}{2} \quad\quad (3.227)$$

N = number of turns, and
D_1 and D_2 are the inside and outside diameters of the coil, respectively, m.

L_{12} can be calculated with formula 3.211.

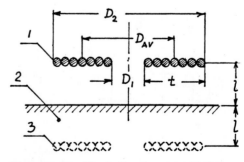

Figure 3.45 Proximity element with flat spiral coil.

Inductance L as a function of distance l is expressed by extracting a part of the imaginary term from formula 3.225:

$$L = (L_0 - L_{12}) + \frac{1}{\mu_0 \sigma \beta \omega} \left(- \frac{\partial L_{12}}{\partial l} \right). \tag{3.228}$$

Sensitivity S_l is

$$S_l = \frac{\partial L}{\partial l} = - \left[\frac{\partial L_{12}}{\partial l} + \frac{1}{\mu_0 \sigma \beta \omega} \left(\frac{\partial^2 L_{12}}{\partial l^2} \right) \right]. \tag{3.229}$$

It follows from these two formulas that reading the inductance depends on the properties of the target's material (μ and σ). An adequate correction must be made to take this into account. When the partial derivatives are calculated, L_{12} can be approximated by a polynomial or any other convenient function of l.

Radio frequencies (0.1 to 20MHz) are usually used to feed the coil and measure inductance.

Thickness-of-metal Sensitive Elements

Skin effects and the effects of eddy currents are employed for measuring the thickness of metal foils, wires, coatings, and so on. For these measurements, the transduction element is a device similar to that described in the preceding paragraph. A variation of the measurand causes a change of the coil inductance due to the redistribution of currents in the target.

Calculations show that the density of eddy currents J is at maximum J_0 at the surface of the conductor and is decreased below the surface (Fig. 3.46):

$$\frac{J}{J_0} = e^{-z\sqrt{\pi f \mu \sigma}} \tag{3.230}$$

where Z = distance from the surface, m.

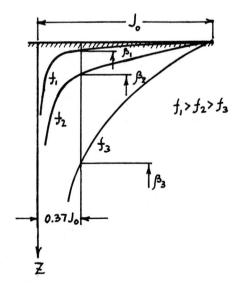

Figure 3.46 Change of eddy currents versus distance from surface of conductor and frequency of currents.

The chart in Figure 3.47 [30] shows the values of "skin depth" as a function of the target material's resistivity and relative permeability for the frequency of 1MHz. With an increase in frequency, the depth of field penetration becomes smaller. Calculations show that an electroconductive target with a thickness 8 to 10 times larger than β has a current density of 37% of that at the surface for distance $Z = \beta$, and only 5% for distance $Z = 3\beta$. Table 3.9 lists the values of β versus the frequencies for four materials [31].

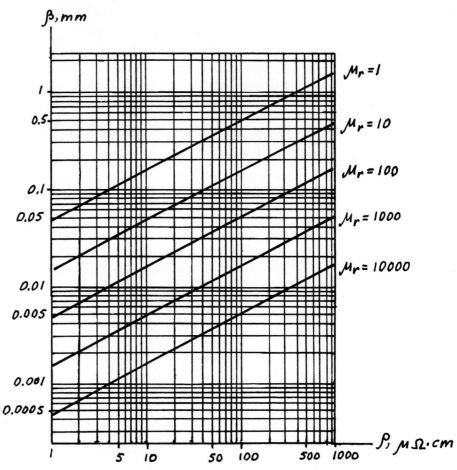

Figure 3.47 "Skin depth" as function of target material resistivity and permeability (adopted from [30]).

TABLE 3.9

"Skin Depth" as a Function of Frequency (Adopted from [31])

	"Skin depth" β in mm			
f Hz	Copper $\sigma = 58.8 \times 10^6$S/m $\mu_r = 1$	Aluminum $\sigma = 34.4 \times 10^6$S/m $\mu_r = 1$	Steel $\sigma = 7.19 \times 10^6$S/m $\mu_r = 100$	Lead $\sigma = 4.52 \times 10^6$S/m $\mu_r = 1$
0.1×10^3	6.7	8.6	1.9	24
0.6×10^3	2.7	3.5	0.77	9.8
1.0×10^3	2.1	2.7	0.60	7.6
6.0×10^3	0.85	1.1	0.24	3.5
10×10^3	0.67	0.86	0.19	2.4
60×10^3	0.27	0.35	0.077	0.98
0.1×10^6	0.21	0.27	0.060	0.76
1×10^6	0.067	0.086	0.019	0.24
10×10^6	0.021	0.027	0.006	0.076

As is clear from the table, even at audio frequencies the depth of current penetration is rather small. It becomes negligible with an increase of frequency up to 1MHz.

In order to measure the thickness of foil, the frequency of excitation should not be very high. Otherwise, the eddy currents become concentrated at the surface, making the sensitivity to thickness negligible.

The reverse conditions exist in measurements of the thickness of a thin wire (Fig. 3.48). At high frequencies, the currents are concentrated along the circumference of the wire, and the electrical characteristics of the element depend mostly on the diameter rather than on the conductivity of the material.

The analytical presentation of a transfer characteristic is complicated for this case, and an experimental evaluation of the characteristic is most practical. Figure 3.48 shows two functions: induction L and quality factor Q of the coil versus diameter d of a copper wire for one particular design [31]. The element constitutes a 10mm-diameter coil (Fig. 3.48) with 60 turns of a 1.0mm-diameter wire. The changes of L and Q are quite linear for a wide range of wire diameters. The frequency of excitation is 10MHz. The sensitivity of the characteristics can be determined by taking slopes from the chart.

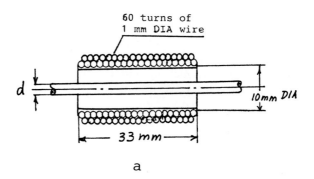

a

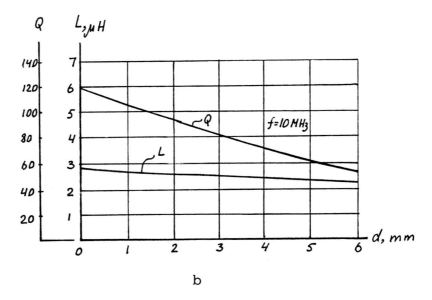

b

Figure 3.48 Characteristics of coil for measuring thickness of wire.
a = schematic diagram of coil, b = inductance and quality factor versus diameter of wire.

The characteristics for copper and aluminum practically coincide. They are very close for brass wire. This device allows continuous measurement without contact with the wire when it is drawn through the coil.

Magnetostrictive Elements

There are four basic magnetostrictive effects.

1. Magnetic induction in a ferromagnetic material is changed under stress (Villary effect).
2. A transient voltage is created between the ends of a wire that is twisted in a longitudinal magnetic field (Weithem effect).
3. The length of ferromagnetic material is changed with a change of longitudinal magnetic field (Joule effect).
4. A rod with electrical current is twisted when subjected to a longitudinal magnetic field. The inverse effect is the axial magnetization of a current-carrying wire when it is undergoing twisting (Wiedemann effect).

These effects are realized in the constructions of the elements that convert compressing, tensile, bending, or torsional forces into stresses in magnetostrictors and into a change of inductances or mutual inductances of coils producing measurable signals. Inverse conversion, such as converting electrical values into mechanical or acoustical signals, is also performed with magnetostrictors.

The magnetostrictive effect is particularly strong in nickel, some alloys (primarily nickel), and ferrites. As a rule, the substance that reveals a direct magnetostrictive effect has a pronounced inverse effect.

Magnetostriction λ $(l_1,\ m_1,\ n_1)$ (elastic strain produced by magnetization of material) for a cubic crystal and arbitrary direction are given [32] by the formula

$$\lambda(l',\ m',\ n') = \lambda_{(100)} + 3(\lambda_{(111)} - \lambda_{(100)})(l'^2 m'^2 + m'^2 n'^2 + n'^2 l'^2) \quad (3.231)$$

where $\lambda_{(100)}$ and $\lambda_{(111)}$ = cube-edge and cube-diagonal magnetostriction coefficients (change in length/unit length),

$l',\ m',\ n'$ = direction cosines.

Strain is anisotropic, and strain tensor coefficients U's are written as follows:

$$U_{xx} = \frac{3}{2}\lambda_{(100)}l'^2, \qquad U_{yy} = \frac{3}{2}\lambda_{(100)}m'^2, \qquad U_{zz} = \frac{3}{2}\lambda_{(100)}n'^2,$$

$$U_{xy} = \frac{3}{2}\lambda_{(111)}l'm', \qquad U_{yz} = \frac{3}{2}\lambda_{(111)}m'n', \qquad U_{xz} = \frac{3}{2}\lambda_{(111)}l'n'. \quad (3.232)$$

For polycrystalline structures, the average saturation magnetostriction λ_s is

$$\lambda_s = \frac{2}{5}\lambda_{(100)} + \frac{3}{5}\lambda_{(111)}. \quad (3.233)$$

Determination of the coefficients from crystalline data is not very tractable mathematically, and measurements give more accurate values.

The relevant information [32] on some magnetostrictive materials is provided in Table 3.10.

Magnetostriction can be negative or positive. Materials with negative magnetostriction contract with magnetization; compressive stress increases permeability,

TABLE 3.10

Saturation Magnetostriction for Some Materials (Adopted from [32])

Material	Density ρ kg/m^3	Young's modulus E N/m^2	Saturation magnetostriction λ_s
Nickel	8.9×10^3	200×10^9	-28
49 Co, 49 Fe, 2 V	8.2×10^3	220×10^9	-65
Iron	7.9×10^3	210×10^9	$+5$
50 Ni, 50 Fe	8.2×10^3	200×10^9	$+28$
87 Fe, 13 Al	6.7×10^3	—	$+30$
95 Ni, 5 Co	8.8×10^3	200×10^9	-35
Cobalt	8.8×10^3	200×10^9	-50
$CoFe_2O_4$	5.3×10^3	—	-250
$Ni_{0.42}Cu_{0.49}Co_{0.01}Fe_2O_4$	5.3×10^3	160×10^9	—

and tensile stress decreases it. Materials having positive magnetostriction extend under magnetization; tensile stress increases permeability, and compressive stress decreases it. The change in length does not depend upon the sign of the field, only on the composition of the material. In order to make a device sensitive to the direction of the field, an offset of the field can be used. The change of magnetic properties under stress is accompanied by a change of slope (permeability) of the *B-H* curve, shape, and area of the hysteresis loop, together with the change of the magnitude of the residual induction. For instance (Fig. 3.49, [33]), permeability of nickel is decreased with stress, and a reverse process takes place in permalloy-68. The slope of the *B-H* curve is larger, more rectangular, and smaller in width.

The element has several basic structures, outlined in Figure 3.50.

At (*a*), force *F* develops stresses in core 1, built of solid ferromagnetic or of a stack of laminations. The induction of coil 2 is sensitive to the stress that is a product of force.

In a similar construction in (*b*), the change of permeability induced by force *F* alters the coupling coefficient between two coils 2 and 3. Coil 2 is fed by stable voltage U_1, and voltage U_2 at secondary coil 3 is proportional to the force.

The element (*c*), in which anisotropic properties of magnetostrictive material are employed, carries two sets of coil 1 and 2 at right angles to each other and at 45° to the load axis. The coils are wound in the four symmetrically and closely placed holes of a magnetostrictive parallelepiped 3. At zero load, an alternating flux created by coil 1 does not induce any voltage in coil 2. In (*d*), the lines of the field do not cross turns of coil 2 due to the isotropic properties of the material and symmetrical shape of the core. Under load, the field is deformed. This causes a change in the flux distribution, shown in (*e*), so that the flux now cuts through the secondary winding and induces an ac voltage proportional to the load. The phase of the voltage is changed by 180° with the change of the sign of force *F*.

A similar concept is utilized in a torque-sensitive element (*f*). Compression (σ_c) and tension (σ_t) stresses are developed in the magnetostrictive cylindrical member 1 as it is twisted by torsional moment *M*. The axes of maximum stresses form a 45° angle with the axis of the member. The change of the inductances of two coils (*g*), whose flux links the stressed area, can be a measure of stress or torque. A differential structure (*h*) is formed of two halves, and each of them is the simple circuit shown in (*a*). As force is applied at the middle of the core, one of the halves is compressed, and the other one is stretched (in another modification, one half is not exposed to the stress at all). In order to increase the output signal or to measure the resultant force, several elements can be stacked in a column or in a row, as is shown in (*i*). In an element based on the Wiedemann effect (*j*), current *i* flows along a rod made of an electroconductive and magnetostrictive material. If the rod is unstressed, the lines

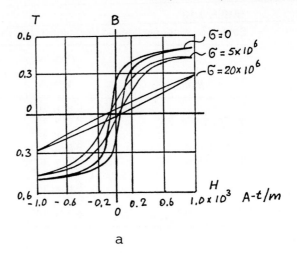

a

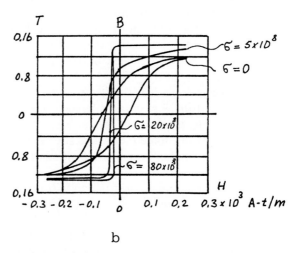

b

Figure 3.49 Change of B-H curves under stress
(stresses σ are given in N/m²). a = curves for nickel,
b = curves for permalloy-68 (adopted from [33]).

of the magnetic field are circular, due to the current *i*. When the rod is twisted by a
torsional moment *M*, the symmetry of the field is disturbed, and voltage *e* is induced
in the coil. This voltage is proportional to the moment.

The sensitivity of the magnetostrictive material to stress S_σ or to strain S_ϵ is
evaluated by the following ratios:

$$S_\sigma = \frac{\Delta\mu}{\mu} \cdot \frac{1}{\sigma} \quad \text{and} \quad S_\epsilon = \frac{\Delta\mu}{\mu} \cdot \frac{1}{\epsilon} \tag{3.234}$$

where μ = permeability of unstressed member, H/m;
        $\Delta\mu$ = change of permeability due to stress σ or strain ϵ, H/m; and
        σ = stress, N/m².

A construction of the element usually contains a simple closed-loop and
zero-gap magnetic circuit to avoid variation of the inductance due to a change of

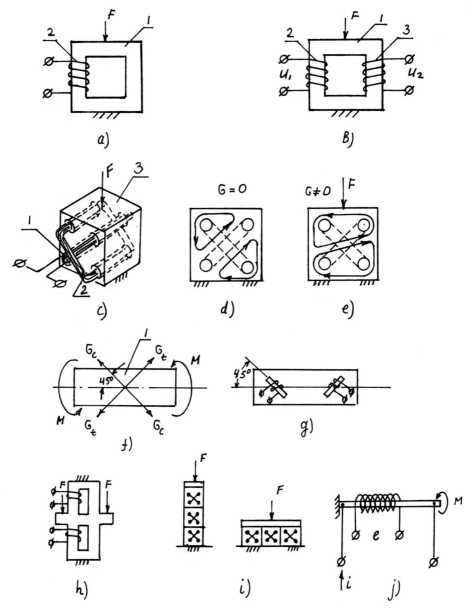

Figure 3.50 Structures with magnetostrictive elements. a = single-coil element, b = two-coil element, c, d, and e = element with core having anisotropic properties, f = torque-sensitive element, g = location of coils for element in f, h = differential structure, i = stacking of elements, j = Wiedemann-effect element.

gaps. Even a small change of the gaps, for instance, between the laminations, can give a drift of reluctance comparable to or greater than that due to the magneto-strictive effect. A calculation of this inductance can be made with formula 3.137 (for $\delta = 0$).

The relative change of the inductance $\Delta L/L$ for the deformed core under load is

$$\frac{\Delta L}{L} = \frac{\Delta \mu}{\mu} + \frac{\Delta A}{A} - \frac{\Delta l}{l} \cdot \frac{1}{\left[1 + \left(\dfrac{\Delta L}{L}\right)\right]^2}. \tag{3.235}$$

Since

$$\frac{\Delta A}{A} = -\nu \frac{\Delta l}{l}, \tag{3.236}$$

the sensitivity S_L of inductance to the strain is written:

$$S_L = \frac{\Delta L/L}{\Delta l/l} = \frac{\Delta L/L}{\epsilon} = S_\epsilon - (\nu + 1) \tag{3.237}$$

where ν = Poisson's ratio.

Having $\nu + 1 \ll S_\epsilon$, the change of the inductance as a function of S_σ and S_ϵ is calculated as follows:

$$\Delta L = L\epsilon S_\sigma = \frac{N^2 \mu A S_\epsilon}{l} \epsilon = \frac{N^2 \mu A S_\sigma}{l} \sigma. \tag{3.238}$$

If the core has the shape of a bar for which force F can be expressed as a product,

$$F = \sigma A, \tag{3.239}$$

the transfer characteristic $\Delta L = f(F)$ is obtained from equation 3.238:

$$\Delta L = \frac{N^2 \mu S_\sigma}{l} F. \tag{3.240}$$

The sensitivities greatly depend on material composition, thermal treatment, type of stress, magnitude of the field, frequency of excitation, and stress distribution in the loaded structure. For instance, if the frequency of field excitation is high, and the skin effect is taken into account, S_σ and S_ϵ are reduced and can be fractions of the computed values. A notable reduction of the sensitivity to force can occur if not all the parts conducting the flux in the magnetic circuit are under stress. In practical designs, $-1.1 \leq S_\sigma \leq +1.8$—for compression, and $+0.6 \leq S_\sigma \leq +2.6$—for tension where S_σ is given in $\%/\text{M}$ (N/m^2).

The sensitivity is at maximum when the operating point is chosen close to the maximum of permeability in the *B-H* curve.

In a magnetostrictive torsion element (Fig. 3.51), the application of a torque generates compressive and tensile normal stresses σ. They are expressed for any elementary rectangular parallelepiped on the surface of the element as follows [34]:

$$\sigma = \sigma_x \cos^2 \psi + \sigma_y \sin^2 \psi + \tau_x \sin 2\psi \tag{3.241}$$

where σ_x and σ_y = normal stresses at the faces of the parallelepiped (x and y are directions along the parallelepiped's axis) and are perpendicular to it, N/m^2;

ψ = angle between the directions of stresses σ and the x-axis, °; and

τ_x = shear stress parallel to the x-axis, N/m^2.

If the element *ABCD*, shown in Figure 3.51, is isolated from a twisted rod by two pairs of axial and transverse sections, only shear stresses can be found on its faces. This means

$$\sigma_x = \sigma_y = 0, \qquad \tau = \tau_x \qquad \text{and} \tag{3.242}$$

$$\sigma = \tau \sin 2\psi. \tag{3.243}$$

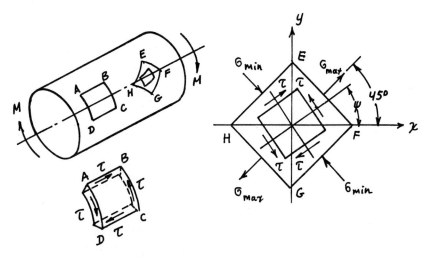

Figure 3.51 Stress state of twisted bar.

The maximum of normal stresses $\sigma_{\max}$ is at

$$2\psi = 90° \quad \text{or} \quad \psi = 45° \tag{3.244}$$

$$\sigma_{\max} = |-\sigma_{\max}| = |\tau_{\max}| = \frac{M}{W_P} \tag{3.245}$$

where M = torque, N · m, and
  W_P = section modulus, m³.

All the elements of the twisted rod are in the state of pure shear, which is not homogeneous since the shear stress varies across the radius of the cross section.
The stress at any distance ρ from the center of the deformed bar is

$$\tau = \frac{M\rho}{J_P} \tag{3.246}$$

where J_P = polar moment of inertia of the section, m⁴.
If the rod has a solid circular section of diameter D, then

$$J_P = \frac{\pi D^4}{32}. \tag{3.247}$$

For a cylinder with an inner central hole of diameter d and outer diameter D, the polar moment is

$$J_P = \frac{\pi}{32}(D^4 - d^4). \tag{3.248}$$

For any shape of the cross section, W_P is calculated by taking the maximum distance $\rho = \rho_{\max}$:

$$W_P = \frac{J_P}{\rho_{max}}. \tag{3.249}$$

A torque-sensitive part should not necessarily be made of a magnetostrictive material. Sleeve 1 of this material (Fig. 3.52) can be welded to cylinder 2 and serve

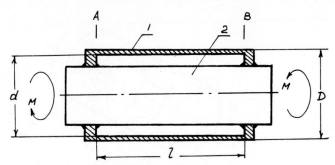

Figure 3.52 Torque-sensitive element with magnetostrictive sleeve.

as a signal-forming component since the shear stress in the sleeve can be a measure of the torque. In order to prove it, we shall first write the equation for the moments, which states that the overall moment M between sections A-A and B-B has two components that are transmitted by the cylinder (M_c) and by the sleeve (M_s):

$$M = M_c + M_s. \tag{3.250}$$

The angles of twist along length l for the cylinder φ_c and for the sleeve φ_s in radians are

$$\varphi_c = \frac{M_c l}{G_c J_c}, \qquad \varphi_s = \frac{M_s l}{G_s J_s} \tag{3.251}$$

where G_c and G_s = shear moduli for the cylinder and sleeve, respectively, N/m^2; and

J_c and J_s = polar moments of inertia for the cylinder and sleeve, respectively, m^4.

The shear modulus for the material is

$$G = \frac{E}{2(1 + \nu)}, \tag{3.252}$$

where E = Young's modulus, N/m^2, and

ν = Poisson's ratio, dimensionless.

Since $\varphi_c = \varphi_s$ and G_c is assumed to be equal to G_s, the following ratio is obtained from formula 3.247:

$$\frac{M_c}{M_s} = \frac{J_c}{J_s}. \tag{3.253}$$

Substituting M_c from formula 3.250 in formula 3.253, we will find that the moment M_s is directly proportional to M and can be its measure:

$$M_s = \frac{J_s}{J_s + J_c} M. \tag{3.254}$$

The output signal from the element is directly proportional to the maximum stresses σ_s or τ_s in the sleeve (formula 3.245):

$$\tau_s = \sigma_s = \frac{M_s}{W_s} \tag{3.255}$$

where W_s = section modulus of sleeve, m³:

$$W_s = \frac{2J_s}{D}. \tag{3.256}$$

Combining formulas 3.248 and 3.254–3.256, we obtain $\tau_s = f(M)$, which can be directly used to form a transfer characteristic of the element:

$$\tau_s = \frac{0.5D}{J_c + \dfrac{\pi}{32}(D^4 - d^4)} M. \tag{3.257}$$

All of these relationships can easily be modified for elements with more complicated shapes. In this case, formulas of strain and stress from the theory of strength of materials can be used.

Sources of sound and ultrasound, solid-state filters, and delay lines with frequencies extending through audio to RF range are built with simple shape resonators (Fig. 3.53) [35]. Magnetostrictive bars are excited for longitudinal or torsional vibrations. This excitation results in mechanical or acoustical energy radiation in the media or energy transmission to the other members for special functions, such as

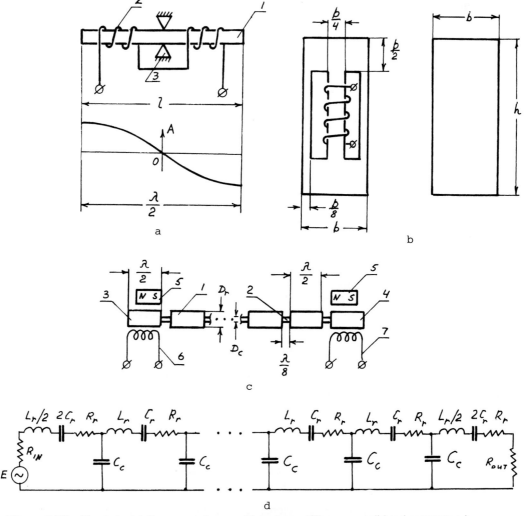

Figure 3.53 Magnetostrictive resonators and band-pass filter. a = solid-rod resonator, b = transformer core-like resonator, c = band-pass filter arrangement, d = filter equivalent circuit.

filtering. A solid rod 1 (a) with an excitation coil 2 and support 3 in the middle, resonates at frequency f, which is defined as

$$f = \frac{nc}{2l} = \frac{n}{2l}\sqrt{\frac{E}{\rho}} \qquad (3.258)$$

where $n = 1,2,3, \ldots$ = number of harmonics (integer numbers);
$\quad l$ = length of the resonator, m;
$\quad c$ = velocity of propagation of vibrations in the resonator's material, m/s;
$\quad E$ = Young's modulus of the resonator's material, N/m^2; and
$\quad \rho$ = density of the resonator's material, kg/m^3.

If the rod is not loaded at the ends, it typically has a node at the middle and a maximum amplitude A at the ends of the bar as is indicated in (a). The bar is periodically stretched and contracted along its length. When the frequency of vibromotive exciting force coincides with the natural mechanical frequency of the bar, it resonates, and the change of length becomes maximum. This amplitude reaches at the end 0.1×10^{-3} the resonator's length. The lowest (fundamental) frequency f' of vibration is at $n = 1$. In this case,

$$f' = \frac{c}{2l}. \qquad (3.259)$$

The length of the bar is equal to the half of the oscillation wave length λ (a):

$$l = \frac{c}{2f'} = \frac{\lambda}{2}. \qquad (3.260)$$

Powerful oscillations can be obtained using a transformer core-like resonator (b). The resonance frequency for this core at longitudinal vibrations along height h is

$$f = \frac{1}{2h}\sqrt{\frac{E}{1 + \dfrac{b}{2h}}}. \qquad (3.261)$$

A resonance band-pass filter (c) is composed of resonator 1 (cylinders or plates), coupler 2, magnetostrictive excitation 3, and pickup (4) parts. It also contains permanent magnets 5 for biasing the field and coils for excitation (6) and pickup (7) of signal. The filter is excited by an input ac voltage or the current feeding coil 6. At the frequency of mechanical resonance, which is calculated with formula 3.258, the pickup part 4 undergoes maximum deformations, inducing increased voltage in coil 7. The quality factor of the electromechanical system for audio and ultrasonic frequencies is quite high (up to 10 000). This allows the construction of filters with high discrimination, which can be increased by iteration of the filter's cells. Calculating the filter's characteristic is simplified by using an equivalent circuit (d) with the following analogies.

Mechanical force	Electrical voltage
Displacement of layers along the resonator's length	Charge
Velocity of layers along the resonator's length	Current
Mechanical damping	Resistance
Mass	Inductance
Compliance of material (displacement-to-force ratio)	Capacitance

Depending on the weight of these parameters, they can be introduced or omitted when the circuit is designed. The characteristic impedances for the cylindrical resonator Z_r and for the coupler Z_c are products of their areas and factor $\sqrt{E\rho}$:

$$Z_r = \frac{\pi D_r^2}{4}\sqrt{E_r \rho_r}, \qquad Z_c = \frac{\pi D_c^2}{4}\sqrt{E_c \rho_c} \qquad (3.262)$$

where D_r and D_c = diameters of resonator and coupler, respectively, m;

E_r and E_c = Young's moduli for resonator and coupler materials, respectively, N/m^2; and

ρ_r and ρ_c = densities of resonator and coupler materials, respectively, kg/m^3.

The unit for Z_r and Z_c is kg/s.

A bandwidth of the filter is found from the expression for quality factor Q_f for the filter and from the standard relationships [36] for the half-power frequencies:

$$Q_f = \frac{f}{\Delta f} = \frac{2\pi f L_r}{R_r + 2R_c}, \qquad (3.263)$$

$$\Delta f = f_1 - f_2 \qquad (3.264)$$

where f = resonance frequency, Hz;

L_r = equivalent inductance of resonator, kg;

R_r and R_c = equivalent resistances of resonator and coupler, respectively, kg/s; and

f_1 and f_2 = half-power points, Hz.

The smaller diameter of the coupler, the higher Q_f. In the equivalent circuit, it corresponds to an increase of capacitances C_c. Table 3.11 gives relevant relationships and equivalent circuits for three typical lengths of the filter parts $\lambda/2$, $\lambda/4$, and $\lambda/8$. Table 3.12 presents several characteristics of several materials for the resonators and couplers. The data summarized in these tables, along with the formulas in this paragraph, are helpful for calculating the filter. In the design, the lengths of the parts can vary; for instance, the length of the coupler can be $\lambda/4$. It should be noted that the structure of the filter discussed here can be adapted not only for the conversion of one electrical value to another, but also for a response to some other physical quantities, such as viscosity of media, density, and so on. These quantities can change the damping of the system, effective mass, or elastic characteristic. The vibrators can be shaped as discs or plates, and vibrations can be torsional or transverse. The basic ideas in calculation remain the same, but the expressions for evaluation of the parameters of vibration should reflect the specifics of vibration. For instance, the frequency f for the torsional vibration of a bar with a node at its middle depends not only on the length of the bar l, its density ρ, and modulus of elasticity E, but also on Poisson's ratio v, as is presented in the following formula (the values of n and c are the same as in formula 3.258):

$$f = \frac{nc}{2l} = n\frac{\sqrt{\dfrac{E}{\rho}}}{\sqrt{2(1+v)}} \cdot \frac{1}{2l}. \qquad (3.265)$$

A magnetostrictive element with an attached mass and coil fed with direct current allows the measurement of acceleration. Under acceleration, the mass

TABLE 3.11

Equivalent Circuits of Resonators

Part of filter	Electrical analog	Relationships
Resonator with $l = \lambda/2$ and free ends		$\omega C = \dfrac{2}{\pi} \cdot \dfrac{1}{Z_r}$ $\omega L = \dfrac{\pi}{2} \cdot Z_r$ $\omega = 2\pi f$
Resonator with $l = \lambda/2$ loaded from one end		$\omega C = \dfrac{2}{\pi} \cdot \dfrac{1}{Z_r}$ $\omega L = \dfrac{\pi}{2} Z_r$
Coupler with $l = \lambda/8$		$\omega C = \dfrac{1}{Z_c}$
Coupler with $l = \lambda/4$		$\omega L = Z_c$ $\omega C = \dfrac{1}{Z_c}$

develops force, compressing or stretching the element. As a result, the ac voltage induced in the coil is proportional to the derivative of acceleration with respect to time. Integrating this voltage in a simple electrical cell gives an output proportional to the acceleration. It is notable that the magnetostrictive body of the elements carries simultaneously three functions: it operates as a part of a converter, as a spring, and as a seismic mass. The following formulas illustrate this concept:

$$F = m\,\frac{d^2x}{dt^2}, \qquad \sigma = k_1 F = k_2\,\frac{d^2x}{dt^2}, \qquad B = k_3\sigma = k_4\,\frac{d^2x}{dt^2},$$

$$U = k_5\,\frac{dB}{dt} = k_6\,\frac{d^3x}{dt^3}, \qquad U_{\text{out}} = k_7 \int U dt = k_8\,\frac{d^2x}{dt^2} \qquad (3.266)$$

where x, m, F, σ, B, U, and U_{out} are displacement, mass, force, stress, flux density, induced voltage, and output voltage, respectively. k_1 to k_8 are constant coefficients.

TABLE 3.12

Characteristics of Materials for Resonators and Couplers (Adopted from [35])

Material	ρ kg/m^3	$\sqrt{E/\rho}$ m/s	$\sqrt{E\rho}$ kg/(m^{2}s)	Q
Fused silica	2.20×10^3	6.04×10^3	12.6×10^6	10 000
Duralumin	2.81×10^3	5.07×10^3	14.3×10^6	8 000
Invar	8.62×10^3	4.01×10^3	34.6×10^6	5 000
Beryllium copper	8.26×10^3	3.73×10^3	30.8×10^6	4 000
Fernico	8.19×10^3	4.08×10^3	33.4×10^6	3 500
Brass	8.54×10^3	3.64×10^3	31.1×10^6	2 500
Phosphor bronze	8.90×10^3	3.52×10^3	31.4×10^6	2 000
Stainless steel	7.94×10^3	4.97×10^3	39.5×10^6	1 500

Several considerations should be taken into account in designing the magnetostrictive element. The maximum stress in the material of the deforming core should not be higher than 0.2 to 0.1 of the elastic limit. This means that the range for the change of μ is approximately $20\% \leq \Delta\mu/\mu \leq 40\%$ for basic materials. The maximum stress for nickel and permalloy is 20N/mm^2 and 80N/mm^2, respectively. This measure reduces mechanical hysteresis and instabilities. In addition to this, training by cyclical load (10^4 cycles) along with thermal treatment and aging also improves the characteristic. The exercise is provided with a magnetic field whose induction is close to saturation and a maximum stress in the cycle approximately 6 to 7 times lower than the elastic limit.

The length of a force-sensitive core along the direction of force should be larger than its width to avoid effects from transverse stresses, which increase the nonlinearity of the transfer characteristic. The sensitivity of the element with a solid core is approximately half of that for the laminated, but it is more accurate since its mechanical characteristics are more stable (there are no gaps in the core). It is expedient to alternate the position of laminations in the stacked core to have a perpendicular orientation of direction of rolling between the laminations (Fig. 3.54). This reduces the effects of anisotropy of the material.

A change of temperature of the element's parts gives a variation of sensitivities S_σ and S_ϵ, permeability, and ohmic resistance of the coil. Depending on the material, the temperature errors can be positive or negative. Sensitivity change reaches 0.2 to 1 percent per 1°C for a noncompensated device and 0.02 to 0.05 percent per 1°C for an element furnished with a compensator.

This element should not be regarded as a highly accurate device. The overall accuracy that can be obtained is between 0.1% and 2%. It is primarily intended for rough measurements, where its durable, strong, and simple construction is important. In terms of interface with a source of force to be measured, the element is almost an ideal receiver of signals. Its mechanical impedance, which can be qualified as a force-to-displacement ratio, is very high. The maximum deformation of the core does not exceed 0.1mm, whereas the force can reach $50 \times 10^6\text{N}$. A high stiffness of the sensing member assures a high natural frequency of the mechanical system and ability to treat input mechanical signals with frequency reaching 60kHz. This characteristic is combined with a low electrical impedance, high power, and high voltage at output. According to the data, which can be found in literature [15, 37], the output power can be up to 10w of 60-Hz current at full mechanical load. For many applications, the system with the element does not need any amplifiers for recording the signals.

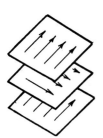

Figure 3.54
Stacking of laminations in force-sensitive core. Arrows denote direction of rolling during fabrication of material.

Other Elements

Variable *inductances with a sliding* or *"capacitive" contact* between coil 1 and moving wiper 2 are shown in Figure 3.55, along with equivalent circuits to the right of the diagrams. If the turns of the coils are evenly distributed along the length of the coil, inductance L_x, measured between terminals 3 and 4 in (*a*), will be a quadratic function of displacement *x*:

$$L_x = kx^2 \tag{3.267}$$

where *k* is constant depending on the construction of the inductor. The "capacitive" contact in (*b*) is a plate 2 that moves at close proximity to the turns and forms with them a constant capacitance. This capacitance *C* does not depend on the position of the plate along the winding. Due to this arrangement, a friction contact is

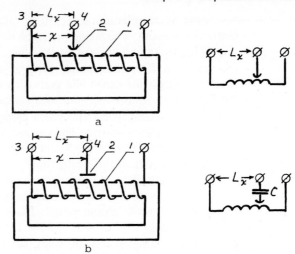

Figure 3.55 Variable inductance element with moving contact. a = version with sliding contact, b = version with "capacitive contact."

eliminated, and that is always desirable. The effective inductance "seen" from the terminals is

$$L_x = kx^2 - \frac{1}{\omega^2 C}. \tag{3.268}$$

Designs with variation of a number of turns are not as popular as those with a change of reluctance, and the concept with a sliding contact has a limited application.

A *temperature-sensitive inductor* is built with a core of ferromagnetic material whose properties are highly dependent on temperature; for example, permalloy and ferrite with high permeability. Since there are many inexpensive specialized components for measuring temperature (thermistors, platinum detectors, and so on), the inductive element can be found chiefly in the compensation for temperature circuits, where having an identical element in a balanced circuit is desirable. One of the attractive features of this element is its simple application in temperature-to-frequency converters. Indeed, a variation of inductance in an elementary tank circuit can easily modulate the frequency of an oscillator and simplify the subsequent digital conversion.

A *discrete inductive element* (Fig. 3.56) produces a pulsating change of inductance in response to the position of the moving armature 1 with respect to core 2 with coil 3. The motion of the armature can be rotational (*a*) or translational (*b*). In this element, the magnitude of the inductance during the pulses is not as crucial for operation as a pulse-repetition period or a number of cumulated pulses. The speed of motion of the armatures can be easily evaluated by standard digital counters.

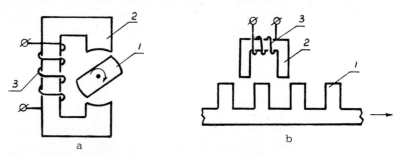

Figure 3.56 Discrete inductive element. a and b = elements with rotational- and translational-motion armatures.

A *pressure-sensitive inductive element* is illustrated in Figure 3.57. A ferromagnetic diaphragm 1 in (*a*) is part of a single magnetic circuit including a pot-type core 2 with winding 3. A differential element (*b*) is composed of two pots 1 and 2 and two windings 3 and 4 around diaphragm 5. Due to the difference in pressures $P_1 - P_2$, the diaphragm is deflected, the gap in the magnetic circuit is changed, and the inductance of the coil (or coils) is changed as well. The calculation of the inductance change versus pressure is similar to that for the absolute pressure capacitance element (see Chapter 3). Inductance L of the single element is a function of the variable reluctance of the gap R_g, whereas reluctance of the magnetic path R_m remains constant.

$$L = \frac{N^2}{R_g + R_m} \approx \frac{N^2}{R_g} = N^2 T_g \qquad (3.269)$$

where N = number of turns, and
$\quad T_g$ = permeance of the gap, H.

Since $R_g \gg R_m$, the inductance is primarily controlled by a variable gap as is reflected in the reduction of formula 3.269. A relationship for T_g is obtained by integrating the equation for the elementary permeancy dT_g, which is formed (Fig. 3.58) by an elementary circular gap d-y at radius x. The area of this gap is $2\pi x \cdot dx$; the permeability of space is μ_0. Therefore, dT_g is written as follows:

$$dT_g = \frac{\mu_0 2\pi x}{d - y} dx = \frac{\mu_0 2\pi x}{d - y_0 \left[1 + \left(\dfrac{x}{R}\right)^2\right]^2} dx. \qquad (3.270)$$

Figure 3.58 shows a part of the element with the diaphragm in a steady (1) and in a deformed-under-pressure (2) condition. It is assumed that the edges of the diaphragm are fixed, and its deflection y at radius x is described by equation 5.60. In this equation, the deflection at center y_0 is a function of difference P between two pressures P_1 and P_2 and is given in formula 5.18. Integrating (formula 3.270) with respect to x and with limits r_1 and r_2, the value of T_g can be calculated. r_1 and r_2 are two arbitrary radii of the central piece 3 facing the diaphragm:

$$T_g = \int_{r_1}^{r_2} dT_g. \qquad (3.271)$$

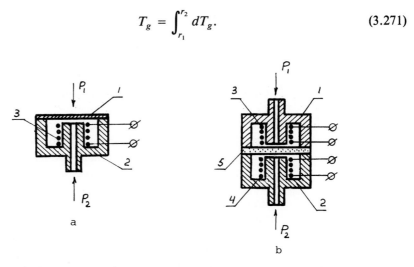

Figure 3.57 Pressure-sensitive inductive elements. a and b = single- and differential-type pressure-sensitive elements.

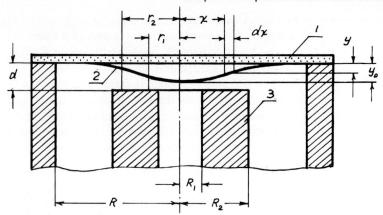

Figure 3.58 Change of gap in pressure-sensitive inductive element.

The inductance in question is determined as follows:

$$L = N^2 \int_{r_1}^{r_2} dT_g = 2\pi N^2 \mu_0 \int_{r_1}^{r_2} \frac{x}{\left[d - y_0 \left[1 + \left(\dfrac{x}{R} \right)^2 \right]^2 \right]} dx$$

$$= \frac{\pi N^2 \mu_0 R^2}{2\sqrt{2} dy_0} \left(\ln \frac{R^2 - r_1^2 - R^2\sqrt{d/y_0}}{R^2 - r_1^2 + R^2\sqrt{d/y_0}} - \ln \frac{R^2 - r_2^2 - R^2\sqrt{d/y_0}}{R^2 - r_2^2 + R^2\sqrt{d/y_0}} \right). \qquad (3.272)$$

If the central hole is not present in the construction (for example, the gas inlet is on the side wall), $r_1 = 0$ and $r_2 = R_2$ are substituted in formula 3.272, which gives the following:

$$L = \frac{\pi N^2 \mu_0 R^2}{2\sqrt{2} dy_0} \left[\ln \frac{1 - \sqrt{\dfrac{d}{y_0}}}{1 + \sqrt{\dfrac{d}{y_0}}} - \ln \frac{1 - \sqrt{\dfrac{d}{y_0} - \left(\dfrac{r_2}{R} \right)^2}}{1 + \sqrt{\dfrac{d}{y_0} - \left(\dfrac{r_2}{R} \right)^2}} \right]. \qquad (3.273)$$

For the configuration of the central piece shown in Figure 3.58, the inductance is calculated with formula 3.272 by substituting $r_1 = R_1$ and $r_2 = R_2$. The performance of the differential element is analyzed by separate evaluations of the inductances of each half when the diaphragm is deflected from the neutral position.

Electromagnetic Forces and Nonlinear Dynamics

A force that is developed between two parts forming a gap of the element's structure has a twofold interest. First, this force must be taken into account if it is comparable with other forces in the force-balance equations. Second, the force is an important characteristic of a driver in a force-balance and vibrating-element transducer.

The force along the lines of magnetic flux is a partial derivative of the energy stored in the gap with respect to the displacement toward each other of the poles forming the gap:

$$F = -\frac{\partial W}{\partial x}, \qquad W = \frac{1}{2} i^2 L, \qquad W = \frac{1}{2} \phi^2 R \qquad (3.274)$$

where F = electromechanical force, N;
$\quad\ \ W$ = energy stored in the gap, J;
$\qquad i$ = current in the coil, A;
$\qquad L$ = inductance of the coil, H;
$\qquad \phi$ = magnetic flux in the gap, Wb; and
$\qquad R$ = reluctance of the gap, 1/H.

The expressions for forces are found by taking derivatives of W:

$$F = -\frac{1}{2} i^2 \frac{dL}{dx} \quad \text{and} \quad F = \frac{1}{2} \phi^2 \frac{dR}{dx} \tag{3.275}$$

where x is a variable separation of the poles, m.

If the flux does not contain an offset, the force always tends to attract the poles and to reduce the distance between them. The change of the direction of the flux flowing through the gap does not change the trend of attraction. The presence of the negative signs in the formulas for forces indicates that they are the forces of attraction that tend to decrease x and the energy stored in the gap. However, minuses are often dropped in calculations, as they will be here in some instances.

If the coil is excited by an alternating current $i = i_m \sin \omega t$, the force is calculated by substituting i in formula 3.275 (we take, as an example, a simple system of one core, one coil, and one gap as in Figure 3.28):

$$F = \frac{1}{2} \cdot \frac{dL}{dx} i_m^2 \sin^2 \omega t = \frac{1}{4} \cdot \frac{dL}{dx} i_m^2 - \frac{1}{4} \cdot \frac{dL}{dx} i_m^2 \cos 2\omega t \tag{3.276}$$

The first term at the right side of this equation is an average force, and the second is an alternating force with a doubled frequency. Quite often in transducer design, the force must have the same frequency as the excitation current or voltage. In this case, a dc component is added to the excitation current, or a permanent magnet is included in the magnetic circuit to create a necessary bias in the flux ϕ_0 through the gap. Therefore, for the linear approximation of function $\phi = f(i)$ at constant μ, the total flux ϕ through the gap is

$$\phi = \phi_0 + \phi_m \sin \omega t \tag{3.277}$$

where the periodic component is due to the alternating current in the winding. Substituting ϕ in formula 3.275, we find that the force has three components:

$$F = \frac{1}{2} \cdot \frac{dR}{dx} \left(\phi_0^2 + \frac{1}{2} \phi_m^2 \right) + \frac{dR}{dx} \phi_m \phi_0 \sin \omega t - \frac{1}{4} \frac{dR}{dx} \phi_m^2 \cos 2\omega t. \tag{3.278}$$

The first component represents an average, time-independent force; the second and the third are forces with frequencies of ω and 2ω. The second harmonic force does not contribute significantly to the overall force if the following ratio is small: $(\phi_m/4\phi_0) \ll 1$. Increasing the bias flux helps to discriminate the second harmonic, which is frequently a source of distortion in signal conditioning.

The calculation of forces is divided into several steps.

1. The equivalent circuit of the magnetic system is drawn and, if it is possible, reduced by the combination of magnetic paths, neglecting reluctances of the ferromagnetic parts, and so on.

2. One of the formulas from 3.274 or 3.275 is selected. If the circuit contains one coil and one gap, formula $F = f(L, i, x)$ is most convenient for application. Multigap and multicoil systems are more easily calculated using the function: $F = f(R, \phi, x)$. For a multicoil system with an inductive interaction between the coils (the magnitudes of the mutual inductances and self-inductances are comparable), the most appropriate formula is $F = f(W, x)$, which is used to calculate the energy stored in the gap. For instance, in the simple systems in Figure 3.59, the following relationships can be used for two and three coils, respectively:

$$W = \frac{L_1 i_1^2}{2} + \frac{L_2 i_2^2}{2} + L_{12} i_1 i_2 \qquad (3.279)$$

$$W = \frac{L_1 i_1^2}{2} + \frac{L_2 i_2^2}{2} + \frac{L_3 i_3^2}{2} + L_{12} i_1 i_2 + L_{13} i_1 i_3 + L_{23} i_2 i_3 \qquad (3.280)$$

where L_i and L_j = self-inductances, H ($i = 1, 2, 3$ and $j = 1, 2, 3$ are coil identifications);

L_{ij} or L_{ji} = mutual inductances, H; and

i_i or i_j = currents in coil i or j, A.

3. Currents, fluxes, reluctances, and inductances, which are needed for substitutions in the formulas, are calculated. The reluctances or inductances that are substituted in the formulas for force must be introduced as functions of the gap or the area of the gap. For calculating the average forces of ac excitation, the root-mean-square values of current can be substituted for i in formula 3.275.

As an example, we will calculate an average force that is exerted on the diaphragm of the pressure-sensitive element (Fig. 3.57a) when all the parameters of the magnetic system are defined, including voltage feeding the coil, $U = U_m \sin \omega t$. During operation, the element's output signal is proportional to the resultant outside force applied to the diaphragm. This force has two components. One of them depends on the pressure to be measured; the other one is defined by the electromechanical force developed due to the magnetic flux flowing through the gap. The latter force must be small to be neglected, or it has to be reproducible to secure the accuracy of transduction. It can also be compensated by a counteracting force as it is in a differential element.

The following steps illustrate the calculation.

1. An equivalent magnetic circuit of the element (Fig. 3.60a) is a series connection of a source of magnetomotive force NI, reluctances of magnetic parts (R_1 through R_4), and reluctance of gap R_g. We assume, as is often done, that $R_g \gg (R_1 + R_2 + R_3 + R_4)$. Thus, the equivalent circuit is reduced to that shown in (b). We also assume that the ohmic resistance of the coil is much smaller than its reactance.

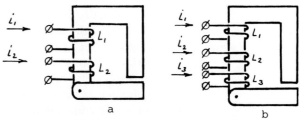

Figure 3.59 Multiply excited systems. a and b = two- and three-coil systems.

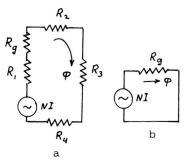

Figure 3.60 Equivalent circuit for pressure-sensitive element. a = complete equivalent circuit, b = reduced equivalent circuit.

2. Since we have a one-coil and one-gap circuit, the formula with the current and inductance's derivative can be taken:

$$F_{av} = \frac{1}{2} I_{RMS}^2 \frac{dL}{dx} \qquad (3.281)$$

where F_{av} = average force, N, and
I_{RMS} = root-mean-square current, A.

3. The current is found as a ratio of voltage U across the coil to its impedance Z:

$$i = \frac{U}{Z}. \qquad (3.282)$$

4. As we assumed, the impedance is purely reactive:

$$Z = \omega L. \qquad (3.283)$$

where ω = angular frequency, rad/s.

5. The inductance is defined only by the reluctance of the gap:

$$L = \frac{N^2}{R_g} = \frac{N^2 A \mu_0}{x} \qquad \text{since} \qquad R_g = \frac{x}{\mu_0 A} \qquad (3.284)$$

where x = gap, m;
A = area of the gap, m^2; and
N = number of turns.

6. The instantaneous and RMS values of the current are

$$i = \frac{U_m R}{\omega N^2} \sin \omega t, \qquad I_{RMS} = \frac{U_m R}{\sqrt{2} \omega N^2}. \qquad (3.285)$$

7. Differentiating L in 5, we obtain (the minus sign is dropped)

$$\frac{dL}{dx} = \frac{N^2 A \mu_0}{x^2}. \qquad (3.286)$$

8. Finally, the average force is found by substituting I_{RMS} and dL/dx in formula 3.281:

$$F_{av} = \frac{1}{4} \cdot \frac{U_m^2}{N^2 \omega^2 A \mu_0}. \qquad (3.287)$$

A solenoid (Fig. 3.61) with coil 1 and pull-in core 2 can be used when a force should be generated in a system with a long travel of the core. It also has an application when a small change of the force within the limited part of the travel is required. One of the representative characteristics of the solenoid is shown in Figure 3.62. Force F is introduced as a function of a relative position of core x_1/l for one

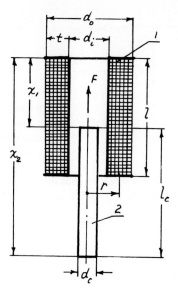

Figure 3.61   Solenoid with pull-in core.

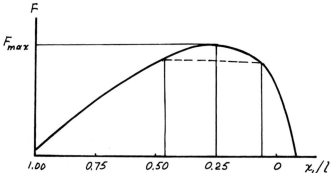

Figure 3.62   Pull-in force of solenoid versus relative position of core.

specific construction (generally the shape of the curve varies depending on the geometry of the parts). It can be seen in the graph that the maximum of the characteristic is at approximately $x_1/l = 0.25$. Around this maximum (dashed line) the change of force is not large.

A calculation [26] of the force can be provided with the formula given below:

$$F = k_1 \frac{Ar}{l_c} \mu_0 \mu_a \frac{(IN)^2}{4} [f_1(y_1) + f(m - y_1) - f(y_2)$$

$$- f(m - y_2)][\varphi(y_2) - \varphi(m - y_2) - \varphi(y_1) + \varphi(m - y_1)]. \qquad (3.288)$$

The following list explains the meaning of different terms in the formula and the methods of their calculation (see also Fig. 3.61). The magnitude of force F is computed in newtons.

$$1. \quad k_1 = \frac{l_c}{l_c - \dfrac{d_c}{2}} \qquad\qquad\qquad (3.289)$$

where l_c = core's length, m, and
$\quad\quad\ d_c$ = core's diameter, m.

2. $A = \dfrac{\pi d_c^2}{4}$ = core's area, m^2.   (3.290)

3. $r = \dfrac{d_0 + d_i}{4}$ = mean radius of winding, m.   (3.291)

4. $\mu_0 = 4\pi 10^{-7}$ = permeability of air, H/m,

5. $\mu_a = \dfrac{\lambda^2}{2.72 lg\lambda - 0.69}$ = relative average permeability of core,   (3.292)
dimensionless,

$\lambda = \dfrac{l_c}{d_c}.$

6. I = current in coil, A.

7. N = number of turns.

8. $f_1(y_1) = \dfrac{y_1}{\sqrt{1 + y_1^2}}, \qquad y_1 = \dfrac{x_1}{r},$ (3.293)

where x_1 = distance from the core's top to the winding's top, m.

9. $f(m - y_1) = \dfrac{m - y_1}{\sqrt{1 + (m - y_1)^2}}, \qquad m = \dfrac{l}{r},$ (3.294)

where l = length of winding, m.

10. $f(y_2) = \dfrac{y_2}{\sqrt{1 + y_2^2}}, \qquad y_2 = \dfrac{x_2}{r},$ (3.295)

where x_2 = distance from the core's bottom to the winding's top, m.

11. $f(m - y_2) = \dfrac{m - y_2}{\sqrt{1 + (m - y_2)^2}}.$ (3.296)

12. $\varphi(y_2) = \sqrt{1 + y_2^2}.$ (3.297)

13. $\varphi(m - y_2) = \sqrt{1 + (m - y_2)^2}.$ (3.298)

14. $\varphi(y_1) = \sqrt{1 + y_1^2}.$ (3.299)

15. $\varphi(m - y_1) = \sqrt{1 + (m - y_1)^2}.$ (3.300)

Note: Parameters λ, y's, f's, φ's, and m are dimensionless.

Calculation with formula 3.288 gives a good correspondence between computed and experimental data (the average discrepancy is 6%).

The equations of statics and dynamics should take into account the electromechanical force when the behavior of the element is analyzed. The analysis is provided with linearization of nonlinear differential equations describing performances of mechanical and electrical parts. As an example, we will take for analysis a single pressure-sensitive element (Fig. 3.63) and model it by the lumped-parameter system. During operation, diaphragm 1 moves translationally along the x-axis. The characteristics essential for the analysis of dynamics of the diaphragm can be introduced by lump parameters. They are mass m, friction coefficient b, and spring factor k. External force F represents a resultant due to summing the pressure applied to the diaphragm. The electrical circuit of the element is composed of variable inductance L, connected in series with ohmic resistance of coil r. Ac voltage U is applied to the coil that develops current i. It is assumed as before that the reluctance

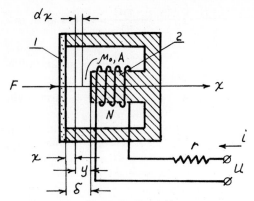

Figure 3.63 Pressure-sensitive element as model for analysis of nonlinear dynamics.

of the magnetic circuit is defined only by the reluctance of the gap between diaphragm 1 and central part 2. Practically any other inductive element having moving parts and characterized by mass, elasticity, friction, and electromagnetic excitation can be reduced to a similar system. In the following analysis, we will write mechanical and electrical equations, qualify their nonlinearity, and show how they are linearized for solution.

The force-balance equation includes several terms.

1. Force F is generated by pressure.
2. Mechanical force F_e results from the magnetic field created by the coil. F_e can be considered as a force externally applied to the diaphragm.
3. A combination of the forces of inertia, friction, and elastic resistance all oppose the external forces. Taking into account the chosen direction for F, the equation of motion is written as follows:

$$m \frac{d^2x}{dt^2} + b \frac{dx}{dt} + kx = F + F_e. \tag{3.301}$$

The equation satisfying the electric circuit is

$$U = ir + \frac{d}{dt}(Li). \tag{3.302}$$

Inductance of coil L is a function of displacement x:

$$L = \frac{N^2}{R(x)} = \frac{N^2 \mu_0 A}{\delta - x} = \frac{B}{y} \tag{3.303}$$

where $B = N^2 \mu_0 A$ and $y = \delta - x$ (y is a variable separation of poles). We assume that fringing, leakage, and reluctance of magnetic parts are neglected, and $R(x)$ is the reluctance of the gap. Using operator (D) notation and substituting the expression for force (equation 3.275), equation 3.301 is rewritten as follows:

$$(mD^2 + bD + k)x + \frac{1}{2} i^2 \frac{dL}{dy} = F. \tag{3.304}$$

Equation 3.302 is modified with the same notation:

$$ri + L_0 \frac{di}{dt} + i_0 \frac{dL}{dt} = U \qquad \text{or} \qquad (3.305)$$

$$ri + L_0 Di + i_0 DL = U \qquad (3.306)$$

where L_0 and i_0 are the inductance and current corresponding to a steady equilibrium point. This point is also characterized by coordinate x_0.

Equations 3.304 and 3.306 are nonlinear in terms of independent variables x, L, and i. They can be solved numerically or by linearization and use of small perturbations around the variables. In order to find the relationships between the perturbations, we will first differentiate equation 3.304, then equation 3.306. The differentiation is given with details that should provide a better understanding of certain aspects of linearization.

$$d[(mD^2 + bD + k)x] + d\left(\frac{1}{2} i^2 \frac{dL}{dy}\right) = dF \qquad (3.307)$$

We denote

$$i^2 \frac{dL}{dy} = J,$$

$$d\left(\frac{1}{2} J\right) = \frac{1}{2}\left[\frac{\partial J}{\partial i} di + \frac{\partial J}{\partial \left(\frac{dL}{dy}\right)} d\left(\frac{dL}{dy}\right)\right], \qquad (3.308)$$

$$\frac{dL}{dy} = -\frac{B}{y^2} = -\frac{B}{(\delta - x)^2} \qquad (3.309)$$

(see formula 3.303),

$$\frac{\partial J}{\partial i} = 2i\left(\frac{dL}{dy}\right)\bigg|_{x=x_0} = -2i_0 \frac{B}{(\delta - x_0)^2}, \qquad (3.310)$$

$$\frac{\partial J}{\partial \left(\frac{dL}{dy}\right)} = i^2, \qquad \frac{\partial J}{\partial \left(\frac{dL}{dy}\right)}\bigg|_{x=x_0} = i_0^2, \qquad (3.311)$$

$$d\left(\frac{dL}{dy}\right) = \frac{2B}{y^3} dy, \qquad d\left(\frac{dL}{dy}\right)\bigg|_{x=x_0} = -\frac{2B}{(\delta - x_0)^3} dx. \qquad (3.312)$$

Substituting all the derivatives obtained in formula 3.307 and replacing differentials dF, di, and dx by small perturbations ΔF, Δi, and Δx, respectively, we obtain the first linearized equation:

$$\left[mD^2 + bD + k - \frac{L_0 i_0^2}{(\delta - x_0)^2}\right]\Delta x - \frac{L_0 i_0}{\delta - x_0}\Delta i = \Delta F. \qquad (3.313)$$

A similar differentiation is made for equation 3.306:

$$d(ri + L_0 Di + i_0 DL) = dU \qquad (3.314)$$

$$rdi + L_0 Ddi + i_0 DdL = dU \qquad (3.315)$$

$$dL = \frac{\partial L}{\partial y} \, dy = \frac{B}{(\delta - x)^2} \, dx, \qquad dy = -dx, \qquad dL \bigg|_{x=x_0} = \frac{B}{(\delta - x_0)^2} \, dx \quad (3.316)$$

$$\frac{dL}{dx}\bigg|_{x=x_0} = \frac{B}{(\delta - x_0)^2}. \quad (3.317)$$

Substituting the derivatives in formula 3.315 and replacing di and dx by Δi and Δx, a linearized electrical equation is obtained:

$$\Delta i(r + L_0 D) + \frac{L_0 i_0 D}{\delta - x_0} \, \Delta x = \Delta V. \quad (3.318)$$

Equations 3.313 and 3.318 are two simultaneous linear differential equations. They can be presented in a matrix form as follows:

$$\left[\begin{array}{cc} \left[mD^2 + bD + k - \dfrac{L_0 i_0^2}{(\delta - x_0)^2} \right] & -\dfrac{L_0 i_0}{\delta - x_0} \\[4mm] \dfrac{L_0 i_0 D}{\delta - x_0} & r + L_0 D \end{array} \right] \left[\begin{array}{c} \Delta x \\[4mm] \Delta i \end{array} \right] = \left[\begin{array}{c} \Delta F \\[4mm] \Delta U \end{array} \right]. \quad (3.319)$$

The equations are solved by standard methods applicable for linear differential equations.

Temperature Instabilities

The sources of temperature instabilities are changes of size, changes of the magnetic properties of ferromagnetic materials, and changes in electrical conductivity. We will consider the instabilities of an element for measurement displacement, which is the most common case. It is convenient to evaluate the drift of the element's impedance rather than inductance. This drift δZ_t is found [38] from the given relationship:

$$\delta Z_t = \frac{1}{Z} \cdot \frac{\partial Z}{\partial t} = \frac{1}{1 + Q_f^2} \left\{ [\alpha_w + a_w + 0.5(\delta \mu_{ct} + \alpha_c)] \right.$$
$$\left. + Q_f^2 \left[\delta \mu_{ct} + a_c + \frac{\delta l_{gt} + \delta l_t}{\dfrac{A_g}{l_g \mu_c} \sum \dfrac{l_c}{A_c} + 1} \right] \right\} \quad (3.320)$$

where Z = impedance, Ω;
$\partial Z / \partial t$ = relative sensitivity of the impedance to temperature, $\Omega/°C$;
Q_f = quality factor, dimensionless (formulas 3.167–3.170);
α_w and α_c = temperature coefficients of resistivity for the materials of winding and core, respectively, $1/°C$;
a_w and a_c = coefficients of temperature expansion, $1/°C$;
$\delta \mu_{ct}$ = temperature coefficient of relative permeability, $1/°C$;
δl_{gt} = relative temperature change of the length of the gap, $1/°C$ [$\delta l_{gt} = \Delta l_g/(l_g \Delta t)$ where Δl_{gt} and Δt are changes of the gap and temperature, respectively];
δl_t = relative temperature change of the lengths of the armature (moving part) and core in the direction of change of the area of the gap, $1/°C$;
μ_c = relative permeability of core's material, dimensionless;

l_g and l_c = lengths of gap and core, respectively, m; and
A_g and A_c = areas of gap and core, respectively, m^2.

δl_{gt} and δl_t must be substituted in formula 3.320 with the corresponding sign, taking into account the direction of the deformation.

The order of magnitude of Q_f for the frequency range of 60 to 20 000Hz is between 1 and 20. It is low for solid cores and higher for laminated cores made of transformer steel and permalloy. The highest Q_f is for ferrite cores.

When δl_{gt} is calculated, the lengths of the parts composing a dimensional loop, coefficients of temperature expansion, and changes of temperature must be considered. As an example, a calculation of δl_{gt} is given for a simple single inductive displacement element (Fig. 3.64) that is composed of three major elements forming the loop: housing 1, core with coil 2, and armature 3. This element undergoes heating in such a manner that the temperature changes of housing, core, and armature are Δt_h, Δt_c, and Δt_a, respectively. In this case, Δl_{gt} is

$$\Delta l_{gt} = \alpha_h \, l_h \, \Delta t_h - \alpha_c \, l_c \, \Delta t_c - \alpha_a \, l_a \, \Delta t_a \qquad (3.321)$$

where α_h, α_c, and α_a = coefficients of temperature expansion for housing, core, and armature, respectively, 1/°C, and
l_h, l_c, and l_a = lengths of housing, core, and armature, respectively, m.

A similar calculation can be provided for the case when the area of the gap is affected by the variation of temperature. In Table 3.13, the parameters [38] useful for calculating the temperature instabilities are given for some materials.

It follows from the data in this table that the instability due to the variation of resistivity is greater than that due to permeability. If the structure does not contain gaps, the change of dimensions due to the temperature expansion can be ignored, and formula 3.320 is reduced to the following:

$$\delta Z_t^1 = \frac{1}{1 + Q_f^2} (\eta \alpha_c + Q_f^2 \cdot \delta \mu_{ct}) \qquad (3.322)$$

where

$$\eta = \frac{1}{\nu(1 + R_c/R_0)} + \frac{0.5(1 + 1/\epsilon)}{1 + R_0/R_c}, \qquad (3.323)$$

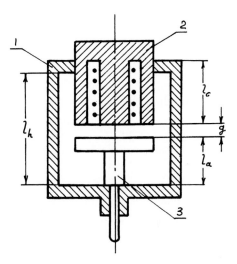

Figure 3.64 Scheme of inductive displacement element for calculation of temperature drifts.

TABLE 3.13

Characteristics of Some Materials Used in Inductive Elements (Adopted from [38])

Material	a, 1/°C From	a, 1/°C To	α, 1/°C From	α, 1/°C To	ρ, $\Omega \cdot$ mm²/m From	ρ, $\Omega \cdot$ mm²/m To	$\delta\mu_{ct}$, 1/°C From	$\delta\mu_{ct}$, 1/°C To
Copper	16.5×10^{-6}	17.1×10^{-6}	0.0039		0.0175	0.0182	—	—
Purified iron	11.7×10^{-6}	12.2×10^{-6}	0.006	0.0065	0.10	0.15	0.032	
Commercial iron	10.6×10^{-6}	12.2×10^{-6}	0.0062		0.10	0.20	0.029	
Silicon iron	12.0×10^{-6}		0.0008	0.003	0.25	0.60	0.002	0.007
45 Permalloy	6.0×10^{-6}	9.0×10^{-6}	0.003	0.004	0.45	0.90	0.0006	0.0015
76 Permalloy	11.0×10^{-6}	13.0×10^{-6}	0.003	0.004	0.50	0.65	0.0009	
Ni-Zi ferrites	6.4×10^{-6}	10.0×10^{-6}	—	—	1×10^{7}	1×10^{12}	0.0008	0.007
Mn-Zn ferrites	6.4×10^{-6}	10.0×10^{-6}	—	—	0.5×10^{6}	1×10^{6}	0.0012	0.009

$$\nu = \alpha_c/\alpha_w, \qquad \epsilon = \alpha_c/\delta\mu_{ct} \tag{3.324}$$

R_0 = resistance of winding, Ω (formula 3.153), and
R_c = effective resistance of core, Ω (formulas 3.161 and 3.164).

Calculations show that the temperature-induced error in the solenoid element is substantially lower than that in the element with the closed magnetic circuit. The error is also smaller in the differential system, where the principle of compensation for different adverse factors is suitable for temperature compensation. Unpredictable heat propagation and temperature gradients cause the errors and should be eliminated.

Features

Besides the temperature errors, several imperfections can affect the characteristics of accuracy.

The nonlinearity of the transfer characteristic is increased, as a rule, with an increase of the range of measurement. Using the differential construction substantially improves the linearity.

Outside electromagnetic fields induce voltage in the coils and can change the permeability of the core.

The induced voltage e [V] is

$$e = 2\pi f N A_w B_m \tag{3.325}$$

where f = frequency, Hz;
N = number of turns in winding;
A_w = cross-sectional area of winding, m²; and
B_m = peak value of induction, T.

In rooms with power equipment, B_m is between 1.0×10^{-5}T and 0.5×10^{-4}T. The element should be shielded. Inside the shield, the field intensity B_{ms} [T] is

$$B_{ms} = B_m \exp\left(-t\sqrt{\pi f \mu \sigma}\right) \tag{3.326}$$

where t = thickness of shield, m;
μ = permeability of shield's material, H/m; and
σ = electrical conductivity of shield's material, S/m.

The induction calculated with formula 3.326 can be substituted in formula 3.325 in order to find whether voltage induced in the coil is acceptable for the normal operation of the device.

Usually the change of the permeability due to the influence of the outside field is very small since this field's induction constitutes a very small fraction of the induction at which the core operates (0.1T to 1.0T).

Stresses in cores due to an assembly imperfection, mounting, or temperature deformation cause a change of permeability because all ferromagnetic materials are magnetostrictive. The range of change of the relative permeability per unit of stress for the typical core materials lies between -0.001 $1/(N/mm^2)$ and -0.07 $1/(N/mm^2)$. Permalloys are most sensitive to stresses.

Some residual stresses are present in the wire of the coil due to tension during winding. Additional stresses can exist when the element is heated. These factors give relative changes of resistance δR_w and impedance δZ_σ, which are expressed as follows:

$$\delta R_w = 17.0 \times 10^{-6} \frac{\Delta F}{d} \quad \text{and} \quad \delta Z_\sigma = \frac{R_w}{Z} \delta R_w \qquad (3.327)$$

where ΔF = variation of wire tension force, N;
  d = diameter of wire, mm;
  R_w = resistance of wire, Ω.

Quite often the element's performance is affected by inaccuracy in assembly, technology, and calibration. Side and end plays in moving parts are typical sources of error. Close parameter tolerances in the design of the magnetic system and winding can cause difficulties in interchanging parts and in compensating for temperature.

The major factors defining aging are the wear of the parts due to friction and a change with time of the magnetic properties of the ferromagnetics. Accelerated, artificial aging under load and heat is helpful in stabilizing the element's characteristics.

The element's dynamic characteristics are improved with a reduction of the moving mass and increase of the restraining spring force. For an individual design, a compromise in the construction and in characteristics of parts must be found. As a rule, the improvement of the dynamic response reduces the sensitivity in static measurements, and vice versa.

It is desirable to have a signal conditioner close to the element to reduce adverse voltage in the connecting cable. This voltage is usually induced from the surrounding electromagnetic fields. The magnitude of the voltage [38] is given below.

$$e = \mu_0 f I_c l \cdot ln\left\{\sqrt{\left[1 + \left(\frac{a_c}{h}\right)^2\right]\left[1 + \left(\frac{a_s}{h}\right)^2\right]}\right\} \qquad (3.328)$$

where e = voltage induced, V;
  μ_0 = permeability of air, H/m;
  f = frequency of current in a power cable, 1/s;
  I_c = current in the power cable, A;
  l = length of the power cable parallel to the signal-carrying cable, m;
  a_c = distance between the conductors in the power cable, m;
  a_s = distance between the conductors in the signal-carrying cable, m; and
  h = distance between the power and signal-carrying cable, m.

The inductive element is usually part of a displacement-, force-, or torque-measuring transducer. It also serves devices in which different mechanical quantities, such as acceleration, pressure, and flow rate, are first converted to displacement and then to a change of inductance. Transducers including the element have high accuracy and resolution. They are suitable for analog and digital conversion. The electrical circuit of the element is convenient for unification.

The characteristics of the element vary widely and depend on application. For instance, the displacement transducer with an inductive element allows the measurement of displacement in a wide range: 6×10^{-6}m to 10m. The resolution can be 10×10^{-9}m and the accuracy can reach 0.1%. The loading force (the force applied to the object of measurement) is usually between 0.001N and 1.0N.

Designs of the element are constantly being improved by the development of devices that have a wider range of the measurand, and better linearity, accuracy, and dynamic response.

chapter 4

Transformer, Electrodynamic, Servo, and Resonant Elements

This chapter unites in one group several electromechanical elements that, despite the differences in composition and mode of operation, contain a wound part. This part manifests itself as an inductor of electromotive force or a source of mechanical force. The exceptions are some of the resonant elements based on the piezoelectric properties of materials. Among the four devices described in this chapter, servo and resonant ones are characterized by extremely high accuracy of conversion and stability in operation.

TRANSFORMER ELEMENTS

A transformer element has a structure similar to that of an inductive element, it converts similar measurands, and its principles of operation are based on the same electromagnetic concepts. Unlike an inductive element, whose output is variable impedance, the transformer develops voltage at the output.

In this section, we briefly consider the performance of several typical transformer transducing circuits and refer the reader to the elements of theory given in the foregoing chapter if a more detailed analysis is needed.

Single-core Transformers

The simplest model of a transformer is a model (Fig. 4.1a) containing a movable armature and a C-shaped core with two windings. The primary coil is excited by current I_1. The element is intended for the measurement of the displacement of the armature varying gap δ. When this gap is changed, the reluctance of the magnetic circuit R and mutual inductance L_{12} between two coils are also changed, causing a change of voltage U_2 in the secondary coil. Neglecting ohmic resistance of the coil, U_2 is calculated as follows:

$$U_2 = j\omega L_{12}I_1. \tag{4.1}$$

The following standard relationships for the given magnetic circuit can be used to derive the expression for voltage U_2 as a function of gap δ and the parameters of the circuit:

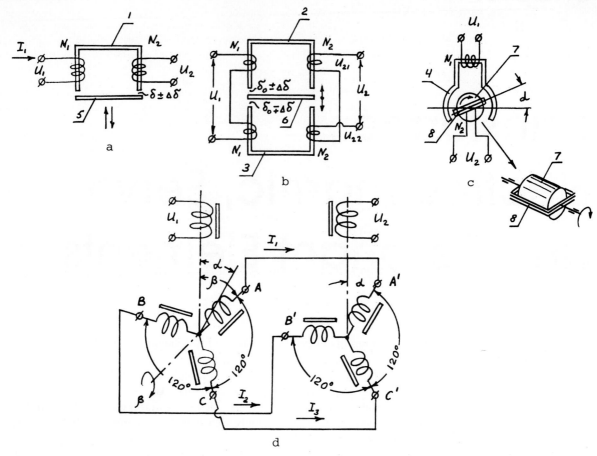

Figure 4.1 Transformer elements. a = element with single core, b = differential transformer, c = transformer with rotating coil, d = synchro transformers; 1 to 4 = *C*-shaped cores, 5 and 6 = movable armatures, 7 = ferromagnetic cylinder, 8 = rotating coil.

$$I_1 = \frac{M_1}{N_1}, \qquad L_{12} = \frac{N_1 N_2 \Phi}{M_1} = \frac{N_1 N_2}{R}, \qquad R = \frac{M_1}{\Phi}, \qquad R = \frac{2\delta}{\mu_0 A}. \quad (4.2)$$

These formulas are correct if we assume that the flux leakage is negligible, fringing effects in the gap can be neglected, and the magnetic reluctance of the circuit is defined by the gap (the permeability of the ferromagnetic part is much larger than that of air).

The combination of formulas in 4.1 and 4.2 gives

$$U_2 = j\omega N_1 N_2 I_1 / R = j\omega N_1 N_2 I_1 \mu_0 A / 2\delta \qquad (4.3)$$

where M_1 = magnetomotive force at the primary, A-turns;

N_1 and N_2 = number of turns of the primary and secondary, respectively;

A = area of gap, m²;

R = reluctance, 1/H;

Φ = magnetic flux, Wb;

δ = gap, m;

ω = angular frequency of current, rad/s; and

μ_0 = permeability of air, H/m.

For a higher sensitivity, the current rather than the voltage excitation of the transducer is preferable.

Differential Transformers

A differential transformer (Fig. 4.1b) contains four windings. The primary windings have a series-aiding connection and the secondary windings are connected in an opposing sense. The displacement of the armature provides an increase in the impedance of one primary (Z_1) and a decrease in another one (Z_2). Thus, the total impedance of the coils connected in series ($Z_1 + Z_2$) remains approximately constant, as well as current I_1, caused by voltage U_1:

$$I_1 = \frac{U_1}{Z_1 + Z_2} \approx \frac{U_1}{2j\omega L_0} = \frac{U_1 \delta_0}{j\omega N_1^2 \mu_0 A} \quad (4.4)$$

where L_0 = inductance of one primary at the neutral position of the armature, H;
N_1 = number of turns of one primary; and
δ_0 = gap at the neutral position of the armature, m.

As before, we assume that the reluctance of the metal portion of the magnetic circuit is neglected. Assuming that there is no electromagnetic interaction between the upper and lower parts of the magnetic circuit in Figure 4.1b, the voltage at output U_2 can be calculated as a difference between two voltages U_{21} and U_{22} induced at each secondary:

$$U_2 = U_{21} - U_{22} = (j\omega N_1 N_2 \mu_0 A / 2\delta_1)I_1 - (j\omega N_1 N_2 \mu_0 A / 2\delta_2)I_1 = \frac{N_2}{N_1} U_1 \frac{\Delta\delta}{\delta_0}. \quad (4.5)$$

This expression is obtained by using formulas 4.3 and 4.4, substituting $\delta_1 = \delta_0 - \Delta\delta$ and $\delta_2 = \delta_0 + \Delta\delta$, and neglecting $(\Delta\delta)^2$, where $\Delta\delta$ is the displacement of the armature from the neutral position.

Rotating-coil Transformers

A rotating-coil transformer (Fig. 4.1c) for the measurement of angular displacements contains a rotating coil that is placed in the gap of a magnetic system formed by a core with rounded poles and a cylindrical ferromagnetic insertion for reduction of reluctance.

Voltage U_1 applied to the coil is balanced by the voltage induced in the coil and by the voltage drop across the resistance of the coil. The latter component is neglected, as it was before:

$$U_1 = j\omega N_1 \phi_1 \quad (4.6)$$

where ϕ_1 = total flux circulating in the magnetic circuit, Wb.

The construction of the device is such that the fraction of the flux ϕ_2 inducing the electromotive force in the movable winding changes proportionally to the coil's angular displacement α (it is taken in radians):

$$\phi_2 = 2\phi_1 \frac{\alpha}{\pi} \quad (4.7)$$

Therefore, voltage U_2 across the coil terminals is

$$U_2 = j\omega N_2 \phi_2 = 2\frac{\alpha}{\pi} \cdot \frac{N_2}{N_1} U_1. \quad (4.8)$$

Unlike the constructions in Figure 4.1a and b, this element does not employ the variation of the reluctance by the moving part, but rather a change in the linkage of the flux with the turns in the secondary.

Synchro-transformers

Synchro-transformers are used to measure relative displacements between two rotating objects. The measuring or control system with synchro-transformers includes two rotor-stator pairs. One part of this pair has three sets of windings that are placed 120° apart at a rotor, as shown in Figure 4.1d. Another part is a single-phase coil attached to the stator. There can be a reverse arrangement of this structure: three coils are in the stator and one coil in the rotor. In terms of principles of operation, these two structures are identical. One of the stators is energized with an ac voltage U_1 and induces voltages in the three-winding segment of the rotor (A, B, and C). These voltages are in phase but their amplitudes depend on the angular position α of the rotor with respect to the stator. The rotor-stator pair experiencing the angular displacement is a transmitter. The wires of the transmitter are connected to another similar pair that is a receiver or control transformer. The voltages from the transmitter ("received" by the receiver) develop magnetic fluxes in the receiver rotor. This flux induces voltage U_2, which is proportional to the displacement angle ($\beta - \alpha$). If the receiver's rotor is held at a reference angle, for example α, voltage U_2 will indicate the deviation from this angle. The following calculations describe this feature in more detail.

The voltages E_a, E_b, and E_c, induced in the three segments (A, B, and C) of the transmitter, are

$$E_a = E_m \cos \beta, \qquad E_b = E_m \cos (\beta - 120°), \qquad \text{and}$$

$$E_c = E_m \cos (\beta - 240°) = E_m \cos (\beta + 120°) \tag{4.9}$$

where E_m = peak value of voltages, V.

Note that these voltages have the same phase but different amplitude.

Voltages E_a, E_b, and E_c create currents in the receiver coils:

$$I_a = \frac{E_a}{2Z}, \qquad I_b = \frac{E_b}{2Z}, \qquad I_c = \frac{E_c}{2Z} \tag{4.10}$$

where Z = impedance of one coil, Ω.

The magnetic fluxes established by the currents induce voltages in the single-phase coil. They are

$$E_a' = KI_a \cos \alpha, \qquad E_b' = KI_b \cos (\alpha - 120°), \qquad \text{and}$$

$$E_c' = KI_c \cos (\alpha - 240°) = KI_c \cos (\alpha + 120°) \tag{4.11}$$

where K = proportionality factor.

The composite electromotive force U_2 in the coil will be

$$U_2 = E_a' + E_b' + E_c' = K[I_a \cos \alpha + I_b \cos(\alpha - 120°) + I_c \cos(\alpha + 120°)]. \tag{4.12}$$

Substituting the expressions for the currents in equation 4.10 gives

$$U_2 = \frac{KE_m}{2Z} [\cos \beta \cos \alpha + \cos(\beta - 120°) \cos(\alpha - 120°)$$

$$+ \cos(\beta + 120°) \cos(\alpha + 120°)] \tag{4.13}$$

Using trigonometric identities, the terms in the brackets are reduced to:

$$\cos \beta \cos \alpha + 2 \cos^2 120° \cos \beta \cos \alpha + 2 \sin^2 120° \sin \beta \sin \alpha$$

$$= \frac{3}{2} (\cos \beta \cos \alpha + \sin \beta \sin \alpha) = \cos(\beta - \alpha) = \cos \theta \qquad (4.14)$$

Expression 4.13 is written now as

$$U_2 = U_m \cos \theta \qquad (4.15)$$

where $U_m = 3KE_m/4Z$ is a constant, and $\theta = \beta - \alpha$ is a displacement angle.

It is seen from formula 4.15 that output voltage U_2 does not depend on the reference positions of rotors α, but is defined by the displacement angle only. U_2 is at maximum for $\theta = 0°$ and equal to zero when $\theta = 90°$. It is more practical to have a 90°-angle offset between two rotors: $\theta = 90° + \theta'$. In this case, a zero-voltage output corresponds to the aligned rotors:

$$U_2 = U_m \sin \theta'. \qquad (4.16)$$

Weak-field Sensors

One of the interesting devices (Fig. 4.2) employing the concepts of variation of magnetic characteristics and induction of voltages is the electromagnetic compass or sensor of very weak magnetic fields (it is also called a "ferrosond"; see the description of fluxgate devices in Appendix 1, Scheme 186).

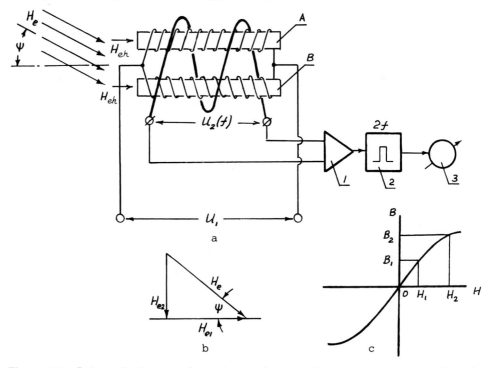

Figure 4.2 Schematic diagram of even-harmonic magnetic compass. a = measuring system, b = components of earth's magnetic field, c = B-H curve approximation; A and B = saturating reactors, f = frequency, H_e = intensity of earth's magnetic field, H_{e1} and H_{e2} = two components of H_e, H_{eh} = horizontal component of H_e, U_1 and U_2 = excitation and pickup voltages, ψ = angle between direction of field and axis of reactors, 1 = amplifier, 2 = filter, 3 = meter.

It is constructed of two magnetic reactors aligned parallel and close to each other. Two excitation coils wound on each reactor have a parallel opposing connection. One pickup coil is wound around two reactors. The cores of the coils are made of a material, such as permalloy, that have a highly nonlinear magnetization characteristic. Excitation of the reactors by ac voltage U_1 does not induce a voltage in the pickup coil since the magnetic fluxes in each reactor compensate each other. The construction and shielding of the element are such that when the reactors placed parallel to the earth's surface are exposed to a magnetic field of intensity H_e (Fig. 4.2b), one of the components of the field H_{e1} establishes a steady flux in the cores. This flux, combined with the fluxes caused by voltage U_1, creates a voltage in the pickup coil that is characterized by the presence of a number of even harmonics. The second-harmonic component is the most powerful one. The amplitudes of the harmonics are proportional to the permanent field intensity. Only one factor can originate the voltages of even harmonics: that is the permanent magnetic field. In other words, in the absence of this field, no signal containing even harmonics is generated in the pickup coil. When rotated, the system can search for the direction to the earth's poles by measuring the maximums in harmonics and be well protected at the same time against various adverse distortions.

As shown in Fig. 4.2, the system includes, besides the reactors, an amplifier, a filter passing voltages of even (usually second) harmonics, and a meter that indicates the deviation of the reactors' axis from the north-south direction.

The analysis [39] illustrating the regularities in the arising of signals carrying even harmonics is performed with the introduction of the *B-H* curve by the hyperbolic-sine function:

$$H = \alpha \sinh \beta B \tag{4.17}$$

where H = field intensity, A-turn/m, and
$\quad\ B$ = flux density, T.

α and β are factors that can be found from the *B-H* curve (Fig. 4.2c) by choosing two points on the curve.

The total induction B in the reactor has two components. One of them is originated by the earth's field, B_0, the other one by the sinusoidal voltage excitation, $B_m \sin \omega t$ with peak value B_m:

$$B = B_0 + B_m \sin \omega t. \tag{4.18}$$

The field intensity H corresponding to B is calculated by substituting B into formula 4.17:

$$H = \alpha \sinh \beta(B_0 + B_m \sin \omega t) = \alpha \sinh \beta B_0 \cdot \cosh (\beta B_m \cdot \sin \omega t)$$
$$+ \alpha \cosh \beta B_0 \cdot \sinh (\beta B_m \cdot \sin \omega t). \tag{4.19}$$

Using the standard relationships for hyperbolic functions and introducing the Fouries series through the Bessel functions, the expression for H is found as follows:

$$H = \alpha \cdot \sinh \beta B_0 [J_0(j\beta B_m) + 2\sum_{1}^{\infty} J_{2n}(j\beta B_m) \cos 2n\omega t]$$

$$- \alpha \cdot \cosh \beta B_0 \cdot 2j\sum_{0}^{\infty} J_{2n+1}(j\beta B_m) \cdot \sin (2n + 1)\omega t. \tag{4.20}$$

The steady component and first five harmonics of the last expression for one of the reactors (for example, A) are

$$H_0 = \alpha \cdot \sinh \beta B_0 \cdot J_0(j\beta B_m),$$

$$H_1 = -\alpha \cdot \cosh \beta B_0 \cdot 2jJ_1(j\beta B_m) \cdot \sin \omega t,$$

$$H_2 = \alpha \cdot \sinh \beta B_0 \cdot 2J_2(j\beta B_m) \cdot \cos 2\omega t,$$

$$H_3 = -\alpha \cdot \cosh \beta B_0 \cdot 2jJ_3(j\beta B_m) \cdot \sin 3\omega t,$$

$$H_4 = \alpha \cdot \sinh \beta B_0 \cdot 2J_4(j\beta B_m) \cdot \cos 4\omega t,$$

$$H_5 = -\alpha \cosh \beta B_0 \cdot 2jJ_5(j\beta B_m) \cdot \sin 5\omega t. \tag{4.21}$$

For reactor B, H_0 to H_5 have the same magnitude, but the terms with odd indexes have opposite signs because of the reverse polarity of the coils. When $-B_m$ is substituted into formulas for H (replacing $+B_m$), the terms with odd indexes change signs. This change comes from the formula for the Bessel function of the first kind of order n, where n is a positive integer:

$$J_n(x) = \frac{x^n}{2^n n!} \left(1 - \frac{x^2}{2^2(n+1)} + \frac{x^4}{2^4 \cdot 2!(n+1)(n+2)}\right.$$

$$\left. - \frac{x^6}{2^6 \cdot 3!(n+1)(n+2)(n+3)} + \ldots\right). \tag{4.22}$$

The total field interacting with the pickup coil H_Σ is the sum of the fields from the two reactors. The terms from formula 4.21 with odd indexes are canceled, and the final expression for H_Σ is

$$H_\Sigma = 2\alpha \cdot \sinh \beta B_0$$

$$\cdot [J_0(j\beta B_m) + 2J_2(j\beta B_m) \cdot \cos 2\omega t + 2J_4(j\beta B_m) \cos 4\omega t + \ldots] \tag{4.23}$$

It is obvious from equations 4.21 and 4.23 that as the permanent field is absent ($B_0 = 0$), no even harmonics are present in the field, and, therefore, no even-harmonic voltage is generated in the pickup coil. The measuring system usually extracts the second-harmonic signal since it is most powerful. With the increase in the order of Bessel functions, the magnitudes of the functions are decreased. It is seen from equation 4.23 that the amplitudes of the harmonics are directly proportional to term $J_n(j\beta B_m)$.

Transducers based on this concept have a sensitivity to a magnetic field of better than 1.0×10^{-3}A/m, whereas the earth's field intensity is between 24 and 63A/m. Several figures can characterize the performance of this device. For reactors having a length of 30 to 60mm and excited by a voltage of frequency 1 to 3kHz, the output voltage of second harmonics is between 30 and 50μV per 1×10^{-3}A/m.

As is clear from the analysis that the variations in the excitation voltage, characteristics of materials, temperature, and so on, cannot create a second-harmonic signal. However, for a stable operation of the system, the feeding voltage must be free of second-harmonic components. The application of a band-elimination filter is desirable.

ELECTRODYNAMIC ELEMENTS

The operation of electrodynamic elements is based on the principles originally formulated by Faraday: voltage e [V] developed in a conductor of length l [m] moving at a right angle through a magnetic field of intensity B [T] with velocity V [m/s] is equal to the product:

$$e = BlV. \tag{4.24}$$

The potential difference that arises in a coil of N turns experiencing a time variation of flux ϕ [Wb] is proportional to the speed of the flux change and to the number of turns:

$$e = -N\frac{d\phi}{dt}. \tag{4.25}$$

Moving-coil Elements

We will consider three representative examples utilizing these principles and begin with a description of a device (Fig. 4.3) that is widely used for the conversion of vibratory motion into an electrical signal. In this device, a *coil* of N turns is placed *in* the *circular gap of* a permanent *magnet*. The coil can be driven along the x-axis by an alternating force developed due to the acoustic pressure on a diaphragm attached to the coil (see Fig. 4.3a). It can also be moved by the seismic mass of an accelerometer, vibrometer, and so on.

The active length l [m] of the coil is

$$l = \pi d N \tag{4.26}$$

where d = diameter of coil, m, and
      N = number of turns.

Therefore, the emf induced in the coil is

$$e = \pi B d N \frac{dx}{dt}. \tag{4.27}$$

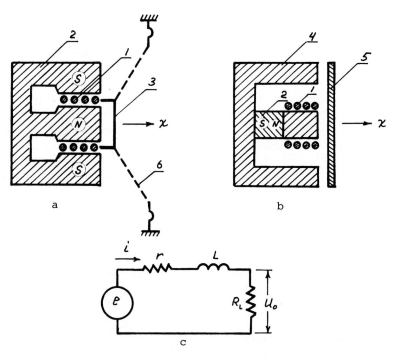

Figure 4.3 Electromagnetic systems. a = moving-coil system, b = moving-armature system, c = electrical equivalent circuit; 1 = coil, 2 = magnet, 3 = driver, 4 = magnetic system, 5 = moving armature, 6 = conical horn, e = emf developed in coil, i = current, r = coil's resistance, L = coil's inductance, R_L = resistance of load, U_0 = output voltage.

Variable-reluctance Elements

Another simple system (Fig. 4.3b) uses modulation of a magnetic flux by a moving armature that changes the reluctance R [1/H] of the magnetic circuit containing a source of magnetomotive force M [A-t]. The variation of the flux causes emf e [V] in a coil that is linked with the flux.

Since $M = \phi R$,

$$e = -N\frac{d\phi}{dt} = -N\frac{d}{dt}\left(\frac{M}{R}\right) = \frac{NM}{R^2} \cdot \frac{dR}{dt}. \qquad (4.28)$$

Assuming that R is a linear function of displacement x [m], and $\Delta R \ll R_0$ (where $R_0 = R$ when displacement x is zero), the emf is reduced as follows:

$$R = R_0(1 + kx), \qquad (k = \text{const}), \qquad dR = R_0 k\,dx \qquad \text{and}$$

$$e = NM\frac{k}{R_0} \cdot \frac{dx}{dt}. \qquad (4.29)$$

The equivalent electrical circuit of the element (Fig. 4.3c) includes resistance r [Ω], inductance of coil L [H], and resistance of load R_L [Ω].

The equations for voltages in the equivalent circuit are

$$e = ir + L\frac{di}{dt} + iR_L \qquad \text{and}$$

$$U_0 = iR_L. \qquad (4.30)$$

The transfer characteristic G (D) for the electrical circuit can be obtained if we substitute e taken from equations 4.27 or 4.29 into equation 4.30. Using the operator notation (D) gives

$$G(D) = \frac{U_0(D)}{x(D)} = S_0\frac{\tau D}{\tau D + 1} \qquad (4.31)$$

where $S_0 = B\pi dNR_L/L$ (for e from equation 4.27) is constant, and $\tau = L/(r + R_L)$ is time constant.

In order to construct the total transfer function of a transducer utilizing this element, the equation of motion for the mechanical system must be written, and the displacement of the moving parts should be introduced as a function of a measurand. For example, the dynamics of the microphone in Scheme 103 is described by the following relation:

$$\frac{x(D)}{P_a(D)} = \frac{A}{mD^2 + bD + k} \qquad (4.32)$$

where $P_a(D)$ = acoustic pressure, N/m^2;
$\quad A$ = effective area of diaphragm, m^2;
$\quad m$ = effective mass of moving system, kg;
$\quad b$ = friction coefficient, N $\cdot$ s/m; and
$\quad k$ = spring factor, N/m.

The combination of equations 4.31 and 4.32 gives the transfer function for the entire system.

Electromagnetic Flowmeters

Among the electrodynamic elements, the electromagnetic flowmeter is one of the most popular instruments (Scheme 115). As electroconductive liquid moves relative to the magnetic field in a pipe (Fig. 4.4), emf is induced in the column of particles between two electrodes at opposite points of the pipe diameter perpendicular to the direction of the field. The basic equation 4.24, giving the voltage induced in a conductor, is valid for the calculation of the signal proportional to the average velocity of the liquid. In this case, the distance between the electrodes (for the round pipe, it is the inner diameter) should be taken as the length of the conductor. For a stationary flow, fluid density, viscosity, and state of turbulence or Reynolds number do not affect the output signal. A volumetric flow rate measurement is possible without obstruction to the flow. In order to eliminate the polarization of electrodes, an alternating magnetic field is used in the instrument. Most liquids used in industry (Table 4.1) have a wide range of electrical conductivities. The formulas given below [40] make it possible to calculate the potential difference at the electrodes as a function of the volumetric flow rate, considering also the influence of the conductivity and dielectric constant on forming the voltage.

$$E = 2ZBQ/\pi R \tag{4.33}$$

$$Z = a + jb = Fe^{j\gamma} \tag{4.34}$$

$$a = \frac{1 + (\omega\epsilon/\sigma_1)^2\epsilon_r(\epsilon_r - 1)}{1 + (\omega\epsilon\epsilon_r/\sigma_1)^2}, \tag{4.35}$$

$$b = \frac{\omega\epsilon/\sigma_1}{1 + (\omega\epsilon\epsilon_r/\sigma_1)^2}, \tag{4.36}$$

$$F = \frac{\{[1 + (\omega\epsilon/\sigma_1)^2\epsilon_r(\epsilon_r - 1)]^2 + (\omega\epsilon/\sigma_1)^2\}^{1/2}}{1 + (\omega\epsilon_r\epsilon/\sigma_1)^2}, \tag{4.37}$$

$$\gamma = \arctan\frac{\omega\epsilon/\sigma_1}{1 + (\omega\epsilon/\sigma_1)^2\epsilon_r(\epsilon_r - 1)}, \tag{4.38}$$

$$Q = 2\pi\int_0^R \rho w d\rho \tag{4.39}$$

where E = voltage across electrodes, V;
$\quad\ \ B$ = flux density, T;
$\quad\ \ Q$ = volumetric flow rate, m³/s;

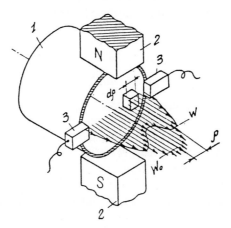

Figure 4.4 Schematic diagram of electromagnetic flowmeter, $d\rho$ = elementary radius, ρ = radius, w = velocity of fluid at radius ρ, w_0 = velocity of fluid at axis of pipe, 1 = pipe, 2 = magnet pole, 3 = electrode.

TABLE 4.1

Electrical Conductivity and Dielectric Constant for Various Fluids (Adopted from [40])

Fluid	Conductivity, σ_1, S/m	Dielectric constant, ε_r, dimensionless
Sulfuric acid		
30%-solution in water	74	82
Sulfuric acid, pure	1.0	84
Milk	0.45	80
Formic acid	5.6×10^{-3}	62
Tap water	5.0×10^{-3}	81
Methyl alcohol	7.2×10^{-4}	31
Distilled water	2.0×10^{-4}	81
Isopropyl alcohol	3.5×10^{-4}	26
Acetic acid	5.0×10^{-5}	7.1
Glycol	3.0×10^{-5}	47
Ammonia liquid	1.3×10^{-5}	16
Acetone	2.0×10^{-6}	30
Aniline	9.0×10^{-7}	7.2
Benzoic acid	3.0×10^{-7}	—
Olive oil	2.0×10^{-13}	3.1
Petroleum	5.0×10^{-15}	2.1

R = interior radius of pipe, m;

ϵ = permittivity of free space (8.85415×10^{-12}), F/m;

ω = angular frequency of the magnetic field, rad/s;

ϵ_r = dielectric constant of the fluid, dimensionless;

σ_1 = conductivity of the fluid, S/m;

ρ = distance (radius) from the center of the pipe to the elementary volume having velocity w, m; and

w = velocity of the elementary volume, m/s.

Analysis shows that even a very low conductivity of liquids (down to 1×10^{-8} S/m) does not attenuate appreciably the emf across the electrodes. Shunting effects in the media adjacent to the electrodes are not significant for the liquids having moderate and high conductivity.

The electromagnetic flowmeter is attractive for measuring the instantaneous flow rate of an alternating flow. However, because of the variation in the diagram of the local velocities (Chapter 5, Sensing and Transduction Elements of Flowmeters), the voltage picked up from the electrodes will not reflect exactly the volumetric flow rate. The output signal will attenuate, and the dynamic error will become significant with the increase in the frequency of the flow alternation.

For the analysis [41] of this error, we will consider the pulsation of liquid caused by a sinusoidal pressure $A(\omega) = Ae^{j\omega t}$ with an angular frequency ω [rad/s]. We assume that the flow is laminar with the distribution of the local velocities given in equation 5.135. In addition to this, the effects associated with the liquid's conductivity are neglected and the magnetic field is created by a permanent magnet. In this case, formula 4.24 can be modified and used to define a differential emf E [V] for an incremental conductor of length $d\rho$ (Fig. 4.4) crossing the magnetic field at distance ρ [m] from the center of the pipe (for the notations, see Chapter 5, Sensing and Transduction Elements of Flowmeters):

$$dE = aBwd\rho \tag{4.40}$$

where a is the constant depending on the construction.

The voltage across the electrodes is the integral:

$$E = 2aB\int_0^R wd\rho = 2aB\int_0^1 wdu, \qquad (u = \rho/R) \qquad (4.41)$$

Substituting w from formula 5.135 into this formula, E is expressed as

$$E = \frac{2aBAe^{j\omega t}}{j\phi\omega}\left[1 - \frac{\int_0^1 J_0(\alpha u j^{3/2})du}{J_0(\alpha j^{3/2})}\right]. \qquad (4.42)$$

Integral $\int_0^1 J_0(\alpha u j^{3/2})du$ is calculated as in formula 5.153. Formula 4.42 is modified by the substitution $\omega\phi = (\alpha^2/R^2)\mu$ ($\alpha = R\sqrt{\omega/\nu}$, $\nu = \mu/\phi$) and by splitting it into two parts:

$$E = K \cdot L. \qquad (4.43)$$

Part K is assumed to be a term that does not depend on α:

$$K = -\frac{2aBAR^2je^{j\omega t}}{\mu}. \qquad (4.44)$$

Part L is a function of α:

$$L = \frac{1}{\alpha^2}\left[1 - \frac{\int_0^1 (\alpha u j^{3/2})du}{J_0(\alpha j^{3/2})}\right]. \qquad (4.45)$$

By substituting the integral from equation 5.153, we will obtain

$$L = \frac{1}{\alpha^2}\left\{H_1(\alpha j^{3/2}) - \frac{\pi}{2}\left[\frac{J_1(\alpha j^{3/2}) \cdot H_0(\alpha j^{3/2})}{J_0(\alpha j^{3/2})}\right]\right\}. \qquad (4.46)$$

The following table reveals numbers for the modulus of L in the range for α from 1 to 10. Values of L_n are normalized and represent the ratio $L_n = L/L_1$ where L_1 is the value of L for $\alpha = 1$ (see p. 255 for calculation of H_0 and H_1).

α	1	2	3	4	5	6	8	10
L	0.1638	0.1368	0.0895	0.0556	0.0363	0.0253	0.0144	0.0103
L_n	1.0000	0.8351	0.5462	0.3394	0.2219	0.1544	0.0880	0.0630

The graph in Fig. 4.5 shows L_n as a function of α. The magnitude of α depends on the frequency of the signal. By taking R, μ, ϕ, B, and A as constants, function $E = f(\omega)$ can be readily computed.

The ability of the flowmeter to emphasize the low-frequency signals can be put to a useful application.

Many liquids whose flow is measured have low conductivity. The internal resistance of the sensor can be high; this means that the signal picked off from the electrodes must be fed into an amplifier having high input resistance.

One of the problems that arises in sensor design is that of generating an unwanted voltage of power source frequency in the signal-carrying wires. The wires connected to the electrodes must pass through the main ac flux in the gap of the electromagnet. The voltage induced in the wires is in quadrature with the signal

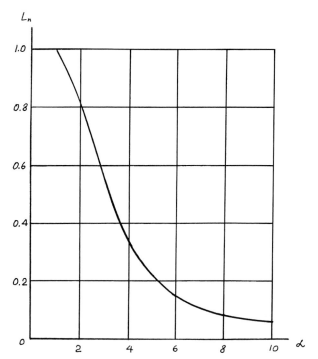

Figure 4.5 Attenuation of signals from electrodes of electromagnetic flowmeter as function of variable α.

voltage. There are several methods of cancelling this noise voltage. The most effective method consists of arranging the wires in the loops to have a geometry such that the voltages generated in different parts of the loops are subtracted.

Nonelectrode Electromagnetic Flowmeters

In a nonelectrode electromagnetic flowmeter (Fig. 4.6), the voltage proportional to the flow rate emerges in signal coils as a result of the interaction of the magnetic flux ϕ created by the excitation coil and the moving fluid. The flux has a radial component ϕ_r directed toward and from the center of the pipe. The layers of electroconductive liquid moving along the pipe cross the lines of flux ϕ_r and create the currents circulating in the concentric paths. These currents are proportional to the

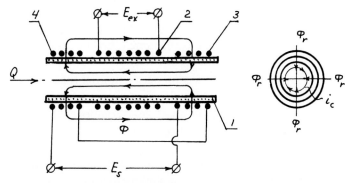

Figure 4.6 Schematic diagram of nonelectrode electromagnetic flowmeter. E_{ex} = excitation voltage, E_s = voltage across signal coils, i_c = circulating current, Q = flow of liquid, Φ = magnetic flux, Φ_r = radial component of magnetic flux, 1 = pipe, 2 = excitation coil, 3 and 4 = signal coils.

flow rate, and they induce proportional electromotive forces in two signal coils that have a series-opposing connection. Due to this connection, the voltages in the coils induced by the main excitation flux are canceled if the structure has a good symmetry.

The electromagnetic flowmeter technique has some application for measurements on uncut blood vessels. Since the walls of the vessels are electroconductive, the electrical signals can be picked up from the electrodes attached to the outside wall of the vessel placed in the magnetic field radiated by a nearby located electromagnet.

If the coil of the element in Figure 4.3a is excited by current I, the coil will experience a pull along the x-axis due to the Lorentz force. The general expression for the force is

$$F = BlI = \pi dNBI \qquad (4.47)$$

where l = length of conductor, m;
$\quad F$ = force, N;
$\quad B$ = flux density, T; and
$\quad I$ = current, A.

The principle of electrodynamic generation of force is quite popular in many devices, but primarily in loudspeakers. The basic construction of the loudspeaker is a modification of the construction shown in Figure 4.3a, where part 3 is provided with conical horn 6, which, being driven by the coil, radiates acoustic waves.

It is common to employ this system as a source of a restoring force in force-balance transducers.

ELEMENTS OF SERVO TRANSDUCERS

The majority of servo transducers are force-balance devices. A block diagram of a transducer (Fig. 4.7) includes several typical components.

1. A sensing element that converts the measurand into force, for example, the spring-suspended seismic mass in an accelerometer, the diaphragm in a pressure meter, the float in a flowmeter, and so on.

2. An error detector that compares the inputs provided by the sensing element and by the element developing a restoring force.

3. An electrical transduction element that converts the difference between the two forces into an analog electrical signal proportional to this difference. It can be an inductive, capacitive, photoelectric or other-type element providing the same function. Usually this element is physically combined with the error detector.

4. A feedforward amplifier that conditions the signal from the error detector and makes it convenient for further treatment.

5. A feedback-loop component that is fed from the output of the feedforward amplifier and controls the force-restoring element.

6. The force-restoring element is an electromechanical device that has an electrical input and force generated at the output. Electromagnetic, electrostatic, and piezoelectric principles are primarily employed in the restoring element.

7. A stabilizing ("equalizing") network, which can be cascaded with or inserted into the forward or feedback loops in order to stabilize the performance of the control loop at a high level of gain.

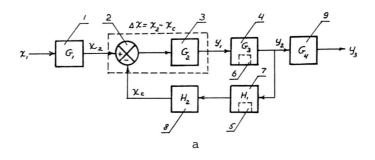

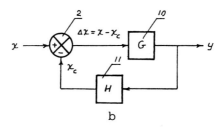

Figure 4.7 Block diagram of force-balance transducer. a = detailed diagram, b = simplified diagram; x and x_1 = measurands, x_2 = measurand modified into force, x_c = compensating force, Δx = error signal (difference between forces), y_1 = electrical signal proportional to difference, y_2 = amplified y_1, y, and y_3 = output signals, G and G_1 to G_4 = transfer functions of cascaded elements, H = transfer function of feedback network (including force generator), H_1 = transfer functions of feedback component, H_2 = transfer function of force-restoring element, 1 = sensing element, 2 = error detector, 3 = transduction element, 4 and 10 = feedforward amplifiers, 5 and 6 = force generator and stabilizing network, 7 = feedback-loop component, 8 = force-restoring element, 9 = output amplifier, 11 = feedback network.

8. An output amplifier that forms the output signal convenient for interfacing with the components that follow in the signal-processing system.

The operation of the transducer is such that the forces caused by the measurands are continuously balanced against those produced by the electrical controlling system. A signal of disbalance ("error") is fed to the amplifier, which in turn feeds back a restoring-force element.

The most important characteristics of this system can be found if we consider a simplified diagram of the sensor (Fig. 4.7b), assuming that the conversions are linear with sensitivities G and H for cells 10 and 11, respectively. For these two cells,

$$y = G\Delta x \quad \text{and} \quad x_c = Hy. \tag{4.48}$$

In the closed loop,

$$x_c = GH\Delta x = GH(x - x_c) \quad \text{or}$$

$$\frac{x}{x_c} = 1 + \frac{1}{GH}. \tag{4.49}$$

Since $GH \gg 1$,

$$x \approx x_c \quad \text{and} \quad y \approx \frac{1}{H}x.$$

It means that, besides x, the output signal is defined mainly by H, which can be very stable and accurate. The degree of accuracy of conversion can be found if we analyze the relative change $\Delta y/y$ of y as a function of the relative changes $\Delta G/G$ and $\Delta H/H$ when $x = $ const. From the last equation,

$$y = \frac{G}{1 + GH} x. \tag{4.50}$$

By taking a full differential of ln y, the value of $\Delta y/y$ is calculated as follows:

$$\frac{\Delta y}{y} = \frac{\Delta G}{G} \cdot \frac{1}{(1 + GH)} - \frac{\Delta H}{H} \cdot \frac{1}{(1 + 1/GH)}. \tag{4.51}$$

If two error components in this formula are random by nature, the mean-square notation can be used for the calculation of the overall error σ due to the unpredictable variations σ_G and σ_H of G and H, respectively:

$$\sigma_1 = \sigma_G \frac{1}{(1 + GH)}, \qquad \sigma_2 = \sigma_H \frac{1}{(1 + 1/GH)} \tag{4.52}$$

$$\sigma = \sqrt{\sigma_1^2 + \sigma_2^2} = \frac{1}{1 + GH} \sqrt{\sigma_G^2 + G^2 H^2 \sigma_H^2} \tag{4.53}$$

It is obvious from this calculation that the increase in gains G and H gives the decrease in errors. However, the high level of G and H leads to the instability of the system, which tends to oscillate. Damping of oscillation is provided by modifying the dynamic characteristics of the system. The insertion of differentiating or integrating RC cells (Table 4.2) in the circuits of amplifiers combined with a negative feedback loop allows manipulation of the time constants and phase shifts between the signals in the system. This provides the necessary degree of damping, and can be illustrated by the following example. Let an integrating network be included into the forward amplifier in Fig. 4.7b having gain K. Its transfer function $G_s(D)$ will be

$$G_s(D) = \frac{\dfrac{K}{1 + DT}}{1 - \dfrac{HKD}{1 + DT}} = \frac{K}{1 + D(T - KH)} = \frac{K}{1 + DT_{eq}} \tag{4.54}$$

where T = time constant of the open-loop system, s, and
      T_{eq} = equivalent time constant, s.

$$T_{eq} = T - KH < T \tag{4.55}$$

Mechanical or electromechanical dampening is also used in the systems' designs.

The characteristics necessary for the analysis of force-balance transducer elements are described by constant coefficients and by differential equations reflecting physical processes. The typical transfer functions G's follow.

$$1. \quad G = K \tag{4.56}$$

This is an equation describing a zeroth-order system. K is time-independent steady-state gain (static gain or simply gain).

$$2. \quad G(D) = \frac{K}{TD + 1} \tag{4.57}$$

TABLE 4.2

Stabilizing Networks for Servo Transducers

Differentiating network		Integrating network	
Circuit	Transfer function	Circuit	Transfer function
	$\dfrac{TD}{TD + 1}$ $T = RC$		$\dfrac{1}{TD + 1}$ $T = RC$
	$\dfrac{T_2 D}{T_1 D + 1}$ $T_1 = (R_1 + R_2)C$ $T_2 = R_2 C$		$\dfrac{T_2 D + 1}{T_1 D + 1}$ $T_1 = (R_1 + R_2)C$ $T_2 = R_2 C$
	$\dfrac{1}{1 + R_1 R_2}$ $\times \dfrac{T_1 D + 1}{T_2 D + 1}$ $T_1 = R_1 C$ $T_2 = \dfrac{R_2 T_1}{R_1 + R_2}$		$\dfrac{1}{1 + R_1 / R_3} \cdot \dfrac{T_2 D + 1}{T_1 D + 1}$ $T_1 = \left(R_2 + \dfrac{R_3 R_1}{R_3 + R_1}\right)C$ $T_2 = R_2 C$
	$-\dfrac{T_2 D}{T_1 D + 1}$ $T_1 = R_1 C$ $T_2 = R_2 C$		$-\dfrac{R_2}{R_1} \cdot \dfrac{1}{TD + 1}$ $T = R_2 C$

D = operator, T = time constant, U_1 and U_2 = input and output voltages, respectively.

This is the model of a first-order system, in other words, the model of the element that can be described by the first-order differential equation.

$$3. \quad G(D) = \frac{K}{T^2 D^2 + 2\zeta TD + 1} \tag{4.58}$$

where ζ = damping factor.

This equation describes a second-order system. The order of the system is defined by the highest power of T in the denominator polynomial.

There are analogies between the elements in different energy domains (electrical, mechanical, thermal, and so on). When a set of transducer components acts in series or in a loop, the output signal of a preceding component must have the nature of the input of the following component.

The analysis of a servo sensor is usually concentrated around the evaluation of its stability when it operates as a small control system. The transfer characteristics of the elements should be known. After the block-diagram of the device is drawn, the analysis can be developed in the following sequence.

1. The transfer function of the open loop is written using operator notation. As an example, for the system in Figure 4.7a, this function will be

$$G(D) = G_2(D) \cdot G_3(D) \cdot H_1(D) \cdot H_2(D). \tag{4.59}$$

2. If the time constant of one of the elements is negligibly small, as compared with the others, it can be introduced by a zeroth-order system, and the transfer function is simplified (for example, $G_3(D) = G_3$).

3. One of the criteria of stability is chosen (Bodo, Nyquist, or Routh-Hurwitz) [42, 43].

4. The Nyquist criterion is appropriate for this analysis. It states that if the open-loop Nyquist plot of a feedback system encircles the point $(-1,0)$ as the frequency ω takes any values from $-\infty$ to $+\infty$, the closed-loop response is unstable. In other words, if the point $(-1,0)$ is not encircled, the response is stable.

5. In order to construct the Nyquist plot, the operator D in the transfer function is replaced by $j\omega$, where $j = \sqrt{-1}$ and ω is angular frequency. In the obtained expression for the transfer function $G(j\omega)$, the real and imaginary parts are separated. For example, if the transfer function is given as in formula 4.58, it turns into the following expression:

$$G(j\omega) = \frac{K(1 - T^2\omega^2)}{(1 - T^2\omega^2)^2 + (2\zeta T\omega)^2} - j\frac{K2\zeta T\omega}{(1 - T^2\omega^2)^2 + (2\zeta T\omega)^2}. \quad (4.60)$$

A Nyquist plot is one of the ways of representing the frequency response characteristics of a dynamic system. It uses the $Im[G(j\omega)]$ as ordinate and $Re[G(j\omega)]$ as abscissa. Figure 4.8 illustrates the forms of the plots for two types of transfer functions:

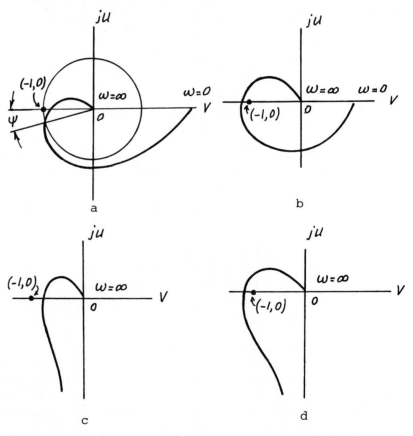

Figure 4.8 Nyquist plots for two types of open-loop transfer functions. a and b = plots for stable and unstable systems, respectively, described by formula 4.61; c and d = plots for stable and unstable systems, respectively, described by formula 4.62; U and V = imaginary and real axis; ψ = phase margin; ω = angular frequency.

Figure 4.8a and b for

$$G(D) = \frac{K}{(T_1D + 1)(T_2D + 1)(T_3D + 1)}. \qquad (4.61)$$

Figure 4.8c and d for

$$G(D) = \frac{K}{D(T_1D + 1)(T_2D + 1)}. \qquad (4.62)$$

6. After the diagram is plotted, the unity circle is drawn (Fig. 4.8a). The angle between the horizontal axis and the line connecting the origin with intersection of the plot and circle is called the phase margin. If this angle is within 30 to 60°, the system operates normally. If it is less than 30°, the stability of the system is not sufficient. If it is larger than 60°, the system's response is considered to be slow.

7. Depending on the determined response, a decision should be made as to whether the stabilizing network must be inserted into the circuit of the transducer.

The dynamic characteristics of different elements that can be used as components of a servo transducer are introduced in the relevant topics. These characteristics should be included into the analysis of the transducer scheme.

A block diagram shown in Figure 4.9 is drawn to illustrate how practical relationships for the force-balance pressure transducer (Scheme 97) can be combined for the analysis of its performance.

In this transducer, pressure P applied to a diaphragm with effective area A is converted into force F, which is compared with the force F_c developed by the force-generating element. The differential force ΔF is applied to the linkage element, causing its motion. The response of this element x is described by a second-order differential equation in which M, B, and K_s stand for the effective mass, friction coefficient, and spring factor, respectively, considering all moving components of the transducer. Motion x is converted into voltage U by the transduction element feeding the current amplifier. This amplifier develops current i, having the level necessary for the excitation of the forcing coil in the electrodynamic element and for feeding the output network, which is introduced in Scheme 97 by a resistor.

As follows from the concept of the force-balance system, its forces do not return the moving parts exactly to null (neutral position), but bring them close to it.

The following set of equations gives the transfer functions and characteristics of transducer elements. Note that only one of them is introduced as a time-dependent function, since it is assumed that the inertia of the mechanical parts constitutes the main factor defining the dynamic response of the system. This assumption is practical and often accepted.

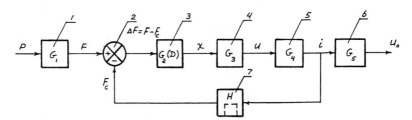

Figure 4.9 Block diagram of a force-balance pressure transducer.
1 = diaphragm, 2 = error detector, 3 = linkage element,
4 = displacement transduction element, 5 = amplifier, 6 = output
network, 7 = force-generating electrodynamic element.

$$G_1 = K_P = \frac{F}{P} = \frac{PA}{P} = A, \qquad \Delta F = F - F_c,$$

$$G_2(D) = \frac{x}{\Delta F} = \frac{1}{MD^2 + BD + K_s}, \qquad G_3 = K_u = \frac{U}{x}, \qquad G_4 = K_i = \frac{i}{U},$$

$$H = K_f = \frac{F_c}{i}, \qquad G_5 = K_0 = \frac{U_0}{i} = R. \qquad (4.63)$$

A combination of these relationships gives the transfer function of the transducer:

$$\frac{U_0}{P} = \frac{ARK_iK_u}{MD^2 + BD + (K_s + K_iK_uK_f)}. \qquad (4.64)$$

For the system without feedback control, the undamped natural frequency ω_n and damping ratio ζ are

$$\omega_n = \sqrt{\frac{K_s}{M}}, \qquad \zeta = \frac{B}{2\sqrt{MK_s}}. \qquad (4.65)$$

They are changed to ω_{n1} and ζ_1 due to the feedback network:

$$\omega_{n1} = \sqrt{\frac{K_s + K_iK_uK_f}{M}}, \qquad \zeta_1 = \frac{B}{2\sqrt{M(K_s + K_iK_uK_f)}}. \qquad (4.66)$$

The increase in ω_n is beneficial since it becomes possible to have a more uniform and stable response of the measuring system in a wider range of measurand time variations. However, ζ is decreased by the same factor, and special measures, like a stabilizing network, can be required to compensate for this decrease.

The major advantage in the use of this system is in a higher accuracy of performance because of the nature of the negative feedback loop: it stabilizes characteristics of the cells that are within the loop.

RESONANT ELEMENTS

A resonator is a core element of digital transducers, which have been intensively developed during the last two decades. The growth of digital systems, and the advent and wide application of minicomputers and microprocessors, have advanced the development of transducers with a digital readout.

A traditional analog transducer coupled with a digital signal processor includes a series of interfaces, such as acceleration to force, force to displacement, displacement to voltage, and voltage to binary date. Each interface contributes errors in conversion. Minimizing the interface effects is always beneficial, and transducers providing a direct measurand-to-output-frequency conversion are more accurate. It should also be noted that the simple- and precision-counting circuits working in conjunction with the frequency-output transducer are not susceptible to the degradation of the characteristics typical for analog circuits, such as changes of the feeding voltages, impedances, gains, and so on.

Frequency-modulating transducers can be classified into two groups. The first group consists of those types in which a single lumped electrical element, such as resistance or capacitance, responds to the measurand by the change of its value. This change is translated into the change of the frequency of an electronic oscillator in which the element is a part of a time-constant controlling circuit.

In this type of conversion, there is at least one intermediate step, and, in addition to this, the oscillation frequency depends on the value of one or more elements that can contribute instabilities.

The second group is comprised of transducers with a direct conversion of the measurand into frequency, and the following paragraphs will be devoted to the elements of these transducers. In spite of the differences in designs, the resonators to be described have one feature in common: the resonant frequency is defined by the inertial, resistive, or elastic properties of the resonating member. Shifting the frequency of the mechanical or acoustical resonance is the most commonly used principle in frequency-modulating sensors. This is why the mechanical and acoustical concepts of conversion are given much attention.

Vibrating Strings

One of the earliest designs of a vibrating-element transducer was a model with a stretched string. The principle of operation of this transducer is illustrated in Figure 4.10. The string is a part of a vibratory system including a pickup, driving element, and feedback-loop amplifier. With the change in tension, the frequency of vibration changes proportionally to the applied force. Electrodynamic or inductive sensing and driving elements are usually employed in the construction. In some designs, a strain-gage effect in the string under alternating deformation also has been used for picking up the signal of displacement.

The frequency of vibration as a function of applied axial force can be obtained from the integration of the one-dimensional wave equation [44], which we will derive having in mind the following assumptions. First, we assume that the string has no flexural stiffness, and it is free to vibrate transversely in the y-direction (Fig. 4.11). Second, the tensile force F in the string remains constant during the vibration in X-Y plane. At certain instants of time, the string is displaced, as shown in (a). An incremental element AB of length dx has, at points A and B, slopes α_1 and α_2, which are small (because of the small magnitude of y) and can be defined as follows:

$$\sin \alpha_1 = \tan \alpha_1 = \partial y/\partial x \quad \text{and} \quad \sin \alpha_2 = \tan \alpha_2 = \frac{\partial y}{\partial x} + \frac{\partial^2 y}{\partial x^2} \, dx. \quad (4.67)$$

The vertical components of force F at A and B are T_1 and T_2, respectively:

$$T_1 = -F \sin \alpha_1 = -F \frac{\partial y}{\partial x}, \quad T_2 = F \sin \alpha_2 = F\left(\frac{\partial y}{\partial x} + \frac{\partial^2 y}{\partial x^2} \, dx\right). \quad (4.68)$$

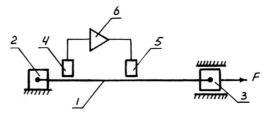

Figure 4.10 Stretched-string element.
F = force, 1 = string, 2 and 3 = fixed and mobile string supports, 4 and 5 = pickup and driving elements, 6 = amplifier.

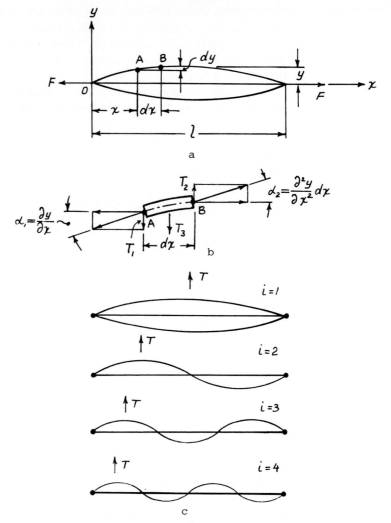

Figure 4.11 Transverse vibration of stretched string.
a = deflection of string, b = diagram of balance forces in
elementary length, c = modes of vibration; T = forces of
excitation at antinodes.

The inertia force T_3 due to the element's motion is

$$T_3 = -dm \cdot \frac{\partial^2 y}{\partial t^2} = -\frac{m}{l} dx \frac{\partial^2 y}{\partial t^2} \qquad (4.69)$$

where dm = mass of the elementary string length dx, kg;
  m = mass of string, kg; and
  l = string's length, m.

The equation of the equilibrium of forces can be written using D'Alembert's principle, which states that the resultant of the external forces and the kinetic reaction acting on a body equals zero:

$$T_1 + T_2 + T_3 = 0 \qquad \text{or} \qquad \frac{\partial^2 y}{\partial t^2} = F \frac{l}{m} \cdot \frac{\partial^2 y}{\partial x^2}. \qquad (4.70)$$

The solution of this equation is introduced in the following form:

$$y = (A_1 \sin \frac{i\pi}{l} x)(B_1 \sin \omega_i t + B_2 \cos \omega_i t) \qquad (4.71)$$

where A_1 = constant depending on the end condition, m;

    B_1 and B_2 = constants depending on the initial conditions, dimensionless; and

        ω_i = frequency of vibration, rad/s.

Expression 4.71 is reduced for these typical conditions: the ends of the string are fixed, the curve shape is described by the function

$$y = y_p \sin (i\pi/l)x \tag{4.72}$$

where y_p is the peak value of the displacement of the sine curve, m, and vibrations are started with the string in its displaced position ($\dot{y} = 0$ at $t = 0$ gives $B_1 = 0$ and $B_2 = 1$).

In this case,

$$y = y_p \left(\sin \frac{i\pi}{l} x \right) \cos \omega_i t. \tag{4.73}$$

The frequency-force relationship is obtained from equation 4.70 in the following form:

$$\omega_i = \pi \cdot i \sqrt{\frac{F}{ml}} \tag{4.74}$$

where i is the order of the mode of vibration ($i = 1, 2, 3, \ldots \infty$; see Fig. 4.11c). The last formula can be written in terms of frequency f_i [Hz], stress in the string σ [N/m^2], and the string's material density ρ [kg/m^3].

$$f_i = \frac{i}{2l} \sqrt{\frac{\sigma}{\rho}} \tag{4.75}$$

If $i = 1$, the vibration is at the fundamental mode, and the corresponding frequency is the first harmonic. The frequencies for the higher values of i are overtones or higher harmonics. Theoretically, there is an infinite number of values of i, and, hence, an infinite number of modes and natural frequencies can be excited. The string-end conditions and the location of excitation determine the mode shape. The excitation, that is driving element, should act at the antinodes, which are the points of maximum displacement. For example, if the string is excited at its mid-point, the odd harmonics (1, 3, 5, and so on) will be excited, but there will be no even harmonics, since they have nodes at the mid-point. The same concept of excitation can be applied to the other vibratory systems characterized by distributed mass and elasticity. If a short string is used in the vibrator and its stiffness should be taken into account, a more exact calculation can be performed with the following formula applicable for the first natural mode of vibration [45].

$$f = \frac{1}{2\pi l^2} \sqrt{\frac{\lambda_1 Er^2 + \lambda_2 \sigma l^2}{\rho}} \tag{4.76}$$

where λ_1 and λ_2 = constants, dimensionless;

        E = Young's modulus of wire material, N/m^2; and

        r = string's radius, m.

The values of λ_1 and λ_2 are taken from the following conditions: $\lambda_1 = 504$ and $\lambda_2 = 11.85$ when $(\sigma l^2/Er^2) \leq 106.5$; $\lambda_1 = 594.5$ and $\lambda_2 = 11$ when $106.5 \leq (\sigma l^2/Er^2) \leq 555.8$; $\lambda_1 = 928$ and $\lambda_2 = 10.4$ when $(\sigma l^2/Er^2) \geq 555.8$.

In the calculations related to string transducers, the initial tension F_0 [N] or initial stress σ_0 [N/m^2] should be a component of the force F or stress σ in the formulas, as shown below.

$$F = F_0 + F_x \quad \text{and} \quad \sigma = \sigma_0 + \sigma_x \tag{4.77}$$

where F_x and σ_x are a force and a stress generated by measurands.

Besides the traditional materials for strings (tungsten alloys, steel, and beryllium copper), metal-coated fused silica and silicon are also of interest. The elastic perfections of these materials are extremely important. Small values of the temperature coefficient of expansion and Young's modulus determine the accuracy of the instrument to a great extent.

The instruments employing the vibrating string as a transduction element are very accurate. For example, gravity acceleration can be measured with an accuracy of approximately 1ppm with such an instrument.

Vibrating Beams

In a vibrating beam element, a measurand (displacement, force, acceleration, pressure, and so on) develops a force against a solid beam whose frequency of oscillation is a function of the applied force [46]. The beam (Fig. 4.12) is made of high-stability age-hardened metal (for example, Ni-Span-C).

The vibrating beam part (Fig. 4.13) includes isolator springs and massive mounting pieces that help localize the vibratory energy within the resonating beam. In the construction, special attention is paid to the joints transmitting or forming the force. These joints are usually welded to eliminate instability, hysteresis, relaxation effects, and so on. A typical driving system includes a capacitive pickup, amplifier, and force-inducing electromagnet.

The natural frequencies of lateral vibration of the beam under an axial load are found by direct solution [47] of the appropriate partial differential equation. The beam is considered to be a distributed elastic system with x-y coordinates as shown in Figure 4.12.

The axial force F [N] can have two components:

$$F = F_0 + F_x \tag{4.78}$$

where F_0 and F_x = pre-load and measurand-induced forces, respectively, N.

The differential equation of the deflection curve under a static lateral load is [44]

$$EI \frac{d^2y}{dx^2} = M - Fy \tag{4.79}$$

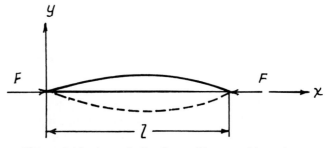

Figure 4.12   Lateral vibrations of beam subjected to axial load.

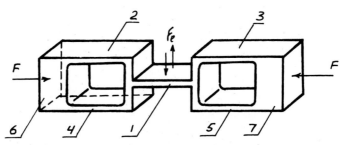

Figure 4.13 Vibrating-beam transduction element. F = force, F_e = alternating force developed by driving electromagnet. 1 = vibrating beam, 2 to 5 = isolator springs, 6 and 7 = massive mounting pieces.

where E = Young's modulus, N/m^2;

$\quad I$ = area moment of inertia of beam's cross section, m^4; and

$\quad M$ = bending moment at x produced by a lateral load, N · m.

The static, lateral load per unit length w [N/m] relates to moment M as

$$\frac{d^2M}{dx^2} = w. \tag{4.80}$$

Differentiating (equation 4.79) twice with respect to x gives

$$EI\frac{d^4y}{dx^4} + F\frac{d^2y}{dx^2} = w. \tag{4.81}$$

This equation is applicable to the vibrating beam. The inertia reaction forces can be substituted for w by using D'Alembert's principle:

$$w = -m\frac{\partial^2 y}{\partial t^2} \tag{4.82}$$

where m = mass per unit length of beam, kg/m.

Thus, equation 4.81 becomes a partial differential equation:

$$EI\frac{\partial^4 y}{\partial x^4} + F\frac{\partial^2 y}{\partial x^2} = -m\frac{\partial^2 y}{\partial t^2}. \tag{4.83}$$

The solution of this equation is found in the following form:

$$y = X(x)\cos \omega t \tag{4.84}$$

where $X(x)$ = peak value of vibration of the beam at the point located at the distance x from the left-hand end, m.

After substituting equation 4.84 into equation 4.83 and canceling cos ωt (it is common to both sides of the equation), we obtain

$$EI\frac{d^4X}{dx^4} + F\frac{d^2X}{dx^2} = m\omega^2X. \tag{4.85}$$

The characteristic equation of 4.85 is

$$EIs^4 + Fs^2 - m\omega^2 = 0 \tag{4.86}$$

Solving the equation for s^2 gives

$$s^2 = -\frac{F}{2EI} \pm \sqrt{\left(\frac{F}{2EI}\right)^2 + \frac{m\omega^2}{EI}}. \tag{4.87}$$

The square root in equation 4.87 is larger than $F/2EI$; this means that there are two real and two imaginary solutions for s:

$$s = \pm s_a \quad \text{and} \quad s = \pm js_b. \tag{4.88}$$

The solution to equation 4.85 is, therefore,

$$X(x) = A \cosh s_a x + B \sinh s_a x + C \sin s_b x + D \cos s_b x \tag{4.89}$$

where A, B, C, D, s_a, and s_b are determined by the end conditions. For the beam fixed at both ends,

$$X(0) = X(l) = 0$$
$$\dot{X}(0) = \dot{X}(l) = 0. \tag{4.90}$$

Now

$$\dot{X}(x) = -s_a A \sinh s_a x + s_a B \cosh s_a x + s_b C \cos s_b x - s_b D \sin s_b x. \tag{4.91}$$

The combination of equations 4.89 and 4.91 gives a set of four linear homogeneous algebraic equations for the conditions $x = 0$ and $x = l$, which can be solved for frequency ω:

$$\omega = \frac{1}{l^2} \sqrt{\frac{EI}{m}} \times \sqrt{H + K \frac{l^2}{EI} F} \tag{4.92}$$

where $H = 496.1$, $K = -12.45$; $I = bh^3/12$ for a rectangular-section beam ($b =$ width [m]; $h =$ thickness [m]); and $m = \rho bh$ ($\rho =$ density of beam's material, [kg/m³]).

In a pressure-measuring transducer,

$$F = A_{ef} \cdot P \tag{4.93}$$

where A_{ef} [m²] = effective area of diaphragm or bellows (see Chapter 5, Elastic
           Elements) connected to the beam by the mechanical linkage that
           converts applied pressure to the axial load, and
       P = applied pressure [Pa].

Here and in the following two cases, we assume that there is no preload ($F_0 = 0$).
    For an accelerometer,

$$F = M_s a \tag{4.94}$$

where M_s = seismic mass, kg, and
       a = acceleration, m/s².

If a displacement is to be measured,

$$F = \frac{bhE}{l} \Delta l \tag{4.95}$$

where Δl[m] = displacement reflected to the beam by (or without) mechanical linkage that converts the displacement to an axial load.

E, b, h, and l in formula 4.92 are temperature-dependent terms. In spite of the use of temperature-stable materials for the resonator, special measures to compensate for temperature drifts are necessary for obtaining highly accurate characteristics. A microprocessor is a relevant tool for this purpose.

For a convenient treatment of signals in a microprocessor, the transfer characteristic of the transducer is introduced by polynomials in which the output, period τ, is a function of pressure P and temperature T:

$$P = A\tau^5 + B\tau^4 + C\tau^3 + D\tau^2 + E\tau + F,$$
$$\tau = a_0 T^4 + b_0 T^3 + c_0 T^2 + d_0 T + e_0,$$
$$C = a_3 T^3 + b_3 T^2 + c_3 T + d_3,$$
$$D = a_2 T^3 + b_2 T^2 + c_2 T + d_2,$$
$$E = a_1 T^3 + b_1 T^2 + c_1 T + d_1. \tag{4.96}$$

where among the letters denoting constants, C, D, and E depend on temperature.

Vibrating Capsules

A vibrating-capsule absolute-pressure transducer is one of the most successful designs [47, 48] among the highly accurate and durable pressure-measuring devices. The following description is related to the design in Figure 4.14 which is composed of an aneroid capsule, two magnetic systems at each side of the capsule, two coils placed in the gaps of the systems, and a feedback amplifier. The capsule is set into oscillation by the excitation network composed of the two coils and the amplifier. The pickup coil produces an electromotive force proportional to the velocity of the

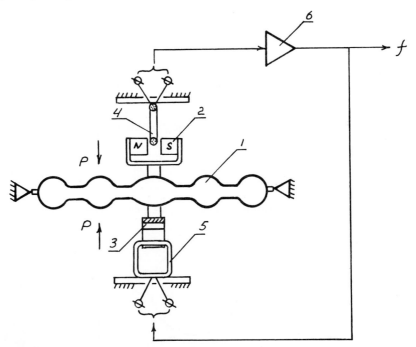

Figure 4.14 Schematic diagram of resonant capsule transducer. f = output frequency, P = applied pressure, 1 = aneroid capsule, 2 and 3 = magnetic systems, 4 and 5 = pickup and driving coil, respectively, 6 = feedback-loop amplifier.

capsule's oscillatory motion, and the driving coil develops the force necessary to maintain the vibrations. The capsule has a pronounced nonlinear pressure-displacement characteristic (see Chapter 5, Elastic Elements). With the change in pressure, the stiffness of the capsule is changed along with the natural frequency of vibration, which is a measure of the applied pressure.

An analysis of the capsule performance is given in [47 and 48]. We will introduce several fragments from these references in order to describe some major features of the vibrating capsule.

The approximate relationship between the applied pressure and the diaphragm's resonant frequency can be determined if the diaphragm is considered as a taut membrane (Fig. 4.15) with no initial stress. It is assumed that the membrane is simply supported around the circumference. As is clear from the principle of transduction, the nonlinear deformation of the diaphragm defines the travel of frequency. According to the large-deflection membrane theory, this nonlinearity is associated with "doming," in other words, with forming a spherical surface under pressure. The length L [m] of the elongated diaphragm diameter measured along the arc of the deflected diaphragm is given by

$$L = R\theta \tag{4.97}$$

where R = spherical radius, m, and
θ = arc angle, rad.

The original value of the diameter is

$$l = 2R \sin \frac{\theta}{2}. \tag{4.98}$$

The stretch ΔL [m] of the diameter l is given by using the expansion for l:

$$\Delta L = \theta R - \left[\theta R - 2R \frac{\theta^3}{8 \cdot 3!} + \ldots\right] \approx \frac{\theta^3 R}{24}. \tag{4.99}$$

The deflection at the center y_0 [m] is determined from the geometry of deformation:

$$y_0 = R\left[1 - \cos\frac{\theta}{2}\right] = R\left[1 - 1 + \frac{\theta^2}{2} - \ldots\right] = \frac{R\theta^2}{2}. \tag{4.100}$$

Therefore, ΔL and the membrane strain ϵ are

$$\Delta L = \frac{\theta y_0}{12} = \frac{y_0 L}{12}, \qquad \epsilon = \frac{\Delta L}{l} = \frac{y_0 L}{12Rl} \approx \frac{y_0}{12R}. \tag{4.101}$$

In writing the last line, we neglected tangential stresses and assumed that $L \approx l$.

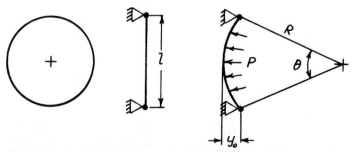

Figure 4.15 Geometrical relationships for simplified model of diaphragm.

The approximate dome pressure-induced stress σ [N/m^2] and strain ϵ are given by

$$\sigma = \frac{PR}{2\delta} \quad \text{and} \quad \epsilon = \frac{\sigma}{E} = \frac{PR}{2E\delta} \tag{4.102}$$

where P = pressure, Pa;
δ = membrane thickness, m; and
E = Young's modulus, N/m^2.

From equations 4.101 and 4.102,

$$\frac{y_0}{12R} = \frac{PR}{2E\delta} \quad \text{and} \quad P = C\frac{y_0}{R^2} \tag{4.103}$$

where C is the constructional constant (in the following formulas the C's with subscripts are constants also).

Combining equations 4.97 and 4.100, we express R as a function of L and y_0, and then substitute R into equation 4.103.

$$R = \frac{L^2}{4y_0}, \quad P = C_1 y_0^3 \tag{4.104}$$

where C_1 = constant.

The spring constant of the diaphragm K [N/m] is proportional to the slope of the pressure-deflection curve; therefore,

$$K = C_2 \frac{dP}{dy_0} = C_3 y_0^2 = C_4 P^{2/3}. \tag{4.105}$$

The natural frequency ω_0 [rad/s] of the diaphragm, when pressurized, is of the form

$$\omega_0 = \sqrt{K/M} = C_5 \sqrt[3]{P} \tag{4.106}$$

where M [kg] is an effective mass of the diaphragm including the attached air mass.

Experiments show substantial agreement with this cubic relationship. The calculations prove that doming is the predominant factor affecting the natural frequency. The second-order effect is the response of the frequency to the change in the attached mass and in the acoustic impedance loading the capsule because of the density-pressure dependency.

If acoustic interaction with the media is not essential, the frequency change due to the variation of only gas density can be evaluated with the following formula [49]:

$$\omega_n = \omega_0\left(1/\sqrt{1 + 0.34\rho_g l/\rho\delta}\right) \tag{4.107}$$

where ω_n and ω_0 = natural frequencies in gas and in vacuum, respectively, rad/s, and
ρ_g and ρ = density of gas and diaphragm material, respectively, kg/m^3.

It can be shown that the hysteresis for the element operating in the frequency domain is less than that in the deflection domain. The diminishing of hysteresis is another valuable feature of this element.

The formulas of Chapter 5, Elastic Elements, establishing relationships between the pressure and displacement of flat and corrugated diaphragms, can be helpful in calculating the transfer functions of the capsules more precisely.

We will illustrate a calculation of the frequency-pressure characteristic, taking as an example a case of the excitation of first-order mode vibrations in a flat diaphragm. The diaphragm is clamped at its edges, and an alternating force is applied at the center of the diaphragm perpendicularly to its plane. The displacement-pressure relationship is described by formula 5.18. The equivalent force F [N] applied to the center of the diaphragm (formula 5.25) due to the acting of pressure P [Pa] when $r = 0$ is

$$F = PA_{\text{ef}} = \frac{1}{3} PA \qquad (4.108)$$

where A_{ef} and A = effective and real areas of diaphragm, m^2.

The slopes of pressure-deflection and force-deflection characteristics, k_P and k_F, respectively, are

$$k_P = \frac{dP}{dy_0} \quad \text{and} \quad k_F = \frac{dF}{dy_0} \qquad (4.109)$$

where y_0 [m] is the deflection of the diaphragm at the center.

If an additional mass M_1 is attached to the center of a flat diaphragm, its effective mass M_{ef} for calculating the fundamental frequency is

$$M_{\text{ef}} = M_1 + 0.3M_2 \qquad (4.110)$$

where M_2 = real mass of the diaphragm, kg.

For corrugated diaphragms, the explicit values of M_{ef} can be obtained from experiments.

The effective mass M_{ef} of the diaphragm without additives is a fraction of its real mass M_2:

$$M_{\text{ef}} = 0.3M_2 = 0.3\rho\delta A \qquad (4.111)$$

where ρ = diaphragm's material density, kg/m^3, and

δ = diaphragm's thickness, m.

Under pressure P, the diaphragm is deflected, and its stiffness at the point of deflection y_0 is k_F; therefore, the equation for the fundamental frequency ω is

$$\omega^2 = \frac{k_F}{M_{\text{ef}}} \approx \frac{k_P}{\delta\rho}. \qquad (4.112)$$

This expression is obtained by substituting M_{ef} from equation 4.111 and differentiating equation 4.108:

$$\frac{dF}{dy_0} = \frac{1}{3} \frac{dP}{dy_0} A. \qquad (4.113)$$

To complete the derivation of $\omega = f(P)$ we should find k_P by differentiating equation 5.18.

Formula 5.18 can be introduced as

$$P = \alpha y_0 (1 + \beta y_0^2) \qquad (4.114)$$

where

$$\alpha = \frac{16E\delta^4 R^4}{3(1 - \nu^2)\delta} \quad \text{and} \quad \beta = \frac{(7 - \nu)(1 + \nu)}{16\delta^2} \qquad (4.115)$$

$$k_P = \frac{dP}{dy_0} = \alpha(1 + 3\beta y_0^2). \tag{4.116}$$

Eliminating βy_0^2 from formulas 4.114 and 4.116 yields

$$y_0 = \frac{3P}{2\alpha + k_P}. \tag{4.117}$$

Substituting equation 4.117 in equation 4.114 gives

$$27\beta \left(\frac{P}{\alpha}\right)^2 = \left(\frac{k_P}{\alpha} - 1\right)\left(\frac{k_P}{\alpha} + 2\right)^2. \tag{4.118}$$

Extracting k_P from this equation and substituting it in equation 4.112 yields

$$27\beta \left(\frac{P}{\alpha}\right)^2 = \left(\frac{\rho\delta\omega^2}{\alpha} - 1\right)\left(\frac{\rho\delta\omega^2}{\alpha} + 2\right)^2. \tag{4.119}$$

The behavior of the capsule is defined not only by the mechanical and acoustical characteristics of the system, but also by the characteristics of the driving network affecting the vibrations and the frequency in the neighborhood of resonance. For these conditions, an analysis of the vibrations will be relevant.

The vibratory motion of the capsule about a static equilibrium is equivalent to that of a linear, single-degree-of-freedom, spring-mass system with damping. A mass, spring, and viscous damping are three lumped mechanical elements. For this analysis, we assume that all mechanical, pneumatic, and acoustic loadings of the capsule are included in these lumped elements. The equation of motion is

$$M \frac{d^2x}{dt^2} + B \frac{dx}{dt} + Kx = F = f \cos \omega t \tag{4.120}$$

where x = displacement, m;
$\quad M$ = equivalent mass, kg;
$\quad B$ = viscous friction coefficient, N $\cdot$ s/m;
$\quad K$ = spring constant, N/m;
$\quad F$ = instantaneous force, N,
$\quad f$ = force peak value, N;
$\quad \omega$ = angular frequency, rad/s; and
$\quad t$ = time, s.

The driving elements of the system, including two coils and an amplifier, can provide an alternating driving force that is proportional either to the displacement or to the velocity of the moving part of the capsule. In the first case, the system operates at the "displacement resonance"; in the second case, at the "velocity resonance."

At the displacement resonance, the relationship between the amplitude of oscillation A and the amplitude of force f is given as

$$A = \frac{f}{\sqrt{(K - M\omega^2)^2 + (\omega B)^2}} \tag{4.121}$$

where ω stands for frequency of oscillation.

The displacement x lags behind the applied force by angle θ:

$$\theta = \arctan \frac{\omega B}{K - M\omega}. \tag{4.122}$$

The angular frequency ω_d, at which the displacement is at maximum:

$$\omega_d = \sqrt{\frac{K}{M} - \frac{B^2}{2M^2}} \tag{4.123}$$

The amplitude A_d at which the resonance takes place is

$$A_d = \frac{f}{B\sqrt{\left(\dfrac{K}{M} - \dfrac{B^2}{4M^2}\right)}}. \tag{4.124}$$

The expression for the displacement is given as

$$x = A \cos(\omega t - \theta) = \frac{f \cos(\omega t - \theta)}{\sqrt{(K - M\omega^2)^2 + (\omega B)^2}}. \tag{4.125}$$

The velocity dx/dt is calculated by differentiation (formula 4.125):

$$\frac{dx}{dt} = -\frac{f\omega \cdot \sin(\omega t - \theta)}{\sqrt{(K - M\omega^2)^2 + (\omega B)^2}}, \qquad \dot{A} = \frac{f\omega}{\sqrt{(K - M\omega^2)^2 + (\omega B)^2}}. \tag{4.126}$$

In order to obtain the frequency at which the velocity amplitude $\dot{A}$ is at maximum, $\partial \dot{A}/\partial \omega$ is found and set to zero. This condition corresponds to $K - M\omega^2 = 0$. Therefore, the frequency ω_v of velocity resonance and the amplitude $\dot{A}_v$ at the resonance are

$$\omega_v = \sqrt{\frac{K}{M}}, \qquad A_v = \frac{f}{B}. \tag{4.127}$$

The velocity lags behind the applied force by angle γ:

$$\gamma = \arctan \frac{M\omega^2 - K}{\omega B}. \tag{4.128}$$

At the velocity resonance $M\omega^2 - K = 0$ and $\gamma = 0$. It is notable that the displacement resonant frequency depends on the viscous friction, which is introduced by the internal friction in the capsule material and by the viscosity of gases. The resonant frequency of the velocity resonance is independent of viscous friction. It is clear that the velocity-resonance mode of operation is preferable for pressure sensing; and the driving circuit must be set up to maintain a zero phase shift between the pickup voltage, which is proportional to the velocity, and the current feeding the driving coil. The force developed by the driving coil is in phase with its current.

For further analysis, we define a quality factor Q as

$$Q = \frac{\omega_v M}{B} \tag{4.129}$$

and insert Q into expression 4.128:

$$\gamma = \arctan\left(Q\frac{\omega}{\omega_v} - \frac{\omega_v}{\omega}\right) = \arctan\frac{(\omega - \omega_v)(\omega + \omega_v)}{\omega_v\omega} = \arctan 2Q\frac{\Delta\omega}{\omega_v} \qquad (4.130)$$

where $\Delta\omega = \omega - \omega_v$ and $(\omega + \omega_v)/\omega\omega_v \approx 2/\omega_v$ since in the neighborhood of resonance $\omega \approx \omega_v$. Relation 4.130 defines the effect of the phase shift on the oscillation frequency. At the resonance $\Delta\omega = 0$, and $\gamma = 0$. The velocity is in phase with the driving force. As follows from formula 4.130, the frequency drift $\Delta\omega/\omega_v$ due to the variation in γ is

$$\frac{\Delta\omega}{\omega_v} = \frac{\tan\gamma}{2Q}. \qquad (4.131)$$

It is obvious that keeping the oscillation frequency equal to the element's resonance frequency is desirable; therefore, the phase shift contributed by the driving electronics must be zero. The increase of Q stabilizes the operation of the resonator since the ratio $\Delta\omega/\omega_v$ becomes smaller. It should be noted that manipulation of the phase shift can be beneficial. For instance, if the resonance system is constructed for detecting the viscous properties of material, and shift of frequency is a measure of viscosity, the sensitivity of the sensor to Q has to be enhanced (Q reflects the dissipation of energy due to friction, which relates to viscosity). In this case, the condition $\tan\gamma \neq 0$ is necessary.

As already mentioned, the resonant capsule transducer is one of the most accurate devices for measurements of pressure. The construction of the transducer does not contain any intermediate elements transmitting pressure or loading the capsule. Free vibrations and a high quality factor assure high resolution, repeatability, and accuracy that can be close to 15ppm *FS*.[*]

Vibrating Cylinders

A change in the frequency of a vibrating cylinder (Scheme 100a) is controlled by pressure (closed-end cylinder), the density of fluid flowing along its walls (opened-end cylinder), and temperature. These physical values cause a variation in the tension of the cylinder and/or change in the elastic properties of its materials. The coupling to the electronic oscillator is via electromagnetic drive and pickup coils that are mounted inside the vibrating cylinder.

The analysis of the cylinder frequency versus load characteristics is considerably more complicated than for the vibrating string or beam. We will review and summarize the calculations presented in [50 and 51] for the pressure-sensitive vibrator, and refer readers to the given sources for collateral studies.

The equations describing the vibratory motion are written and solved using the curvilinear coordinate system, as shown in Figure 4.16a, where the direction of the x-axis is along the cylinder. The y-axis, $y = a\varphi$, is measured clockwise in the circumferential direction. The z-axis is directed inward and normal to the middle surface of the shell. The equations of motion (formulas 4.132) are based on the D'Alembert's principle. They are written for the small displacements of the shell, u, v, and w, in the x, y, and z directions, respectively, assuming that the stress resultants due to pressure are constant.

---

[*] Here and further, *FS* = full scale.

$$
\begin{aligned}
&(\partial^2 u/\partial x^2) + [(1 - \eta)/2a^2](\partial^2 u/\partial \varphi^2) + [(1 + \eta)/2a](\partial^2 v/\partial x \partial \varphi) \\
&- (v/a)(\partial w/\partial x) - [\bar{N}_\varphi(1 - \eta^2)/Eha] \times [(\partial^2 v/\partial x \partial \varphi) - (\partial w/\partial x)] \\
&\qquad\qquad\qquad\qquad - [(1 - \eta^2)/Eh]\rho h(\partial^2 u/\partial t^2) = 0
\end{aligned}
$$

$$
\begin{aligned}
&[(1 + \eta)/2a](\partial^2 u/\partial x \partial \varphi) + [(1 - \eta)/2](\partial^2 v/\partial x^2) + (1/a^2)(\partial^2 v/\partial \varphi^2) \\
&- (1/a^2)(\partial w/\partial \varphi) + (h^2/12a^2) \times [(\partial^3 w/\partial x^2 \partial \varphi) + (\partial^3 w/a^2 \partial \varphi^3)] \\
&+ (h^2/12a^2)[(1 - \eta)(\partial^3 v/\partial x^2) + (\partial^2 v/a^2 \partial \varphi^2)] + [(1 - \eta^2)/Eh] \\
&\qquad\qquad\qquad\qquad \times [\bar{N}_x(\partial^2 v/\partial x^2) - \rho h(\partial^2 v/\partial t^2)] = 0
\end{aligned}
\tag{4.132}
$$

$$
\begin{aligned}
&v(\partial u/\partial x) + (\partial v/a \partial \varphi) - (w/a) - (ah^2/12)\nabla^4 w \\
&- (h^2/12)\{[(2 - \eta)/a](\partial^3 v/\partial x^2 \partial \varphi) + (\partial^3 v/a^3 \partial \varphi^3)\} \\
&+ [a(1 - \eta^2)/Eh]\{\bar{N}_x(\partial^2 w/\partial x^2) + (\bar{N}_\varphi/a)[(\partial v/a \partial \varphi) + (\partial^2 w/a \partial \varphi^2)] \\
&\qquad\qquad\qquad\qquad - \rho h(\partial^2 w/\partial t^2)\} = 0
\end{aligned}
$$

where ∇^4 denotes the operator

$$
[(\partial^2/\partial x^2) + (\partial^2/a^2 \partial \varphi^2)]^2
\tag{4.133}
$$

η = Poisson's ratio, dimensionless;
E = Young's modulus, N/m^2;
a = cylinder radius, m;
h = thickness of cylinder wall, m;
L = length of cylinder; m;
ρ = density of material, kg/m^3;
$\bar{N}_\varphi$ = circumferential force resultant per unit length (due to internal pressure), N/m;
$\bar{N}_x$ = axial force resultant per unit length (due to internal pressure), N/m; and
P = internal pressure, Pa.

The solution of this equation for a simply supported cylinder shows that the cylinder has several vibration modes (Fig. 4.16). Displacements u, v, and w are expressed as the space and time periodic functions:

$$
u = f(\cos \frac{m\pi x}{L}, \cos n\varphi, \cos \omega t)
$$

$$
v = f(\sin \frac{m\pi x}{L}, \sin n\varphi, \cos \omega t)
$$

$$
w = f(\sin \frac{m\pi x}{L}, \cos n\varphi, \cos \omega t)
\tag{4.134}
$$

where m and n are integers.

There are three natural frequencies of the cylinder. The smallest frequency corresponds to hoop modes, and the larger frequencies correspond to tangential motion. The hoop-mode vibrations are utilized for pressure sensing because these types of vibrations are much more sensitive to pressure than the tangential ones.

For the derivation of the frequency-pressure relationship of the hoop-mode vibrations, the tangential forces of inertia can be neglected, which finally gives the following frequency equation:

$$
\frac{\rho \omega_0^2 a^2}{E} = \frac{\lambda^2}{n^2 + \lambda^2} + \frac{\left(\dfrac{h}{a}\right)^2 (n^2 + \lambda^2)^2}{12(1 - v^2)} + \frac{Pa}{2Eh}(\lambda^2 + 2n^2)
\tag{4.135}
$$

where $\lambda = m\pi a/L$. This value of λ corresponds to the solution for the vibrations of the simply supported ends of the cylinder. For clamped ends, $\lambda = (m + 0.3)\pi a/L$. In practical designs, the end conditions define the resonant frequencies that lie

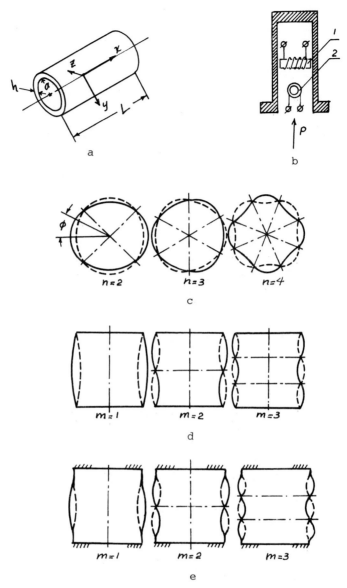

Figure 4.16 Schematic diagrams for vibrating cylinder.
a = cylinder coordinates, b = mutually perpendicular
orientation of driving (1) and pickup (2) coils inside cylinder,
reducing electromagnetic interaction between coils,
c = circumferential vibration forms, d = axial vibration forms
for cylinder with freely supported ends, e = axial vibration
forms for cylinder with fixed ends, m and n = indices of
vibratory modes.

between the frequencies for the simply supported and clamped cases. For example,
if the cylinder has closed ends integral with the cylindrical shell as in Scheme 100a,
the value of λ will be

$$\lambda = [m + 0.3 \exp(-qh/d)] \frac{\pi a}{L}. \tag{4.136}$$

The suitable factor q is determined from experiments; d [m] is the thickness of the
closed end. When d is very small, the end conditions are considered as for the simply
supported ends, while the large d is equivalent to the clamped ends.

In practice, the transfer characteristic is introduced by a polynomial with calibration coefficients k_i.

$$P = k_0 + k_1\omega_0 + k_2\omega_0^2 + k_3\omega_0^3 \tag{4.137}$$

This expression, derived by the expansion of the analytical functions, closely relates to those obtained from the experiments. As for the other types of resonators having a direct contact with the measured media, compensation for gas density and temperature is necessary.

In terms of accuracy, vibrating-cylinder pressure meters belong to the same high-level class of instruments as quartz and resonant capsule transducers. Long-term stability and excellent repeatability are the most attractive qualities of these instruments.

Piezoelectric Resonators

Due to its mechanical and electrical perfections, the *crystalline quartz resonator* has become the core of a number of transducers for measuring different physical quantities. The frequency of vibration of a crystal depends on its strain, effective mass, and temperature, which can be controlled by measurands. In a number of designs, the oscillating crystal is combined with a sensing element, such as bellows, [52, 53] providing initial conversion of the measurand into a force acting upon the crystal.

Crystals can have several kinds of vibration (Fig. 4.17). In general, the vibrating mode of a crystal can be analyzed by using an electrical equivalent circuit (Fig. 4.18). The terms in the equivalent circuit depend on the size of the element and electrodes (see Chapter 9, Piezoelectric Elements), their configuration, the orientation of the element with respect to the crystallographic axes, the method of mounting, mode of oscillation, and so on. These terms are determined by the ac excitation of the crystal and measurement of frequencies and impedances in the characteristic points.

C_0 [F] is the element's electrical capacitance measured at low frequencies. C [F] is the equivalent of compliance and is defined by the mechanical energy stored in the crystal due to the applied electrical energy $[(1/2)\,CU^2]$. L represents the vibrating

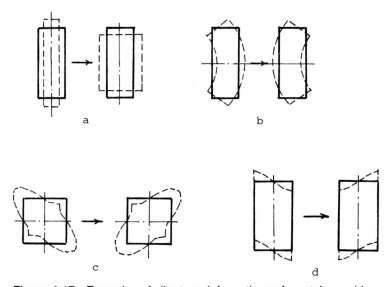

Figure 4.17 Examples of vibratory deformations of crystals used in sensors. a, b, c, and d = longitudinal, flexural, contour, and shear deformations, respectively.

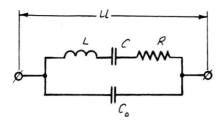

Figure 4.18 Quartz crystal equivalent circuit. C_0 = shunt capacitance, C = motional capacitance, L = motional inductance, R = motional resistance, U = voltage across crystal.

effective mass of the element. This term depends on the mode of vibration. R is an equivalent corresponding to the losses in the moving system.

The crystal impedance Z [Ω] can be written as

$$Z_c = R_c + jX_c, \quad R_c = \frac{RX_0^2}{R^2 + (X + X_0)^2}, \quad X_c = X_0 \frac{R^2 + X(X + X_0)}{R^2 + (X + X_0)^2} \quad (4.138)$$

where R_c and X_c = real and imaginary parts of Z_c, respectively, Ω,

$$X = \omega L - \frac{1}{\omega C} \quad \text{and} \quad X_0 = -\frac{1}{\omega C_0}. \quad (4.139)$$

The circuit is characterized by two resonances (Fig. 4.19): series, at frequency ω_s [rad/s], and parallel, at frequency ω_p. At the series resonance, the reactances of the elements in the LCR-branch of the equivalent circuit are equal, and

$$\omega_s = 1/\sqrt{LC}. \quad (4.140)$$

At the parallel resonance, the reactances of two branches are equal, and

$$\omega_p = 1/\sqrt{LC[C_0/(C + C_0)]}. \quad (4.141)$$

One of the characteristic parameters of crystal is the capacitance ratio $m = C/C_0$. From equation 4.141,

$$\omega_p = \omega_s \sqrt{1 + m}. \quad (4.142)$$

The larger m is, the easier the control of resonance frequencies will be.

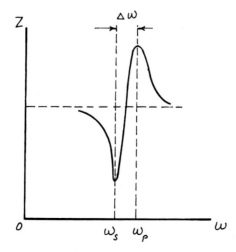

Figure 4.19 Resonant characteristic of piezoelectric crystal.

As mentioned in Chapter 9, the coupling coefficient k for a piezoelectric crystal is defined by the ratio of the electrical energy converted into mechanical to the total electrical energy consumed by the crystal. In terms of the equivalent circuit it is

$$k^2 = \frac{\frac{1}{2} CU^2}{\frac{1}{2} C_0 U^2 + \frac{1}{2} CU^2} = \frac{C}{C_0 + C} = \frac{m}{1 + m}. \qquad (4.143)$$

$k^2 \approx C/C_0 = m$ if $C \ll C_0$. For quartz crystals, typically m is not larger than 1×10^{-2} to 1×10^{-3}.

The relative difference between the frequencies ω_s and ω_p is a function of m:

$$(\omega_p - \omega_s)/\omega_s = m/2. \qquad (4.144)$$

The quality factor Q for piezoelectric materials is quite high: 1×10^4 to 1×10^6 for quartz and 1×10^2 to 1×10^4 for piezoelectric ceramics. Q reflects losses of energy for internal friction in the vibrating material, electrodes, and supports; radiation of energy in the media; and developing heat in the electrical circuit of excitation. In terms of the equivalent circuit, Q is calculated as follows:

$$Q = \frac{1}{R} \sqrt{\frac{L}{C}} = \frac{\omega_s L}{R} = \frac{1}{\omega_s CR}. \qquad (4.145)$$

In piezoresonance sensors, the crystal is connected to the circuit that maintains oscillations at or close to the resonance frequency. Due to the presence of direct and reverse piezoelectric effects, the maximum of the generated signal corresponds to the vibrations of the crystal at the resonance frequency. The vibrations are most powerful if a standing wave is established in the crystal. For longitudinal deformations, the distance between two surfaces reflecting the waves should be a multiple of the half-wave length. The relevant formulas establishing the relationships between the characteristics of vibration and parameters of the crystal are as follows:

$$\lambda = v/f, \qquad v = \sqrt{E/\rho}, \qquad f_u = \frac{n}{2h} \sqrt{\frac{E}{\rho}}, \qquad f_0 = \frac{1}{2h} \sqrt{\frac{E}{\rho}} \quad (4.146)$$

where λ = wave's length, m;
$\quad v$ = sound speed in piezoelectric, m/s;
$\quad E$ = Young's modulus, N/m^2;
$\quad f_u$ = resonant frequency of harmonic number u, Hz;
$\quad n$ = number of half-waves that can be placed between two faces of vibrating body;
$\quad h$ = distance between two faces, m;
$\quad \rho$ = density of material of vibrating body, kg/m^3; and
$\quad f_0$ = resonant frequency of fundamental harmonic, Hz.

For the ideal resonator, $f_0 = f_s = \omega_s/2\pi$. If during the vibrations the crystal has shear deformation over its thickness, the frequency will be

$$f_u = \frac{u}{2a} \sqrt{\frac{E}{\rho} \cdot \frac{1 - \nu}{(1 + \nu)(1 - 2\nu)}} \qquad (4.147)$$

where a = element's thickness, m, and
$\quad \nu$ = Poisson's ratio.

For the radial oscillation,

$$f_u = \frac{J_u}{2\pi r} \sqrt{\frac{E}{\rho(1 - \nu)^2}} \qquad (4.148)$$

where r = crystal's radius, m, and
J_u = constant defined by Bessel functions, dimensionless.

In measurements of temperature, the heating of a crystal causes changes of its size, density, and elastic properties (E). When a force or forces inducing physical quantities (torque, pressure, acceleration, and so on) act upon the crystal, its stress condition, and parameters h, ρ, and ν undergo alternation. A deposition of material on the surface of a vibration, or the change of a covibrating mass, leads to changes in h and in the average value of ρ.

The frequency $f(t)$ [Hz] of crystal as a function of temperature t [C°] can be introduced by an expansion [54]:

$$f(t) = f_0 \left[1 + \sum_{n=1}^{3} \frac{1}{n!} \cdot \frac{\partial^n f}{f_0 \partial t^n} \bigg|_{t=t_0} (t - t_0)^n \right] \qquad (4.149)$$

where f_0 = resonant frequency at temperature t_0, Hz, and

$$\frac{1}{n!} \cdot \frac{\partial^n f}{\partial t^n f_0} = a_n = \text{temperature coefficient of frequency of order } n, 1/°C.$$

For practical calculations in the range of temperatures from -200 to $+200°$C, the order three of the polynomial (formula 4.149) is acceptable:

$$\begin{aligned}
f(t) &= f_0 [1 + \sum_{n=1}^{3} a_n(t - t_0)^n] \\
&= f_0 [1 + a_1(t - t_0) + a_2(t - t_0)^2 + a_3(t - t_0)^3] \qquad (4.150)
\end{aligned}$$

where

$$a_1 = \frac{\partial f}{\partial t f_0} \bigg|_{t=t_0}, \qquad a_2 = \frac{\partial^2 f}{2\partial t^2 f_0} \bigg|_{t=t_0}, \qquad a_3 = \frac{\partial^3 f}{6\partial t^3 f_0} \bigg|_{t=t_0}. \qquad (4.151)$$

Coefficients a's depend on the crystal cut. For example, for Y-cut, a_1, a_2, and a_3 are 92.5×10^{-6}, 57.5×10^{-9}, and 5.8×10^{-12} $1/°C^n$, respectively.

Two types of crystal vibratory deformations are primarily used for the construction of force-converting elements: shear and bending. One of the problems existing in the design of these elements is in the decoupling of the vibratory portion of the crystal from the force-transmitting parts. In order to obtain adequate accuracy in conversion, the vibratory system should have a high level of quality factor. This condition can be achieved if the vibrations are localized within the crystal and there is no leak of the vibratory energy through the parts attached to the element. The contact with the vibratory parts is provided at the places where the amplitude of vibration is minimal or zero. This arrangement is easily realized for the shear-type vibrations but requires special measures for the bending type. In the latter case, the construction incorporates a mechanical suppressing filter blocking the flow of energy outside.

For the shear-type element, the relative change of frequency is approximately equal to the relative change of the characteristic length h:

$$\frac{\Delta f}{f} = \frac{\Delta h}{h}. \qquad (4.152)$$

The elements have a central frequency of 0.3 to 100MHz, thickness of 0.05 to 3mm, and length and width of 3 to 30mm. Quartz crystal can withstand tensile and compression stress of 1×10^2 and $24 \times 10^2 M(N/m^2)$, respectively.

An example of a bending vibratory element [55] is shown in Figure 4.20. It is successfully used in a high-accuracy pressure sensor (Fig. 4.21) and in precision accelerometers [56, 57]. The core of the resonator is a beam. It is excited by the alternating-polarity voltages applied to the set of electrodes, as explained in Figure 4.20, and vibrates at a lateral primary resonant frequency when no axial load is applied. As force F is applied, the frequency of vibration increases with the tension of the beam and decreases with its compression. Being deposited on the beam surface and connected diagonally, the electrodes stimulate the beam to be bent upward for one sense of applied voltage and downward for the opposite voltage. The quartz beam is free to vibrate in the fixed-fixed flexural mode with minimal energy transfer to the outer structure. This condition is possible because the structure on each end of the beam provides compliance for mounting torques from the end-mounting pads. The compliance, combined with the isolation mass, comprises a low-pass mechanical vibration filter (isolator) for the high-frequency quartz vibrating beam. This two-way filter effectively isolates the beam from the external vibrations that

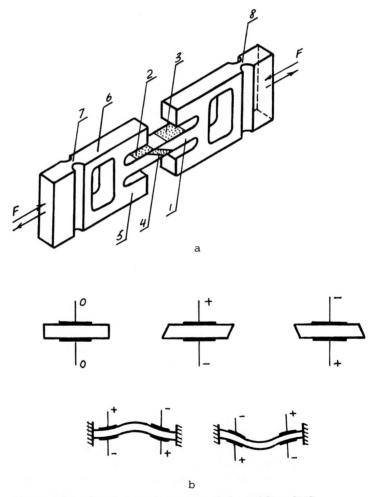

Figure 4.20 Quartz crystal resonator (adopted from [55]).
a = resonator, b = schematic diagram of piezoelectric excitation;
F = applied force, 1 = vibrating beam, 2 and 3 = electrodes,
4 = jumper, 5 = isolator mass, 6 = isolator spring, 7 and
8 = flexure reliefs.

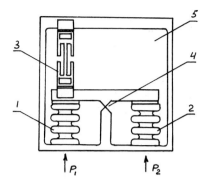

Figure 4.21 One of designs of differential-pressure transducers with quartz transduction element (adopted from [52]). P_1 and P_2 = pressures whose difference is to be measured, 1 and 2 = bellows, 3 = quartz crystal, 4 = pivot, 5 = internal vacuum.

could pass through the support. In order to eliminate the damping effects of air that diminish the quality factor, quartz is placed in a vacuum chamber.

For a beam loaded by an axial force and vibrating in the flexural mode, the force-to-frequency conversion is given by formula 4.92. Sometimes, for convenience, the output is represented in terms of period rather than frequency. For example, the transfer characteristic of one of the models of quartz pressure sensors [55] for measurements of absolute pressure (its construction is similar to that in Figure 4.21 but with one bellows), is written as follows:

$$P = A(1 - \tau_0/\tau) - B(1 - \tau_0/\tau)^2 \qquad (4.153)$$

where
P = applied pressure, Pa;
τ = output period at zero pressure, s;
τ_0 = output period at applied pressure, s; and
A and B = first- and second-order coefficients, Pa (they are found from experiments).

A 100kPa-range pressure transducer is characterized by the following characteristics: τ_0 is nominally about 25μs, corresponding to 40kHz beam natural resonance; and τ_0 changes to 28μs at full-scale pressure. Coefficients A and B in formula 4.153 are approximately 1.22×10^6 and 0.68×10^6, respectively, for pressure P measured in Pa. The characteristics of accuracy of such a transducer are high. For example, repeatability is within 0.005% *FS*. Unlike some other types of resonant pressure transducers, this device is not sensitive to gas density, moisture content, and so on.

The resonators vibrating at shear mode are most sensitive to the change of the mass M [kg] of the vibrating body and are used to determine the thickness of the deposition on their surface. It is easy to show that the frequency change is directly proportional to the change in the added mass Δm [kg] or to the thickness h' [m] and density ρ' [kg/m^3] of the addition:

$$\frac{\Delta f}{f} = -\frac{\Delta m}{M} = -\frac{\rho' h'}{\rho h}. \qquad (4.154)$$

In this formula, ρ and h are the vibrator density and thickness (it is assumed that the vibrator and deposition surfaces are equal). If a crystal is thermally stabilized, the minimum detectable $\Delta m \approx 1 \times 10^{-7} M$. The thickness of the deposition should not exceed 2μm to avoid degradation of the quality factor and accuracy of conversion.

A vibrating crystal exposed to gaseous or liquid media experiences an acoustic impedance that is defined by the losses of the resonator's energy for the radiation of waves and their dissipation in the media. The basic parameters of fluids: pressure, temperature, density, and viscosity, affect the acoustic impedance and, as a result, vibratory characteristics of the resonator. A reading of the resonant frequency or attenuation of vibrations can be useful for measurements of the listed quantities.

A resonator made of quartz is not very sensitive to the applied static field E. For instance, for $E = 1 \times 10^3$V/mm, the relative change of frequency $\Delta f/f$ is not greater than 0.1×10^{-6}. The sensitivity of piezoceramic materials to the field is more pronounced.

A magnetic field also does not significantly affect the crystal oscillation. A magnetic flux with a density of 0.1T gives $\Delta f/f = 0.5 \times 10^{-6}$. Furnishing the crystal with magnetic-material electrodes, such as nickel, appreciably changes the frequency sensitivity to a magnetic field. For example, a field of intensity of approximately 30×10^3 A/m gives a frequency shift $\Delta f/f$ of about 0.1%.

Quartz resonators can withstand a high radiation dose. For instance, for a radiation of 100×10^{12} neutrons per square centimeter-second (n/cm$^2 \cdot$ s), $\Delta f/f$ = 0.2×10^{-6}. However, at radiation of 0.1×10^{18} n/cm$^2 \cdot$ s, quartz resonators become nonoperational. Piezoceramic materials can work at doses exceeding 1×10^{18} n/cm$^2 \cdot$ s.

Nonreversible changes in quartz characteristics during aging can be explained by the degradation of the surface layers of crystal, spreading or recovering of microcracks, stress relaxation after mechanical treatment and assembly, diffusion and redistribution of impurities in the crystal, changes in physical properties, films on the crystal's surface, holders, and so on. Etching and cleaning of the surface, and mechanical and thermal cycling of crystal stabilize its characteristics. As a rule, the aging is most intensive in the first days after the fabrication of the element. After this, the properties become stable gradually. Placing the crystal in a vacuum chamber is a good practice in quartz element design, since processes of degradation of the surface contacting with the media contribute greatly to the instability of the vibratory characteristics. Quite often, the reliability of the mounting elements, such as cements, and different mechanical and electrical interfaces are the factors that mainly define the quality of the resonator.

More information on solid-state resonators can be found in the sections on microsensors in Chapter 9.

Acoustical Resonators

The resonant frequency of a cavity filled with gas can be a measure of the gas temperature, as well as its molecular weight, density, and composition. In this paragraph, we will consider how temperature and gas composition control the frequencies of two resonators that are the cores of transducers described in Appendix 1 and sketched in Schemes 148 and 161. In these transducers, the elastic properties of gases define the velocity of sound, which is expressed as follows:

$$c = \sqrt{\chi \frac{RT}{\mu}} \qquad (4.155)$$

where c = velocity of sound in gas, m/s;
$\quad\quad \chi$ = adiabatic index C_p/C_v (see Chapter 5, Sensing and Transduction Elements of Flowmeters);
$\quad\quad R$ = gas constant = 8.314 J/(mole $\cdot$ K);
$\quad\quad T$ = absolute temperature, K; and
$\quad\quad \mu$ = molecular weight of gas, kg/mole.

If the gas composition is known and all the parameters in formula 4.155 are constant but T, the resonator responding to c can serve as a temperature sensor. On the other hand, if T is constant, and the gas is a mixture of two gases for which μ's and χ's are known, the frequency response to c determines the content of each gas. The following calculation explains how the concept can be realized.

The characteristics of the mixture of two gases should satisfy the equations:

$$PV = \frac{m}{\mu} RT \qquad P_1 V = \frac{m_1}{\mu_1} RT$$

$$P_2 V = \frac{m_2}{\mu_2} RT \qquad P = P_1 + P_2 \qquad (4.156)$$

P, P_1, P_2 = pressures of gas mixtures and partial pressures of each gas, respectively, Pa;

m, m_1, m_2 = masses of gas mixtures and each gas, respectively, kg; and

μ, μ_1, μ_2 = molecular weights of gas mixtures and each gas, respectively, kg/mole.

From the preceding set of equations:

$$\mu = \frac{\mu_1 \mu_2 (m_1 + m_2)}{\mu_1 m_2 + \mu_2 m_1} \qquad \text{and} \qquad (4.157)$$

$$n_1 = \frac{m_1}{m_1 + m_2} = \frac{\mu_1 (\mu_2 - \mu)}{\mu (\mu_2 - \mu_1)} \qquad (4.158)$$

where n_1 = concentration of one of the gases.

Now we will find a relationship between the resonant frequency of the resonator and the velocity of the sound. The method of calculation of resonant frequency for this acoustic system is similar to that used for any dynamic system. First, the equation of the motion of the system is written. Second, the expression for the amplitude of vibration or for the impedance of the system is derived. Third, the resonant frequency is found from the conditions of maximum amplitude or minimum impedance.

A model of a temperature-sensitive element (Scheme 161 in Appendix 1) can be introduced by a Helmholtz resonator [58] (Figure 4.22), which is comprised of a rigid enclosure of volume V [m³] with a small opening of radius r and length L [m]. Plane harmonic acoustic waves from a sound generator are directed inside the cavity. Before the equation of motion is written, we will make several definitions using the following notations:

ρ = density of gas, kg/m³;
A = cross-sectional area of the neck, m²;
x = displacement of volume in the neck, m;

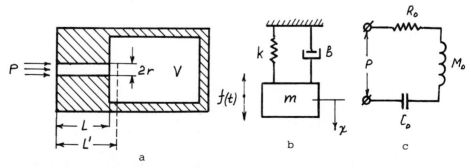

Figure 4.22 Helmholtz resonator as model of transduction element. a = resonator, b = mechanical analogs, c = electrical analogs.

r = neck's radius, m;
L' = neck's effective length, m;
ω = angular frequency of pressure oscillation, rad/s;
P_0 = peak value of harmonic acoustical pressure, Pa;
M_a = acoustical mass, kg/m^4;
R_a = acoustical resistance, kg/m$^4 \cdot$ s;
k_a = acoustical stiffness, kg/m$^4 \cdot$ s^2; and
η = dynamic viscosity, N $\cdot$ s/m^2.

The volume of gas in the neck moves as a unit and can be considered as a mass element of the system (see Chapter 5, Acoustical Elements). The force F_1 [N] needed to drive this mass is

$$F_1 = \rho A L' \frac{d^2x}{dt^2} \quad \text{where} \quad L' = L + 1.48r \qquad (4.159)$$

and L = real length of neck, m.

The mechanical resistance is defined by two components. One of them is radiation resistance, corresponding to energy dissipation at the opening. Another one, absorption resistance, reflects the losses because of viscous friction in the system. These resistances are

$$A^2 \rho \omega^2 / 4\pi c \quad \text{and} \quad A^2 \left[(L/\pi r^3) \sqrt{2\rho\eta\omega} \right]. \qquad (4.160)$$

The force F_2 [N] for overcoming these resistances is assumed to be proportional to the velocity of the motion of the gas masses:

$$F_2 = A^2 \left[\rho\omega^2 / 4\pi c + (L/\pi r^3) \sqrt{2\rho\eta\omega} \right] \frac{dx}{dt}. \qquad (4.161)$$

The elastic reluctance of volume V [m^3] or the effective spring stiffness is defined by the ratio: $\rho c^2 A^2 / V$. Thus, the counteracting spring force F_3 [N] will be

$$F_3 = \frac{\rho c^2 A^2}{V} x. \qquad (4.162)$$

The driving force of the system F_4 [N] is due to harmonic acoustic pressure $P = P_0 \times \sin \omega t$. The force is expressed as pressure times area:

$$F_4 = A P_0 \sin \omega t. \qquad (4.163)$$

Summing all the forces gives the equation of motion:

$$F_1 + F_2 + F_3 = F_4. \qquad (4.164)$$

In order to introduce the motion in the acoustical rather than mechanical domain, the terms of equation 4.164 should be divided by the area A, which modifies this equation as follows:

$$M_a \frac{d^2X}{dt^2} + R_a \frac{dX}{dt} + k_a X = P_0 \sin \omega t \qquad (4.165)$$

where

$$M_a = \frac{\rho L'}{A}, \qquad R_a = \rho\omega^2/4\pi c + (L/\pi r^3)\sqrt{2\rho\eta\omega}, \qquad k_a = \frac{\rho c^2}{V} = \frac{1}{C_a}, \quad (4.166)$$

$X = Ax$ is volume displacement, $dX/dt = (dx/dt)A$ is volume velocity, and $(d^2X/dt^2) = (d^2x/dt^2)A$ is volume acceleration. The acoustical compliance $C_a = 1/k_a$ instead of k_a is often used as a parameter of an acoustic system. This conversion is convenient for the analysis of acoustic systems; it allows us to operate with the lumped mechanical elements of mass, resistance, and stiffness. The use of electrical equivalents of the mechanical circuit (Fig. 4.22c) is helpful for this calculation (see Chapter 5, Acoustical Elements). The final equation of motion, 4.165, corresponds to the equation of motion for the forced oscillation of a mechanical system with damping. Following the standard steps in the analysis of the elementary tank circuit, we can write the equation for the acoustic impedance Z [kg/m$^4 \cdot$ s]:

$$Z = R_a + j(\omega M_a - 1/\omega C_a). \qquad (4.167)$$

The impedance is at minimum (the resonant condition) when $\omega M_a - 1/\omega C_a = 0$. The substitutions for M_a and C_a from formula 4.166 give the resonance frequency ω_0 [rad/s]:

$$\omega_0 = c\sqrt{\frac{A}{L'V}}. \qquad (4.168)$$

In the design of the sensor, the quality factor Q can be of interest. It is

$$Q = \omega\frac{M_a}{R_a}. \qquad (4.169)$$

By substituting c from expression 4.155 into formula 4.168, the ω-T transfer function is determined:

$$\omega_0 = \sqrt{\frac{\chi R T A}{\mu L'V}}. \qquad (4.170)$$

A similar calculation leads to the following frequency—sound velocity relationship for the resonator shown in Scheme 148 [59]:

$$\omega_0 = \frac{\pi c}{L} \qquad (4.171)$$

where L = distance between the sound generator and receiver (microphone), m. A combination of equations 4.155, 4.158, and 4.171 gives the frequency versus gas concentration characteristic:

$$\omega_0 = \frac{\pi}{L}\sqrt{\frac{\chi R T}{\mu_1\mu_2}[n_1(\mu_2 - \mu_1) + \mu_1]}. \qquad (4.172)$$

The relevant parameters of several gases are listed in Table 4.3. Sound velocities for various conditions can be calculated by using formula 4.155. More data on gases can be found in [60 and 61].

TABLE 4.3

Thermodynamic Properties of Selected Gases at Atmospheric Pressure (Source: [60 and 61])

Parameter	T [K]	H_2	T [K]	O_2	T [K]	N_2	T [K]	Air	T [K]	H_2O Vapor
$\chi = C_p/C_v$		1.605		1.401		1.402		1.403		1.324
Molecular weight μ [kg]		0.002		0.032		0.028		0.029		0.020
Density ρ [kg/m³]	200	1.123	200	1.956	200	1.711	273	1.252	380	0.586
	300	0.082	300	1.301	300	1.142	293	1.164	450	0.490
	400	0.061	400	0.975	400	0.854	353	0.968	500	0.440
Dynamic viscosity η [N · s/m²] × 10^{-6}	200	6.813	200	14.85	200	12.95	273	17.456	380	12.71
	300	8.963	300	20.63	300	17.84	293	18.240	450	15.25
	400	10.864	400	25.54	400	21.98	353	20.790	500	17.04

Microwave Cavity Resonators

At microwave frequencies, the resonant frequency of a cavity depends upon the cavity dimensions and the dielectric properties of the material inside the cavity.

The dimensions can be changed by a sensitive element converting the measurand into the displacement of a moving electrode, which changes the cavity dimensions. A number of characteristics of substances, such as density, moisture content, temperature, and solidification during the curing process relate to the dielectric constant, which can be easily measured by sampling the substances in the cavity. The transfer characteristics of the microwave resonator can be obtained by solving the electromagnetic wave equations for a waveguide or resonator.

Various Resonators

A disc supported at the center (Fig. 4.23a) and vibrating at different modes has been used for measuring the elastic properties of the material composing the disc [62]. It is feasible to apply this element for other measurements as well. The resonant frequency of the vibration of the disk is [63]:

$$\omega_0 = \frac{k^2 h}{b^2} \sqrt{\frac{E}{12(1 - \nu^2)\rho}} \tag{4.173}$$

where k = constant dependent on the mode of vibration and the ratio a/b, dimensionless, and

$a, b,$ and h = radius of rigid center (stem), radius of disk, and disk's thickness, respectively, m.

The disk can vibrate at several modes characterized by nodal diameters "n" (Fig. 4.24) and circumferences "s." If a/b is small (less than 0.2) the frequency is only slightly affected by the change in a. The mode of vibration with two mutually perpendicular diameters ($n = 2$) is one of the most stable. The disk should be excited at the antinodal points at its edge, halfway between the nodes.

Torsionally vibrating rods are very immune to the accelerations acting upon the instrument. Among the different configurations, we will name two: a uniform-section rod (*b*) and a ring (*c*). The fundamental frequency of the rod (whose nodal section is at the middle of its length) is given by

$$\omega_0 = \frac{\pi}{L} \sqrt{\frac{G}{\rho}} \tag{4.174}$$

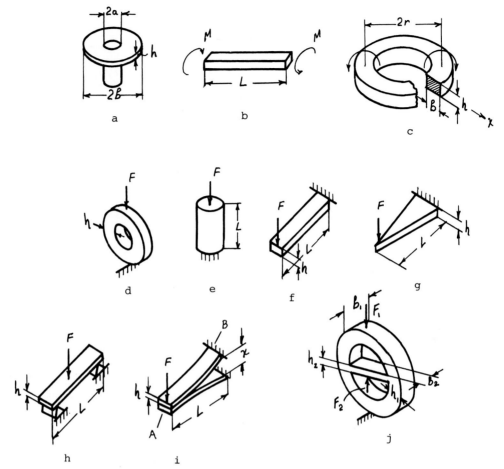

Figure 4.23 Various resonators. a = disk at flexural vibration, b = straight rod at torsional vibration, c = ring at torsional vibration, d = ring at flexural vibrations, e = straight bar at longitudinal vibration, f = uniform-section cantilever at flexural vibrations, g = triangular cantilever at flexural vibrations, h = free-free bar at flexural vibrations, i = fastened rectangular plates at flexural vibrations, j = proving ring with bridge at flexural vibrations.

where G = modulus of rigidity $[E/2(1 + \nu)]$, N/m^2;
   L = rod's length, m; and
   ρ = density of material, kg/m^3.

The ring's frequency is

$$\omega_0 = \sqrt{\frac{EI_x}{\rho r^2 I_p}} \qquad (4.175)$$

where r = ring's radius, m;
   I_x = moment of inertia of cross section about the x-axis (Fig. 4.23c), m^4; and
   I_p = polar moment of inertia of cross section, m^4.

For example, for a circular section of radius R,

$$I_x = \frac{\pi R^4}{4} \qquad \text{and} \qquad I_p = \frac{\pi R^4}{2}. \qquad (4.176)$$

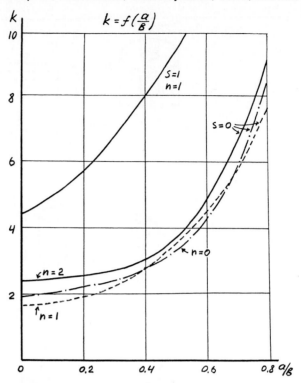

Figure 4.24 Coefficient k in formula 4.173 as function of ratio a/b for different modes of vibration. n = number of nodal diameters, s = number of nodal circumferences.

For a rectangular section of height h and width b,

$$I_x = \frac{bh^3}{12} \quad \text{and} \quad I_p = \frac{bh^3}{16}\left[\frac{16}{3} - 3.36\frac{h}{b}\left(1 - \frac{h^4}{12b^4}\right)\right]. \qquad (4.177)$$

The resonance frequency of a ring fixed at one point (d) and excited by an alternating radial force F directed toward this point is given by

$$\omega_0 = \frac{0.773h}{R^2}\sqrt{\frac{E}{\rho}} \qquad (4.178)$$

where h = ring's thickness, m.

Note that this frequency is independent of the ring's width.

A straight bar fixed at one end and excited by a longitudinal force F at the other end (e) has a resonance frequency of

$$\omega_0 = \frac{1.565}{L}\sqrt{\frac{E}{\rho}} \qquad (4.179)$$

where L = length of bar, m.

Note that this frequency does not depend on the cross-sectional area.

A uniform-section cantilever (f) excited at its end by a transverse force F resonates at the following frequency:

$$\omega_0 = \frac{1.018h}{L^2}\sqrt{\frac{E}{\rho}} \tag{4.180}$$

where L and h = length and thickness of beam, m.

A triangular cantilever of uniform thickness (g), when excited by a transverse force F, resonates at frequency

$$\omega_0 = \frac{1.985h}{L^2}\sqrt{\frac{E}{\rho}}. \tag{4.181}$$

A cantilever of such a shape has a uniform strain along its length when it is loaded by the transverse force at its end.

A free-free bar (h) excited by a force at the middle of its length resonates at frequency

$$\omega_0 = \frac{0.566h}{L^2}\sqrt{\frac{E}{\rho}}. \tag{4.182}$$

Two thin, flat rectangular plates (i) fastened together at one end (A) change their resonant frequency of flexural vibrations when the distance between their edges at the other end (B) varies [64]. It is assumed that these plates are excited by a transverse force F and that the edges at B are rigidly fixed to a massive part whose displacement x controls the stiffness of the vibrating structure and, thus, the resonant frequency. The frequency can be calculated approximately as follows:

$$\omega_0 \approx \frac{2.036h}{L^2}\sqrt{\frac{E}{\rho}\left[1 + \left(\frac{x+h}{h}\right)^2\right]}. \tag{4.183}$$

A proving ring (j) provided with a thin bridge along the diameter can serve as a frequency-output transduction element that responds to the force F_1 [N] applied to the ring as shown in (j). The bridge is excited at the middle of its length by alternating force F_2. The resonant frequency is directly proportional to the force F_3, which causes tension in the bridge [65]:

$$\omega_0 = \frac{1}{L^2}\sqrt{\frac{504EJ_2 - 12F_3L^2}{\rho b_2 h_2}} \tag{4.184}$$

$$F_3 = \frac{0.137F_1 r^3}{J_1\left(0.149\dfrac{r^3}{J_1} + \dfrac{L}{b_2 h_2}\right)} \tag{4.185}$$

where J_1 = moment of inertia of cross section of ring ($b_1 h_1^3/12$), m^4;
   J_2 = moment of inertia of cross section of bridge ($b_2 h_2^3/12$), m^4;
   b_1 and b_2 = widths of ring and bridge, respectively, m;
   h_1 and h_2 = thicknesses of ring and bridge, respectively, m; and
   r = radius of ring's mean circumference, m.

In this section, we have reviewed the representative resonant techniques and structures for transducers. The concept of a measurand-to-frequency conversion is being developed today not only on the macro- but on the micro-level as well. Progress in this field of transducer engineering is quite impressive, resulting in the additional description of microelectronic resonators as introduced in Chapter 9.

chapter 5

Mechanical, Acoustical, and Flowmetering Elements

This chapter primarily deals with sensing elements of transducers used for measurements of pressure, acceleration, sound intensity, and flow. However, some aspects of transduction elements will also be discussed. Elastic elements are constituent parts of many electromechanical transducers and are typical examples of sensing elements. We begin this chapter with a review of the elements which are in common use.

ELASTIC ELEMENTS

A spring element converts force, moment, acceleration, pressure, or temperature into displacement or strain, which are inputs of strain- or displacement-sensitive transduction elements. The calculation of spring elements uses the standard principles and formulas of the theory of elasticity and its application to precision mechanics. In this section, we will review the calculation of several elements [66–69] and recommend the cited literature for deeper study of the subject. We will consider in more detail the issues related to the design of one specific element (the absolute pressure capsule). The material in this section introduces a set of diversified problems typical for the design of sensitive and transduction elements.

The *major characteristic of spring elements* is the relationship between the linear (f [m]) or angular (φ [rad]) deformations and force (F [N]) or torque (T [N · m]). These characteristics can be linear (Fig. 5.1), nonlinear, and nonlinear with a linear portion. The sensitivity S is the ratio of two increments taken at a particular point of the characteristic:

$$S_F = \frac{\Delta f}{\Delta F} \approx \frac{df}{dF} \quad \text{or} \quad S_T = \frac{\Delta \varphi}{\Delta T} \approx \frac{d\varphi}{dT} \tag{5.1}$$

where S_F = sensitivity to force, m/N, and
S_T = sensitivity to torque, rad/N · m.

The reciprocal of S is stiffness or spring factor k. If n elements are connected in series and loaded with one force F (Fig. 5.2a), the sensitivities of the individual elements S_i and entire system S_Σ will be

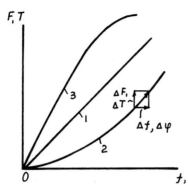

Figure 5.1 Typical characteristics of spring element. F and T = force and torque, f and φ = displacement and angle, 1, 2, and 3 = linear, nonlinear, and partly nonlinear characteristics, respectively.

$$S_1 = \frac{df_1}{dF}; \quad S_2 = \frac{df_2}{dF}, \quad \ldots; \quad S_n = \frac{df_n}{dF}; \quad S_\Sigma = \sum_{i=1}^{n} S_i = \frac{\sum_{i=1}^{n} df_i}{dF}. \quad (5.2)$$

The total stiffness k_Σ, as a function of the elements' stiffness k_i, is

$$k_\Sigma = \frac{1}{\sum_{i=1}^{n} \frac{1}{k_i}}. \quad (5.3)$$

In the parallel connection (Fig. 5.2b), the elements have common displacement f_c:

$$f_c = f_1 = f_2 = \ldots = f_n = F_1 S_1 = F_2 S_2 = \ldots = F_n S_n \quad (5.4)$$

Therefore,

$$F_1 = \frac{f_c}{S_1}, \quad F_2 = \frac{f_c}{S_2}, \quad \ldots, \quad F_n = \frac{f_n}{S_n} \quad \text{and} \quad (5.5)$$

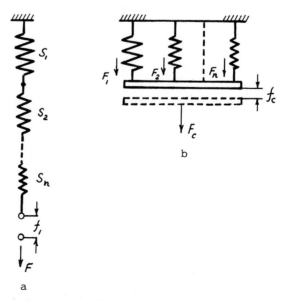

Figure 5.2 Series (a) and parallel (b) connection of spring elements.

common force F_c, common sensitivity S_c, and common stiffness k_c will be as follows:

$$F_c = \sum_{i=1}^{n} F_i = f_c \sum_{i=1}^{n} \frac{1}{S_i}, \qquad S_c = \frac{1}{\sum_{i=1}^{n} \frac{1}{S_i}}, \qquad k_c = \sum_{i=1}^{n} k_i. \qquad (5.6)$$

The most simple construction of the spring element is a *straight rod* (Fig. 5.3), which experiences tension, compression, bending, or twisting under forces F [N], P [N], or twisting moment M [N · m]. This element is usually combined with strain gages fixed on its surface. Typically, the rod can be solid or hollow and have a circular or rectangular cross section.

The formulas describing the strain state of the rod that can be used for practical calculation are as listed.

For tension or compression under force F (Fig. 5.3a),

$$\epsilon_x = F/AE, \qquad \epsilon_y = \nu F/AE, \qquad \Delta l = \epsilon_x \cdot l. \qquad (5.7)$$

For bending under force P (Fig. 5.3b),

$$\epsilon_x = \frac{T}{WE}, \qquad T = P \cdot x, \qquad T_{\max} = Pl, \qquad f = \frac{Pl^3}{3EJ}, \qquad (5.8)$$

$$J = \frac{bh^3}{12} \quad \text{and} \quad W = \frac{bh^2}{6} \quad \text{for a rectangular section,} \qquad (5.9)$$

$$J = \frac{\pi d^4}{64} \quad \text{and} \quad W = \frac{\pi d^3}{32} \quad \text{for a circular section.} \qquad (5.10)$$

For twisting of the circular-section rod under moment M (Fig. 5.3c),

$$\epsilon_s = \frac{Md}{2J_P E}, \qquad \theta = \frac{Ml}{J_P G}, \qquad G = \frac{E}{2(1 + \nu)}, \qquad J_P = \frac{\pi d^4}{32}. \qquad (5.11)$$

For twisting of the rectangular-section rod under moment M,

$$\epsilon_s = \frac{M(3b + 1.8h)}{Eb^2 h^2} \qquad (5.12)$$

at the mid-point of each longer side, and

$$\theta = \frac{Ml}{KG}, \qquad K = \frac{bh}{16}\left[\frac{16}{3} - 3.36\frac{h}{b}\left(1 - \frac{h^4}{12b^4}\right)\right] \qquad (5.13)$$

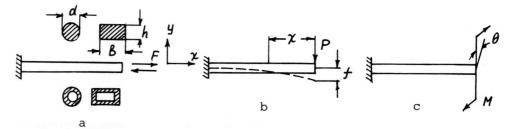

Figure 5.3   Rod spring element under tension-compression (a), bending (b), and twisting (c).

where ϵ_x and ϵ_y = strains in x- and y-directions, respectively;
$\qquad \epsilon_s$ = strain corresponding to the shear stress;
$\qquad A$ = cross-sectional area of rod, m^2;
$\qquad E$ = Young's modulus of material, N/m^2;
$\qquad \nu$ = Poisson's ratio,
$\qquad \Delta l$ = rod's elongation under force F, m;
$\qquad T$ = torque, $N \cdot m$;
$\qquad W$ = section modulus, m^3;
$\qquad x$ = length of arm for calculating torque T, m;
$\qquad T_{\max}$ = maximum value of T, $N \cdot m$;
$\qquad f$ = deflection of rod's end, m;
$\qquad J$ = moment of inertia of the section, m^4;
$\qquad d$ = diameter of rod, m;
$\qquad M$ = twisting moment, $N \cdot m$;
$\qquad J_P$ = polar moment of inertia of the section, m^4;
$\qquad \theta$ = angle of twist at the rod end, rad;
$\qquad G$ = modulus of rigidity of material, N/m^2; and
b and h = width and thickness of rectangular bar, respectively, m.

A round *proving ring* (Fig. 5.4) is also used quite frequently in combination with strain gages that are attached to the inner and outer surfaces of the ring in the places of maximum sensitivity (Scheme 86c). For the arrangement given in Figure 5.4, these places are at the intersections of the horizontal axis with the ring circumferences (at $\alpha = 0$). The strains that are developed on the outer and inner surfaces, ϵ_0 and ϵ_i, respectively, can be calculated with the following formulas:

$$\epsilon_0 = -\frac{r}{Ebh^2}(1.91 - 3\cos\alpha)F, \qquad \epsilon_i = \frac{r}{Ebh^3}(-1.91 + 3\cos\alpha)F. \quad (5.14)$$

In some designs, the ring is combined with a transduction element sensing the changes of the horizontal and vertical diameters, Δx [m] and Δy [m], respectively. They can be calculated as follows:

$$\Delta x = 1.64\frac{r^3}{Ebh^3}F \quad \text{and} \quad \Delta y = -1.79\frac{r^3}{Ebh^3}F \qquad (5.15)$$

where $\qquad r$ = inner radius of ring, m, and
b and h = width and thickness of ring, respectively, m.

Fixed-edged flat diaphragms with or without a bossed center (Fig. 5.5) are used in many pressure transducers. As a pressure difference P [Pa] acts across the diaphragm, the deflection of its center y_0 [m] is calculated as follows:

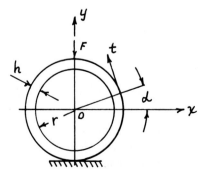

Figure 5.4 Round proving-ring element.

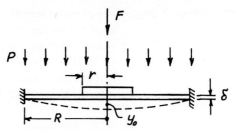

Figure 5.5   Circular bossed diaphragm.

$$\frac{PR^4}{E\delta^4} = A\,\frac{y_0}{\delta} + B\left(\frac{y_0}{\delta}\right)^3 \tag{5.16}$$

where $\quad A = 16/\{3(1 - \nu^2)[1 - \rho^4 - 4\rho^2 ln(R/r)]\}, \tag{5.17}$

$$B = \left[\frac{7 - \nu}{3}(1 - \rho^2 + \rho^4) + \frac{(3 - \nu)^2}{1 + \nu}\rho^2\right]\bigg/\left[(1 - \nu)(1 - \rho^4)(1 - \rho^2)^2\right]$$

R and r = diaphragm and bossed area radii, respectively, m;
$\rho = r/R$;
δ = diaphragm thickness, m; and
y_0 = diaphragm center displacement, m.

If the diaphragm does not contain the bossed area, ($r = 0$ and $\rho = 0$), then expression 5.16 is reduced:

$$\frac{PR^4}{E\delta^4} = \frac{16}{3(1 - \nu^2)} \cdot \frac{y_0}{\delta} + \frac{7 - \nu}{3(1 - \nu)} \cdot \left(\frac{y_0}{\delta}\right)^3. \tag{5.18}$$

Force F [N] applied to the diaphragm center will give displacement at the center, which is calculated as follows:

$$y_0 = \frac{3(1 - \nu^2)}{\pi}\left[\frac{R^2 - r^2}{4R^2} - \frac{r^2 \cdot ln^2(R/r)}{R^2 - r^2}\right]\frac{FR^2}{E\delta^3}. \tag{5.19}$$

The next section, entitled Absolute Pressure Capsules, is devoted to the calculations of a circular diaphragm with a more detailed presentation of its stress and strain state.

Square diaphragms (Fig. 5.6) have become popular in sensor designs because of their wide application in silicon microsensors. The diaphragm, formed as a square plate with fixed edges, has a deflection given by the equation [70–72] similar to that for circular diaphragms:

$$\frac{Pa^4}{E\delta^4} = \frac{4.20}{(1 - \nu^2)} \cdot \frac{y_0}{\delta} + \frac{1.58}{(1 - \nu^2)}\left(\frac{y_0}{\delta}\right)^3 \tag{5.20}$$

where a = half sidelength of diaphragm, m.

A rough approximation of the expression describing the deflection y [m] of the points on the axes of symmetry at the diaphragm (distance r [m] in Figure 5.6) is given as

$$y = y_0\left[1 - 0.9\left(\frac{r}{a}\right)^2\right]^2. \tag{5.21}$$

A *corrugated diaphragm* provides greater deflection than does a flat diaphragm of the same size. If the entire surface of the diaphragm is corrugated, the equation of deflection of the diaphragm center will be

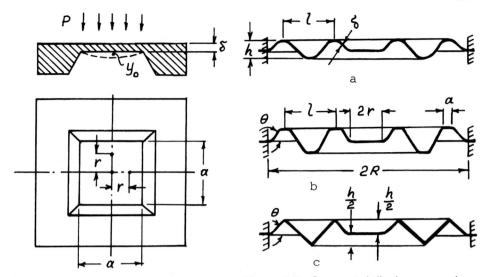

Figure 5.6 Square diaphragm.

Figure 5.7 Corrugated diaphragms. a, b, and c = diaphragms with sinusoidal, trapezoidal, and triangular corrugations, respectively.

$$\frac{PR^4}{E\delta^4} = \frac{2(\alpha + 3)(\alpha + 1)}{3k_1(1 - \nu^2/\alpha^2)} \cdot \frac{y_0}{\delta} + \frac{32k_1}{\alpha^2 - 9}\left[\frac{1}{6} - \frac{3 - \nu}{(\alpha - \nu)(\alpha + 3)}\right]\left(\frac{y_0}{\delta}\right)^3 \quad (5.22)$$

where $\alpha = \sqrt{k_1 k_2}$; k_1 and k_2 are defined by the type of the corrugations (their profiles) and by the deepness of corrugation (Fig. 5.7 and Table 5.1).

A corrugated diaphragm having a bossed center is calculated by taking into account the radius r[m] of the center and specifics of the corrugations (Fig. 5.7).

Coefficients η_1 and η_2 in the following formula are functions of parameters α and $\rho (\rho = r/R)$. They are tabulated [68] and given in Table 5.2.

$$\frac{PR^4}{E\delta^4} = \eta_1 \frac{2(\alpha + 1)(\alpha + 3)}{3k_1\left(1 - \frac{\nu^2}{\alpha^2}\right)} \cdot \frac{y_0}{\delta} + \eta_2 \frac{32k_1}{\alpha^2 - 9}\left[\frac{1}{6} - \frac{3 - \nu}{(\alpha - \nu)(\alpha + 3)}\right]\left(\frac{y_0}{\delta}\right)^3 \quad (5.23)$$

If force F [N] is applied to the diaphragm center, its displacement is described by the following expression:

TABLE 5.1

Coefficients k_1 and k_2 to Formulas 5.22 to 5.24 (Adopted from [73]). Dimensions a, l, and h are revealed in Figure 5.7.

Type of corrugation	k_1	k_2
Sinusoidal with $\theta < 15°$ (Fig. 5.7)	1	$\frac{3}{2}\left(\frac{h}{\delta}\right)^2 + 1$
Trapezoidal	$\frac{l^2 - 2a}{l \cdot \cos\theta} + \frac{2a}{l}$	$\left(\frac{h}{\delta}\right)^2\left[\frac{l - 2a(1 - 3\cos\varphi)}{2\cos\theta}\right] + \left(1 - \frac{2a}{l}\right)\cos\theta + \frac{2a}{l}$
Triangular	$\frac{1}{\cos\theta}$	$\frac{h^2}{R^2 \cos\theta} - \cos\theta$

TABLE 5.2

Coefficients η_1, η_2, η_3, and η_4 to Formulas 5.23 and 5.24 (Adopted from [68])

	$\rho = r/R$															
α	0.2	0.4	0.6	0.8	0.2	0.4	0.6	0.8	0.2	0.4	0.6	0.8	0.2	0.4	0.6	0.8
	η_1				η_2				η_3				η_4			
2	1.10	1.68	4.25	28.3	1.14	1.89	5.21	36.7	1.32	2.72	9.04	76.2	2.36	6.69	26.2	237
4	1.01	1.22	2.33	12.8	1.13	1.75	4.46	30.6	1.12	1.69	4.17	29.0	2.35	6.52	25.3	229
6	1.01	1.11	1.75	7.77	1.13	1.73	4.28	28.7	1.08	1.45	2.89	16.2	2.34	6.49	25.1	227
8	1.00	1.08	1.52	5.53	1.13	1.73	4.22	27.9	1.07	1.36	2.39	10.9	2.34	6.49	25.0	226
10	1.00	1.06	1.40	4.34	1.13	1.72	4.20	27.4	1.06	1.31	2.15	8.31	2.34	6.48	25.0	226
12	1.00	1.05	1.34	3.64	1.13	1.72	4.20	27.2	1.06	1.29	2.01	6.81	2.34	6.48	25.0	225
16	1.00	1.04	1.27	2.90	1.13	1.72	4.18	27.0	1.05	1.26	1.86	5.25	2.34	6.48	25.0	225

$$\frac{FR^2}{\pi E\delta^4} = \eta_3 \frac{(1 + \alpha)^2}{3k_1(1 - \nu^2/\alpha^2)} \cdot \frac{y_0}{\delta} + \frac{\eta_4 k_1}{\alpha^2 - 1}\left[\frac{1}{2} - \frac{1 - \nu}{(\alpha - \nu)(\alpha + 1)}\right]\left(\frac{y_0}{\delta}\right)^3. \qquad (5.24)$$

The force F_d [N] developed by the diaphragm due to applying pressure can be calculated approximately, taking a product of pressure and effective area A_{ef} [m²]:

$$F_d = P \cdot A_{ef} \approx P \cdot \frac{\pi}{3}(R^2 + Rr + r^2). \qquad (5.25)$$

By combining the diaphragm parameters, the required performance can be obtained. Usually, a compromise between the displacement and linearity of the transfer characteristic should be tolerated.

Bellows are soft, pressure-sensitive springs (Fig. 5.8) that have a small effective area and low stiffness. In the construction of transducers, they serve as elements creating a force or displacement but not contributing a significant stiffness in the system. The regular bellows have a number of deep corrugations that constitute a series connection (see beginning of this section) of compliant segments; therefore the resulting stiffness is small. The effective area of the bellows A_{ef} [m²] and force developed F_{ef} [N] are calculated as follows:

$$A_{ef} = \pi(R_1 + R_2)^2/4, \qquad F_{ef} = PA_{ef} \qquad (5.26)$$

where R_1 and R_2 = two characteristic radii of the bellows as in Figure 5.8, [m], and P = pressure difference, Pa.

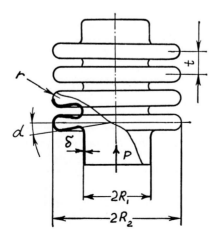

Figure 5.8 Bellows.

A thorough study [52] shows that the effective area can slightly change in response to a variation of pressure. Attention must be paid to this fact, especially in the design of differential pressure transducers where insensitivity to the common-mode pressure of two uniform bellows is important. If one of the ends of the bellows is fixed, the displacement f [m] of the other end is calculated as [67, 74]

$$f = PA_{ef} \frac{1-\nu}{E\delta} \cdot \frac{n}{A_0 + \alpha A_1 + \alpha^2 A_2 + B_0 \left(\dfrac{\delta}{R_1}\right)^2} \qquad (5.27)$$

where

δ = thickness of wall at R_1, m;

$\alpha = (4r - t)/[2(R_2 - R_1 - 2r)]$;

r = radius of curvature of corrugation crest, m;

t = pitch of bellows corrugations, m;

n = number of active corrugations; and

A_0, A_1, A_2, and B_0 = coefficients; they can be found as the functions of parameters $k = R_2/R_1$ and $m = r/R_1$ in Figure 5.9.

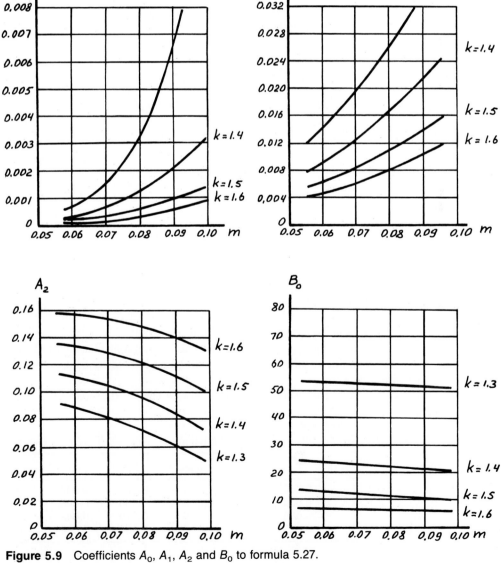

Figure 5.9 Coefficients A_0, A_1, A_2 and B_0 to formula 5.27.

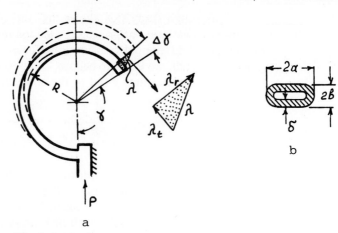

Figure 5.10 Bourdon tube.

A *Bourdon tube* is also an element (Fig. 5.10) creating a large displacement in response to the pressure applied to its inner cavity. In its simple form, it is a flattened, oval-section bent pipe. The pipe is sealed from one end and forms a C-shaped member. Being pressurized from the inlet at the other end, the pipe tends to be straightened, thus deflecting the free end. The relative angular deflection of the thin-wall ($\delta/b < 0.8$) spring is calculated [68] as follows:

$$\frac{\Delta\gamma}{\gamma} = P\epsilon \frac{R^2}{b\delta}\left(1 - \frac{b^2}{a^2}\right)\frac{k_1}{k_2 + A}. \tag{5.28}$$

For the thick-wall tube ($\delta/b > 0.8$) with extended width $b(b > a)$,

$$\frac{\Delta\gamma}{\gamma} = P\epsilon \frac{R^2}{b\delta} \cdot \frac{1 - C_1}{C_1 + \delta^2/(12b^2)} \tag{5.29}$$

where

$$C_1 = \frac{1}{\psi} \cdot \frac{\sin^2\psi + \sinh^2\psi}{\sin\psi \cdot \cos\psi + \sinh\psi \cdot \cosh\psi}, \tag{5.30}$$

$$\psi = \sqrt[4]{3/A} = 1.316a/\sqrt{R\delta}, \qquad A = (R\delta/a^2)^2, \qquad \epsilon = (1 - \nu^2)/E.$$

Dimensions a, b, R, and δ (all in [m]) are revealed in Figure 5.10; coefficients k_1 and k_2 as functions of ratio a/b are given by numbers in Table 5.3; γ and $\Delta\gamma$ (in radians) are the angle of the working part of the tube and the change of this angle, respectively. A radial, tangential, and total displacement of tube free end λ_r, λ_t, and λ (all in [m]), respectively, can be calculated by using geometrical relationships:

$$\lambda_r = (\Delta\gamma/\gamma)R(1 - \cos\gamma), \qquad \lambda_t = (\Delta\gamma/\gamma)R(\gamma - \sin\gamma), \qquad \lambda = \sqrt{\lambda_r^2 + \lambda_t^2}. \tag{5.31}$$

TABLE 5.3

Coefficients k_1 and k_2 as Functions of Ratio a/b (Fig. 5.10) (Adopted from [68])

a/b	k_1	k_2	a/b	k_1	k_2	a/b	k_1	k_2
1	0.750	0.083	4	0.452	0.044	8	0.400	0.042
1.5	0.636	0.062	5	0.430	0.043	9	0.395	0.042
2	0.566	0.053	6	0.416	0.042	10	0.390	0.042
3	0.493	0.045	7	0.406	0.042	∞	0.368	0.042

A deformation of *cylindrical and hemispherical cups* (Fig. 5.11) can be a measure of the pressure difference across the cups' wall.

In the cup element, strains ϵ_x and ϵ_y on the wall, in the x- and y-directions, respectively, are

$$\epsilon_x = \frac{0.5r}{E\delta}(1 - 2\nu)P, \tag{5.32}$$

$$\epsilon_y = \frac{0.5r}{E\delta}(2 - \nu)P \tag{5.33}$$

where r and δ = radius and thickness of cup [m].

As it follows from formulas 5.32 and 5.33, the strains are independent of length *l*.

For the spherical element, strains ϵ's in x- and y-directions are the same and equal to

$$\epsilon = \frac{0.5r}{E\delta}(1 - \nu)P. \tag{5.34}$$

With the development of solid-state sensitive and transduction elements, the *multilayer constructions* containing layers with different elastic properties are typical. The calculation of such elements begins with finding the forces and/or moments causing the deformation of the elements, then involves writing equations balancing the forces and/or moments. The final stage is solving these equations for strains or deformations. A similar approach is used when we analyze various emerging undesirable stresses in the elements of structure having nonuniform elastic characteristics of materials.

A representative example, which we will consider in this section, deals with conditions quite common in the practice of design [75]. A part of a spring element (it can be a cantilever, beam on two supports, diaphragm, and so on) experiences a bending moment (Fig. 5.12). The value of this moment, M [N · m], at a particular section of the element is known or can be determined. Our task is to calculate the strain along the element's axis (for this case, x-axis), taking into account the difference in the coefficients of elasticity of the layers. The statement of this problem is typical for the calculation of structures with strain gages when the gages, substrates, and intermediate layers are sandwiched together. To be more specific in the calculation, we will assume that the substrate and layers are rectangular in section and the latter form a symmetrical structure around the horizontal axis (all relevant dimensions are shown in Figure 5.12).

As a result of deformation, the moment M from the outside sources of mechanical energy is balanced by the moment developed due to the elastic forces in the element.

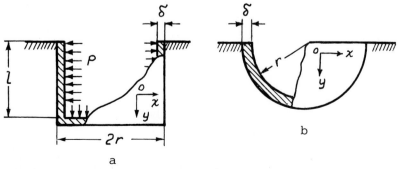

Figure 5.11 Cylindrical (a) and hemisphere (b) cups.

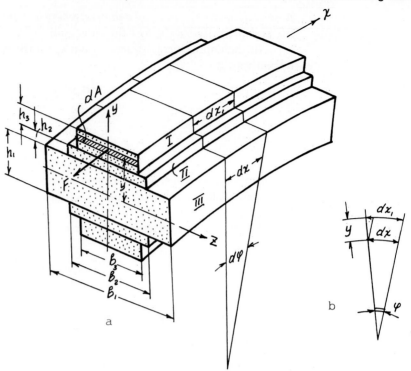

Figure 5.12 Schematic diagram of multilayer structure subjected to bending moment.

For the elementary area dA [m²], the elementary torque dM [N · m] due to the elastic force dF [N] normal to dA will be

$$dM = ydF \qquad\qquad (5.35)$$

where y = distance from the force to the neutral axis (z), m.
The neutral axis lies on the neutral plane (XZ) where fibers of material do not experience stress. Force dF is a resulting force in area dA due to the presence of normal stress σ [N/m²]; therefore,

$$dF = \sigma \cdot dA \qquad\qquad (5.36)$$

where $dA = b \cdot dy$ (b is the width of the area in question; for this case $b = b_3$).
According to Hooke's law and the geometrical relations shown in Figure 5.12,

$$\sigma = \epsilon E, \qquad \epsilon = \frac{dx_1 - dx}{dx}, \qquad dx_1 - dx = yd\varphi, \qquad (5.37)$$

therefore,

$$\epsilon = y\,\frac{d\varphi}{dx}. \qquad\qquad (5.38)$$

Combining formulas 5.35–5.38, the expression for the elementary moment is written as follows:

$$dM = \frac{d\varphi}{dx}\,Eby^2dy. \qquad\qquad (5.39)$$

In order to find the strain ϵ in any plane at distance y from the neutral, we must write integral equations giving the solution for $d\varphi/dx$. These equations should reflect the balance between the applied moment M and the moments obtained as a result of integration over the thicknesses. The integration is performed for each fraction of the section (areas I, II, areas symmetrical to them at another side of the substrate, and area III):

$$M = \int_0^{\frac{h_1}{2}} \frac{d\varphi}{dx} E_1 b_1 y^2 dy + \int_{\frac{h_1}{2}}^{\left(h_2 + \frac{h_1}{2}\right)} \frac{d\varphi}{dx} E_2 b_2 y^2 dy + \int_{h_2 + \frac{h_1}{2}}^{\frac{h_1}{2} + h_2 + h_3} \frac{d\varphi}{dx} E_3 b_3 y^2 dy$$

$$+ \int_{-\frac{h_1}{2}}^{0} \frac{d\varphi}{dx} E_1 b_1 y^2 dy + \int_{-\left(h_2 + \frac{h_1}{2}\right)}^{-\frac{h_1}{2}} \frac{d\varphi}{dx} E_2 b_2 y^2 dy + \int_{-\left(\frac{h_1}{2} + h_2 + h_3\right)}^{-\left(h_2 + \frac{h_1}{2}\right)} \frac{d\varphi}{dx} E_3 b_3 y^2 dy. \tag{5.40}$$

After elementary integration and substitution of $d\varphi/dx$ into formula 5.38, the expression for the strain at any level y is determined as follows:

$$\epsilon = \frac{3}{2} M \frac{y}{L + K + N} \tag{5.41}$$

where

$$K = \frac{E_1 b_1 h_1^3}{8}, \qquad L = E_2 b_2 \left(h_2^3 + \frac{3}{2} h_1 h_2^2 + \frac{3}{4} h_1^2 h_2\right), \qquad \text{and}$$

$$N = E_3 b_3 h_3 \left(h_3^2 + \frac{3}{2} h_1 h_3 + 3h_2 h_3 + \frac{3}{4} h_1^2 + 3h_1 h_2 + 3h_2^2\right). \tag{5.42}$$

With the strains determined, the stresses at each area can be calculated by simply multiplying them by E's.

A similar approach is used when a *bimorph (bimetallic) structure* is analyzed. In its regular form (Fig. 5.13), the bimetal element is constituted of a strip of two

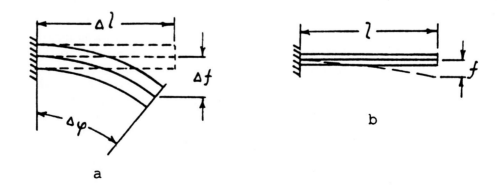

Figure 5.13 Bimetallic temperature-sensitive spring. a = deformation of incremental length Δl, b = deformation of total spring.

rigidly connected (welded) layers of materials with different coefficients of expansion. With the change in temperature, the layers are extended or contracted differently, thus causing bending of the spring. The ratio of the angular deformation $\Delta\varphi$ [rad] to the change in temperature Δt [°C] is

$$\frac{\Delta\varphi}{\Delta t} = \frac{6(\alpha_1 - \alpha_2) \cdot \Delta l}{(E_1\delta_1^2 - E_2\delta_2^2)^2/[E_1E_2\delta_1\delta_2(\delta_1 + \delta_2)] + 4(\delta_1 + \delta_2)} \qquad (5.43)$$

where α_1 and α_2 = temperature coefficients of expansion for two materials;
$\quad\quad \delta_1$ and δ_2 = thicknesses of layers, m;
$\quad\quad E_1$ and E_2 = Young's moduli of layers, N/m²; and
$\quad\quad \Delta l$ = incremental length of element, m.

It is obvious from formula 5.43 that the sensitivity $\Delta\varphi/\Delta t$ is at maximum when

$$E_1\delta_1^2 - E_2\delta_2^2 = 0 \qquad \text{or} \qquad \frac{\delta_1}{\delta_2} = \sqrt{\frac{E_2}{E_1}}. \qquad (5.44)$$

For a spring having such a ratio of thicknesses,

$$\frac{\Delta\varphi}{\Delta t} = (3/2)(\alpha_1 - \alpha_2) \cdot \Delta l/(\delta_1 + \delta_2). \qquad (5.45)$$

Displacement f [m] of the free end of the strip having length l [m] is obtained by taking the integral over the length:

$$f = (3/4)(\alpha_1 - \alpha_2)l^2 \cdot \Delta t/(\delta_1 + \delta_2). \qquad (5.46)$$

The maximum force F_m [N] which can be developed by the spring at its end will be

$$F_m = \frac{(E_1 + E_2)b(\delta_1 + \delta_2)^3}{8l^3} f \qquad (5.47)$$

where b = width of strip, m.

The selection of the *material for an elastic element* is an important step in its design. The plasticity of the material must be appropriate for shaping it, but special treatment, such as hardening, should provide necessary elastic properties. The material should have low mechanical hysteresis and creep. It must be stable over its lifetime. Its temperature drifts have to be small and repeatable. Quite often, in addition to these common requirements, other specific properties are required; for example, chemical resistivity, low magnetic permeability, the ability to work at a high temperature, and so on. Industry offers a wide variety of high-quality materials. It is worth noting that nonmetal spring elements are more widely used in contemporary transducers. The data on some of them can be found in Chapter 5, Absolute Pressure Capsules; and in Chapter 7, Mechanical Characteristics of Semiconductors. Table 5.4 of this section gives the characteristics of several materials. Depending on the variation in impurities and on specifics of the fabrication process, some properties of the materials vary significantly. Direct information from the manufacturer on the characteristics of the chosen material is desirable. Special attention should be paid to the calculation of stresses that must be much lower than the elastic limit. One of the approaches in the calculation of allowable stresses in an element is discussed in the following section.

TABLE 5.4
Characteristics of Several Materials for Elastic Elements

Material	Young's modulus (1×10^9)N/m^2	Coefficient of linear expansion (1×10^{-6})1/°C	Temperature coefficient of Young's modulus (1×10^{-4})1/°C
Tin-phosphor bronze	112	17.1	4.8
Beryllium copper	131	16.6	3.1
Invar	144	8.0	0.2
Ni-Span-C	176	8.0	0.2
German Silver (nickel silver)	126	16.6	4.0
Fused silica	73	0.55	1.7

Absolute Pressure Capsules

This section describes the absolute pressure capsule, which is a representative example of the integration of a sensitive element (pressure-sensitive diaphragm) with a transduction element (variable capacitance). It also illustrates typical stages in analysis and calculation during the design of an element.

Various modifications of the capsule are widely used as the core of the pressure transducers of avionic, industrial, medical, and other instruments [6]. Many of them are based on the deformation of the pressure-sensitive spring element, which is exposed to a vacuum on one side and to pressure on another. The deformation of the element is converted into an electrical signal proportional to the measured pressure.

Capsule Structure. The simplest model of the capsule (Fig. 5.14) includes flexible metal diaphragms 1 and 2, deformed by pressure P to be measured, and housing 3 with an evacuated cavity. The diaphragms are electrically insulated and form two electrodes of a capacitor. The change of the distance between the diaphragms causes a change of the capacitance, providing an adequate output.

Progress in solid-state technology and mechanical treatment of quartz, ceramics, and silicon makes it possible to build the elements [76, 77] with a diaphragm made of nonconductive material. The diaphragm carries a thin metal film or a diffused electroconductive layer (silicon) in order to form a capacitor. One of the versions of this type of capsule is shown in Figure 5.15. Two circular fused-silica, ceramic, or silicon diaphragms 1 and 2 with metal platings (3, 4) and electrical leads (6,7), sealed on ring 5, form a capacitance enveloped in a vacuum. The operation of this capsule is similar to that shown in Figure 5.14.

Requirements for high accuracy and temperature insensitivity have led to the use of a reference capacitance, which works together with a pressure-sensitive capacitance to reduce or eliminate temperature effects on the reading. Success in thermocompensation often requires the expedient location of compensating elements in

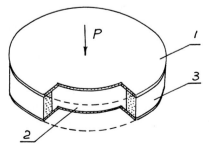

Figure 5.14 Simplest model of capsule with metal diaphragm.

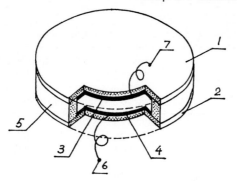

Figure 5.15 Capsule with nonmetal parts and platings.

order to subject both capacitances to the same temperature and to have similar heat-transfer characteristics in the parts forming the capacitances.

In the improved construction (Fig. 5.16), the compensating and pressure-sensitive capacitors are combined and formed of the electrodes on the same members.

Two basic configurations of the capsule are used. In the first type (Fig. 5.16a), two flat circular diaphragms 1 and 2, made of dielectric material with good spring characteristics (fused quartz or special ceramics), are sealed together on ring 3. Chamber 4 is evacuated. Conducting plates 5, 6, 7, and 8 are deposited as a thin metal film on the inner surfaces of the diaphragms. Electrical conductors 9, 10, 11, and 12 are attached to plates to be connected with an electrical circuit. Plates 5 and 6 are circular and are positioned centrally. Plates 7 and 8 are annular and are located at the edges of the diaphragms.

Plates 5 and 6 form a capacitor sensitive to pressure P, which is applied externally, whereas the capacitor formed between plates 7 and 8 is substantially constant because the deformation of the diaphragm is small at the edge. The pressure-sensitive and reference capacitances are made equal when the lowest pressure is measured. Electrostatic shielding, made of a thin metal film on the outside surface of the capsule, protects the capsule against outside fields.

Figure 5.16b illustrates a similar approach with a slight difference. Here diaphragms 1 and 2 are attached to the circular housing 3. Armature 4 is also rigidly fixed on the housing. All parts are made of the same material and form a sealed cavity with a vacuum inside. The internal surfaces of the diaphragms are coated uniformly.

The armature coating is divided into two sections (7, 8 and 9, 10) as described above. The sections can be electrically connected, and they can work in parallel.

Pressure-sensitive and reference capacitors are formed with platings 5 and 6 and corresponding platings on armature 4. Electrostatic shielding covers the outside surface of the capsule.

This structure is more complicated than that in Figure 5.16a, but has several advantages. The absolute value of the capacitance change is larger here if the diaphragm dimensions are the same for both approaches. Connecting the diaphragm platings to the ground is more convenient. The capsule is better protected against noise from the outside field. Both types have a symmetrical configuration and are mechanically balanced to reduce responses to acceleration and vibration.

The capsule can also be built with a differential change of two capacitors; in this construction, when one capacitance is increased under pressure the other is decreased (Fig. 5.16e). The pressure-sensitive diaphragm 1 is connected with diaphragm 2, which is encapsulated inside housing 3. Deformations of diaphragm 1 are transmitted to diaphragm 2 through stem 4. The surfaces of partition 5 and the diaphragms have solid platings 6, 7, 8, and 9 as shown in Figure 5.16e. Applied pressure causes practically equal, opposite-sign changes of capacitance formed by couples of electrodes: 10-11 and 12-13. Like any differential structure, this capsule has an improved electrical performance, but is sensitive to acceleration (especially

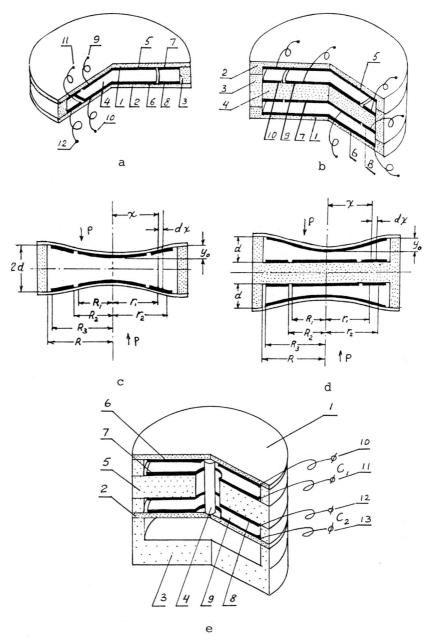

Figure 5.16 Capacitance capsules with two diaphragms and combination of pressure-sensitive and compensating parts. a = structure with only two diaphragms, b = construction with diaphragms and armature, c = radial section of capsule (a), d = radial section of capsule (b), e = capsule with differential capacitances.

perpendicular to the surface of the diaphragm). There are several more complicated constructions that have different mechanical, in addition to electrical, structures.

Capsule Material, Shape, and Size. After choosing a particular configuration of the capsule, the selection of materials is important. If all the parts of the structure are made of the same material, then temperature stresses, which cause instabilities, are reduced or eliminated. One of the materials used for the capsule is fused silica. It has excellent characteristics as a spring material. Internal friction in the material is low, and mechanical hysteresis is negligible. Quartz is chemically resistive, responds to temperature much less than metal, and is stable in time. However, its tensile strength is low, and the diaphragm is relatively thick in order to keep stresses within the allowed limit. Diaphragm deflection is often a small fraction of its thickness.

Silicon, sapphire, alumina, pyroceramic, and several other materials (Chapter 7, Mechanical Characteristics of Semiconductors; Chapter 9, Electromechanical Microsensors) can also be employed for the housing and diaphragms. The quality of the machining and joining of the parts, the reliable bonding of the thin metal film on the surfaces, and the electrical stability of the film are the major factors that define the performance of the device. A deposition of platinum is used to form electrodes. A gold-over-nickel layer is vacuum-deposited and is employed as an outer electrostatic shield.

The diaphragm deformation depends to a considerable extent on the condition of the edge because this is an area of maximum stress. Any imperfection in the attachment of the diaphragm makes the instrument unreliable.

Fused quartz, like any glass material, has microcracks on the surface, which tend to propagate and change the original tuning of the instrument. Thus, it is desirable to etch off the upper layer of the diaphragm using hydrofluoric acid. Etching is also recommended for preparing the surfaces. After the appropriate material for the diaphragm is chosen, tensile strength (σ_{ts}), Young's modulus (E), and Poisson's ratio (ν) become known.

The next stage in capsule analysis is the evaluation of the size of the diaphragm. A diaphragm with a large diameter is desirable because the proportional increase of the capacitance always simplifies signal conditioning (low capacitance causes high electrical disturbance). However, this diameter is limited by the size of the element and should meet special requirements.

Calculation of the diaphragm's thickness is the next step. First of all, the type of the diaphragm (flat or corrugated) and its bonding at the edges should be selected. A flat diaphragm develops a smaller stroke than the corrugated one; however, its performance is more accurate, and it is less expensive to fabricate. Experience shows that a rigid rather than flexible attachment of the diaphragm's edge ensures a repeatable deformation of the structure. As a rule, the diaphragm is shaped like a thin disk for easy machining. In contemporary technology, rectangular diaphragms etched in silicon are widely used for mini- and micro- strain-gage sensors (see Chapter 8, Fabrication of Pressure-sensitive Microstructures; Chapter 9, Semiconductor Strain Gages). The same trend exists in the development of the newest capacitive microsensors. However, all new designs of fused silica or ceramic pressure-sensitive macroelements are shaped like disks. For this reason, the circular diaphragm is chosen for the design that is considered below.

Stressed State of Diaphragm.   Once the shape of the diaphragm and method of its fixing have been chosen, a stress analysis should be performed to calculate the diaphragm's thickness and to obtain a diagram of stresses in the material of the diaphragm.

A cross cut along the diameter of the diaphragm is shown in Figure 5.17a. Under pressure P, applied to the upper surface of the diaphragm, any point with radius x undergoes deflection y. At the center it is y_0. The radius of the diaphragm is R, and its thickness is δ. The stressed state of each elementary cube (Fig. 5.17d) is determined by two stresses. One of them is the radial stress σ_r, measured along a radius, and the other is the tangential stress σ_t, which is perpendicular to σ_r. These two stresses are maximum at the surface, and they are equal to zero at the central line of the diaphragm (broken line in Fig. 5.17d). When the diaphragm is deformed, its elements are under compression or tension, depending on the sign of the radius of curvature (Fig. 5.17e). It is noteworthy that at the points labeled A and B, the radius of the curvature changes its sign, and the stresses are at zero. The following relationships [66] allow calculation of the magnitudes of stresses σ_r and σ_t and deflections y as functions of the major parameters of the diaphragm:

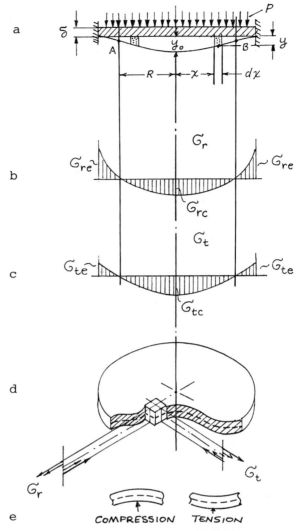

Figure 5.17 Flat circular diaphragm under stress.
a = diaphragm's cut, b = radial stress plot, c = tangential stress plot, d = diaphragm stress state.

$$\sigma_r = \frac{3w}{8\pi m\delta^2}\left[(3m + 1)\frac{x^2}{R^2} - (m + 1)\right], \tag{5.48}$$

$$\sigma_t = \frac{3w}{8\pi m\delta^2}\left[(m + 3)\frac{x^2}{R^2} - (m + 1)\right], \tag{5.49}$$

$$y = \frac{3w(m^2 - 1)}{16\pi E m^2 \delta^3}\cdot\frac{(R^2 - x^2)^2}{R^2}, \tag{5.50}$$

$$w = P\pi R^2 \tag{5.51}$$

where σ_r = stress at the surface in radial direction, N/m^2;
   σ_t = stress at the surface in tangential direction, N/m^2;
   m = reciprocal of Poisson's ratio ν, that is, $m = 1/\nu$, dimensionless;
   P = pressure, Pa;
   E = Young's modulus, N/m^2;
   R = diaphragm's radius, m;
   δ = diaphragm's thickness, m; and
   y = deflection at radius x, m.

By substituting $x = R$ and $x = 0$ into formulas 5.48 and 5.49, the characteristic stresses at the edge (σ_{re}, σ_{te}) and at the center (σ_{rc}, σ_{tc}) can be found:

$$\sigma_{re} = \frac{3w}{4\pi\delta^2}, \qquad \sigma_{te} = \frac{3w}{4\pi m\delta^2}, \qquad \sigma_{rc} = \sigma_{re} = \frac{3w(m+1)}{8\pi m\delta^2}. \qquad (5.52)$$

The deflection in the center y_0 is defined by substituting $x = 0$ into formula 5.50:

$$y_0 = \frac{3w(m^2 - 1)R^2}{16\pi E m^2 \delta^3}. \qquad (5.53)$$

This is a linear approximation of y_0. In accurate calculations, more exact values for y_0 can be found with the equation 5.18 [66].

The resulting or equivalent stress σ_{eq} must be found in order to calculate the thickness. In this case, a two-dimensional stressed state is taken into account. According to the distortion-energy concept for a two-dimensional state [67], the equivalent stress for this case is defined:

$$\sigma_{eq} = \sqrt{\sigma_r^2 + \sigma_t^2 - \sigma_r \sigma_t}. \qquad (5.54)$$

Since the maximum stresses are developed at the edge of the diaphragm (Fig. 5.17b and c; formulas 5.52), the maximum of σ_{eq} is calculated with formula 5.54 by substituting: $\sigma_r = \sigma_{re}$ and $\sigma_t = \sigma_{te}$.

Hence,

$$\max \sigma_{eq} = \sqrt{\sigma_{re}^2 + \sigma_{te}^2 - \sigma_{re}\sigma_{te}}. \qquad (5.55)$$

Taking values for σ_{re} and σ_{te} from equations 5.52 and substituting them into equation 5.55 leads to the following:

$$\max \sigma_{eq} = 0.75 \frac{PR^2}{\delta^2} \sqrt{1 - \nu + \nu^2} = 0.695 \frac{PR^2}{\delta^2}. \qquad (5.56)$$

Following from this expression for $\nu = 0.17$ (fused silica),

$$\delta = 0.834R \sqrt{\frac{P}{\max \sigma_{eq}}}. \qquad (5.57)$$

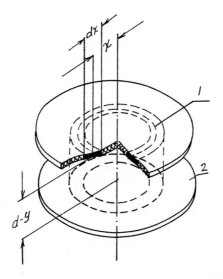

Figure 5.18   Incremental annual capacitor.

For a material like fused silica, allowable max σ_{eq} should be taken at least 1.5 times lower than the damaging stress (which is $60.0 \times 10^6 \text{N/m}^2$ for fused silica) in order to ensure a proper margin of security against failure.

The maximum stroke max y_0 at the center of the diaphragm can be found if the known maximum pressure P_{max} is substituted in formula (5.18):

$$\max y_0 = \frac{3P_{max}(m^2 - 1)R^4}{16Em^2\delta^3}. \tag{5.58}$$

The distance between the diaphragm and the stationary plate d should be slightly larger than max y_0 to avoid mechanical and electrical contacts between the parts at maximum pressure.

Capsule Transfer Characteristic. The transfer characteristic estimates relations between the capacitance and pressure to be measured. It is obtained by integrating the equation for the differential capacitance dC, which is formed by an elementary circular electrode 1 and plate 2 (Fig. 5.18). The electrode has a mean radius x, width dx, and distance $(d - y)$ from the plate. This is a model for incremental capacitance in the structure shown in Figure 5.16d.

$$dC = \epsilon \frac{2\pi x \cdot dx}{d - y} \tag{5.59}$$

where ϵ = permittivity, F/m, and
$2\pi x \cdot dx$ = area of the elementary electrode, m^2.

From formulas 5.50 and 5.18, y can be expressed as a function of y_0 (which is by itself a function of P):

$$y = y_0 \left[1 - \left(\frac{x}{R}\right)^2\right]^2. \tag{5.60}$$

Hence,

$$dC = \epsilon \frac{2\pi x}{d - y_0\left[1 - \left(\frac{x}{R}\right)^2\right]^2} dx. \tag{5.61}$$

The capacitance of any capacitor C, formed by the annular electrodes with arbitrary radii r_1 and r_2 and the plate, can be found by integrating equation 5.61 with respect to x:

$$C = 2\pi\epsilon \int_{r_1}^{r_2} \frac{x}{d - y_0\left[1 - \left(\frac{x}{R}\right)^2\right]^2} dx. \tag{5.62}$$

In order to solve the integral, the integrand's denominator D is modified and some substitutions are made:

$$D = d - y_0\left[1 - \left(\frac{x}{R}\right)^2\right]^2 = \frac{1}{R^4}\left[(d - y_0)R^4 + 2y_0R^2x^2 - y_0x^4\right]$$

$$= \frac{1}{R^4}(a_1 + 2b_1z + c_1z^2) \tag{5.63}$$

where

$$z = x^2, \qquad a_1 = (d - y_0)R^4, \qquad b_1 = y_0 R^2, \qquad c_1 = -y_0 \qquad dx = \frac{1}{2x} dz. \quad (5.64)$$

Finding C permits the solution of the following integral:

$$\int \frac{dz}{a_1 + 2b_1 z + c_1 z^2} = \frac{1}{2\sqrt{b_1^2 - a_1 c_1}} \times \ln \frac{c_1 z + b_1 - \sqrt{b_1^2 - a_1 c_1}}{c_1 z + b_1 + \sqrt{b_1^2 - a_1 c_1}} \quad (5.65)$$

This integral gives real values if

$$b_1^2 - a_1 c_1 > 0. \quad (5.66)$$

Substitutions a_1, b_1, and c_1 into formula 5.66 give

$$y_0^2 R^4 + (d - y_0)R^4 y_0 = y_0 d \cdot R^4$$

$y_0 d \cdot R^4$ is always larger than zero.

Finally, substituting $z = x^2$ in formula 5.65 and computing the integral within limits r_1 and r_2 yield

$$C = \int_{r_1}^{r_2} \epsilon \pi R^4 \frac{1}{2\sqrt{b_1^2 - a_1 c_1}} \cdot \ln \frac{c_1 x^2 + b_1 - \sqrt{b_1^2 - a_1 c_1}}{c_1 x^2 + b_1 + \sqrt{b_1^2 - a_1 c_1}}$$

$$= \frac{\epsilon \pi R^2}{2\sqrt{d y_0}} \left(\ln \frac{R^2 - r_2^2 - R^2\sqrt{d/y_0}}{R^2 - r_2^2 + R^2\sqrt{d/y_0}} - \ln \frac{R^2 - r_1^2 - R^2\sqrt{d/y_0}}{R^2 - r_1^2 + R^2 \sqrt{d/y_0}} \right). \quad (5.67)$$

This formula can be modified by substitutions: $\gamma = y_0/d$ = relative deflection of the center of the diaphragm, $\rho_1 = r_1/R$ and $\rho_2 = r_2/R$ = relative radii for calculation of the capacitance:

$$C = \frac{\epsilon \pi R^2}{2d \sqrt{\gamma}} \ln \frac{\left[\gamma(1 - \rho_1^2) + \sqrt{\gamma}\right]\left[\gamma(1 - \rho_2^2) - \sqrt{\gamma}\right]}{\left[\gamma(1 - \rho_1^2) - \sqrt{\gamma}\right]\left[\gamma(1 - \rho_2^2) + \sqrt{\gamma}\right]} \quad (5.68)$$

where C can be calculated as the signal capacitance C_s (most sensitive to pressure) or the reference capacitance C_r (less sensitive to pressure), depending on the value of the relative radii ρ_1 and ρ_2, which are substituted in formula 5.68.

The following conditions for r_1 and r_2 are necessary if formula 5.68 is used:

$$0 \leq r_1 \leq R, \qquad 0 \leq r_2 \leq R, \qquad r_2 > r_1.$$

Expression 5.68 is convenient for calculation if $0 < \gamma \leq 1$. For $\gamma = 0$ (zero deflection of the diaphragm), the following relation is more simple:

$$C = \frac{\epsilon \pi R^2}{d} (\rho_2^2 - \rho_1^2). \quad (5.69)$$

The two formulas are immediately applicable for calculating the values of the capacitances for the most common cases such as when one of the platings belongs to the diaphragm deformed under pressure, and another is on the part remaining flat. By a simple assigning of specific values to the terms in the formulas, the calculations for the different modifications of the capsule can be performed. For instance, to calculate C_s and C_r for half of the capsule in Figure 5.16b, the following substitutions must be made for C_s: $r_1 = 0$, $r_2 = R_1$, or $\rho_1 = 0$, $\rho_2 = R_1/R$; for C_r: $r_1 = R_2$, $r_2 = R_3$, or $\rho_1 = R_2/R$, $\rho_2 = R_3/R$.

For the capsule with two working sides, if they are in a parallel connection, the total capacitance is doubled: $2C_s$ and $2C_r$. For the calculations related to the version in Figure 5.16a, the physical gap between two diaphragms $2d$ must be divided by 2 to obtain an equivalent gap for substitution in formula 5.67 and 5.68. In this case, capacitances C_s' and C_r' are calculated from the following simple relationships: $C_s' = C_s/2$ and $C_r' = C_r/2$ where C_s and C_r are defined in formulas 5.67 and 5.68.

Selection of the magnitudes of R_1, R_2, and g depends on the design ($g = R_2 - R_1$). If C_s and C_r are chosen equal to each other, R_1 and R_2 are found by equating the areas for C_s and C_r at $P = 0$:

$$R_1 = \frac{-g + \sqrt{2R_3^2 - g^2}}{2} \tag{5.70}$$

$$R_2 = R_1 + g. \tag{5.71}$$

Some departures of the experimental data from the calculated data may occur in the event of imperfections in the geometry or assembly of parts and variations of the physical parameters of materials. It is difficult to define precisely the diaphragm radius R (to the inside edge of the seal) if the diaphragm is fused. Exact values of the elasticity modulus and Poisson's ratio should be taken for a given lot of material.

The structures in Figures 5.14–5.16 are well protected against vibration. First, the natural frequency of the diaphragms is usually higher than the frequencies of the specified vibrations. Second, the structure itself inherently quenches the vibrations of the diaphragms. When the vibrations are oriented perpendicularly to the diaphragm surface (the most sensitive direction), each diaphragm is stimulated to move in the same direction. However, for the capsule to be excited by an intense vibration, a phase angle of 180° must be present between the two disturbing forces applied to each diaphragm. Since this angle is usually small, the structure tends to damp oscillations.

Dynamic Characteristics of the Diaphragm.   The natural frequency and the damping coefficient are the most essential dynamic parameters of the diaphragm. The latter is usually obtained from experiments. For a flat diaphragm rigidly fixed at the edge, the natural frequency f_i of the vibration's mode number i is defined as follows [44]:

$$f_i = \frac{A_i}{2\pi R^2} \sqrt{\frac{E\delta^2}{12(1 - \nu^2)m_v}} \tag{5.72}$$

where R, E, δ, and ν have been defined above;
   m_v = density of the diaphragm's material, kg/m^3; and
   A = factor depending on the mode of vibration, dimensionless.
   A_i is equal to 10.21, 39.78, and 88.90 for the first, second, and third modes
      of vibration, respectively.

When the capsule is connected to a pneumatic or hydraulic system, the change of the volume at the diaphragm due to its deformation can be important for the evaluation of the dynamic characteristics of the system. This change of volume ΔV is between the initial diagram plane and its deflected surface. It is determined by integrating the volume of the elementary ring dV (Fig. 5.17a):

$$dV = 2\pi x \cdot dx \cdot y. \tag{5.73}$$

Substituting y from formula 5.60 and taking the integral between 0 and R gives

$$\Delta V = \frac{1}{3} \pi R^2 y_0 \qquad (5.74)$$

where ΔV = change of the volume, m^3.

The pertinent characteristic for the measuring system is its volume change per unit pressure change $\Delta V/\Delta P$. Usually, the smaller this ratio, the better the dynamic characteristics of the system. This ratio is readily calculated if expressions 5.51, 5.53, and 5.74 are combined, and P in formula 5.51 is substituted by ΔP:

$$\frac{\Delta V}{\Delta P} = \frac{\pi R^6 (m^2 - 1)}{16 E m^2 \delta^3}. \qquad (5.75)$$

Temperature Drifts. When the basic calculations are completed, the values of R, R_1, R_2, R_3, δ, d, E, and ν become known. The temperature characteristics of all these parameters are also known. They are usually described by a linear function of temperature:

$$A = A_0(1 + \alpha_i t) \qquad (5.76)$$

where A = value of the parameter at the specific temperature;
$\quad A_0$ = value of A at the reference temperature; and
$\quad t$ = departure of the temperature from that at which A_0 is measured, and α_i is the temperature coefficient related to the parameter in question.

As a rule, the most significant errors are brought about by temperature. In order to predict these errors, formulas 5.18, 5.67, and 5.69 can be effectively used. All the arguments in the formulas are substituted by the corresponding linear functions of temperature using the linear approximation in formula 5.76. Then variations are assigned to the temperature while the other terms remain constant. As a result, the computed values of C_s and C_r as functions of temperature allow the evaluation of instabilities and the development of measures for diminishing them. Computer simulation is helpful in this calculation.

Signal conditioners usually provide an output proportional to the ratio C_s'/C_r', which is a function of pressure to be measured. The error Δ due to temperature can be evaluated by the following ratio:

$$\Delta = \left(\frac{C_s}{C_r} - \frac{C_s'}{C_r'} \right) \Big/ \frac{C_s}{C_r} \qquad (5.77)$$

where C_s and C_r are taken for the initial conditions. C_s' and C_r' are the signal and reference capacitances that have experienced a temperature drift.

Sensitivity to Acceleration. In spite of special measures of protection, the capsule may be sensitive to acceleration due to a departure of the real characteristics of specimens from the ideal. This departure occurs primarily due to various tolerances. The diaphragm undergoes the maximum deflections when acceleration is directed perpendicularly to its surface. This deflection should be compared with that which is a response to pressure to be measured. If a unit of the diaphragm surface A is affected by acceleration a, the equivalent pressure P_s is calculated:

$$P_s = m_s a \qquad (5.78)$$

P_s = equivalent pressure, Pa;

m_s = mass of 1m^2 of the diaphragm, kg/m^2; and

a = acceleration, m/s^2.

The error caused by acceleration can be evaluated with formula 5.77 by assigning to P the value of P_s, found in formula 5.78, and using this result in formulas 5.18, 5.53, and 5.67–5.69.

Some Characteristics of Design. The following constants describe the physical parameters of fused silica [77,78]. They can be used to calculate different constructional and functional characteristics of the capsule.

$$\text{Mass density } \rho = 2.202 \times 10^3 \text{ kg/m}^3;$$

$$\text{Elasticity modulus } E = 73 \times 10^9 \text{N/m}^2 \text{ at } 25°C;$$

$$\text{Poisson's ratio } \nu = 0.17 \text{ at } 25°C;$$

$$\text{Tensile strength } \sigma_{ts} = 58 \times 10^6 \text{N/m}^2;$$

$$\text{Compressive strength } \sigma_{cs} = 1340 \times 10^6 \text{N/m}^2;$$

$$\text{Temperature coefficient of expansion } \alpha_l = 0.55 \times 10^{-6} \text{ 1/°C};$$

$$\text{Temperature coefficient of elasticity modulus } \alpha_E = 167.8 \times 10^{-6} \text{ 1/°C; and}$$

$$\text{Temperature coefficient of Poisson's ratio } \alpha_\nu = 78.8 \times 10^{-6} \text{ 1/°C}.$$

The last three values are defined for the temperature range of 0 to 150°C. In addition to these characteristics, the permittivity for vacuum $\epsilon = 8.85 \times 10^{-12}$F/m.

Elastic characteristics for other materials used for diaphragms can be found in Chapters 7 and 9. The values of capacitances calculated with the formulas of this paragraph are obtained in farads when the dimensions are substituted in meters.

Fused silica is regarded as a stable material because of its low expansion coefficient. However, its Young's modulus instability is relatively high. It is known from practice that the zero drift for a fused silica-type transducer is low because it is usually governed by the linear expansion of material. However, the sensitivity, which is directly defined by the diaphragm stiffness, is significantly affected by the temperature drift of the elasticity modulus.

The choice of the capsule's structure and constructional characteristics varies depending on the specifications of a particular instrument. A transducer that incorporates this capsule is usually furnished with one more capsule that is not exposed to the pressure to be measured but is employed to compensate for temperature and other environmental factors. The following parameters describe one of the practical elements based on scheme of Figure 5.15.

The diameter of the diaphragm $2R$ is between 30 and 50mm; the thickness of the diaphragm δ is between 1 and 2mm; and the gap between two diaphragms or between one diaphragm and a stationary plate, d or $2d$, is from 0.1 to 0.2mm.

The maximum deflection of the diaphragms in the center is between 0.03 and 0.08mm.

The safety coefficient is between 1.5 and 2 (when tensile stress is taken as a reference).

The capacitance in a vacuum is from 50 to 100pF. Its change at full-rated pressure is between 30 and 60 percent.

The temperature coefficient of a zero shift (without any compensation and only for the signal capacitance) is less than $\pm 0.1\%$ FS/100°C.

The sensitivity change for the uncompensated capsule is approximately 2%/100°C.

When the capsule works in a measuring system that has adequate compensation, an accuracy between 0.02 and 0.01% FS for the temperature range between $-55°C$ and $+70°C$ can be achieved.

The capsules shown in Figure 5.16 have approximately the same basic dimensions (they are slightly thicker). Over the pressure range, the change of the signal capacitance C_s varies between 50 and 200%, whereas the reference capacitance change is between 5 and 10%.

INERTIAL-MASS ELEMENTS

In this section, we will consider the functions of the elements that relate to accelerometers and seismometers (Figure 5.19). The essential component of these instruments is a mass m [kg], which is suspended inside the case by springs with a spring factor k [N/m]. The mass is connected to a damper having friction coefficient b [N · s/m] and symbolically designated by a dashpot. The mass links to a sensor that allows the measurement of the relative-to-the-case motion $z = (X - x)$, where X is a displacement of the object, and x is a displacement of the mass.

The behavior of the instrument is described by the equation of motion:

$$m \frac{d^2x}{dt^2} + b\left(\frac{dx}{dt} - \frac{dX}{dt}\right) + k(x - X) = 0 \qquad (5.79)$$

This equation written in terms of z and X is

$$m \frac{d^2z}{dt^2} + b \frac{dz}{dt} + kz = -m \frac{d^2X}{dt^2}. \qquad (5.80)$$

It can be assumed for our analysis that the motion of the vibrating object is sinusoidal with peak value X_0 [m]:

$$X = X_0 \sin \omega t. \qquad (5.81)$$

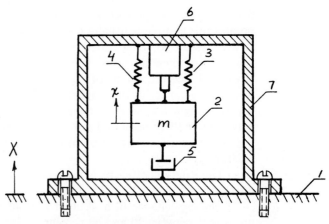

Figure 5.19 Schematic diagram of vibration measuring instrument. X = motion to be measured, x = motion of seismic mass, 1 = moving support, 2 = seismic mass, 3 and 4 = springs, 5 = dashpot (damper), 6 = motion sensor, 7 = supporting case.

Differentiating X and substituting in formula 5.80 gives

$$m \frac{d^2z}{dt^2} + b \frac{dz}{dt} + kz = mX_0\omega^2 \sin \omega t. \qquad (5.82)$$

The steady-state solution of this equation will be of the following form:

$$z = z_0 \sin (\omega t - \varphi) \qquad (5.83)$$

where

$$z_0 = \frac{mX_0\omega^2}{\sqrt{(k - m\omega^2)^2 + (b\omega)^2}} = \frac{X_0 \left(\frac{\omega}{\omega_n}\right)^2}{\sqrt{\left[1 - \left(\frac{\omega}{\omega_n}\right)^2\right]^2 + \left(2\zeta \frac{\omega}{\omega_n}\right)^2}}, \qquad (5.84)$$

$$\varphi = \arctan \frac{\omega b}{1 - m\omega^2} = \frac{2\zeta \frac{\omega}{\omega_n}}{1 - \left(\frac{\omega}{\omega_n}\right)^2} \qquad (5.85)$$

where $\omega_n = \sqrt{k/m}$ = natural frequency of undamped oscillation, rad/s;
$\zeta = b/b_c$ = damping factor, dimensionless; and
$b_c = 2m\omega_n$ = critical damping, kg/s.

Equation 5.84 is graphically presented in Figure 5.20. The useful range of frequencies for the seismometer and accelerometer is indicated in the graph with reference to the relative frequency ω/ω_n. The seismometer is usually designed for operating at a high ratio ω/ω_n. With the increase in ω/ω_n, the relative displacement

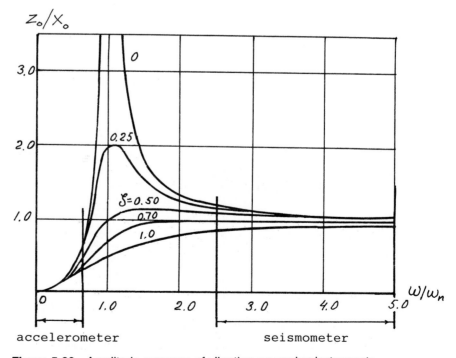

Figure 5.20 Amplitude response of vibration-measuring instruments.

z_0 becomes equal to X_0; in other words, the mass tends to remain stationary while the case moves with the object whose vibrations are measured. If the sensor provides differentiating of the relative displacement (piezoelectric or electrodynamic type), the output of the transducer will be read as the velocity of the moving object. The useful range for accelerometers is $0 \leq \omega/\omega_n \leq 0.4$ (sometimes $0 \leq \omega/\omega_n \leq 0.6$). When the ratio ω/ω_n approaches zero, the value of z_0 becomes equal to the amplitude of the vibratory acceleration of the object times the constant factor:

$$z_0 = \frac{X_0 \omega^2}{\omega_n^2}. \tag{5.86}$$

Note that

$$\frac{d^2 X}{dt^2} = -X_0 \omega^2 \sin \omega t. \tag{5.87}$$

The value of

$$N = 1 \bigg/ \sqrt{\left[1 - \left(\frac{\omega}{\omega_n}\right)^2\right]^2 + \left(2\zeta \frac{\omega}{\omega_n}\right)^2} \tag{5.88}$$

in expression 5.84 determines the error in the reproduction of the measured acceleration (Fig. 5.21). The more it differs from unity, the greater the error. When the moving system of the accelerometer is not damped ($\zeta = 0$), the value of $N = 1/[1 - (\omega/\omega_n)^2]$ changes rapidly with the increase in the frequency. However, when ζ is between 0.65 and 0.70, the drop of $[1 - (\omega/\omega_n)^2]$ is compensated, to a certain extent, by the growth of $2\zeta(\omega/\omega_n)^2$.

For a faithful reproduction of a complex-wave periodic process and a saving of the shape of the wave, all harmonic components must have phase shifts proportional to their frequencies. Theoretically, shift φ must be linearly proportional to the ratio ω/ω_n. Practically, this condition is nearly satisfied for $\zeta = 0.70$ when $\varphi \approx (\pi/2) \times (\omega/\omega_n)$ (Fig. 5.22). For example, if the system has $\zeta = 0.7$, and the object's periodic motion is described by the equation

$$X = X_1 \sin \omega_1 t + X_2 \sin \omega_2 t + X_3 \sin \omega_3 t, \tag{5.89}$$

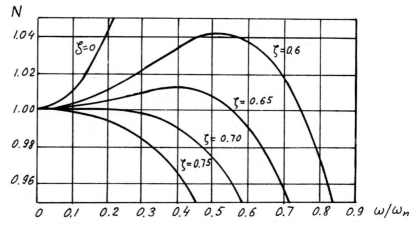

Figure 5.21   Factor N characterizing accelerometer error versus frequency. Note that plateau zone of curve for condition $\zeta = 0.7$ is longest.

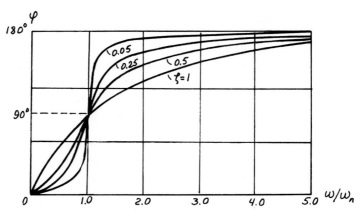

Figure 5.22   Phase shift between readout sinusoidal signal and vibratory sinusoidal wave of object.

the phase shift corresponding to the frequencies will be approximately

$$\varphi_1 = (\pi/2) \times (\omega_1/\omega_n), \qquad \varphi_2 = (\pi/2) \times (\omega_2/\omega_n), \qquad \text{and}$$

$$\varphi_3 = (\pi/2) \times (\omega_3/\omega_n). \tag{5.90}$$

In equation 5.89, X_1, X_2, and X_3 = peak values of the harmonics of frequencies ω_1, ω_2, and ω_3, respectively, m.

According to formulas 5.83 and 5.86, the output z of the accelerometer is

$$z = z_1 \sin(\omega_1 t - \varphi_1) + z_2 \sin(\omega_2 t - \varphi_2) + z_3 \sin(\omega_3 t - \varphi_3)$$

$$= \frac{1}{\omega_n^2}\left[\omega_1^2 X_1 \sin \omega_1\left(t - \frac{\pi}{2\omega_n}\right) + \omega_2^2 X_2 \sin \omega_2\left(t - \frac{\pi}{2\omega_n}\right)\right.$$

$$\left. + \omega_3^2 X_3 \sin \omega_3\left(t - \frac{\pi}{2\omega_n}\right)\right] \tag{5.91}$$

where z_1, z_2, and z_3 = peak values of relative displacements (in m), calculated with formula 5.86.

The real acceleration of the object, whose motion equation is given by formula 5.89 is

$$\frac{d^2X}{dt^2} = -(\omega_1^2 X_1 \sin \omega_1 t + \omega_2^2 X_2 \sin \omega_2 t + \omega_3^2 X_3 \sin \omega_3 t). \tag{5.92}$$

It is obvious from comparing the expressions 5.91 and 5.92 that there is no distortion in the reproduction of the real acceleration. The differences between the two functions are the amplitude factor $1/\omega_n^2$ and time shift of $\pi/2\omega_n$, which do not distort the reproduction of the acceleration in the measuring system.

The increased requirements for accuracy and ranges in the measurement of acceleration have stimulated the development of servo-type accelerometers. Their principles are discussed in pp. 180–186. Some attention is given to the solid-state designs introduced in the beginning of Chapter 9.

The spring factors of simple elastic elements and their fundamental frequencies can be found in Chapter 4, Resonant Elements; and Chapter 5, Elastic Elements. Some aspects of acoustic damping are discussed in the section of Chapter 5 entitled Acoustical Elements.

SENSING AND TRANSDUCTION ELEMENTS OF FLOWMETERS

In addition to the electromagnetic flowmeter analysis (Chapter 4, Electrodynamic Elements) we will present calculations for several mechanical flow-sensing elements [79] converting the flow to be sensed into mechanical quantities. These elements are usually combined with a transduction element or with a transducer forming an electrical signal. At the end of this section we will review several types of ultrasonic elements.

A *turbine wheel* installed in a pipeline (Scheme 110) rotates at a speed proportional to the flow rate. The number of revolutions of the rotor per second n is directly proportional to the average velocity of the fluid w [m/s] and inversely proportional to the turbine blade pitch l [m]:

$$n = k\frac{w}{l} \tag{5.93}$$

where k = constructional constant, dimensionless.

Since the volumetric flow rate Q [m³/s] is

$$Q = wA \tag{5.94}$$

where A = area of pipe, m²,

$$n = \frac{kQ}{Al}. \tag{5.95}$$

For accurate measurements, it is desirable to have straightening flow vans mounted before the turbine. The pipe before and after the turbine must be straight at lengths $10D$ and $5D$, respectively (D is the diameter of the pipe).

A *differential-pressure centrifugal-type (elbow) sensing element* (Scheme 109e) contains two pressure ports, and the pressure difference between these two ports is a measure of the flow:

$$Q = 0.22R^2 \sqrt{\frac{r \cdot \Delta P}{R \cdot \sigma}} \tag{5.96}$$

where ΔP = pressure difference, [mm H₂O];
R = radius of tube, m;
r = average radius of pipe elbow, m; and
σ = fluid density with respect to water, dimensionless.

In the derivation of the transfer characteristic of an *orifice-plate element* (Scheme 109a; Fig. 5.23a) we assume that the liquid is not compressible, the velocities in areas I and II are uniform with respect to the pipe radius, and they are parallel to the pipe radius.

According to the law of conservation of energy,

$$\int w \cdot dw = -g \int \frac{dP}{\gamma_1}. \tag{5.97}$$

For the flow between areas I and II, solution of this equation gives

$$P_1' - P_2' = \frac{\gamma_1}{2g}(w_2'^2 - w_1^2) \tag{5.98}$$

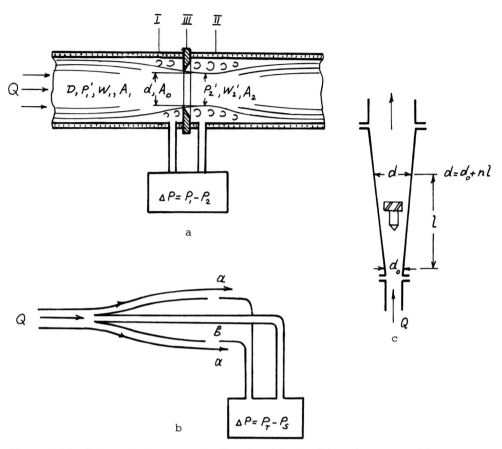

Figure 5.23 Schematic diagrams of orifice plate (a), Pitot (b), and rotametric (c) sensing elements.

where P'_1 and P'_2 = absolute pressures in areas I and II, respectively, Pa;

γ_1 = specific gravity of fluid, N/m³, ($\gamma_1 = \phi g$, ϕ = fluid density [kg/m³], g = gravitational acceleration, 9.807 m/s²); and

w_1 and w'_2 = average fluid velocities in areas I and II, respectively, m/s.

Due to the flow continuity,

$$A_1 w_1 = A_2 w'_2. \tag{5.99}$$

Defining μ as a coefficient of the contraction of the stream $\mu = A_2/A_0$ and taking the ratio $m = A_0/A_1$, the combination of the last equations gives

$$w_1 = \mu w'_2 \frac{A_0}{A_1} = \mu m w'_2 \quad \text{and} \tag{5.100}$$

$$w'_2 = \frac{1}{\sqrt{1 - \mu^2 m^2}} \sqrt{\frac{2g}{\gamma_1}(P'_1 - P'_2)} \tag{5.101}$$

where A_0, A_1, and A_2 = areas of orifice, pipe, and contracted stream at temperature t, respectively, m².

The theoretical value of w'_2 from equation 5.100 should be corrected for obtaining the real velocity w_2. It can be done by using coefficient ζ and taking into account two facts:

1. Some losses of energy exist due to the viscosity of the fluid,
2. The pressures for the measurements, P_1 and P_2, are evaluated immediately at the orifice plate.

Therefore, the expression 5.101 is rewritten as follows:

$$w_2 = \frac{\varsigma}{\sqrt{1 - \mu^2 m^2}} \sqrt{\frac{2g}{\gamma_1} (P_1 - P_2)}. \tag{5.102}$$

Weight flow rate G [N/s] is determined as follows:

$$G = w_2 \gamma_1 A_2 = \mu w_2 \gamma_1 A_0 = \frac{\mu \varsigma}{\sqrt{1 - \mu^2 m^2}} A_0 \sqrt{2g\gamma_1(P_1 - P_2)}. \tag{5.103}$$

Coefficients ς and μ are experimentally determined together and, being combined, form coefficient α:

$$\alpha = \frac{\mu \varsigma}{\sqrt{1 - \mu^2 m^2}}. \tag{5.104}$$

Now the basic equations for the flow rates are

$$G = \alpha A_0 \sqrt{2g\gamma_1(P_1 - P_2)} \quad \text{and}$$

$$Q = \frac{G}{\gamma_1} = \alpha A_0 \sqrt{\frac{2g}{\gamma_1} (P_1 - P_2)}. \tag{5.105}$$

Similar equations can be used for describing gas and steam flow rate as a function of pressure drop, but the possible change in γ_1 under pressure and temperature must be accounted for. It is possible to assume that the process is adiabatic and the gas state is described by the following equation:

$$\frac{P}{\gamma^\chi} = C \tag{5.106}$$

where P = pressure in the pipe, Pa;
$\quad \gamma$ = specific weight of gas or steam, N/m³;
$\quad C$ = constant; and
$\quad \chi$ = adiabatic index.

Adiabatic index χ is the ratio of the specific heats: $\chi = C_p/C_v$. For monatomic, diatomic, and polyatomic gases, χ is 1.67, 1.40, and 1.33, respectively. Table 5.5 gives the numerical values for χ and γ of some gases used in industry. Note that γ is calculated by multiplying densities by g, which is taken as equal to 9.807 m/s². Substituting

$$\frac{1}{\gamma_1} = \left(\frac{C}{P}\right)^{\frac{1}{\chi}} \tag{5.107}$$

into equation 5.97 and taking the integral,

$$\int_{w_1}^{w_2'} w\,dw = -\, gC^{\frac{1}{\chi}} \int_{P_1'}^{P_2'} P^{-\frac{1}{\chi}} \cdot dP, \tag{5.108}$$

TABLE 5.5

Specific Weights γ and Adiabatic Index χ for Gases (Adopted from [79])

Gas	Chemical formula	γ at 20°C and 760mm Hg N/m³	χ at 20°C dimensionless
Air	—	11.768	1.40
Oxygen	O_2	13.014	1.40
Nitrogen	N_2	11.386	1.40
Hydrogen	H_2	0.819	1.41
Chlorine	Cl_2	29.313	1.34
Carbon monoxide	CO	11.386	1.40
Carbon dioxide	CO_2	17.996	1.30
Sulfur dioxide	SO_2	26.645	1.26
Hydrogen sulphide	H_2S	13.916	1.32
Ammonia	NH_3	7.022	1.31
Methane	CH_4	6.531	1.31
Acetylene	C_2N_2	10.660	1.23
Ethylene	C_2H_4	11.474	1.24
Ethane	C_2H_6	12.347	1.21
Water steam	H_2O	—	1.31

we obtain

$$w_2'^2 - w_1^2 = 2gC^{\frac{1}{\chi}} \cdot \frac{\chi}{\chi - 1} P_1'^{\frac{\chi-1}{\chi}} \left[1 - \left(\frac{P_2'}{P_1'} \right)^{\frac{\chi-1}{\chi}} \right]. \tag{5.109}$$

The coefficient of the contraction for gases μ_χ will differ from that for liquids (μ) because of the ability of gases to change volume under pressure, and its dependence on the ratio P_1/P_2. Therefore, the equation of continuity will be

$$\gamma_1 w_1 A_1 = \gamma_2 w_2' A_2 = \gamma_2 w_2' \mu_\chi A_0. \tag{5.110}$$

Solving this equation for w_1, we will obtain

$$w_1 = \mu_\chi w_2' \frac{\gamma_2}{\gamma_1} \cdot \frac{A_0}{A_1} = \mu_\chi m w_2' \left(\frac{P_2'}{P_1'} \right)^{\frac{1}{\chi}}. \tag{5.111}$$

For P_1' and γ_1, equation 5.107 is written as

$$C^{\frac{1}{\chi}} = \frac{P_1'^{\frac{1}{\chi}}}{\gamma_1}, \tag{5.112}$$

and expression 5.109 is modified as

$$w_2'^2 - w_1^2 = 2g \frac{\chi}{\chi - 1} \cdot \frac{P_1'}{\gamma_1} \left[1 - \left(\frac{P_2'}{P_1'} \right)^{\frac{\chi-1}{\chi}} \right]. \tag{5.113}$$

Combining equations 5.111 and 5.113 and keeping in mind the sense of factor ζ for the real condition of fluid flow, the equations for the flow rates are obtained as follows:

$$G = \alpha \cdot \epsilon \cdot A_0 \sqrt{2g\gamma_1(P_1 - P_2)} \qquad \text{and} \tag{5.114}$$

$$Q = \alpha \cdot \epsilon \cdot A_0 \sqrt{\frac{2g}{\gamma_1}(P_1 - P_2)}. \tag{5.115}$$

In these equations, coefficient ϵ reflects a correction for gas compressibility:

$$\epsilon = \frac{\alpha_\chi}{\alpha}\sqrt{\frac{1 - \mu_\chi m^2}{1 - \mu_\chi^2 m^2\left(\frac{P_2}{P_1}\right)^{\frac{2}{\chi}}}} \times \sqrt{\frac{P_1}{P_1 - P_2} \cdot \frac{\chi}{\chi - 1}\left[\left(\frac{P_2}{P_1}\right)^{\frac{2}{\chi}} - \left(\frac{P_2}{P_1}\right)^{\frac{\chi+1}{\chi}}\right]} \tag{5.116}$$

where

$$\alpha_\chi = \frac{\mu_\chi \zeta}{\sqrt{1 - \mu_\chi^2 m^2}}. \tag{5.117}$$

The last equations are valid for the Venturi tube and for a tube with a nozzle (Scheme 109b), for which the following two conditions are accepted: $\mu \approx \mu_\chi \approx 1$, and $\alpha_\chi \approx \alpha$. Coefficient ϵ can be calculated using formula 5.116.

When the orifice plate is used, the recommended value of m lies between 0.4 and 0.5 and should not exceed 0.7.

A *Pitot tube* (Fig. 5.23b) gives the velocity of fluid by the measurement of the difference between the impact, or total (P_T), and static (P_S) pressures in a flow. The outer pipe of this element has openings at a wall parallel to the direction of flow. The openings are far enough back so that the velocity and pressure outside the openings have a free-stream value, which is translated in the differential-pressure meter as a static pressure. The opening of the inner pipe is at a right angle to the stream and accepts the total pressure developed by the moving particles of fluid. Applying Bernoulli's equation to points a and b, we obtain

$$P_T - P_S = \frac{\gamma_1}{2g} w^2 \qquad \text{or} \tag{5.118}$$

$$w = \xi \sqrt{\frac{2g}{\gamma_1}(P_T - P_S)} \tag{5.119}$$

where P_T and P_S = total and static pressures, Pa, and
ξ = coefficient determined experimentally ($\xi = 0.98 - 0.99$); it depends on the constructional features, e.g., location of intake holes, their dimensions, etc.; the remaining terms as denoted before.

For gases at high velocities, ($a > w > 0.2a$, where a is sound velocity in given gas) w is calculated by using the following formula:

$$w = \xi \sqrt{2g \frac{\chi}{\chi - 1} RT_1\left[1 - \left(\frac{P_S}{P_T}\right)^{\frac{\chi-1}{\chi}}\right]} \tag{5.120}$$

where R = universal gas constant = 8.314 J/mole $\cdot$ K = 1.986 cal/mole $\cdot$ K, and
T_1 = temperature at the stagnation point, K.

The stagnation point is at the front of the aerodynamic body placed into a high-speed gas stream. At this point, the stream is split, the speed of gas is zero, and therefore, the temperature is at maximum.

A Pitot tube can be used for the measurements of Q and G. For this purpose, the area A of the pipeline is divided into n equal parts and the local speeds (w_1, w_2, w_3, . . . , w_n) are measured for each part. Thus, flow rates G and Q are calculated as follows:

$$G = w_1 \frac{A}{n} \gamma_I + w_2 \frac{A}{n} \gamma_I + \ldots + w_n \frac{A}{n} \gamma_1, \quad \text{and} \quad Q = \frac{G}{\gamma_1}, \quad (5.121)$$

where w's can be calculated using formula 5.119

$$G = \xi \frac{A}{n} \sqrt{2g\gamma_1} \left(\sqrt{P_{T1} - P_S} + \sqrt{P_{T2} - P_S} + \ldots + \sqrt{P_{Tn} - P_S} \right), \quad (5.122)$$

where P_{T1}, P_{T2}, . . . , P_{Tn} are total pressures corresponding to velocities w_1, w_2, . . . ,w_n.

A *rotameter-type sensing element* contains a float in a uniformly tapered tube (Fig. 5.23c). An upward flow causes the float to reach equilibrium height proportional to the flow rate. At this height, the weight of the float is in balance with the resulting force developed by the flow. This state is described by the equation

$$F_f = F \frac{\gamma_f - \gamma_1}{\gamma_1} = F_s = (P_1' - P_2') \cdot A_f \quad (5.123)$$

where F_f = float weight as it is suspended in the flow, N;
F = real weight of float, N;
γ_f = specific weight of the float material, N/m³;
γ_1 = specific weight of the fluid, N/m³;
F_s = resulting force on float, N;
P_1' = full pressure on the float acting on the area of float upward, Pa;
P_2' = pressure on the float acting downward, Pa; and
A_f = area of float in upward-downward direction, m².

Using formula 5.119 the difference of pressures $P_1' - P_2'$ can be expressed as a function of the fluid velocity w in the round gap between the float and pipe:

$$P_1' - P_2' = \xi \frac{w^2 \cdot \gamma_1}{2g}. \quad (5.124)$$

Combining this equation with equation 5.123, we will find that

$$w = \sqrt{\frac{2gF(\gamma_f - \gamma_1)}{\xi \gamma_1 A_f \gamma_1}}. \quad (5.125)$$

Assigning A_0 [m²] to the area of the round gap and denoting $\psi = \sqrt{1/\xi}$, the weight flow rate G [N/s] is calculated as

$$G = w\gamma_1 A_0 = \psi A_0 \sqrt{2gF \left(\frac{\gamma_f - \gamma_1}{\gamma_1} \right) \frac{\gamma_1}{A_f}}. \quad (5.126)$$

The volumetric flow rate Q [m³/s] is

$$Q = \psi A_0 \sqrt{\frac{2g}{\gamma_1 A_f} F \left(\frac{\gamma_f - \gamma_1}{\gamma_1} \right)} \quad (5.127)$$

A_0 is defined by the geometry of the float and pipe:

$$A_0 = \frac{\pi}{4}\left[(d_0 + nl)^2 - d_f^2\right]$$ (5.128)

where d_0 = inner diameter of the tapered tube at the lower end, m;
n = the change of the inner diameter of the tube per unit of length, m/m;
l = displacement of the float with respect to the lower end, m; and
d_f = the maximum diameter of the float, m.

The magnitude of coefficient ψ depends on the shape of the float and the viscosity of the fluid. Assuming that ψ is constant, it becomes obvious from formulas 5.126–5.128 that G or Q is simply defined by the gap A_0 or, finally, by the height l. Therefore, if the position of the float is monitored by an electrical transduction system, a reading of the flow rate becomes possible.

If a rotameter is calibrated for the measurement of the volume flow rate of the liquid having specific weight γ_1, and measurements are provided on the other liquid having a different value of γ_1', the rate will be C_1 times greater, where $C_1 = \sqrt{\gamma_1/\gamma_1'}$. Similarly, for the weight flow rate, the correcting coefficient is $C_2 = \sqrt{\gamma_1'/\gamma_1}$. These factors are applicable for the measurements of gas flow. However, one more cofactor C_3 depending on the gas pressure and temperature must be used:

$$C_3 = \sqrt{\frac{P_1 T_1'}{P_1' T_1}}$$ (5.129)

where P_1 and P_1' = absolute pressures at the calibration and measurement conditions, respectively, Pa, and
T_1 and T_1' = temperatures at the calibration and measurement conditions, respectively, K.

A *target-type flow-sensitive element,* described in this paragraph, is one more example of a transducer component in which sensing and transduction functions are combined. The analysis [41] of this device should be provided simultaneously for sensing and transduction portions of the element. As in the preceding cases of this section, the emphasis is on the hydrodynamical aspects of the problem. A strain-gage transduction element (see Chapter 2, Strain-gage Elements) is chosen due to its small size and mass. This makes it convenient to combine the element with a light, flow-sensitive target for measurements of instantaneous variations of pulsating flow.

The device for analysis is shown in Figure 5.24. A thin, round stem 1 is attached to a flexible plate 2. The stem passes through the opening in pipe 3, which carries liquid. Two strain gages 4 are bonded on each side of the plate. They can also be diffused layers in the silicon plate (Fig. 9.32). The real construction is sealed, and there is no leak from the inside of the pipe. Due to the drag force, the stem is deflected proportionally to the volume rate of the flow, causing deformation of strain gages. This deformation or electrical signal produced by the strain gage is a measure of the flow rate. If the diameter of the stem is much smaller than the diameter of the pipe, the drag force F can be assumed to be directly proportional to the velocity w [41]:

$$F = \psi w$$ (5.130)

F = drag force per unit of stem's length, N/m;
w = velocity of liquid, m/s; and
ψ = coefficient, N · s/m².

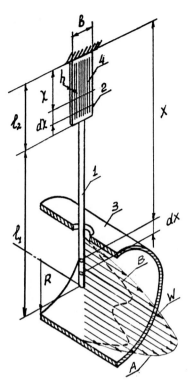

Figure 5.24 Schematic representation of strain gage flow-rate-sensitive element.

In Figure 5.24, l_2, b, and h denote the length, width, and thickness of plate 4, respectively. l_1 is the length of stem 1. X is the coordinate of the local velocity w, and x is the coordinate of the cross-sectional area where the deformation is determined.

The goal of the following analysis is to derive the transfer characteristic for this device; in other words, to find the strain in the plate as a function of the flow velocity. It is assumed that the cantilever, composed of the stem and plate, has a quick response to a change of velocity and does not contribute any amplitude or phase distortion to the transfer function. The analysis is concentrated on the influence of the steady and pulsating parameters of the flow on the transfer characteristic. The solution of this problem includes two steps [41]. In the first step, the local velocities of liquid are found by integrating differential equations of the motion of an incremental volume of liquid. In the second step, the equation of stresses in the elementary length of the plate is written. In this equation, the moments bending the plate are introduced as functions of drag forces and velocities. It is assumed in the following derivations that a steady flow is free of turbulences and that particles of a pulsating flow have the dominating velocity along the pipe. This assumption is true for some viscous liquids (for instance, in blood circulation).

For a steady flow, the equation of motion [80] of an elementary volume is written as

$$\frac{d^2w}{d\rho^2} + \frac{1}{\rho} \cdot \frac{dw}{d\rho} + \frac{1}{\mu} \cdot \frac{\Delta P}{\Delta l} = 0 \tag{5.131}$$

A pulsating flow is described by another equation:

$$\frac{\partial^2 w}{\partial \rho^2} + \frac{1}{\rho} \cdot \frac{\partial w}{\partial \rho} - \frac{1}{\nu} \cdot \frac{\partial w}{\partial t} = -\frac{A}{\mu} e^{j\omega t}. \tag{5.132}$$

For the purpose of analysis, the oscillations in the pipe are assumed to be caused by a sinusoidal function $A(\omega)$:

$$A(\omega) = Ae^{j\omega t} \tag{5.133}$$

where $A = \Delta P/\Delta l$ = peak value of $A(\omega)$, Pa/m;
       ΔP = pressure drop along the pipe, having length Δl, Pa;
       Δl = length of the pipe, m; and
       ω = angular frequency, rad/s.

The solution [81] of equation 5.131 for velocity w is expressed as follows:

$$w = \frac{\Delta P}{\Delta l} R^2 \frac{1 - u^2}{4\mu}. \tag{5.134}$$

The solution [81] of equation 5.132 is

$$w = \frac{Ae^{j\omega t}}{\varphi j\omega}\left[1 - \frac{J_0(\alpha u j^{3/2})}{J_0(\alpha j^{3/2})}\right] \tag{5.135}$$

where w = local velocity, m/s;
           ρ = distance (radius) from the center of the pipe to the elementary volume, having velocity w, m;
           μ = dynamic viscosity of liquid, kg/m · s;
           ν = kinematic viscosity of liquid ($\nu = \mu/\varphi$), m²/s;
           φ = density of liquid, kg/m³;
           u = ρ/R = relative radius, dimensionless;
           α = $R\sqrt{\omega/\nu}$ = parameter of pulsations, dimensionless;
           j = $\sqrt{-1}$;
           R = radius of pipe, m; and
$J_0(\alpha j^{3/2})$ and
$J_0(\alpha u j^{3/2})$ = Bessel functions of zero order and complex argument.

It should be noted that the diagram of the velocities for the steady flux (curve A in Fig. 5.24) is parabolic, whereas it has a complicated curvature for the pulsating flow (curve B). The shape of this curve depends on the frequency of pulsation and viscosity of liquid.

By substituting velocities w from equations 5.134 and 5.135 in expression 5.130, the drag force at any radius ρ can be found. It is obvious that the strain gages are subjected to the strain, which is proportional to the flow velocity. The stem sums the bending moments of the drag forces that are applied to the stem. The moments are transmitted to the strain gages when the plate is bent. In order to find the relationship between the flow velocity and the strain, the moments should be integrated along the length of the stem from the axis of the pipe to its wall.

The longitudinal strain $\epsilon(x)$ at the surface of the bent plate is given by the standard formula of the Strength of Materials [66]:

$$\epsilon(x) = \frac{M(x)}{IE} \cdot \frac{h}{2} \tag{5.136}$$

where $\epsilon(x)$ = longitudinal strain $\epsilon(x)$ at the surface of the bent plate for the section with coordinate x, dimensionless;
    $M(x)$ = bending moment at the section with coordinate x, N · m;
       E = modulus of elasticity, N/m²;

I = moment of inertia of the section of the plate with respect to the neutral axis, m^4; and

$h/2$ = distance from the neutral of the plate to the most remote point (half of the plate's thickness), m.

The average strain ϵ of the strain gage is

$$\epsilon = \frac{1}{l_2}\int_0^{l_2}\epsilon(x)dx = \frac{h}{2EIl_2}\int_0^{l_2}M(x)dx. \tag{5.137}$$

It is assumed that the strain gages occupy the entire length of the plate.

The differential bending moment $dM(x)$ in the section having coordinate x due to drag force with coordinate X is represented in the following form (Fig. 5.24):

$$dM(x) = F(X - x)\cdot dx = \psi w(X - x)\cdot dx. \tag{5.138}$$

The bending moment $M(x)$ due to all drag forces applied to the stem is the integral of $dM(x)$ with respect to the length of the stem:

$$M(x) = \psi\int_{a_1}^{a_2}w(X - x)dx \tag{5.139}$$

where the limits of integration are

$$a_1 = l_1 + l_2 - R \quad\text{and} \tag{5.140}$$

$$a_2 = l_1 + l_2. \tag{5.141}$$

Hence, the equation for ϵ is written in the following form:

$$\epsilon = \frac{N}{l_2}\int_{a_1}^{a_2}\int_0^{l_2}w(X - x)\,dX\cdot dx \tag{5.142}$$

where

$$N = \frac{\psi h}{2EI}. \tag{5.143}$$

It is convenient for further analysis to make a substitution:

$$x = l_1 + l_2 - Ru, \text{ where } 0 \le u \le 1 \tag{5.144}$$

Equation 5.142 can then be rewritten as

$$\epsilon = \frac{NR}{l_2}\int_0^1\int_0^{l_2}w(l_1 + l_2 - Ru - x)\,dudx =$$

$$NR\left[\left(l_1 + \frac{l_2}{2}\right)\right]\int_0^1 wdu - R\int_0^1 wu\cdot du. \tag{5.145}$$

For the steady flow,

$$\int_0^1 wu\cdot du = \frac{1}{2}\cdot\frac{Q_S}{\pi R^2} \quad\text{and}\quad \int_0^1 w\cdot du = \frac{4}{3}\cdot\frac{Q_S}{\pi R^2}. \tag{5.146}$$

For the pulsating flow,

$$\int_0^1 wu \cdot du = \frac{1}{2}\frac{Q_P}{\pi R^2} \quad \text{and} \quad \text{(5.147)}$$

$$\int_0^1 w \cdot du = \frac{Q_P}{\pi R^2} \cdot \frac{1 - [J_0(\alpha j^{3/2})]^{-1} \cdot \int_0^1 J_0(\alpha u j^{3/2}) \cdot du}{1 - [\alpha j^{3/2} \cdot J_0(\alpha j^{3/2})]^{-1} \cdot 2J_1(\alpha j^{3/2})} \quad \text{(5.148)}$$

where Q_S and Q_P = volume flow rates for steady and pulsating flows, respectively, m^3/s.

The expressions for average strains corresponding to the steady and pulsating flows (ϵ_S and ϵ_P, respectively) are obtained by taking the integrals determined in formulas 5.146–5.148 and substituting them in formula 5.145:

$$\epsilon_S = \frac{NQ_S}{\pi R}\left[\frac{4}{3}\left(l_1 + \frac{l_2}{2}\right) - \frac{1}{2}R\right] \quad \text{(5.149)}$$

$$\epsilon_P = \frac{NQ_P}{\pi R}\left[T\left(l_1 + \frac{l_2}{2}\right) - \frac{1}{2}R\right] \quad \text{(5.150)}$$

where

$$T = \frac{1 - [J_0(\alpha j^{3/2})]^{-1} \cdot \int_0^1 J_0(\alpha u j^{3/2})du}{1 - [\alpha j^{3/2}J_0(\alpha j^{3/2})]^{-1} \cdot 2J_1(\alpha j^{3/2})}. \quad \text{(5.151)}$$

The solution of the integral

$$\int_0^1 J_0(\alpha u j^{3/2}) \cdot du$$

is known [82]:

$$\int_0^1 J_0(\alpha y) \cdot dy = \frac{1}{2}\pi y[J_1(ay) \cdot H_0(ay) - J_0(ay) \cdot H_1(ay)] + yJ_0(ay). \quad \text{(5.152)}$$

For the case analyzed here, $y = 1$ and $a = \alpha j^{3/2}$; therefore,

$$\int_0^1 J_0(\alpha u j^{3/2})du = \frac{1}{2}\pi[J_1(\alpha j^{3/2}) \cdot H_0(\alpha j^{3/2}) - J_0(\alpha j^{3/2}) \cdot H_1(\alpha j^{3/2})]$$

$$+ J_0(\alpha j^{3/2}) \quad \text{(5.153)}$$

where H_0 and H_1 are Struve functions of zero and the first order, respectively. J_1 is a Bessel function of the first order and complex argument, and a is a factor at variable y.

The data necessary for calculating the integral given in formula 5.153 can be found in mathematical handbooks [83].

$J_0(\alpha j^{3/2})$ and $J_1(\alpha j^{3/2})$ are numerically introduced as follows [82]:

$$J_0(\alpha j^{3/2}) = \text{ber}_0(\alpha) + j\text{bei}_0(\alpha) = M_0 e^{j\theta_0(\alpha)} \quad \text{(5.154)}$$

$$J_1(\alpha j^{3/2}) = \text{ber}_1(\alpha) + j\text{bei}_1(\alpha) = M_1 e^{j\theta_1(\alpha)} \quad \text{(5.155)}$$

where ber's and bei's are real and imaginary parts, respectively, of the Bessel functions. $M_0(\alpha)$, $M_1(\alpha)$, $\theta_0(\alpha)$, and $\theta_1(\alpha)$ are magnitudes and angles of functions (see also formula 4.22 for calculations of J_0 and J_1).

The necessary figures for Struve functions can be found as real (ster) and imaginary (stei) components:

$$H_0(\alpha j^{3/2}) = \text{ster}_0(\alpha) + j\text{stei}_0(\alpha)$$

$$H_1(\alpha j^{3/2}) = \text{ster}_1(\alpha) + j\text{stei}_1(\alpha). \tag{5.156}$$

A set of values of Struve functions has been tabulated [82], and some numbers are given in the following Table 5.6.

The transfer characteristic of the element for a steady flow $\epsilon_S = f(Q_S)$ is readily computed with formula 5.149. For the pulsating flow, characteristic $\epsilon_P = f(Q_P)$, given by formula 5.150, is a function of parameter α, which depends on the frequency of pulsation and on the viscosity of liquid. After calculating the term T, ϵ_P can be introduced as a complex number:

$$\epsilon_P = Re + jIm \tag{5.157}$$

where Re and Im are the real and imaginary parts of ϵ_P, respectively. Its amplitude-frequency response is analyzed by tabulating modulus $|\epsilon_P|$ and the phase angle θ of ϵ_P as a function of parameter α or the angular frequency ω:

$$|\epsilon_P| = \sqrt{(Re)^2 + (Im)^2} \tag{5.158}$$

$$\tan\theta = \frac{Im}{Re}. \tag{5.159}$$

Practical calculations of $|\epsilon_P|$ and θ show that the transfer characteristic has attenuation and a lagging phase angle with an increase of parameter α or frequency ω.

Ultrasonic Elements

Originally, the ultrasonic flow-rate meter was designed as a linear speed-measuring instrument. However, great progress in the development of the concept and construction has led to the creation of ultrasonic volume and mass flow-rate meters.

In the measurements by means of acoustic waves, the characteristics of wave propagation are evaluated in order to detect the rate of flow. A rough error can occur if the linear velocity reading is used as a measure of volume or mass flow. One of the typical sources for this error is a change in the fluid local velocities pattern because of the change in the nature of the flow. One simple calculation illustrates this case. Let us assume that we have a circular pipe with running liquid (Fig. 5.25) whose flow changes from laminar to turbulent. This change makes the pattern of velocities

TABLE 5.6

Struve Function for Parameter α (Adopted from [82])

	$\text{Ster}_0(\alpha)$	$\text{Stei}_0(\alpha)$	$\text{Ster}_1(\alpha)$	$\text{Stei}_1(\alpha)$
0	0.0000	0.0000	0.0000	0.0000
2	−1.2315	0.4416	0.2247	−0.8230
4	−2.4319	−2.6671	3.2088	−1.8070
6	7.2578	−8.9265	8.5194	7.4797
8	34.9607	20.9189	−21.0363	32.5157
10	−56.4158	138.7957	−131.2420	−59.4713
20	−114 775.23	47 489.35	−44 583.74	−113 602.52

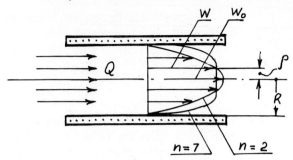

Figure 5.25   Patterns of local velocities of flows.

flatter and can be modeled by the variation in the power n of the parabola describing the pattern:

$$w = w_0(1 - u^n), \qquad u = \rho/R \qquad (5.160)$$

where w = local velocity at distance (radius) ρ from pipe's axis, m/s;
  w_0 = axial velocity, m/s; and
  R = pipe's radius, m.

Power n varies usually from 2 for laminar flow, to 7 for turbulent flow. The relation between the local velocity and volumetric flow rate Q [m³/s] can be readily found by taking the integral:

$$Q = 2\pi \int_0^R w\rho \cdot d\rho = 2\pi w_0 R^2 \int_0^1 (1 - u^n)u \cdot du = \pi w_0 R^2 \frac{n}{n + 2}$$

$$= \pi R^2 \frac{n}{(1 - u^n)(n + 2)} w. \qquad (5.161)$$

The average velocity w_{av} [m/s] is

$$w_{av} = \frac{1}{R} \int_0^R w dR = \frac{1}{R} w_0 R \int_0^1 (1 - u^n)du = w_0 \frac{n}{n - 1} \qquad \text{and} \qquad (5.162)$$

Q, expressed in terms of w_{av}, is written as

$$Q = \pi R^2 \frac{n - 1}{n + 2} w_{av}. \qquad (5.163)$$

It is seen from formulas 5.161 and 5.163 that, if the transduction element intended for volumetric measurements responds to the local or average velocity only, the reading of Q will not be correct because of the variation in the n-dependent ratio. For example, if the velocity w_0 is measured at the pipe's axis ($\rho = 0$, and $u = 0$), and n varies from 2 to 7, the variation in the reading of Q reaches 56%.

There are several approaches in the arrangement of the sensing part of the flowmeter [84, 85]. One of them is the most attractive; it is a clamp-on transducer of the transit-time ("wide beam") flowmeter. The performance parameters of this instrument are superior to that of any other type of ultrasonic flowmeter, including the Doppler, the clamp-on "shear mode" transit-time, and the inserted "wetted transducer." The same elements of this transducer, which radiate and receive waves, are clamped on the outside surface of a pipe. They do not disturb the flow stream and do not take energy from it. The flow of all types of liquids, except heavily aerated ones, can be measured. In order to have a mass flow-rate measurement, the instrument incorporates a digital computer, which treats data about the liquid density along with its velocity. There is a known relationship between the sonic velocity in liquid and its density, and measurement of this velocity is intrinsic to the flow-measurement process. Due to computer application, it is easy to add to the measuring system a temperature sensor. This feature makes it possible to measure not only the volumetric and mass flow, but also the energy flow along the pipe. At present, four types of flowmeters are in common use: clamp-on wide-beam transit-time, clamp-on shear-mode transit-time, wetted (inserted) transit-time, and clamp-on Doppler transducers. The waves are usually generated and received by piezoceramic elements that are in direct (wetted) or indirect (through the pipe's wall) contact with the liquid.

The diagram in Figure 5.26 shows a two-element *system based on the Doppler effect*. The measurement of flow is provided by detecting the frequency shift imposed on an ultrasonic carrier wave reflected from particles moving with the flow stream. A transmitter at one side of the wall propagates continuous ultrasonic waves into fluid. Particles of fluid reflect some of the energy to the receiver fixed at the other side of the pipe. The difference Δf [Hz] between the frequency of the transmitted signal f_t and frequency of the received signal f_r is given [86, 87] as

$$\Delta f = f_t - f_r = \frac{2f_t \cos \theta}{V_w} V_F \qquad (5.164)$$

where V_w = propagation velocity for the wedge material, m/s;
$\quad\ V_F$ = linear flow velocity, m/s; and
$\quad\ \theta$ = angle between the direction of the sound beam and pipe's axis, degree.

It is noteworthy that there is no term in formula 5.164 representing sound velocity in the fluid. Analysis shows that the changes in this velocity are compensated by similar changes in $\cos \theta$; and angle θ should be interpreted as the angle defined by the wedge.

This method is recognized as one of the most sensitive among the ultrasonic principles and is characterized by a linear output-input conversion in a wide range of measurands. However, there are two problems associated with this principle: reading depends on the flow profile and on the sonic conductivity and reflectivity of liquid.

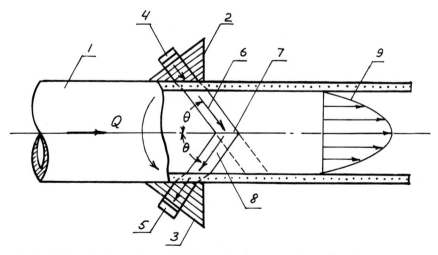

Figure 5.26 Clamp-on Doppler ultrasonic flowmeter. Q = flow, θ = beam angle, 1 = pipe, 2 and 3 = wedges, 4 and 5 = transmitter and receiver, respectively, 6 = transmitting beam, 7 = zone of Doppler reflection, 8 = Doppler reflection, 9 = flow profile (adopted from [84 and 85]).

Changes in liquid chemistry, line pressure, and cavitation can greatly affect the performance of the transducer.

A two-element wetted (inserted) transducer (Fig. 5.27) contains two crystals alternatively working as transmitters and receivers. Ultrasonic pulses are directly injected into the flow stream, first upstream and then downstream. The same principle is realized in a construction combining one transmitter and two receivers (Scheme 117a) at equal distances from the transmitter. Flow is measured by detecting the difference in transit time when the velocities of pulse propagation and flow are added and subtracted due to the alteration of the ultrasonic beam. The reading of the difference in the "upstream and downstream frequencies" is proportional to the flow velocity. This difference is readily found from the following simple relationships:

$$t_1 = L/(V_S + V_F \cos \theta), \qquad t_2 = L/(V_S - V_F \cos \theta),$$

$$\Delta f = 1/t_1 - 1/t_2, \quad \text{and} \quad \Delta f = (2V_F \cos \theta)/L \qquad (5.165)$$

where t_1 and t_2 = signal travel time in the direction and against the direction of flow, s;

L = the separation distance of transmitter and receiver, m;

V_S and V_F = velocity of sound propagation in the medium and linear flow velocity, m/s; and

Δf = frequency difference of signal, Hz.

As is clear from formula 5.164, Δf is independent of V_S and conversion is linear. However, the flow has distortion at the "wells" of elements. The angle between the ultrasonic beam and the flow stream adjacent to the well is uncertain. This angle varies with liquid viscosity and flow velocity, generating errors in measurements. Filling the "wells" with patching material does not solve the problem. This filling produces beam refraction, diverting the beam from its intended arrival point and deteriorating the passage of signals.

The operation of the clamp-on shear-mode transit-time ultrasonic flowmeter is similar to that of the wetted type with one difference: the system does not contain "wells." However, the beam refraction in this construction makes measurements sensitive to the sonic velocity of liquid, which is a function of liquid chemistry and temperature. Besides this, the calibration factor of the instrument is a function of the sonic propagation velocity in the parts of the transducer.

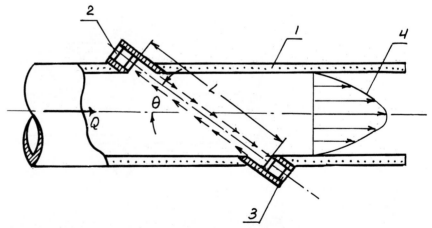

Figure 5.27   Wetted transducer transit-time ultrasonic flowmeter. Q = flow, θ = beam angle, 1 = pipe, 2 and 3 = transmitting and receiving elements, 4 = flow profile (adopted from [84 and 85]).

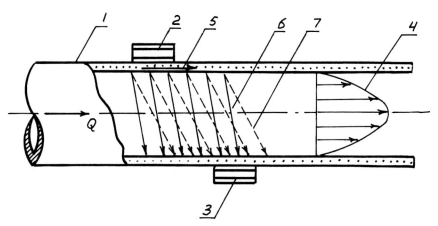

Figure 5.28 Wide-beam ultrasonic flowmeter. Q = flow, 1 = pipe, 2 and 3 = transmitting and receiving elements, 4 = flow profile, 5 = axial beam injection, 6 and 7 = low and high liquid sonic velocity beams, respectively (beam covers receive element for any liquid) (adopted from [84 and 85]).

A wide-beam system (Fig. 5.28) contains a clamp-on transmitting element that excites a natural mode of sonic propagation in the pipe wall, stimulating the waves to move along the main axis of the pipe. Due to this motion, a wide beam is continuously radiated in the liquid and travels away from the transmitting element. The receiving element is excited by the propagating waves regardless of the refraction angle of the beam. Analysis [85] shows that the penetration of the beam through the entire flow profile permits simple-profile compensation. A similar approach is utilized in the design of gas flow-rate measuring instruments.

ACOUSTICAL ELEMENTS

When measurements are associated with an alternating pressure that is transmitted via gas communication lines, consideration of the acoustical characteristics of waveguides, and of gas's elastic, inertial, and resistive characteristics is necessary. These characteristics also play an important role when the motion of gas in a small-volume transducer cavity, such as in an acoustical labyrinth, causes changes in the dynamics of a pressure transduction element. The operation of acoustical filters used as parts of resonant-type transducers is also guided by the mentioned characteristics. This section is devoted to the analysis of typical problems related to the issues just listed. See the section in Chapter 4 on acoustical resonators for more information on the subject.

It is convenient to analyze a transducer's acoustic dynamics by applying electrical analogies. In this method, differential equations having the same form as those of the system under consideration are set. Then the two systems are considered to behave in a similar manner. By interchanging the parameters and variables, the responses of the analogous systems are investigated. In terms of analysis, great mathematical strides have been made in the electric-circuit field; and by setting up electrical equivalents, one can take advantage of this progress. Combined with the computer technique, the method of analogies is a valuable tool for solving a good number of transducer problems. Two distinct electrical analogies are in use. They are known as duals of each other; that is, their behavior is described by the same kind of differential equations. The first kind is a force-voltage analogy. It is based on Kirchhoff's voltage law, which states that at each instant of time the algebraic sum of the voltages around a closed loop in a network is equal to zero. Kirchhoff's current law is the basis for the so-called force-current analogy. According to this law, at any given instant, the sum of all the currents flowing toward a point is equal to the sum of values of all currents flowing away from the point. The principles of using these two analogies are quite similar, and we will concentrate our attention on the former type.

In the force-voltage analogy, the corresponding couples of the mechanical and electrical systems are force (F) – voltage (U); velocity (v) – current (i); mass (m) – inductance (L); friction coefficient (b) – resistance (R); and compliance (c) – capacitance (C). These equivalents are originated by two differential equations for elementary mechanical and electrical systems (Fig. 5.29):

$$m\frac{dv}{dt} + bv + \frac{1}{c}\int v\,dt = F \qquad \text{and} \tag{5.166}$$

$$L\frac{di}{dt} + Ri + \frac{1}{C}\int i\,dt = U. \tag{5.167}$$

Force F applied to the mechanical system (Fig. 5.29a) is opposed by three forces F_1, F_2, and F_3 developed due to three mechanical impedances Z_m, Z_b, and Z_c created by mass, friction, and elasticity (Fig. 5.29b). These impedances form a circuit with a parallel connection of elements. In the electrical analogue described by equation 5.167, voltage U should also be divided between three components modeling forces F_1, F_2, and F_3. These components are voltage drops U_1, U_2, and U_3, which must be formed by a series (not a parallel) connection of the circuit elements (c).

Now we can turn to an acoustic system (d). First, let us assume that a certain volume of gas having mass m is forced to move with velocity v in a manifold by a force F applied to a piston. It is relevant to describe the characteristics of the motion of the system in terms of pressure and volume velocity, as is commonly accepted for an hydraulic or acoustical system. Force F in equation 5.166 is replaced by the product $F = PA$, where A is the area of manifold. Velocity v is introduced as a ratio: $v = Q/A$, where Q is volume velocity. Since the motion of gas is accompanied by the dissipation of energy, and since the gas is compressible and inertial, the nature of the terms in the differential equation for the forces will be the same. Now making substitutions for F and v in equation 5.166, we obtain

$$\left(\frac{m}{A^2}\right)\frac{dQ}{dt} + \left(\frac{b}{A^2}\right)Q + \left(\frac{1}{cA^2}\right)\int Q\,dt = P \tag{5.168}$$

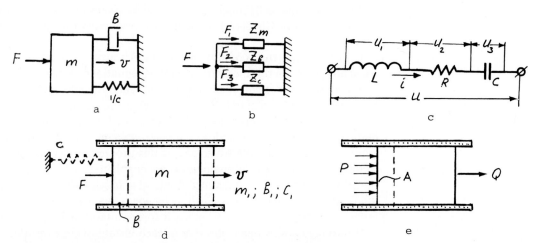

Figure 5.29 Mechanical, acoustical, and electrical analog circuits. a and b = mechanical models, c = electrical analog, d and e = acoustical models.

where P = acoustical pressure, Pa;

Q = volume velocity, m^3/s;

m/A^2 = acoustical mass, kg/m^4;

b/A^2 = acoustical resistance or friction coefficient, kg/s $\cdot$ m^4; and

cA^2 = acoustical compliance, m^{4}s^2/kg.

Thus, we have obtained an acoustic derivative of a force-voltage analogy that is a pressure-voltage analogy (*e*). A comparison of equations 5.168 and 5.166 shows that terms m/A^2, b/A^2, and $1/cA^2$ are the analogues of L, R, and $1/C$, respectively. This system is convenient for the calculation of a device containing acoustic elements only. In transducers, quite often an acoustical system is combined with a mechanical one. In this case, the analog circuits should reflect this fact, and the presentation of physical parameters must be done in terms of one system. A simple example of such a system is an elastic diaphragm that is exposed to a pressure at one side and to elastic-acoustic counteraction at the other. An approximate presentation of this system can be given by the diagram in Figure 5.29d if it is completed by a spring attached to the piston. Force F applied to the diaphragm is opposed by the diaphragm-generated forces due to its mechanical parameters m, b, and c; and by the gas, due to its acoustical parameters m_1, b_1, and c_1. In order to solve the problem for diaphragm motion (velocity), we can build an equivalent circuit, and write and solve the force equations. However, before doing this, it is necessary to reduce the parameters to one system. In choosing the mechanical system for calculation, the acoustic parameters must be converted: m_1A^2, b_1A^2, and c_1/A^2. Constructing the equivalent circuits requires some skill. In the following topics, representative examples of drawings of the equivalent circuits and of some elementary calculations are given. Reading the relevant parts of books on acoustics can be helpful for deeper study of the subject.

An *acoustical labyrinth* is a standard part of any microphone. If properly designed, the labyrinth makes it possible to have a flat frequency characteristic of output, or to depress the passage of sound within the prescribed frequency band. A typical arrangement of the microphone interior [21] is shown in Figure 5.30a. There are three inner cavities that form, along with the diaphragm, the compliances of the system. The orifices represent masses and resistances. Mass and resistance of the diaphragm are also included into the system. Figure 5.30b illustrates the acousto-mechanical structure of the system. The principles of constructing of the system are explained in the following items.

1. At the microphone input, the excitation force $F = PA$ is developed due to the existence of acoustic pressure at the front of the protection grid. A is the area of all input holes in the grid.
2. This force is applied to the mass m_1 in the capillaries of the grid and, at the same time, to the resistance r_1, which is also located in the same capillaries. This resistance is created by viscous friction, radiation of sound waves, and by other losses.
3. Practically, mass m_1 is incompressible; therefore, force F is considered to be applied directly to the volume that is enclosed between the grid and diaphragm and that has compliance c_1.
4. The combination of m_1, r_1, and c_1 forms nodal point 1 (see *b*).
5. The pressure developed at the diaphragm surface acts upon it. The diaphragm itself makes nodal point 2 where m_2, r_2, and c_2 are jointed.
6. The same pressure must overcome the impedance created by the elasticity of the volume at the diaphragm back; therefore, the stiffness of this volume $1/c_3$ must be added to the diaphragm stiffness $1/c_2$ (formula 5.3).

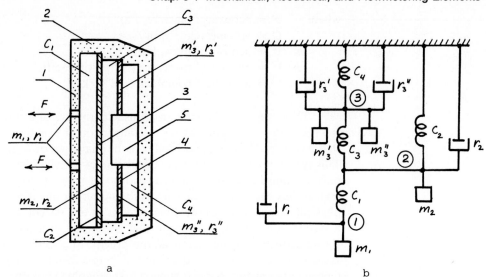

Figure 5.30 Acoustical labyrinth of microphone (a) and its acousto-mechanical circuit (b) (adopted from [21]). 1 = protection grid, 2 = housing, 3 = diaphragm, 4 = partition, 5 = part of transduction element, F = force equivalent to acoustic pressure, m_1 = air mass in capillaries of protection grid, m_2 = mass of diaphragm, m_3' and m_3'' = air masses in orifices of partition, c_1 = compliance of volume before diaphragm, c_2 = compliance of diaphragm, c_3 = compliance of volume between diaphragm and partition, c_4 = compliance of volume between partition and backplate, r_1 = resistance of capillaries of grid, r_2 = resistance of diaphragm (represents active losses), r_3' and r_3'' = resistance of orifices in partition.

Adding the stiffness corresponds to the parallel connection of two elastic elements.

7. The acoustic pressure developed between the diaphragm and partition acts upon masses m_3', m_3'', and resistances r_3' and r_3'', which are concentrated in the partition holes. Since the mass in these holes is incompressible, the pressure is transferred to the volume at the backplate directly. Therefore, the nodal point 3 is formed by the connection of five elements: m_3', m_3'', r_3', r_3'', and c_4.

Outlining this structure helps in drawing the electrical analog circuit (Fig. 5.31), which can be completed in three steps:

1. The applied force and all components of nodal point 1 are arranged in a loop with the voltage (represented by F) applied to the loop.
2. Then the closed loops with elements of nodal points 2 and 3 are drawn similarly. The elements common to the adjacent loops are included in the circuit of each loop. In our case, they are c_1 and c_3.
3. Finally, the loops are combined by joining them at the places having the common element and by showing one of the common elements at the juncture.

Further analysis can include the calculation of voltages and currents in this circuit, which simply characterizes the mechanical and acoustical function of the microphone. By varying the magnitudes of the circuit parameters, the optimal performance can be found.

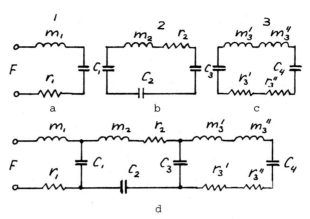

Figure 5.31 Construction of electrical analog circuit. a, b, and c = steps in drawing loops, d = final outline of analog circuit; 1, 2 and 3 = loops corresponding to nodal points in Figure 5.30b.

A *pressure transmitting line* (Fig. 5.32) is used for remote measurements of pressure, especially in cases where direct contact of the transducer with the agent to be measured is not possible or not desirable (high temperature, chemically aggressive fluid, etc.). The line is usually introduced by a pipe filled with liquid or gas. Compression or decompression of this fluid at the remote end is transmitted to the sensitive element of the transducer. Because of the hydraulic or acoustical impedance of the pipe, the pressures at the pipe have some differences that are a source of measurement errors. A convenient method of calculating [88, 89] these differences is the use of electrical analogies and the presentation of the pipe and the transducer cavity as a system described by the lumped acoustic capacitance C [m$^4 \cdot$ s^2/kg], inductance L [kg/m^4], and resistance R [kg/m$^4 \cdot$ s]:

$$C = \frac{V}{\rho v^2}, \qquad L = \frac{4l\rho}{3\pi r^2}, \qquad R = \frac{8\eta l}{\pi r^4} \qquad (5.169)$$

where V = transducer cavity volume, m^3;
　　　ρ = density of fluid, kg/m^3;
　　　v = sound velocity, m/s;
　　　l = pipe length, m;
　　　r = pipe inner diameter, m; and
　　　η = dynamic viscosity of fluid, N $\cdot$ s/m^2.

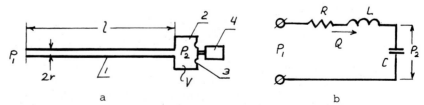

Figure 5.32 Schematic view of pressure-transmitting system (a) and its electrical equivalent (b). *C, L,* and *R* = electrical equivalents, *l* and *r* = pipe length and radius, P_1 and P_2 = pressures at input and at transducer, Q = fluid flow, V = transducer cavity volume, 1 = pipe, 2 = transducer cavity, 3 = diaphragm, 4 = transducer.

For air at room temperature, $v = 343.3$, $\rho = 1.164$, and $\eta = 18.240 \times 10^{-6}$ (see also Table 4.3).

The equivalent capacity for the cavity is usually larger than that for the pipe; and the simplified equivalent circuit (Fig. 5.32b) contains three elements: R and L for the pipe and C for the cavity. In our equivalent circuit, the analogs of pressures are voltages P_1 (D) and P_2 (D) (D is operator). Current Q represents the fluid volumetric flow rate. P_2 (D) is calculated as a voltage drop across C:

$$P_2(D) = P_1(D) \frac{1/(DC)}{1/(DC) + R + DL}. \qquad (5.170)$$

The transfer characteristic $G(j\omega) = P_2(j\omega)/P_1(j\omega)$ is

$$G(j\omega) = \frac{1}{1 + j\omega CR - \omega^2 CL} \qquad (5.171)$$

where ω = angular frequency of oscillation of P_1, rad/s.
The amplitude-frequency relationship can be found from the calculation of modulus $|G|$:

$$|G| = \frac{1}{\sqrt{(1 - \omega^2 LC)^2 + \omega^2 R^2 C^2}} = \frac{1}{\sqrt{\left[1 - \left(\dfrac{\omega}{\omega_0}\right)^2\right]^2 + 4\zeta^2\left(\dfrac{\omega}{\omega_0}\right)^2}} \qquad (5.172)$$

where ω_0 is the natural frequency of oscillation of the hydraulic or acoustic system (pipe and cavity); and ζ is the damping of the system.

$$\omega_0 = \frac{1}{\sqrt{LC}} = \sqrt{\frac{2\pi r^2 v^2}{4lV}}, \qquad \zeta = \frac{1}{2} RC\omega_0 = \frac{2\eta}{v\rho r^2}\sqrt{\frac{3lV}{\pi}} \qquad (5.173)$$

The phase shift between P_2 and P_1 is calculated as

$$\varphi = \arctan\left(-\frac{\omega RC}{1 - \omega^2 LC}\right) = \arctan\left[-\frac{2\zeta \dfrac{\omega}{\omega_0}}{1 - \left(\dfrac{\omega}{\omega_0}\right)^2}\right]. \qquad (5.174)$$

For a pipe of small diameter, L can be neglected. In the calculation of the system, the "weight" of each term is evaluated by comparing the resistance R with reactances ωL and $1/\omega C$.

Acoustical Filters

In some designs, a low-pass acoustic filter is inserted in the resonant transducer inlet to prevent the transducer from acoustic interaction with similar devices through the gas carrying communications. Use of the filter is also desirable in cases where an unpredictable change in the acoustic load from the outside components of a measuring system can adversely affect the resonator vibration. One of the simple models of such a filter is shown in Fig. 5.33. It includes two rigid enclosures introducing acoustical compliances C_1 and C_2 and pipes giving masses M_1, M_2, and M_3. The pressure at the filter inlet is P_1 and the acoustic load at the outlet is Z_L. In the calculation of the filter, we are usually interested in the attenuation of the alternating pressure applied

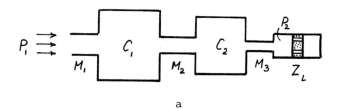

a

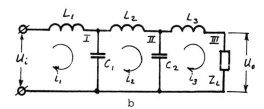

b

Figure 5.33 Low-pass acoustical filter. a = acoustical coupling of filter elements, b = electrical equivalent circuit (resistive elements are ignored in this circuit).

to the inlet for a specific range of pressure frequencies. The problem is conveniently solved by using a voltage-pressure analogy (b) with the calculation of voltage drop U_0 across impedance Z_L as a function of input voltage U_i. A mesh circuit analysis is helpful for this kind of calculation. We assign clockwise direction to the mesh currents of each loop. The subscript at the current indicates the number of the mesh. The system of voltage mesh equations in matrix notation is written as

$$\begin{bmatrix} Z_{11} & Z_{12} & Z_{13} \\ Z_{21} & Z_{22} & Z_{23} \\ Z_{31} & Z_{32} & Z_{33} \end{bmatrix} \times \begin{bmatrix} i_1 \\ i_2 \\ i_3 \end{bmatrix} = \begin{bmatrix} U_i \\ 0 \\ 0 \end{bmatrix}. \tag{5.175}$$

In this Z-matrix, the impedances with the uniform subscripts (Z_{ii}) are total impedances of each mesh (Z_{11} − of mesh I, Z_{22} − of mesh II, and Z_{33} − of mesh III). Z's having different subscripts (Z_{ik}) stand for the impedances between the meshes (e.g., Z_{23}). All Z_{ik}'s are taken negative. The values of impedances in operator notation are

$$Z_{11} = DL_1 + \frac{1}{DC_1}, \quad Z_{12} = Z_{21} = -\frac{1}{DC_1}, \quad Z_{22} = DL_2 + \frac{1}{DC_1} + \frac{1}{DC_2},$$

$$Z_{23} = Z_{32} = -\frac{1}{DC_2}, \quad Z_{33} = DL_3 + \frac{1}{DC_2} + Z_L, \quad Z_{13} = Z_{31} = 0. \tag{5.176}$$

Voltage U_0 is a product:

$$U_0 = i_3 Z_L. \tag{5.177}$$

Current i_3 is calculated by taking the ratio of two determinants:

$$i_3 = \frac{\begin{vmatrix} Z_{11} & Z_{12} & U_i \\ Z_{21} & Z_{22} & 0 \\ Z_{31} & Z_{32} & 0 \end{vmatrix}}{\begin{vmatrix} Z_{11} & Z_{12} & Z_{13} \\ Z_{21} & Z_{22} & Z_{23} \\ Z_{31} & Z_{32} & Z_{33} \end{vmatrix}}. \tag{5.178}$$

To conclude the calculation, operator D is replaced by $j\omega$ in the expression for U_0, and the characteristic $U_0 = f(\omega)$ for $\omega = $ var is determined.

It is not desirable to have the filter's resonance frequency close to that of the transduction resonator, since instabilities of the filter's resonant characteristics (thermal and mechanical) can track the transducer output.

chapter 6

Heat-exchange Elements

This chapter is devoted to several traditional elements widely used to measure temperature and fluid velocity or its composition. Description of solid-state devices for measurements of temperature can be found in Chapter 10.

TEMPERATURE-SENSITIVE ELEMENTS

In this section, we will discuss thermocouples and metal resistors sensitive to temperature. Semiconductor temperature sensors will be considered among the solid-state devices in Chapter 10, under "Solid-state Temperature Sensors" and "Fiber-optic Sensors."

A thermocouple (Fig. 6.1) develops an electromotive force proportional to the temperature difference between its hot and cold junctions. The thermocouples feature a wide temperature span (from near absolute zero to $+2700°C$), low cost, simplicity, and versatility.

Three effects are associated with forming the voltage in a thermocouple:

1. The *Thomson emf* is originated if the temperature of one end of a homogeneous conductor with predominating electron conductivity is raised above that of the other end. The hot end becomes positive with respect to the cold end.

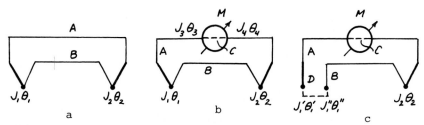

Figure 6.1 Connections of thermocouples. a = circuit with two thermocouples, b = insertion of meter into circuit, c = insertion of intermediate material into circuit; A to D = different materials, J's = junctions, M = meter, $θ$'s = temperatures.

2. The *Peltier emf* is developed at the junction of the ends of two dissimilar materials if the other ends of the materials are kept at a temperature different from that at the junction.

3. The *Seebeck effect* is a combination of these two phenomena. The rising of the voltages can be explained by the increase of density of free-charge carriers with temperature and by the diffusion of the excess electrons from the hot end to the cold end. In addition to this, the different materials have different free-carrier densities, and when two conductors are joined, electrons tend to diffuse through the junction. The conductor with the higher electron density loses electrons and acquires a positive charge.

The contact voltage is the major contributor to the potential for the common materials of thermocouples. Thermoelectric sensitivity is usually given relative to a standard material such as platinum at a specified temperature, because the voltage developed by a thermocouple is not a linear function of temperature. The following is an enumeration of these materials, with sensitivities in $\mu V/°C$ at 0°C given in the parentheses: bismuth (-72), constantan (-35), nickel (-15), platinum (0), mercury ($+0.6$), carbon ($+3$), aluminum ($+3.5$), lead ($+4$), silver ($+6.5$), copper ($+6.5$), gold ($+6.5$), tungsten ($+7.5$), iron ($+18.5$), nichrome ($+25$), germanium ($+300$), silicon ($+440$), tellurium ($+500$), and selenium ($+900$). The sensitivities for any combinations of the materials can be obtained by taking the difference of the sensitivities of the materials forming the couple. For example, the sensitivity of nichrome versus constantan is $25 - (-35) = 60\mu V/°C$.

A number of materials are standardized for common usage, and the characteristics of some of them are given in Table 6.1 [1, 90].

In constructions of thermocouples, two materials A and B (Fig. 6.1a) are involved in a basic thermoelectric circuit. They form two junctions J_1 and J_2 operating at temperatures θ_1 and θ_2. Junction J_2 can be considered as a reference. A current in the loop proportional to the difference in temperatures is defined by the difference between the emf's developed at the junctions and by the resistance of the circuit. In a practical connection, the loop must be broken (Fig. 6.1b) and the wires from meter M are inserted in the circuit. These wires of material C create two more junctions: J_3 and J_4 at temperatures θ_3 and θ_4. The measuring system will not be affected by these junctions if the emf produced by A–C is inherently small and if $\theta_3 = \theta_4$. Quite often, in circuits of compensation, one more material D (Fig. 6.1c) is introduced at one of the junctions, thus forming two new junctions J_1' and J_1'' at temperatures θ_1' and θ_1''. If these temperatures are uniform, no additional emf will affect the performance of the circuit. Maintaining a constant temperature at the places of connection is mandatory when a reference-junction technique is utilized (Fig. 6.2). The measuring setup contains an isothermal terminal block that has a resistive temperature sensor (usually semiconductor). Terminals from the thermocouple are rigidly attached to

TABLE 6.1

Characteristics of Thermocouples

Type	Composition, percent	Range of application, °C	emf*, mV
Copper/constantan	100 Cu/60 Cu 40 Ni	-200 to $+300$	4.24
Iron/constantan	100 Fe/60 Cu 40 Ni	-200 to $+1100$	5.28
Chromel/constantan	90 Ni 10 Cr/55 Cu 45 Ni	0 to $+1100$	6.30
Chromel/alumel	90 Ni 10 Cr/94 Ni 2 Al 3 Mn 1 Si	-200 to $+1200$	4.10
Platinum/platinum rhodium	100 Pt/90 Pt 10 Rh	0 to $+1450$	0.64
Carbon/silicon carbide	100 C/100 SiC	0 to $+2000$	17

*Emf is developed at 100°C; reference junction is at 0°C.

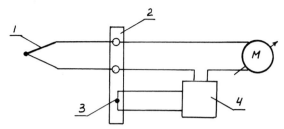

Figure 6.2 Isothermal block reference connection. M = voltmeter, 1 = thermocouple, 2 = isothermal block, 3 = temperature sensor, 4 = compensating circuit.

the block and their temperature remains precisely uniform with the temperature of the sensor. The compensation circuitry develops a voltage that is combined with the voltage from the junction. The algebraic sum of these two voltages is measured and is proportional to the measuring temperature.

There are several sources of errors associated with measurements by thermocouples. Voltaic effects take place when moisture in porous insulation is combined with salt or acid, creating a difference in potential in the dissimilar materials. Extraneous junctions are sources for the parasitic emf. Equalizing the temperature of different currents carrying conductors and eliminating random junction effects is important. Processes of material oxidation are more intensive at high temperatures. Corrosion and degradation of material cause changes in electrical characteristics (e.g., resistance). If temperature is monitored by a contact sensor, an undesirable transmission of heat energy from the ambient to the junction can cause heating or cooling of wires by radiation and can, as a result, lead to significant errors in reading the temperature. A sink of heat through the wires of a thermocouple to or from the area or parts of the junction's attachment is possible, especially if the wires are heavy and the parts are light.

Contemporary resistive temperature sensors are fabricated of thin wire, strips, and films made primarily of platinum, nickel, and copper. The practical temperature ranges for these materials are, for platinum, -258 to $1000°C$; for nickel, -150 to $300°C$; and for copper, -200 to $120°C$. Elements made of platinum run uniform and are high in resistance and stable in value but expensive. At a high temperature, platinum can be contaminated by the vapors of metals (e.g., iron); in a reduction environment, it becomes brittle and electrically unstable. Nickel's chemical stability, temperature coefficient, and resistivity are high, but the element does not run uniform and requires individual adjustment. Copper-made elements are uniform in their characteristics, stable in value, and inexpensive but low in resistance. They cannot be used for high temperatures because of the oxidation and possible destruction of the elements' integrity. Some more characteristics of Pt, Ni, and Cu along with other materials used for thermoresistors are indicated in Table 6.2.

The resistance of a conductor as a function of temperature is described by linear or nonlinear equations with constant coefficients valid for a particular temperature range. These coefficients are defined not only by the impurities of materials but also by the method of fabrication of the conductors. For platinum, the resistance as a function of temperature between 0 and 850°C [91] is written

$$R_\theta = R_0(1 + A\theta + B\theta^2) \tag{6.1}$$

where R_0 and R_θ = resistances of conductor at 0 and $\theta°C$, Ω, respectively;
$\qquad A = 3.90784 \times 10^{-3} 1/K$; and
$\qquad B = -0.578408 \times 10^{-6} 1/K^2$.

TABLE 6.2

Characteristics of Thermoresistive Materials at 20°C

Material	Resistance-temperature coefficient 1×10^{-3} 1/K	Resistivity $\Omega \cdot$ mm²/m	Melting temperature °C
Platinum	3.91	0.105	1773
Copper	4.28	0.017	1083
Nickel	6.3 to 6.6	0.068	1455
Tungsten	4.82	0.055	3410
Molybdenum	4.57	0.052	2630
Rhenium	3.11	0.211	3170
Rhodium	4.57	0.047	1960
Graphite	0.02	46.0	3870
Platinum and 20 Rhodium	0.21	0.160	1900

For the range 0 to $-200°C$,

$$R_\theta = R_0[1 + A\theta + B\theta^2 - C(\theta - 100)\theta^3] \tag{6.2}$$

where A and B are as given for formula 6.1 and

$$C = -4.481924 \times 10^{-12} 1/K^3.$$

All these coefficients allow accurate calculation if the ratio of resistances taken at 100 and 0°C is 1.385.

Similar formulas for nickel at temperatures from -50 to $+180°C$ are

$$R_\theta = R_0(1 + A\theta + B\theta^2 + C\theta^3) \tag{6.3}$$

where $A = 5.47 \times 10^{-3} 1/K$;
$B = 0.639 \times 10^{-5} 1/K^2$; and
$C = 0.69 \times 10^{-8} 1/K^3$.

The resistance versus temperature characteristic of copper is usually introduced by a linear function in the range of -50 to $+180°C$:

$$R_\theta = R_0(1 + \alpha_\theta \theta) \tag{6.4}$$

where $\alpha_\theta = 4.26 \times 10^{-3} K^{-1}$.

If the resistance $R_{\theta1}$ of a copper conductor at temperature θ_1 is known and the resistance $R_{\theta2}$ at temperature θ_2 should be found, the following formula can be helpful:

$$R_{\theta2} = R_{\theta1} \frac{(1 + \alpha_\theta \theta_2)}{(1 + \alpha_\theta \theta_1)}. \tag{6.5}$$

Thermoresistive elements having different physical forms should be fabricated in such a manner as to avoid stresses in wires and unpredictable changes of resistance because of strain-gage effects. A wound coil is a typical construction of the element. A very small element exhibits a rapid response to the change of temperature, but it has small resistance (low output signal). Relatively large coils have a large time constant. Resistance thermometers are usually parts of a Wheatstone bridge. Temperature differences are readily measured when the elements are introduced in the adjacent arms of the bridge. When an element is used, a special arrangement of leads (Fig. 6.3) has to be made for compensation for the variation in lead resistance. The

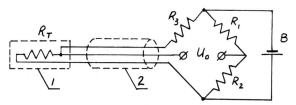

Figure 6.3 Lead arrangement for lead-resistance compensation. B = battery, R_1 to R_3 = resistors of bridge, R_T = thermoresistor, 1 = three-wire cable, 2 = sensor.

compensation is usually accomplished by the addition of leads into two adjacent bridge arms. The uniform change of their resistances does not significantly affect the bridge output. Thermoelectric emf should not affect the measuring circuit and can be eliminated by utilizing ac excitation. The heat-transfer conditions adjacent to the element largely define self-heating and the maximum current that can be tolerated. Empirical formulas can be recommended for rough calculations. One of them is

$$I = 2d^{1.5}\Delta\theta^{0.5} \text{ (for } \theta \text{ up to 750°C)} \qquad (6.6)$$

where I = current, A;
   d = diameter of wire, mm; and
   $\Delta\theta$ = acceptable change in θ, °C.

If the wire is in still air, $\Delta\theta = 5I^2/d^2$.

The heat transfer through the leads should be minimized by using small-diameter wire, insulations, and so on.

Besides self-heating, heat-transfer characteristics from media to the element should be studied to verify whether the measurement is adequate. It should be understood that the temperature transduction element senses the temperature of its parts (e.g., case), and, depending on the heat-transfer conditions, this temperature can differ from that to be examined.

As an example, we will consider one of the common cases of gas-stream temperature measurement when a thermoresistive sensor is placed in a pipe carrying gas (Fig. 6.4). The sensor is oriented along the radius of the pipe and thermally isolated from the pipe wall. Our task is to find the difference between the sensor and gas temperatures and define the factors affecting this difference.

The equation balancing powers developed in the sensor and exchanged with gas is

$$(\theta_s - \theta_g)\bar{h}_1 A = I^2 R \qquad (6.7)$$

where

$$\bar{h}_1 = \frac{Nu'_D \cdot K}{D}, \quad Nu'_D = 0.8\sqrt{Re'}, \quad Re' = \frac{vl'}{v_g}, \quad \text{and} \quad l' = 0.5\pi d; \qquad (6.8)$$

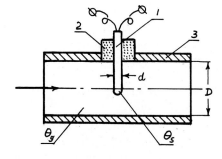

Figure 6.4 Thermoresistive sensor in gas-carrying pipe. D = diameter of pipe, d = diameter of sensor, θ_g = temperature of gas, θ_s = temperature of sensor, 1 = sensor, 2 = thermal insulation, 3 = pipe.

θ_s and θ_g = sensor and gas temperatures, respectively, °C;
$\bar{h}_1$ = heat transfer coefficient, W/m² · K;
A = sensor's area, m²;
I = current in thermoresistor, A;
R = thermoresistor's resistance, Ω;
Nu'_D = Nusselt number, dimensionless;
K = thermal conductivity, W/m · K;
D = diameter of pipe, m;
Re' = the Reynolds number, dimensionless;
v = average linear speed of gas, m/s;
l' = characteristic length, m; and
ν_g = kinematic viscosity of gas, m²/s.

Equation 6.7 is written with the assumption that thermal losses through thermal insulation and radiation are negligible.

From equation 6.7 the difference in the temperatures is

$$\theta_s - \theta_g = \frac{I^2 R}{\bar{h}_1 A}. \tag{6.9}$$

In terms of the measurement accuracy, the change of this difference $\Delta(\theta_s - \theta_g)$ is important:

$$\Delta(\theta_s - \theta_g) = \frac{2IR}{\bar{h}_1 A}\Delta I + \frac{I^2}{\bar{h}_1 A}\Delta R - \frac{I^2 R}{A\bar{h}_1^2}\Delta\bar{h}_1 - \frac{I^2 R}{\bar{h}_1 A^2}\Delta A. \tag{6.10}$$

It is obvious from equations 6.9 and 6.10 that an increase in $\bar{h}_1$ and A and a decrease in I and R are desirable. The dependence of reading on gas velocity can be controlled by the variation of the same parameters, since $\bar{h}_1 = f(v)$.

The measuring of temperature by thermoresistive elements is an inertial process. The equation describing this process can be obtained by equating the expressions for the elementary energies during the heat exchange between the sensor and medium:

$$\rho_s V_s C_p \cdot d\theta_s = \bar{h}_1 A(\theta_m - \theta_s)dt \qquad \text{or} \tag{6.11}$$

$$\tau \frac{d\theta_s}{dt} + \theta_s = \theta_m \tag{6.12}$$

where τ = time constant, s;
t = time, s;
ρ_s = sensor's material density, kg/m³;
V_s = sensor's volume, m³;
C_p = specific heat for the sensor's material, J/kg · K; and
θ_m = medium temperature, °C (the other terms as before).

Equation 6.12 describes the behavior of a first-order system. If the sensor is encased, and one more heat-transfer process takes place (from the case to the sensor), the second-order equation can be applicable:

$$\tau^2 \frac{d^2\theta_s}{dt^2} + 2\zeta\tau \frac{d\theta_s}{dt} + \theta_s = \theta_m. \tag{6.13}$$

The parameters τ and ζ depend on the constructional and material characteristics of the sensor. It is clear that the response of a sensor modeled by equation 6.13

is slower than that of a sensor modeled by equation 6.12. In most cases, modeling by equation 6.12 is satisfactory for the evaluation of dynamic responses.

In operator notation, equation 6.12 is written

$$\frac{\theta_s(D)}{\theta_m(D)} = \frac{1}{\tau D + 1} \tag{6.14}$$

where

$$\tau = \frac{\rho_s V_s C_p}{\bar{h}_1 A}. \tag{6.15}$$

If an abrupt change of the media temperature from θ_O ("old") to θ_N ("new") takes place, the change in the sensor body temperature will be

$$\theta_s = \theta_N - (\theta_N - \theta_O)e^{-(t/\tau)}. \tag{6.16}$$

This formula can be modified for the calculation of the time that is needed for the sensor body to reach a specific temperature θ_s.

$$t = -\tau ln \frac{\theta_s - \theta_N}{\theta_O - \theta_N}. \tag{6.17}$$

If the media temperature changes gradually and linearly with respect to time ($\theta_m = \varphi t$), equation 6.12 is rewritten

$$\tau \frac{d\theta_s}{dt} + \theta_s = \varphi t \tag{6.18}$$

where φ = constant, °C/s. The solution of this equation is

$$\theta_s = \varphi t + \varphi \tau (1 - e^{-(t/\tau)}). \tag{6.19}$$

For $t \gg \tau$, $\theta_s = \varphi(t - \tau)$ and the difference between θ_m and θ_s is independent of time:

$$\theta_m - \theta_s = \varphi \tau. \tag{6.20}$$

An exponential change of temperature θ_m with time constant τ_m and $\theta_m = \theta_O$ at $t \to \infty$ gives

$$\theta_m = \theta_O(1 - e^{-(t/\tau_m)}), \tag{6.21}$$

$$\tau \frac{d\theta_s}{dt} + \theta_s = \theta_O(1 - e^{-(t/\tau_m)}). \tag{6.22}$$

At moment t, the sensor body temperature will be

$$\theta_s = \theta_O \left(1 - \frac{\tau_m e^{-(t/\tau_m)} - \tau e^{-(t/\tau_m)}}{\tau_m - \tau}\right). \tag{6.23}$$

For the conditions $\tau_m \gg \tau$ and $t \gg \tau$,

$$\theta_s = \theta_O \left(1 - \frac{\tau_m}{\tau_m - \tau} e^{-(t/\tau_m)}\right). \tag{6.24}$$

For all these cases, the error of dynamic responses can be calculated if the differences between θ_m and θ_s are taken for the given moment t. As is clear from the analysis, the improvement of the heat exchange between the sensor and media is always beneficial for decreasing the errors. In this regard, the increase in the magnitudes of the terms in the denominator of expression 6.15 and the decrease of its numerator are helpful when decisions are made on the materials, sizes, and shapes of the constructional elements.

HOT-WIRE ELEMENTS

The cooling effect of a fluid on the resistance of a current-heated wire is used as a measure of the fluid's velocity or composition. The heat loss from the wire depends upon the flow rate, thermal conductivity, and specific heat of the fluid. There are two methods of measurement. In one of them, current through the wire is maintained constant. The voltage drop is monitored and is used as a measure of cooling. A second approach is to measure the current when the voltage drop across the element is held constant. The output of hot-wire anemometers is nonlinear. Transfer characteristics of the elements (output versus flow rate) are usually determined experimentally. A typical relation is

$$U_o = \sqrt{A + BV^n} \tag{6.25}$$

where
$$U_o = \text{output voltage, V;}$$
$$A, B, \text{ and } n = \text{constants; and}$$
$$V = \text{fluid velocity, m/s.}$$

The behavior of the anemometer inserted in a pipe is described by the same relationships obtained for the thermoresistive sensor, which was discussed in section "Temperature-sensitive Elements." A thermocouple heated by an ac current is also utilized in thermoanemometers. A dc voltage at the thermocouple terminals is a function of the flow rate.

The concentration of gas components in a mixture of two known gases can also be measured by a hot-element sensor [8]. Thermal conductivities ν of gases vary (see Table 6.3) [8]. For dry air, $\nu = 2.38 \times 10^{-3} \text{J/m} \cdot \text{s} \cdot \text{K}$.

ν for the mixture of two gases can be defined as

$$\nu = g_1\nu_1 + g_2\nu_2 \tag{6.26}$$

where ν_1 and $\nu_2 = \nu$'s for gases 1 and 2, respectively, J/m $\cdot$ s $\cdot$ K; and
g_1 and $g_2 = $ gas concentrations as fractions of unity for gases 1 and 2, respectively.

TABLE 6.3

Relative Thermal Conductivities ν/ν_{air} and Temperature Coefficients of Thermal Conductivity α_ν for Gases

Gas	ν/ν_{air}	α_ν [1/K]
Air	1.000	0.00253
Argon	0.685	0.00311
Oxygen	1.015	0.00303
Nitrogen	0.998	0.00264
Hydrogen	7.130	0.00261
Carbon oxide	0.964	0.00262

Since $g_1 = 1 - g_2$, measuring ν is sufficient for detecting ν_1 and ν_2; therefore, the concentrations are

$$g_1 = \frac{\nu - \nu_2}{\nu_1 - \nu_2} \quad \text{and} \quad g_2 = \frac{\nu_1 - \nu}{\nu_1 - \nu_2}. \tag{6.27}$$

If the gas to be tested is forced to flow with a stable speed through a cell (Fig. 6.5) having a thin, heated wire in a pipe, cooling of the wire (depending on the gas concentration) changes the wire resistance. This resistance is a measure of the concentration.

It should be noted that formula 6.26 gives an approximate relationship for the gas mixture. More exact evaluation of the gas content can be done by calibration of the instrument using gas samples.

Taking the construction shown in Figure 6.5 as a model, we can derive a transfer characteristic for the element. Fourier's law of heat conduction states that the rate of heat flow through a substance is proportional to the area normal to the direction of the flow and to the negative of the temperature rate change with distance along the direction of the flow. It means that the heat lost from the elementary surface of the wire is

$$dQ = -\nu \cdot dA \frac{d\theta}{dr} \tag{6.28}$$

where dQ = elementary heat flow rate from the surface, J/s;
dA = elementary surface, m^2; and
$d\theta/dr$ = the gradient of the temperature along radius r, K/m.

The temperature distribution along the radius is found by taking the integral of equation 6.28 with respect to the radius:

$$\theta = \theta_p - \frac{1}{2\pi\nu} \cdot \frac{dQ}{dl} \, ln \frac{r_w}{r_p} \tag{6.29}$$

where θ_p = temperature at wall, K;
r_w = radius of wire, m;
r_p = inner radius of pipe, m; and
l = wire length, m.

By taking one more integral of equation 6.29 with respect to length l, the heat loss rate from the entire wire is obtained:

$$Q = \frac{(\theta_p - \theta_w)2\pi\nu l}{ln \dfrac{r_w}{r_p}}. \tag{6.30}$$

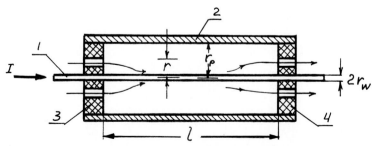

Figure 6.5 Hot-wire element. I = current, l = length of wire, r = variable radius, r_p and r_w = pipe and wire radii, 1 = hot wire, 2 = pipe, 3 and 4 = wire supporting insulators with inlets for gas.

Assuming that the wire has resistance R and is heated by current I, the equation of the power balance is $Q = I^2R$, and

$$R = \frac{1}{I^2} \cdot \frac{(\theta_p - \theta_w)2\pi l}{ln \dfrac{r_w}{r_p}} \nu.$$ (6.31)

Resistance R is a function of temperature:

$$R = R_0(1 + \alpha\theta_w)$$ (6.32)

where R_0 = resistance of wire at 0°C, Ω; and
  α = resistance-temperature coefficient of wire material, 1/K.

Extraction θ_w from equation 6.32 and substitution into equation 6.31 gives

$$R = \frac{\theta_p + \dfrac{1}{\alpha}}{\dfrac{I^2}{2\pi l\nu} ln \dfrac{r_w}{r_p} + \dfrac{1}{\alpha R_0}}.$$ (6.33)

In this transfer characteristic, r_w, r_p, l, R_0, and α are constants. If I and θ_p are stabilized during the measurement, R simply defines ν or the concentration in question.

One of the most important conditions for accurate measurements is maintaining constant gas speed and correctly choosing the temperature of wire. For example, for measurements of CO_2 concentration, the temperature of the wire should not be higher than 100–120°C because of the rapid increase in α_ν with temperature. In a mixture with air, this coefficient becomes almost equal to that of air and it affects the accuracy of the method.

It is obvious that the basic relationships illustrating performance of the wire element can be used without substantial changes for the devices containing silicon or metal-film elements.

chapter 7

Characteristics of Solid-state Transducing Devices

The rapid and successful development of the technology of solid-state devices has made possible the creation of a new class of measuring devices—solid-state sensors. These sensors can be introduced in four categories:

1. sensors based on semiconductor effects in materials,
2. sensors employing non-semiconductor effects in solids,
3. sensors in which semiconductor and nonsemiconductor principles are combined, and
4. Sensors in which one of the above devices is combined with a computing element.

Among these devices, silicon sensors and microstructures based on micromachining (μ machining) have the potential to revolutionize the field of sensors and create a new industry of silicon microstructures.

Micromachining is primarily the three-dimensional sculpting of silicon using well-developed or modified semiconductor batch-processing technology. Solid-state sensors were originated many years ago, but the first publication in 1954 [92] on the piezoresistive properties of germanium and silicon stimulated extensive research and development of different semiconductor sensors and especially strain gages, which became pioneers in the microsensor business.

The wide application of microprocessors whose inputs receive signals from sensors has created the need for low-cost, high-performance, and mass-produced sensors. It is natural that IC-batch processing techniques used for manufacturing many electronic components, including microprocessors, attracted sensor designers because of their compatibility with microprocessor technology and their low cost.

As the initially expensive and complicated technology of silicon strain gages was improved over the years, doors were opened to many new types of sensors for measuring different physical quantities.

Today, the following applications are typical for solid-state sensors: automotive, medical, aerospace, military, process control, industrial control, consumer, test, communication, factory automation, heating, ventilation, air-conditioning, and computer peripherals. Depending on their complexity, sensors are divided into several categories which are specifically termed by the manufacturers. The main categories are:

A sensor die is usually an inexpensive, micromachined silicon chip produced in high volumes as a commodity product.

A sensor is defined as a first-level packaged sensor die that is provided with basic compensation and normalization of characteristics.

A transducer is a fully packaged, compensated, and calibrated sensor that can be provided with a signal conditioner and is convenient for use by an end user.

A smart transducer is a packaged sensor combined with a microprocessor to provide digital compensation for different adverse factors, precise calibration, and sometimes remote communication and control.

We will begin this part of the book devoted to solid-state devices with a review of some general aspects of solid physics and descriptions of basic processes in the fabrication of structures that are quite common for many solid-state sensors; then we will consider several groups of the most popular sensors.

SOLIDS IN SENSORS

Electroconductivity of Solids

Many solid-state sensors are based on the effects of change of conductivity or generation of charges in solids. In spite of their wide variety, these sensors have one quality in common. As a physical quantity affects a solid transduction element, the charged microparticles in the element are displaced or stimulated to flow, thus producing the electrical output signal.

In terms of electrical properties, the elements can be conductors, semiconductors, and insulators.

The atoms of these substances consist of a central *nucleus* (Fig. 7.1) surrounded by *orbiting electrons,* in a model of a *planetary atom*. The electrons carry a negative charge (1.60219×10^{-19}C). The nucleus is composed of positive *protons* combined with *neutrons* having no charge. The mass of each particle in the nucleus is approximately 1836 times the mass of the electron (it is 9.1095×10^{-31}kg).

The charges of protons and electrons are equal to each other.

The atoms of various substances are composed of the same particles—protons, neutrons, and electrons. The difference in properties of the substances depends on the number and arrangement of the particles, combination of atoms, and other factors.

The electrons can occupy only certain *orbital* rings or *shells* around the nucleus. Each shell contains a limited number of electrons, and they move at a fixed distance from the nucleus. The outer-shell electrons (*valence electrons*) largely determine the electrical properties of an atom. Normally, an atom is electrically neutral since the number of electrons and protons is in balance. The atom turns into a positive ion if it loses electrons and into a negative ion if it gains some electrons. The electrons that are close to the nucleus are held in the orbits with large forces and are bound tightly to the atom. The electrons in the outermost orbits experience smaller forces of attraction from the nucleus; therefore, they can be easily displaced, thus producing ions. The electrons in the outer orbit (*valence shell*) determine the electrical characteristics of an atom. For example, a copper atom (Fig. 7.2) contains only one electron in the valence shell with a weak attachment to the nucleus. It can easily be removed from its orbit and drift between the atoms. This is why copper is one of the best conductors.

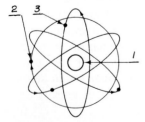

Figure 7.1 Planetary atom. 1 = nucleus, 2 and 3 = orbiting electrons.

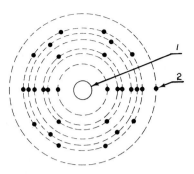

Figure 7.2 Copper atom. 1 = nucleus,
2 = one electron in valence shell.

A silicon atom (Fig. 7.3) has four electrons in the outer shell but can contain four more. In other words, the valence shell has four vacancies or holes which can be filled with electrons. If a certain amount of energy is applied to an atom, it can release its electrons. However, in silicon, the link of valence electrons with the nucleus is stronger than in copper. Easily detached electrons in copper form a mass of electrons, called *electron gas,* drifting about between the copper atoms and carrying a negative charge. Since the atoms are positively charged (with losing negative particles), they electrostatically attract the electron gas (Fig. 7.4), creating a bonding force termed *metallic bonding* (for metals). As an electrical field is applied to metal, the electron gas is given motion, which makes the material electroconductive.

The bonding arrangement in silicon is different. Silicon's atoms are close to each other, and outer-shell electrons orbit in the valence shell of two atoms, filling the holes of neighboring atoms (Fig. 7.5) and developing a bonding force known as *covalent bonding*. Due to this arrangement, all of the valence shells are filled, there are no charge carriers in the material, and pure (intrinsic) silicon behaves like an

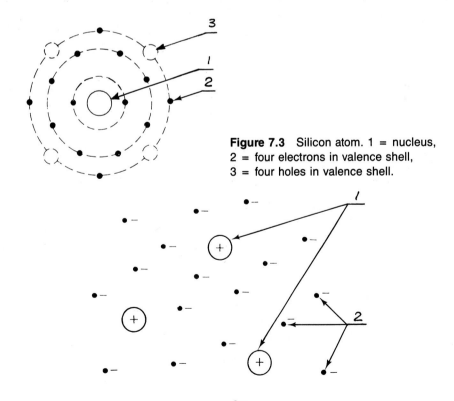

Figure 7.3 Silicon atom. 1 = nucleus,
2 = four electrons in valence shell,
3 = four holes in valence shell.

Figure 7.4 Atomic bonding in conductors.
1 = positive ions, 2 = free electrons.

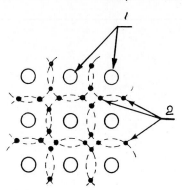

Figure 7.5 Covalent bonding in semiconductors. 1 = nuclei, 2 = shared valence electrons.

insulator. Indeed, at very low temperatures, silicon does not conduct current; but when some electrons are given thermal energy with a rise in temperature, electrons become mobile, a small current can flow through the silicon, and it behaves as a semiconductor. The energy required to break such a covalent bond for silicon (*energy gap*) is about 1.1eV.

In certain materials, if the electrons in covalent bonds are very strongly attached to their atoms, the current is zero; such materials are then said to be insulators. There are insulators in which the electrons can leave one valence orbit and be accepted into the orbit of another atom. Atoms become ionized by forming positive and negative ions because of the loss and gain of electrons. This process results in the formation of an electrostatic attraction between the ions (ionic bonding), thus holding the material together. There are no free electrons in the material, and the current cannot flow, but negative and positive ions can be arranged in groups.

If, under certain circumstances, an electron leaves a covalent-bond atom, it forms a *hole,* which can be filled by a neighboring electron. The vacancies can move from ion to ion in the material. This motion can be regarded as a flow of positive charges ("*flow of holes*"). In the theory of semiconductors, it is common and convenient to consider both electron and hole flow.

Electrical Characteristics of Semiconductors

The motion of electrical particles in a semiconductor is stimulated by an electrical field that develops a force on particles.

$$F = q\epsilon = m \frac{dv}{dt} \tag{7.1}$$

where F = force, N;
    q = particle's charge, C;
    ϵ = field intensity, V/m;
    m = particle's mass, kg;
    v = particle's velocity, m/s; and
    t = time, s.

When a unit charge moves against the field, the work for the charge displacement along the x-axis is

$$V = -\int_{x_1}^{x_2} \epsilon \, dx \tag{7.2}$$

where V = work done or difference in potentials between points with coordinates x_1 and x_2, V. The potential energy for the particle with charge q is

$$U = qV \qquad (7.3)$$

where U = potential energy, J.

According to the law of conservation of energy, the total energy W of a particle moving in a field remains constant.

$$W = U + \frac{mv^2}{2}. \qquad (7.4)$$

If this particle leaves one point (x_1) with velocity v_0 and moves against the field toward another point (x_2) having potential V with respect to the first point, the equation of total energy can be written

$$\frac{mv_0^2}{2} = qV + \frac{mv^2}{2}. \qquad (7.5)$$

It is worth noting that the particle's speed at the second point does not depend on the distribution of the field along the particle's path but only on the difference in potentials between the points. The particle can reach the point x_2 if its original energy $W_2 = mv_0^2/2 \geq qV$ (otherwise v is imaginary). For an intermediate value of energy W_i (Fig. 7.6), the particle comes to rest at coordinate x_i and it can never "penetrate" in the shaded area between x_i and x_2. At point P_1, the particle acts as if it met a barrier after which the direction of its motion was altered (backward, along the field). This barrier is called *the potential-energy barrier* or *potential hill*.

An electron volt (eV), rather than the joule (J), is used as the unit of energy in the analysis of semiconductor devices. It is originated from formula 7.3 by setting q equal to the charge of an electron ($\sim 1.6 \times 10^{-19}$C) and the difference in the potentials—to 1V. Therefore, $1\text{eV} = 1.6 \times 10^{-19}$J.

As mentioned above, electrons in metal form electron gas and move freely, colliding with heavy, almost-stationary atoms. The average distance between collisions is termed the *mean free path*. The direction of motion is random and it does not provide an oriented flow of carriers. As an electrical field is introduced to the material, the electrons obtain an additional acceleration a. They still collide with ions and bounce off, but attain an average *drift speed* v in the direction opposite to the field, thus creating a directed flow of electrons which is a current. The acceleration is defined as force over mass:

$$a = \frac{F}{m} = \frac{q\epsilon}{m}. \qquad (7.6)$$

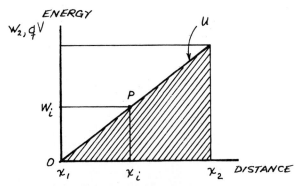

Figure 7.6 Potential-energy barrier.

The speed at a time between collisions is *at*; therefore, the average speed of electrons is proportional to the field intensity ϵ or

$$v = \mu\epsilon \tag{7.7}$$

where μ $[m^2/V \cdot s]$ is known as the *mobility* of the electrons.

If a certain volume of a material or medium contains n electrons per cubic meter, the charge per cubic meter is $nq = \rho$ (charge density). If we multiply ρ by a volume speed of electrons Q, the magnitude of the current I, flowing due to the field, is found in

$$I = \rho Q = \rho vA \tag{7.8}$$

where $Q = A \cdot v$ $[m^3/s]$ and A $[m^2]$ is a material's cross-sectional area crossed by a stream of electrons.

The current density J $[A/m^2]$ is obtained by taking the ratio of current to cross-sectional area and substituting v extracted from equation 7.7:

$$J = \rho v = \rho\mu\epsilon = nq\mu\epsilon = \sigma\epsilon \tag{7.9}$$

where $\sigma = nq\mu$ is the conductivity of the material. The conductivity is measured in siemens per meter (S/m).

It is clear from the expression for σ that the larger the number of electrons capable of moving in the material the higher the conductivity. For conductors $n \approx 1 \times 10^{28}$, and for insulators $n \approx 1 \times 10^7$, electrons/m^3. For semiconductors, n lies between these two values and can be greatly varied by adding to the intrinsic semiconductor a small amount of trivalent or pentavalent atoms. Such a semiconductor is called *extrinsic, doped,* or *impure.* As a dopant with five valence electrons is added (for example *Sb* to *Si*), it impregnates the crystal lattice (Fig. 7.7) and displaces some of the silicon atoms, forming unbound electrons that can be carriers of current. The energy required to move the unbound electrons out of the atoms is only about 0.05eV for silicon. The pentavalent impurities (antimony, phosphorus, and arsenic) create negative carriers. They are called *donor* or *n-type* impurities. Along with the increase in the number of electrons (*majority carriers*), the number of holes (*minority carriers*) in the material is decreased.

When a trivalent impurity (boron, gallium, or indium) is added to the intrinsic semiconductor, it becomes capable of accepting electrons (Fig. 7.8) because only three covalent bonds are filled and one free bond constitutes a hole. These dopants are termed acceptors or p-type impurities. When acceptors are added, the concentration of free electrons is decreased. In this case, the majority carriers are holes, and minority carriers are electrons.

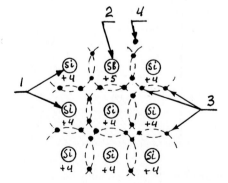

Figure 7.7 Extrinsic silicon with atom displaced by atom of antimony. 1 = nuclei of Si, 2 = nucleus of Sb, 3 = valence electrons, 4 = unbound electron.

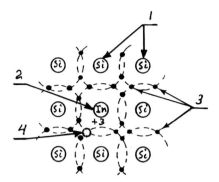

Figure 7.8 Extrinsic silicon with atom displaced by atom of indium. 1 = nuclei of Si, 2 = nucleus of In, 3 = valence electrons, 4 = hole.

The conductivity of materials can be controlled by a very small amount of dopants, which makes it possible to obtain materials either with electron or hole conductivities. They are called n-*type,* or p-*type,* respectively.

The mass-action law states that the product of the free negative and positive concentrations is constant and independent of the donor or acceptor impurity content:

$$np = n_i^2 \tag{7.10}$$

where *n, p,* and n_i are electron, hole, and intrinsic concentrations, respectively.

It can be easily proven that in an n-type material, the free-electron concentration is approximately equal to the density of donor atoms (N_D), and in a p-type material, the hole concentration is approximately equal to the density of acceptor atoms (N_A)

$$n \approx N_D \quad \text{and} \quad p \approx N_A \tag{7.11}$$

Combining expressions 7.10 and 7.11, we can find the concentration *p* of holes in an n-type semiconductor:

$$p = \frac{n_i^2}{N_D}. \tag{7.12}$$

For a p-type semiconductor, the concentration of free electrons is

$$n = \frac{n_i^2}{N_A}. \tag{7.13}$$

It is possible to control or totally change the conductivity of the material by adding donors to a p-type crystal or acceptors to an n-type material.

In an intrinsic semiconductor, the concentrations of free electrons and holes are in equilibrium. Due to thermal agitation, new electrons are created constantly and recombine with holes. At the same time, some existing hole-electron pairs disappear as a result of an opposing process. On the average, a hole and an electron exist for a certain time before recombination. This time is known as a *mean time* for holes (τ_p) and for electrons (τ_e).

When an electrical field affects a semiconductor, the positive and negative carriers of charges move in opposite directions. Since the charges differ in signs, they produce a common current. The current density is calculated the same way as it is for formula 7.9, but for *n* and *p* carriers having mobilities μ_n and μ_p, respectively:

$$J = (n\mu_n + p\mu_p)q\epsilon = \sigma\epsilon \tag{7.14}$$

where n and p = concentrations of free electrons and holes, respectively, electrons/m^3 and holes/m^3; and

σ = conductivity of a semiconductor, S/m.

$$\sigma = (n\mu_n + p\mu_p)q. \tag{7.15}$$

The density of hole-electron pairs increases with temperature, causing the increase in the conductivity.

For an intrinsic semiconductor [93]:

$$n_i = \sqrt{A_0 T^3 exp\,(-E_{G0}/kT)} \tag{7.16}$$

where A_0 = constant;

T = temperature, K;

E_{G0} = energy gap at 0K (zero kelvin), eV; and

k = Boltzmann constant, eV/K ($k = 8.620 \times 10^{-5}$ eV/K).

The energy gap $E_G(T)$ is a linear function of temperature.

For silicon [94],

$$E_G(T) = 1.21 - 3.60 \times 10^{-4} \times T. \tag{7.17}$$

For germanium [95],

$$E_G(T) = 0.785 - 2.23 \times 10^{-4} \times T \tag{7.18}$$

The mobility depends on temperature, and at temperatures between 100 and 400K, it falls off approximately as the inverse square of the temperature.

High electric field intensities (larger than 1×10^3V/cm) affect mobility, which varies approximately as the inverse square root of the intensity.

The motion of carriers in a semiconductor occurs not only due to the accelerating force from an electrical field but also because of the diffusion of carriers. There can be appreciable spatial differences in the concentration of uniform carriers in a crystal. This non-uniformity leads to the migration of carriers from an area of high concentration to an area of small concentration. The process is purely statistical and is determined by a larger number of randomly moving particles that move toward the area with the smaller concentration. If electrons move along the x-axis and the concentration gradient along this axis is dn/dx, the diffusion electron-current density is

$$J_n = qD_n \frac{dn}{dx}. \tag{7.19}$$

A similar expression for the diffusion hole-current density is written

$$J_p = -qD_p \frac{dp}{dx} \tag{7.20}$$

where J_n and J_p = diffusion electron- and hole-current densities, respectively, A/m^2;

D_n and D_p = diffusion constants for electrons and holes, respectively, m^2/s; and

J_p is negative because it is taken for holes.

In practical problems, usually one or the other transport process predominates. If both the potential and concentration gradient exist simultaneously, the total current density will be as follows:

$$J = J_n + J_p = q\left(n\mu_n\epsilon + D_n \frac{dn}{dx}\right) + q\left(p\mu_p\epsilon - D_p \frac{dp}{dx}\right). \qquad (7.21)$$

The mobility and diffusion constants for a moderately doped semiconductor material are dependent. They are related by an equation called the Einstein equation:

$$D_n = \frac{\bar{k}T}{q}\,\mu_n, \qquad D_p = \frac{\bar{k}T}{q}\,\mu_p \qquad (7.22)$$

where $\bar{k}$ = Boltzmann constant expressed in J/K ($\bar{k} = 1.381 \times 10^{-23}$ J/K). The ratio $\bar{k}T/q$ is given in volts and equals 0.026V at room temperature.

Comparing the electrical properties of metals and semiconductors, we can state that metals have a good conductivity that decreases with temperature, a single charge carrier (electrons), a large carrier density, a fixed carrier density, a drift current, and no diffusion current. Semiconductors are distinguished by fair conductivity which increases with temperature, bipolar charge carriers (electrons and holes), moderate carrier density which varies with temperature and number of dopants, and non-fixed carrier concentration controlled by the amount of dopants, drift and diffusion currents.

Several electrical properties of three popular semiconductors are summarized and given in Table 7.1.

Semiconductor Junctions

So far we have considered the properties of semiconductors with a uniform distribution of impurity atoms. A number of sensors employ these materials—for example, strain gages, thermistors, and photoconductive cells. In these linear devices, the electrical current is directly proportional to the voltage applied to the device (Fig. 7.9). Among contemporary sensors, we can find devices that have a nonlinear current-voltage characteristic, for example, temperature-sensitive diodes, photo-sensitive diodes and transistors, and pressure-sensitive transistors. There is a pronounced trend to employ a metal-oxide-silicon field-effect transistor for sensing physical quantities. Furthermore, many transducers incorporate diodes and unipolar, bipolar, and metal-oxide transistors, which are used for signal conditioning and are integrated with other transducer elements. All of these devices contain a p-n junction which will be described briefly in the following sections.

A p-n junction can be introduced by a single-crystal bar (Fig. 7.10a) or a planar structure (Fig. 7.10b) with predominantly p-type impurities at one side of the junction and n-type impurities at the other side. The majority carriers for the p and n sides are holes and electrons, respectively. If these two parts were separated, they would be electrically neutral because of the balance of the carriers. However, since the two sides are in contact, some charges travel through the border between the sides. This diffusion is caused by a sharp carrier concentration gradient at the junction. As holes move toward the n-region, they leave behind negatively charged

TABLE 7.1

Electrical Properties of Silicon, Germanium, and Gallium Arsenide

Property	Symbol	Unit	Si	Ge	GaAs
Intrinsic carrier concentration at 300K	n_i	cm^{-3}	1.5×10^{10}	2.5×10^{13}	9×10^6
Electrons' mobility	μ_n	cm^2/V·s	1350	3900	8500
Holes' mobility	μ_p	cm^2/V·s	480	1900	400
Energy gap at 300K	E_G	eV	1.11	0.67	1.43
Electrons' diffusion constant	D_n	cm^2/s	34	98	212
Holes' diffusion constant	D_p	cm^2/s	12	48	10

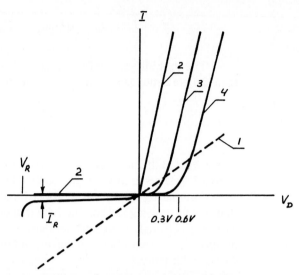

Figure 7.9 Current-voltage characteristics.
I = current through junction (diode), I_R = reverse-leakage current ($0.01\mu A$ for Si), V_D = voltage across junction (diode), V_R = reverse breakdown, 1 = linear element, 2 = ideal junction (diode), 3 = germanium, 4 = silicon.

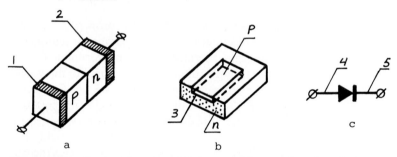

Figure 7.10 P-n junction. a = single-crystal bar p-n junction diode, b = planar p-n junction (p-type impurity is diffused into n-type material), c = diode symbol; 1 and 2 = metal contacts, 3 = junction, 4 = anode lead, 5 = cathode lead.

acceptor ions. Migrating electrons leave behind positively charged donor ions. This motion results in the creation of a retarded electrical field in the area next to the junction. The field tends to restrict the migration and establish an equilibrium state.

The region of the junction is depleted of mobile charges and is called *the depletion zone, region, the space-charge region,* or *the transition region.* The thickness of this zone is within a fraction of a micrometer. The voltage corresponding to the field across the junction is called the *barrier voltage.*

If we provide the junction with two metal contacts, as shown in Figure 7.10, the semiconductor can serve as a diode. One of the major functions of a diode is its use as a unidirectional conductor of current. The mechanism of this conductance is illustrated in Figure 7.11. In a diode having no outside connections, steady-state conditions are established with a certain width of the depletion zone Z_1 (Fig. 7.11a). As the positive and negative terminals of a battery are applied to the n- and p-sides of the diode, respectively, some changes in the charge distribution occur. The mobile electrons in the n material are attracted to the positive terminal, and the holes in the p material are attracted to the negative terminal. These carriers move away from the junction and deplete a wide area around the junction (depletion zone Z_2 in Figure

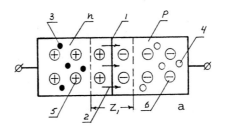

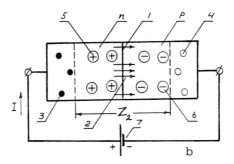

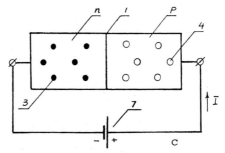

Figure 7.11 P-n junction. a = p- and
n-type semiconductors in contact,
b = reverse-biased p-n junction,
c = forward-biased p-n junction;
I = current, *n* = n-type material,
p = p-type material, Z_1 and
Z_2 = depletion zones, 1 = junction,
2 = retarding electric field, 3 = electrons,
4 = holes, 5 = positively charged donor
ions, 6 = negatively charged acceptor
ions, 7 = battery.

7.11b). In other words, a larger electric field barrier to the diffusion across the
junction is created. This barrier withstands the motion of carriers through the junction, and a very small current results. The current flows because a small number of
hole-electron pairs are generated as a result of thermal agitation. This current
increases with temperature and is called the diode's *reverse saturation current* or
reverse leakage current. The voltage applied in the direction shown in Figure 7.11b
is called *the reverse* or *blocking bias*.

A p-n junction biased in the forward direction is illustrated in Figure 7.11c.
Applying the plus terminal of a battery to the p material and minus terminal to the
n material will repel the carriers from the terminals and propel them toward the
junction. The width of the depletion zone decreases along with the electric field

barrier. Now the carriers can pass the barrier much more easily. This condition corresponds to a significant increase in the current, and the diode becomes electro-conductive. After passing the depletion zone, the carriers are neutralized with carriers of the opposite type. The battery supplies carriers to the terminals continuously. The carriers follow toward the junction; thus, the current circulates in the loop of the forward-biased diode. In order to maintain this circulation, the applied voltage should be as large as the barrier voltage, which is about 0.3V for a germanium junction and about 0.6V for a silicon junction. With the voltage increase, the barrier is practically removed and the only limit for the current is the resistance of the semiconductor material. If the diode were ideal, its characteristics would be a straight line close to the I-axis (Fig. 7.9) for positive voltages V_D and a straight line along the V_D-axis for negative voltages V_D. In a forward-biased diode, the current rises appreciably if the voltage applied to the diode is of the order of the barrier voltage.

The relation between the current passing through the diode and the leakage current is given by the *diode law*:

$$I = I_R \left(exp\, \frac{qV_D}{\bar{k}T} - 1 \right)$$

(7.23)

where I = current passing the diode, A;
   I_R = reverse leakage current, A;
   q = electron's charge, C;
   $\bar{k}$ = Boltzmann constant, J/K; and
   T = absolute temperature, K.

The reverse leakage current is of the order of a few hundredths of a microampere for silicon and up to $50\mu A$ for germanium. With the increase in the thermally generated intrinsic carriers, I_R also increases.

The reverse-biased junction can withstand *a back-bias voltage* up to a certain limit called *the reverse-breakdown voltage* (V_R in Figure 7.9). At this limit, the reverse current abruptly increases and, if it is not restricted, it can destroy the diode.

BIPOLAR TRANSISTORS

The description of the p-n junction provides a foundation for understanding *the bipolar semiconductor junction transistor* which is a semiconductor triode. There are two basic models of the transistors: n-p-n (Fig. 7.12) and p-n-p (Fig. 7.13) types. The structure of the n-p-n type (Fig. 7.12a) is built of three sandwiched blocks, which are provided with leads and called *collector* (C), *base* (B), and *emitter* (E). Collector and emitter blocks are made of n material and the base is fabricated from p material. The transistor model can be introduced by back-to-back connected diodes (Fig. 7.12b). It should be noted that, physically, these two diodes are combined in one structure that can function if the diodes have a common base made as a very thin p layer between two more massive n blocks. Figure 7.12c shows the transducer's circuit symbol, and Figure 7.12d reveals the connection of a transistor to two voltage sources to illustrate the principle of operation. It is clear from the model in Figure 7.12b that, if the battery V_{BE} were not connected, current I_C would not flow (the C-B diode is reverse-biased). As the battery V_{BE} is connected to the B-E diode in a forward-bias sense, electrons begin to flow from the emitter to the base (note that the electrons' flow is opposite to the direction of the current).

In the base, the electron flow is split into two streams. A small fraction of the flow is directed to the base terminal and forms the *base current*. The large fraction of the flow passes the thin base region and, being accelerated by the large voltage

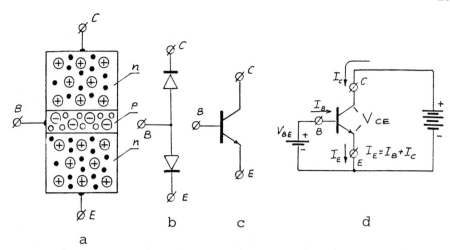

Figure 7.12 N-p-n bipolar transistor. a = transistor's structure, b = transistor's diode model, c = transistor's symbol, d = electrical connection; B, C, and E = base, collector, and emitter, respectively, n and p = n- and p-type material, I_B, I_C, and I_E = base, collector, and emitter currents, respectively, V_{BE} and V_{CE} = base-emitter and collector-emitter voltages.

V_{CE}, flows through the terminal C and constitutes *the collector current*. This mechanism can be explained in more detail. The forward bias of the emitter-base junction reduces the barrier from the electrical field at the junction, as was explained above. The electrons from the emitter region can easily reach the base, which is very thin and lightly doped. There are a limited number of holes in the base region available for recombination, and the current drawn from the base lead is small. The absolute majority of electrons diffuse through the base toward the base-collector junction, which is reverse-biased by voltage V_{CE}. This voltage is rather large. The electric field acting in the depletion zone of the base-collector junction has the magnitude and orientation across the junction to accelerate electrons toward the collector. Here they continue to move toward the positive collector lead and represent the main current between the collector and emitter. Because some holes also take part in forming the currents in the transistor, the device employs two poles of carriers and is called *a bipolar transistor*. The operation of a p-n-p transistor (Fig. 7.13) can be

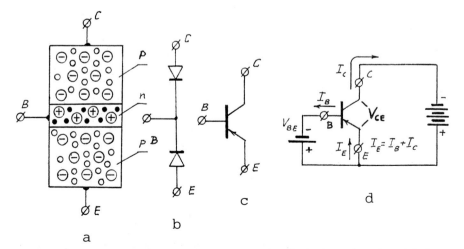

Figure 7.13 P-n-p bipolar transistor. a = transistor's structure, b = transistor's diode model, c = transistor's symbol, d = electrical connection. All notations are as in Figure 7.12.

described identically to that of the n-p-n device, if the roles played by the electrons and holes are reversed along with the polarity of the batteries and the directions of the currents.

Among the transducer's various electrical characteristics, the forward-current transfer ratio is one of the most important. It is denoted by β (beta) or h_{fe} and is defined as a ratio of the current passing through the transistor's collector I_C to the base current I_B:

$$\beta = h_{fe} = \frac{I_C}{I_B}. \tag{7.24}$$

Depending on the type of the transistor, β is usually between 30 and 150.

The transistor's input and output electrical characteristics can be given as graphs. *The input* or *base characteristics* are a plot of a base current I_B as a function of the base-emitter voltage V_{BE}. Virtually, it is a characteristic of a forward-biased p-n junction (Fig. 7.14). The output or collector characteristics (Fig. 7.15) are primarily defined by the base current (I_B through I_{BN}) whose increase gives a corresponding increase in the output. For small values of V_{CE}, the current I_C is also

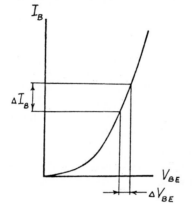

Figure 7.14 Input or base characteristic. I_B = base current, V_{BE} = base-emitter voltage, ΔI_B and ΔV_{BE} = small variations in current and voltage, respectively.

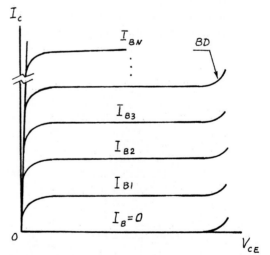

Figure 7.15 Output or collector characteristics. *BD* = transistor breakdown, I_B, I_{B1}, I_{B2}, and I_{BN} = different values of base current controlling I_C ($I_B < I_{B1} < I_{B2} < I_{B3} < I_{BN}$), I_C = collector current, V_{CE} = collector-emitter voltage.

small, but greatly sensitive to the voltage (in the saturation region). In the operating region, between saturation and the transistor's breakdown, the collector currents are determined almost entirely by the base current which is controlled by V_{BE}. It is clear from this description that a transistor is a current- or a voltage-to-current amplifying device. The small variations ΔV_{BE} in V_{BE} cause changes ΔI_B in the current I_B, thus creating a relatively large change in I_C.

FIELD-EFFECT TRANSISTORS (FET's)

Field-effect transistors belong to the category of electronic devices that are controlled by voltage (at this point, we are not considering the FET's response to other physical quantities). The description "controlled by voltage" indicates the fact that the main applications of FET's are circuits in which FET's are placed at the input of the circuit to provide a high input impedance and low current consumption.

The schematic structures and symbols of n- and p-channel junction field-effect transistors are given in Figures 7.16 and 7.17.

In the n-type transistor, a block of an n-type semiconductor contains, at its side, a region of a p-type material forming a junction. This is why the transistor is called the junction FET or JFET. Leads are attached to the top and bottom of the bar and they are called *the drain* and *the source,* respectively. The third lead is connected to the p-material, and it is called the *gate.* As with the p-n junction of a diode, the arrow at the gate on the FET's symbol indicates the easy-current-flow direction. The drain and source are not distinguishable in the transducer's symbol because they are usually interchangeable in many applications.

When a voltage V_{DS} is applied between the drain and source (Fig. 7.16a), a drain current I_D is established and flows through the channel formed by the

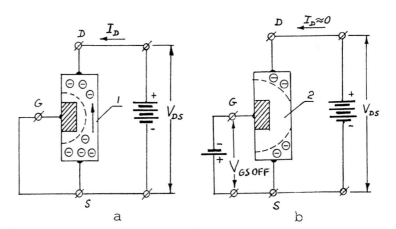

Figure 7.16   N-channel FET. a and b = transistor's structures at conducting and pinched-off condition, respectively; *D, G,* and *S* = drain, gate, and source, respectively, 1 = n-type channel, 2 = depletion region.

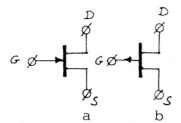

Figure 7.17   n- and p-channel FET symbols, a for n-type and b for p-type; *D, G,* and *S* as in Figure 7.16.

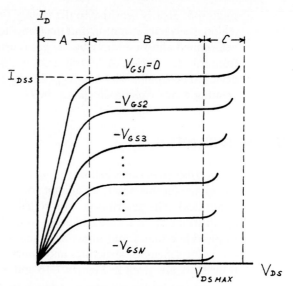

Figure 7.18 Output characteristics of n-channel JFET. *A, B,* and *C* = ohmic, saturation, and breakdown regions, respectively, I_D = drain current, I_{DSS} = saturation current, V_{DS} = drain-source voltage, V_{GS} = gate-source voltage, $V_{DS\ max}$ = breakdown voltage, V_{GSI} = through $-V_{GSN}$ = different values of V_{GS} controlling I_D (magnitudes of negative voltages are increased from V_{GS1} to V_{GSN}).

n-material. Electrons come to the source, pass the channel, and are "drained" at the drain. If voltage V_{GS} is established at the gate (Fig. 7.16b) and it is negative with respect to the source, the diode becomes reverse-biased, and a very small leakage current ($\sim 1 \times 10^{-9}$A) flows through the gate. At the same time, if voltage V_{GS} is rather large, the depletion region is increased and crosses the bar (*b*). Since no carriers are present in the region, the channel becomes non-conductive (the FET is "pinched off") at the so-called pinch-off voltage ($V_{GS\ OFF}$). Depending on the FET type, this voltage varies from about -2 to -15V. With the change in voltage V_{GS} between $V_{GS\ OFF}$ and zero, the FET's characteristics vary as shown in Figure 7.18. The output characteristics have three regions: ohmic or nonsaturation, constant current or saturation, and breakdown. Current I_D is controlled primarily by the voltage V_{GS} in the saturation region; this is why the FET is usually considered a voltage-to-current amplifying device.

At voltage $V_{DS\ max}$, current I_D is increased sharply and can destroy the device; therefore, the operating voltage V_{DS} should not exceed $V_{DS\ max}$. For a particular FET, the maximum current the device will pass (I_{DSS}) is called *the saturation current* (see Fig. 7.18).

In a p-channel FET, the polarities of voltages V_{DS} and V_{GS} must be reversed as compared to an n-channel FET. The principles of operation and the nature of the basic characteristics are similar for these two types of FET's.

METAL-OXIDE-SILICON FET'S (MOSFET'S)

In a MOSFET, a thin layer of silicon dioxide is placed between the gate and the channel. This insulation diminishes the current through the gate to very small values (down to 1×10^{-14}A). Because of the insulation, the MOSFET is also called an insulated-gate FET or IGFET. The device exists in two basic versions: depletion-type and enhancement-type (Fig. 7.19 and Fig. 7.20). Each of them can have either an n or p channel.

The operation of *the depletion-type MOSFET* and *JFET* are very similar. In the n-channel version, the drain, channel, and source regions are made of n-type material, which is in contact with a substrate fabricated of p-type material. A layer of insulation covers the channel region. This layer and the regions of the drain, source, and substrate are metallized. Usually, the source and substrate are connected

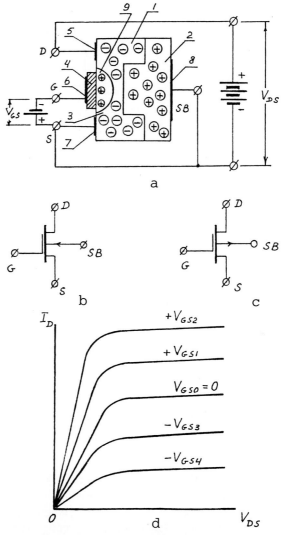

Figure 7.19 Depletion-type MOSFET.
a = structure of n-channel device, b = n-channel device's symbol, c = p-channel device's symbol, d = output characteristics; D, G, S, and SB = drain, gate, source, and substrate, respectively, I_D = drain current, V_{DS} and V_{GS} = drain-source and gate-source voltages, respectively, V_{GS0} through V_{GS4} = different values of V_{GS} controlling I_D ($|V_{GS4}| > |V_{GS3}|$ and $V_{GS2} > V_{GS1}$), 1 = n-material, 2 = p-material, 3 = channel, 4 = insulation, 5, 6, 7, and 8 = metallizations, 9 = area with holes induced by gate's potential.

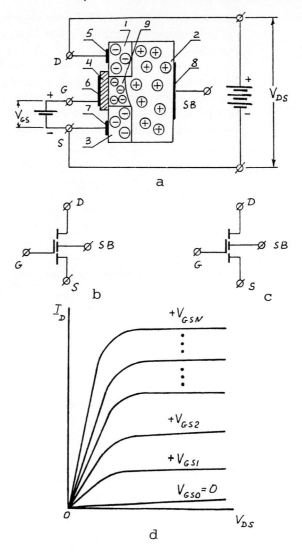

Figure 7.20 Enhancement-type MOSFET.
$V_{GS0} < V_{GS1} < V_{GS2} < \ldots < V_{GSN}$, 1 and 3 = two
n-material regions, 9 = induced n-type channel. All
other notations are as in Figure 7.19.

inside the transistor package. If the substrate has an outer lead, it is connected to the base lead outside the transistor's housing.

When a dc voltage V_{DS} is applied between the source and drain and the gate potential is zero (the gate is connected to the source), a certain current will flow through the channel because of the conductivity of the n-material. Now suppose the gate is supplied with a potential that is negative relative to the source. This potential will repel the majority carrier electrons from the channel's region next to the gate and will attract holes to this region. N-type carriers conducting electricity through the channel become *depleted,* and the current I_D is decreased. The reverse mechanism takes place when the gate is made positive. The channel is now enlarged into the p-material substrate, more electrons flow through the channel, and the current is increased. This fact is reflected in the set of the output characteristics shown in Figure 7.19d.

In an n-channel *enhancement-type MOSFET* (Fig. 7.20), two separate n-channel regions are formed in a p-type substrate. These two regions are provided with metallization and constitute the source and drain of the transistor. Metallization and insulating material forming the gate are placed over the area of p-material between the regions of the drain and source. The base is also metallized and connected to the source.

When the positive terminal of the power-supplying battery is connected to the drain and a negative terminal connected to the source, the equivalent circuit of the transistor is a back-to-back connection of two p-n diodes (we assume that the gate's potential is zero with respect to the source). The diode at the drain region is reverse-biased and conducts a very small current. When the gate potential is made positive, it induces a negative charge at the oxide-substrate interface because of the capacitorlike gate structure. The capacitor is formed by two electrodes: the gate's metallization and conducting p-type substrate, and the dielectric silicon-dioxide spacer between the electrodes. The negative charges in the p-type material are minority carrier electrons. They are collected at the oxide-silicon interface along with ionized acceptors. With the increase in the gate's potential and in the number of electrons concentrated in the channel, it becomes converted to the n-type conductor, thus permitting majority electrons to flow between the source and drain. This *enhancement* in the drain-to-source current (it can reach 1×10^9) is reflected in the name of this transistor. The output characteristics of the device controlled by the gate-to-source voltage are shown in (*d*).

In the complimentary, p-channel MOSFET's, the structure of the transistors remains the same, but the materials are replaced by the complimentary types (n's become p's and p's become n's). The polarity of the voltages and directions of the currents are reversed.

A wide variety of references are available for the further study of transistors. The basics described in the preceding sections can be sufficient for an understanding of the sensors' fundamentals. However, some topics need more detailed studies (such as those within a standard college program on electronics).

CRYSTAL STRUCTURES OF SEMICONDUCTORS

The semiconductor materials that are used for the fabrication of sensors are introduced by three modifications:

1. *A single crystal* that is characterized by a repeated arrangement in a regular manner of some atoms within a unit cell.
2. *A polycrystalline material* that contains only portions of the regular structure (the portions, "crystallites," are aligned arbitrarily).
3. *An amorphous material* that contains the arbitrarily aligned crystallites reduced to the size of a unit cell.

The majority of solid-state sensors are fabricated from single-crystal materials, such as silicon, germanium, gallium arsenide, and from several other semiconductors. Silicon is, by far, the material most commonly used in these sensors. Its atoms are organized in a lattice having several axes of symmetry (*A, B, C, D, E,* and *F* in Figure 7.21). The lattice is cubic (diamond-type) and composed of unit cells, whose crystallographic axes *X, Y,* and *Z* are defined by the edges of the cell. The axes are perpendicular to each other. The size of the cell is defined by the lattice parameter *a*.

The orientation of any plane in the crystal is identified by Miller indices, which are three integers determined in the following manner: The intercepts of a plane on the three crystallographic axes are expressed as fractions of the crystal's parameters; then, the reciprocals of these fractions, reduced to integral proportions, are taken.

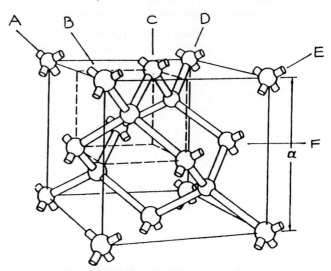

Figure 7.21   Cell of silicon crystal lattice. *A* through *F* = axes of symmetry. a = lattice parameter (a = 5.43Å for Si and a = 5.65Å for Ge).

Figure 7.22 illustrates a reference system to define the Miller indices. *X, Y,* and *Z* are chosen as crystallographic axes, and *ABC* is a unit plane, which is characterized by angles α, β, and γ, and the ratio *OA : OB : OC.* For the cubic crystal $\alpha = \beta = \gamma = 90°$, and *OA : OB : OC* is 1 : 1 : 1. Miller indices of another plane *HKL* are integers proportional to *OA/OH, OB/OK,* and *OC/OL.* For instance, if a plane (Fig. 7.22b) has intercepts at points 6, 1, and 2, the reciprocals will be 1/6, 1/1, and 1/2; integral proportions will be 1, 6, and 3, and the plane will be identified as (1, 6, 3). In the cubic crystal, directions are defined by the perpendiculars to the particular plane and are identified by the same numbers as the plane but put in brackets. If the direction is negative, it is marked by a minus sign above the numbers. As an example, a positive direction of the *z*-axis is [001] and negative [00$\bar{1}$]. Figure 7.23 shows the plots of several planes and directions for the silicon crystal. As discussed later in more detail, the physical properties of different crystalline structures greatly depend on the orientation of the die that is cut from the single-crystal ingot. In the fabrication of a precision sensor, it is important to secure a correct

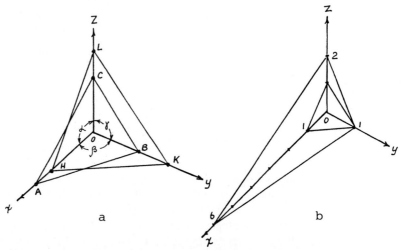

Figure 7.22   Reference system used to define Miller indices.

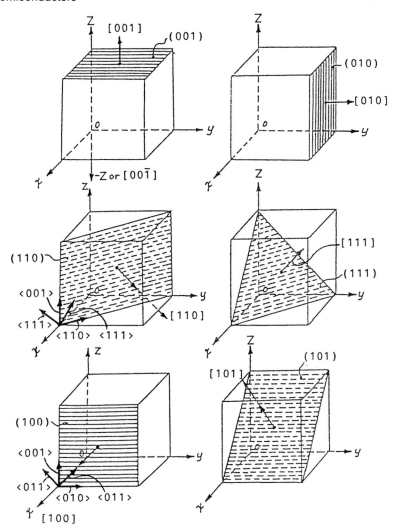

Figure 7.23 Some planes and directions for silicon (cubic) crystal. Angle bracket denotations are explained on p. 382.

orientation of the die's planes with respect to the crystal axes. Depending on the principle of fabrication and operation of a sensor, a two-dimensional orientation of the crystal can be required (plane and direction). It is obvious that the elements made of amorphous materials are free of these requirements.

MECHANICAL CHARACTERISTICS OF SEMICONDUCTORS

Among the different semiconductors, silicon is used most often in solid-state electromechanical sensors, not only because of its specific electrical characteristics but also due to its excellent mechanical properties. Silicon's frequent usage requires that particular attention be given to silicon in the following discussion.

Since 1972, micromechanical devices fabricated from silicon have become commercially established. It was proven [96] that silicon can be regarded as a high-precision, high-strength, and high-reliability material, especially applicable wherever miniaturized precision mechanical devices must be fabricated in large quantities.

Table 7.2 presents the basic mechanical properties of silicon and some other materials given for comparison. The data relates to a single-crystal silicon (SCS)

TABLE 7.2

Mechanical Characteristics of Silicon and Some Other Materials (Extraction from [96])

Materials	Yield strength (1×10^9)N/m^2	Young's Modulus (1×10^9)N/m^2	Density g/cm^3	Thermal conductivity W/cm · °C	Thermal Expansion (1×10^{-6})1/°C
Diamond*	53	1035	3.5	20	1.0
SiC*	21	700	3.2	3.5	3.3
TiC*	20	497	4.9	3.3	6.4
Al$_2$O$_3$*	15.4	530	4.0	0.5	5.4
Si$_3$N$_4$*	14	385	3.1	0.19	0.8
Iron	12.6	196	7.8	0.803	12.0
SiO$_2$ (fibers)	8.4	73	2.5	0.014	0.55
Si*	7.0	190	2.3	1.57	2.33
Steel (max strength)	4.2	210	7.9	0.97	12.0
W	4.0	410	19.3	1.78	4.5
Stainless steel	2.1	200	7.9	0.329	17.3
Mo	2.1	343	10.3	1.38	5.0
Al	0.17	70	2.7	2.36	25.0

* single crystal

quite frequently utilized in sensor fabrication. The Young's modulus of silicon is very close to that of nickel and stainless steel and is well above that of many glasses. The hardness of silicon is close to that of quartz and chromium. The tensile yield strength is almost three times higher than that of stainless-steel wire. At room temperature, metals yield by deforming inelastically, but silicon yields by fracturing. Silicon tends to cleave along crystallographic planes, especially if some stresses are oriented along the planes. A careful treatment of the silicon wafer's edges during the fabrication cycle is important because mechanical damage often starts from the edge. Rough handling of a die (resulting in, for example, scratches on the surface) causes fractures in brittle material such as SCS. Concentrated stresses and cleavage can be caused by high-temperature treatment, bulk imperfections, and deposition of different films on the surface of an SCS element. It is obvious that the strength of the element depends also on its crystallographic orientation and geometry. In order to preserve the intrinsically strong structure of SCS, several measures are recommended during construction:

1. The silicon element should have small bulk, surface, and edge crystallographic defects.

2. Friction and abrasion at SCS surfaces must be minimized.

3. Mechanical processing (sawing, grinding, scribing, and polishing) of an SCS element should be minimized or eliminated, especially for the stress-exposed surfaces. Etching is preferable; it is necessary to remove some types of imperfections of the mechanical treatment if still present in the process.

4. Because anisotropic etching forms sharp corners and edges (see following text), which become places with high concentrations of stresses, a subsequent anisotropic etching is recommended to round such corners.

5. Direct mechanical contact to the silicon can be prevented by a thin-film coating of the SCS's surfaces (chemical vapor depositions of SiC or Si$_3$N$_4$ are applicable).

6. It is preferable to use low-temperature processing. Differently doped silicon and deposited materials on surfaces have variations in thermal

coefficients of expansion and develop a nonuniform deformation under heating.

Passivating the surface with the thin film not only protects the element against undesirable mechanical and chemical interactions but also lowers fatigue which is usually initiated at defects in the surface. It is proven that an intergrain diffusion of H_2O into an unprotected surface increases the fatigue rate. In polycrystalline materials, various surface irregularities and grain boundaries can be sources of stress concentration. In this regard, the crystalline perfection of SCS is advantageous. One method of increasing the strength of vitreous materials is to create compressing stresses at the surface. Depositing the Si_3N_4 film on the Si surface provides the same effect. The underlying silicon surface undergoes compression, and the structure becomes stronger. Besides this, the wear-resistant quality of Si_3N_4 is very high, and the combination of Si_3N_4-on-Si is quite practical.

In many applications, knowledge of a semiconductor's elastic characteristics is important. Typical calculations of the deformation of a spring element can be provided with standard methods of the theory of elasticity if the relationship between stress and strain in the element is known. For the isotropic polycrystalline and amorphous materials, this relationship is described by Hooke's law given in its simplest form:

$$\sigma = \epsilon E \tag{7.25}$$

where σ = stress, N/m^2;
$\quad \epsilon$ = strain, dimensionless; and
$\quad E$ = Young's modulus, N/m^2.

When stress or strain is calculated with formula 7.25, E is taken as a constant parameter independent from the directions of σ or ϵ.

In crystal materials, whose elastic properties are anisotropic, the relation between σ's and ϵ's is more complex and depends greatly on the spatial orientation of these quantities with respect to the crystallographic axes (a review of Appendix 2 is advised as preparation for reading this material). For this case, Hooke's law is written with two formulas [97, 98]:

$$\sigma_{ij} = C_{ijkl} \cdot \epsilon_{kl} \quad \text{and} \quad \epsilon_{ij} = S_{ijkl} \cdot \sigma_{kl} \tag{7.26}$$

where σ_{ij} and σ_{kl} = stress tensors of rank 2, N/m^2;
$\quad \epsilon_{ij}$ and ϵ_{kl} = strain tensors of rank 2, dimensionless;
$\quad C_{ijkl}$ = stiffness-coefficient tensor of rank 4, N/m^2; and
$\quad S_{ijkl}$ = compliance-coefficient tensor of rank 4, m^2/N.

Generally, for crystals, C_{ijkl} and S_{ijkl} are each composed of 36 coefficients. Due to the symmetry, however, only 21 coefficients are left.

The tensor presentation in equation 7.26 can be reduced to matrices

$$\sigma_m = \sum_{n=1}^{6} C_{mn}\epsilon_n \quad \text{and} \quad \epsilon_m = \sum_{n=1}^{6} S_{mn}\sigma_n \tag{7.27}$$

or to products

$$[\sigma] = [C] \cdot [\epsilon] \quad \text{and} \quad [\epsilon] = [S] \cdot [\sigma]. \tag{7.28}$$

Components of tensors C_{ijkl} and S_{ijkl} are substituted by elements of matrix C_{mn} and S_{mn}, respectively. Conversion of indices ij to m and kl to n are provided using the following scheme:

$$11 \rightarrow 1, 22 \rightarrow 2, 33 \rightarrow 3, 23 \text{ and } 32 \rightarrow 4, 13 \text{ and } 31 \rightarrow 5, 12 \text{ and } 21 \rightarrow 6$$

$$C_{ijkl} \rightarrow C_{mn} \text{ and } S_{ijkl} \rightarrow S_{mn} \text{ when } m \text{ and } n = 1, 2, 3;$$

$$2 S_{ijkl} \rightarrow S_{mn} \text{ when } m \text{ or } n = 4, 5, 6;$$

$$4 S_{ijkl} \rightarrow S_{mn} \text{ when } m \text{ and } n = 4, 5, 6;$$

$$\sigma_{ij} \rightarrow \sigma_m \text{ when } m = 1, 2, 3; \text{ and}$$

$$\epsilon_{ij} \rightarrow \epsilon_m \text{ when } m = 4, 5, 6. \tag{7.29}$$

A simplified form of the stiffness-coefficient and compliance-coefficient matrices with the reduced indices gives the following two expressions:

$$C_{mn} = \begin{bmatrix} C_{11} & \cdots & C_{16} \\ \cdot & & \cdot \\ \cdot & & \cdot \\ \cdot & & \cdot \\ C_{61} & \cdots & C_{66} \end{bmatrix} \quad \text{and} \quad S_{mn} = \begin{bmatrix} S_{11} & \cdots & S_{16} \\ \cdot & & \cdot \\ \cdot & & \cdot \\ \cdot & & \cdot \\ S_{61} & \cdots & S_{66} \end{bmatrix} \tag{7.30}$$

These relationships are substantially simplified for cubic-lattice crystals. If the vector of stress is oriented along the [100]-axis, the matrices 7.30 are written

$$C_{mn} = \begin{bmatrix} C_{11} & C_{12} & C_{12} & 0 & 0 & 0 \\ C_{12} & C_{11} & C_{12} & 0 & 0 & 0 \\ C_{12} & C_{12} & C_{11} & 0 & 0 & 0 \\ 0 & 0 & 0 & C_{44} & 0 & 0 \\ 0 & 0 & 0 & 0 & C_{44} & 0 \\ 0 & 0 & 0 & 0 & 0 & C_{44} \end{bmatrix} \tag{7.31}$$

$$S_{mn} = \begin{bmatrix} S_{11} & S_{12} & S_{12} & 0 & 0 & 0 \\ S_{12} & S_{11} & S_{12} & 0 & 0 & 0 \\ S_{12} & S_{12} & S_{11} & 0 & 0 & 0 \\ 0 & 0 & 0 & S_{44} & 0 & 0 \\ 0 & 0 & 0 & 0 & S_{44} & 0 \\ 0 & 0 & 0 & 0 & 0 & S_{44} \end{bmatrix} \tag{7.32}$$

The formulas for conversion of C_{mn} to S_{mn} and back are

$$C_{11} = \frac{S_{11} + S_{12}}{(S_{11} - S_{12})(S_{11} + 2S_{12})}, \quad S_{11} = \frac{C_{11} + C_{12}}{(C_{11} - C_{12})(C_{11} + 2C_{12})}, \tag{7.33}$$

$$C_{12} = \frac{-S_{12}}{(S_{11} - S_{12})(S_{11} + 2S_{12})}, \quad S_{12} = \frac{-C_{12}}{(C_{11} - C_{12})(C_{11} + 2C_{12})}, \tag{7.34}$$

$$C_{44} = \frac{1}{S_{44}}, \quad S_{44} = \frac{1}{C_{44}}. \tag{7.35}$$

The main elastic constants for materials are determined as follows:

$$\frac{1}{S_{11}} = E = \text{Young's modulus};$$

$$-\frac{S_{12}}{S_{11}} = \nu = \text{Poisson's ratio; and}$$

$$\frac{1}{S_{44}} = G = \text{shear modulus } [G = E/2(1 + \nu)].$$

E and G can be determined experimentally from measurements of the resonance frequency of the longitudinal vibrations of a bar ($\sim$ 30mm long and 20mm^2 in the cross-sectional area) cut in the proper direction [98]:

$$E = \frac{4\rho l^2 f^2}{[1 + (\Delta l/l)]n^2}, \qquad G = \frac{4.77\rho l^2 f^2}{[1 + (\Delta l/l)]n^2} \qquad (7.36)$$

where E and G = Young's and shear moduli, respectively, N/m^2;
$\quad\rho$ = density of material, kg/m^3;
$\quad l$ = length of bar at room temperature, m;
$\quad\Delta l$ = change of l with deviation of temperature from room temperature, m;
$\quad f$ = frequency, Hz; and
$\quad n$ = number of harmonic.

The temperature characteristics of E and G are usually determined for the cubic crystals oriented along [100], [110], and [111] directions. Several temperature characteristics for silicon and gallium arsenide are illustrated in Figure 7.24 [99, 100]. As is seen from the graphs, the magnitudes of E and G gradually decrease with temperature.

Table 7.3 illustrates stiffness and compliance coefficients for a number of semiconductors. These coefficients can be used in calculations of elastic constants for the arbitrary-oriented element whose axes are x_i' ($i = 1, 2, 3$).

$$x_i' = p_i x + q_i y + r_i z \ (i = 1,2,3) \qquad (7.37)$$

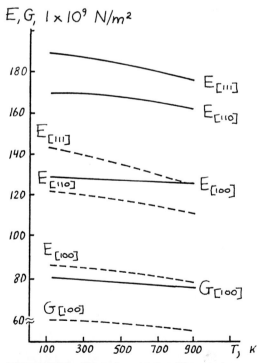

Figure 7.24 Young's modulus (E) and shear modulus (G) as function of absolute temperature (T) for silicon (—) and gallium arsenide (---) specimens oriented along [100], [110], and [111] axes (adopted from source [101]).

TABLE 7.3

Stiffness and Compliance Coefficients for Semiconductors (Adopted from [97]).

Factor	C_{11}	C_{12}	C_{44}	S_{11}	$-S_{12}$	S_{44}
Semiconductor	1×10^9 [N/m^2]			1×10^{-12} [m^2/N]		
Ge	129.2	47.9	67.0	9.64	2.60	14.9
Si	165.7	63.9	79.6	7.68	2.14	12.6
GaAs	117.6	52.7	59.6	11.77	3.64	16.76
InAs	83.4	45.4	39.5	19.46	6.86	25.30
InSb	67.2	36.6	30.2	24.40	8.61	33.10
GaSb	88.5	40.4	43.2	18.80	4.97	23.20
AlSb	89.4	44.2	41.6	16.60	5.48	24.00
ZnS	98.1	62.7	44.8	20.30	7.90	22.25
CdTe	53.3	36.5	20.4	4.15	1.68	49.05
HgTe	53.6	36.6	21.1	4.17	1.69	47.40
PbS	127.0	29.8	25.0	10.10	2.31	40.00
PdTe	108.0	7.7	13.4	9.35	0.62	73.50
SnTe	109.3	2.1	9.7	9.16	0.17	103.00
Pb$_{0.5}$Sn$_{0.5}$Te	181.0	3.1	13.0	5.82	0.85	76.90

The axes form angles with the x [100], y [010], and z [001] axes (Fig. 7.23). The direction cosines of the angles are p_i, q_i, and r_i.

Young's modulus and Poisson's ratio for an arbitrary direction i (p_i, q_i, r_i) and for an orthogonal-to-this-direction j (p_j, q_j, r_j) are determined by the relationships

$$E'_i = 1/S'_{ii}, \qquad \nu'_{ij} = -S'_{ij}/S'_{ii} \tag{7.38}$$

where $i, j = 1', 2', 3'$.

The shear modulus is

$$G'_k = \frac{1}{S'_{kk}} \tag{7.39}$$

where $k = 4', 5', 6'$.

The stiffness and compliance coefficients for arbitrary crystallographic directions are found in the following equations [101]:

$$C'_{11} = C_{11} + C_c(p_1^4 + q_1^4 + r_1^4 - 1),$$
$$C'_{12} = C_{12} + C_c(p_1^2 p_2^2 + q_1^2 q_2^2 + r_1^2 r_2^2),$$
$$C'_{14} = C_c(p_1^2 p_2 p_3 + q_1^2 q_2 q_3 + r_1^2 r_2 r_3),$$
$$C'_{44} = C_{44} + C_c(p_2^2 p_3^2 + q_2^2 q_3^2 + r_2^2 r_3^2), \tag{7.40}$$

$$S'_{22} = S_{11} + S_c(p_2^4 + q_2^4 + r_2^4 - 1),$$
$$S'_{13} = S_{12} + S_c(p_1^2 p_3^2 + q_1^2 q_3^2 + r_1^2 r_3^2),$$
$$S'_{14} = 2S_c(p_1^2 p_2 p_3 + q_1^2 q_2 q_3 + r_1^2 r_2 r_3),$$
$$S'_{56} = 4S_c(p_1^2 p_2 p_3 + q_1^2 q_2 q_3 + r_1^2 r_2 r_3),$$
$$S'_{55} = S_{44} + 4S_c(p_3^2 p_1^2 + q_3^2 q_1^2 + r_3^2 r_1^2) \tag{7.41}$$

where

$$C_c = C_{11} - C_{12} - 2C_{44} \text{ and } S_c = S_{11} - S_{12} - \tfrac{1}{2}S_{44}. \tag{7.42}$$

TABLE 7.4

Analogous Coefficients (They are calculated as C's and S's in Formulas 7.40 and 7.41).

Coefficients	C'_{11}	C'_{12}	C'_{14}				C'_{44}
Analogous	C'_{22}	C'_{13}	C'_{36}	C'_{35}	C'_{34}	C'_{46}	C'_{55}
			C'_{25}	C'_{26}	C'_{16}	C'_{45}	
Coefficients	C'_{33}	C'_{23}	C'_{24}	C'_{36}	C'_{15}	C'_{56}	C'_{56}

Coefficients	S'_{22}	S'_{13}	S'_{14}				S'_{56}	S'_{55}
Analogous	S'_{11}	S'_{12}	S'_{36}	S'_{25}	S'_{24}	S'_{35}	S'_{46}	S'_{44}
Coefficients	S'_{33}	S'_{23}	S'_{26}	S'_{34}	S'_{16}		S'_{45}	S'_{66}

All other coefficients are calculated as explained in Table 7.4. It should be kept in mind that $C'_{ij} = C'_{ji}$ and $S'_{ij} = S'_{ji}$.

When the analogous coefficients are calculated with formulas 7.40 and 7.41, the direction cosines for a particular coefficient are determined by the subscripts on the term. Each new subscript of the coefficient represents two old subscripts (see Appendix 2) by the following conversions:

$$1 \rightarrow 11,\ 2 \rightarrow 22,\ 3 \rightarrow 33,\ 6 \rightarrow 12,\ 5 \rightarrow 13,\ 4 \rightarrow 23.$$

For example, the coefficient subscript 36 expands to 3312 and the geometrical factor is

$$p_3 p_3 p_1 p_2 + q_3 q_3 q_1 q_2 + r_3 r_3 r_1 r_2 = p_1 p_2 p_3^2 + q_1 q_2 q_3^2 + r_1 r_2 r_3^2 \quad (7.43)$$

The relationships given in the sets of formulas 7.38 through 7.42 assist in the calculation of the elastic constants for semiconductors and in the application of the elasticity theory to semiconductor stress problems.

We will consider a representative example with the calculation of Young's modulus and Poisson's ratio for a silicon strip (Fig. 7.25) that is cut from a crystal in such a way that the strip's plane is in the (001)-plane of the material (see Fig. 7.23) and its axis is displaced from the y-axis by 45°. For convenience in notations, the axes x, y, and z are denoted as 1, 2, and 3, respectively, and the axes indicating the orientation of the strip are $1'$, $2'$, and $3'$.

1. The formulas for E and ν given in 7.38 can be rewritten for our case as follows:

$$E'_1 = 1/S'_{11} \quad \text{and} \quad \nu'_{12} = -S'_{12}/S'_{11}. \quad (7.44)$$

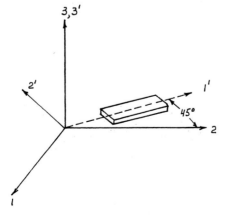

Figure 7.25 Silicon strip in plane 1-2 and oriented along $1'$-axis that makes 45° angle with 2-axis.

2. As we can see from Table 7.4, the coefficient S'_{11} can be calculated by using the formula for S'_{22} taken from 7.41 and by modifying the geometrical factor. Since the subscripts are 11, which expands to 1111, the geometrical factor in the formula for S'_{11} will be

$$p_1p_1p_1p_1 + q_1q_1q_1q_1 + r_1r_1r_1r_1 - 1 = p_1^4 + q_1^4 + r_1^4 - 1; \text{ therefore,}$$

$$S'_{11} = S_{11} + (S_{11} - S_{12} - \tfrac{1}{2}S_{44})(p_1^4 + q_1^4 + r_1^4 - 1). \quad (7.45)$$

3. The similar definitions for S'_{12} allow us to state that S'_{12} is an analogous coefficient for S'_{13}, the expansion for subscripts 12 is 1122, and this yields

$$p_1p_1p_2p_2 + q_1q_1q_2q_2 + r_1r_1r_2r_2, \qquad \text{and} \qquad (7.46)$$

$$S'_{12} = S_{12} + (S_{11} - S_{12} - \tfrac{1}{2}S_{44})(p_1^2p_2^2 + q_1^2q_2^2 + r_1^2r_2^2). \quad (7.47)$$

4. In order to find the direction cosines, we will state that p_1, q_1, and r_1 are cosines of the angles (β_{11}, β_{12}, β_{13}) between the axes 1' and 1, 1' and 2, and 1' and 3, respectively; p_2, q_2, and r_2 are the cosines of the angles (β_{21}, β_{22}, β_{23}) between the axes 2' and 1, 2' and 2, and 2' and 3, respectively. It follows from the expansion of formula 7.37 that

$$X'_1 = p_1x + q_1y + r_1z$$

$$X'_2 = p_2x + q_2y + r_2z$$

$$X'_3 = p_3x + q_3y + r_3z \qquad (7.48)$$

Note that X'_1, X'_2, X'_3, x, y, and z in these formulas correspond to axes in Figure 7.25 marked by 1', 2', 3', 1, 2, and 3, respectively. The matrices for the angles are written

$$\begin{bmatrix} \beta_{11} & \beta_{12} & \beta_{13} \\ \beta_{21} & \beta_{22} & \beta_{23} \\ \beta_{31} & \beta_{32} & \beta_{33} \end{bmatrix} = \begin{bmatrix} 135° & 45° & 90° \\ 135° & 135° & 90° \\ 90° & 90° & 0° \end{bmatrix}. \qquad (7.49)$$

The matrices of the direction cosines are

$$\begin{bmatrix} p_1 & q_1 & r_1 \\ p_2 & q_2 & r_2 \\ p_3 & q_3 & r_3 \end{bmatrix} = \begin{bmatrix} -\sqrt{2}/2 & \sqrt{2}/2 & 0 \\ -\sqrt{2}/2 & -\sqrt{2}/2 & 0 \\ 0 & 0 & 1 \end{bmatrix}. \qquad (7.50)$$

5. The values for S_{11}, S_{12}, and S_{44} can be taken from Table 7.3:

$$S_{11} = 7.68 \times 10^{-12}\text{m}^2/\text{N}, \qquad S_{12} = -2.14 \times 10^{-12}\text{m}^2/\text{N}, \qquad \text{and}$$

$$S_{44} = 12.6 \times 10^{-12}\text{m}^2/\text{N}.$$

Substituting the numbers into equations 7.44, 7.45, and 7.47 gives

$$S'_{11} = \{7.68 + (7.68 + 2.14 - \tfrac{1}{2}12.6) [(-\sqrt{2}/2)^4 + (\sqrt{2}/2)^4 - 1]\}$$
$$\times 10^{-12} = 5.92 \times 10^{-12}\text{m}^2/\text{N},$$

$$S'_{12} = \{-2.14 + (7.68 + 2.14 - \tfrac{1}{2}12.6) [(-\sqrt{2}/2)^2(-\sqrt{2}/2)^2$$
$$+ (\sqrt{2}/2)^2(-\sqrt{2}/2)^2 \times 10^{-12}\} = -0.38 \times 10^{-12}\text{m}^2/\text{N},$$

$$E' = \frac{1}{5.92 \times 10^{-12}} = 0.17 \times 10^{12}\text{N/m}^2, \qquad \text{and}$$

$$\nu' = -\frac{(0.38 \times 10^{-12})}{5.92 \times 10^{-12}} = -0.064.$$

After similar calculations are performed, the diagrams showing the values of E and ν for different directions can be drawn. Figures 7.26 through 7.29 display E and ν for Ge and Si in planes (100) and (110) as functions of direction (see reference [101]).

Calculations show that E, G, and ν are constant for any direction that is in the (111) plane. In other words, a plate lying in this plane can be considered as having isotropic elastic properties.

Finally, we should note that a review of the physics of photoelectric, thermoelectric, and magnetoelectric phenomena in solids is also important for understanding a sensor's performance, and the relevant topics will be given in the sections related to the particular type of device.

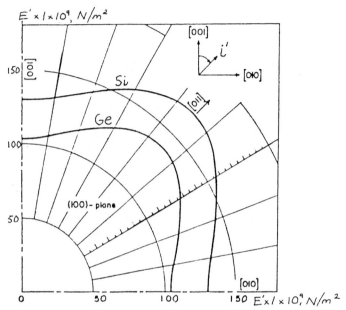

Figure 7.26 Young's modulus as function of direction in (100) plane (adopted from [101]).

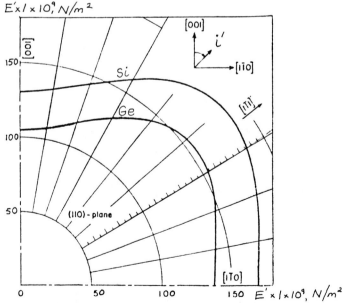

Figure 7.27 Young's modulus as function of direction in (110) plane (adopted from [101]).

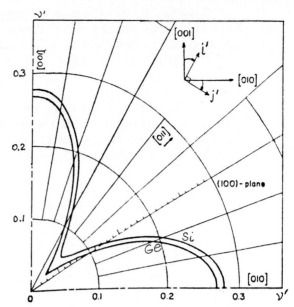

Figure 7.28 Poisson's ratio as function of direction in (100) plane. Both *i′* and *j′* are in (100) plane (adopted from [101]).

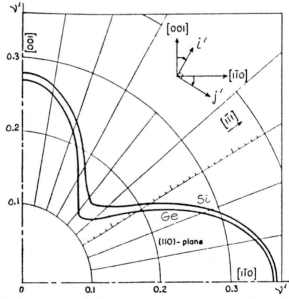

Figure 7.29 Poisson's ratio as function of direction in (110) plane. Both *i′* and *j′* are in (110) plane (adopted from [101]).

chapter 8

Fabrication of Microsensors

We begin the description of microsensor fabrication with the presentation of the processes typical for the electronic industry and adopted for sensors. Then we will consider specific operations that are chiefly applicable for sensors.

BASIC OPERATIONS IN FABRICATION OF MICROSENSORS

A wide variety of technologies is employed in the fabrication of sensors. Standard operations such as machining, stamping, welding, and heat treatment are typical examples. In the following paragraphs, we will discuss the specific processes that are used in the fabrication of the sensing and transduction elements of sensors. These elements constitute the cores of the sensors and primarily dictate their performance. However, it should be noted that there are no "standard" processes since manufactures often incorporate proprietary features. The basic processes described here, however, are quite common for many devices.

As mentioned before, silicon is the leading material for solid-state sensors. The semiconductor industry has gained great experience in the methods of treatment of silicon as a material for devices providing purely electrical conversion (diodes and transistors, for example). These methods are largely used for the fabrication of sensors with the addition of specific techniques necessary for the sensors of physical quantities. In our review of the processes, we will emphasize these techniques. More detailed descriptions of the processes can be found in the specialized literature on the fabrication of semiconductor materials, VLSI's and IC's [102–104].

Generally speaking, the basic processing steps that are used for the fabrication of diodes, transistors, and integrated circuits are applicable in forming microelectronic sensors.

There are three main types of processes that can be independent or combined with each other:

1. fabrication of a semiconductor transduction element,
2. shaping of sensing and/or transduction elements operating on nonsemiconductor principles, and

3. fabrication of elements for signal conditioning (diodes, transistors, resistors, capacitors). These elements can be regarded as part of a transducer when they are integrated with other transducer components in one monolithic or hybrid structure.

Purification of Silicon

The fabrication begins from the preparation of materials, and the first step is *the purification of silicon*. The raw material for silicon is silicon dioxide (SiO_2) and various silicates. For the growth of semiconductor-grade crystals, the electrically active impurities should be removed. Reduction of SiO_2 to Si is performed with carbon in an electrode arc furnace; as a result, a metallurgical-grade silicon (98 percent purity) is obtained. After this, the Si is converted, by reaction with HCl, to the chemical compound of Si, H, and Cl, for example, $HSiCl_3$. Then, the interaction of this compound with H_2 at temperatures between 1000 and 1200°C gives a semiconductor-grade polycrystalline silicon.

Growth of Silicon Crystals

The growth of single-crystal silicon is performed by using the Czochralski method and by the float-zone technique. In the Czochralski process, polycrystalline silicon and dopant are placed into a quartz crucible (Fig. 8.1), which is put into a furnace heated in an inert-gas atmosphere to a temperature in excess of the silicon melting point (1420°C).

A small piece of crystalline silicon called *a seed crystal* is dipped into the silicon melt; then it is slowly pulled out of the melt with a solidified mass of silicon that keeps a crystallographic orientation of the seed crystal. Both the crucible and crystal are rotated in opposite directions, and the crystal with the growing ingot of silicon is continuously pulled out of the crucible, reaching a length of the order of one meter. The diameter of the ingot is between 0.1 and 0.15m.

In *the float-zone process* (Fig. 8.2), a rod of polycrystalline silicon is placed in a pumped-down quartz tube. A small seed crystal is attached to one of the ends of

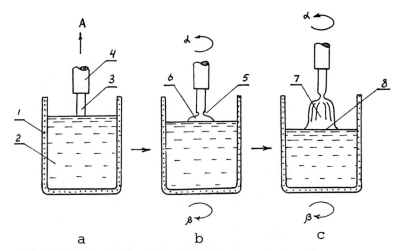

Figure 8.1 Czochralski crystal growth. a = crystal setting, b = forming of crystal at shoulder, c = ingot pulling; *A* = direction of pull, α and β = directions of rotation of shaft and crucible, respectively, 1 = quartz crucible, 2 = silicon melt, 3 = seed crystal touching melt surface, 4 = pulling shaft, 5 = neck region, 6 = shoulder, 7 = silicon ingot, 8 = liquid-solid interface.

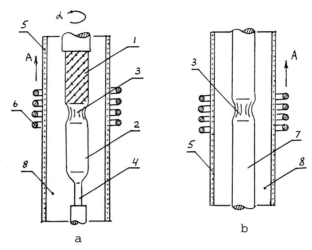

Figure 8.2 Vacuum float-zone crystal growth and refining. a = crystal growth system, b = refining system; A and α = directions of translational motion of coil and rotational motion of silicon, respectively, 1 = polycrystalline silicon rod, 2 = single-crystal silicon, 3 = molten zone, 4 = seed crystal, 5 = quartz tube, 6 = RF induction heating coil, 7 = silicon ingot, 8 = vacuum.

the rod. The tube with silicon is placed inside a radio-frequency (RF) induction coil that induces eddy currents in the silicon, thus raising its temperature above the melting point. The rod moves slowly along the coil's axis, exposing new areas to the RF radiation. The molten zone passes along the length of the rod, and when it starts at the seed end, the melt silicon resolidifies as a crystallographic continuation of the seed crystal. Surface-tension effects prevent the rod from breaking off at the molten zone.

A similar process is used for *zone refining* of silicon to purify the material. Impurities in silicon have a tendency to stay in the liquid phase; this is why the impurities can be segregated and moved to one of the ends of the rod when the molten zone is progressively shifted along the rod.

This process is used for obtaining a high-resistivity material since a high degree of purification can be achieved. In a number of sensor material processes [105], a high-purity silicon is required to create structures sensitive to various physical values.

Ingot Slicing

The next step is *ingot trimming and slicing*. The extreme top and bottom portions of the ingot are cut off (Fig. 8.3) and a cylindrical surface is ground to an exact diameter; as a rule, it is 75, 100, or 125mm. A flat is also ground along the cylinder's length. The flat is necessary for finding the crystallographic orientation (X-ray technique is used) and for establishing mechanical settings in the following operations.

An inner-diameter, diamond-embedded, stainless-steel saw blade is used for slicing 0.6 to 0.7mm-thick ingot wafers.

In order to improve the rough surface of the wafer after the slicing operation, the wafer undergoes *polishing and cleaning,* which make the surfaces smooth, planar, and parallel. These qualities are required for the photolithographic process. After lapping and mechanical polishing, the surfaces are lightly etched to remove possible vestiges of the mechanical treatment. Finally, the wafers are cleaned, rinsed, and dried; then they're ready for a number of processing steps.

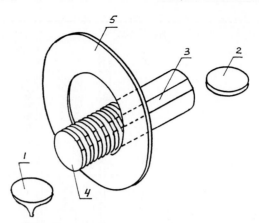

Figure 8.3 Slicing ingot. 1 and 2 = removed top and bottom portions, 3 = ground orientation flat, 4 = cut-off wafer, 5 = blade.

Diffusion

Diffusion is widely used in the fabrication of various sensors when it is necessary to introduce dopant impurities into the surface region of the wafer. The impurities penetrate into the silicon at a wafer temperature between 900 and 1200°C and at the rate of approximately $1\mu m/h$. The penetration depth lies between 0.3 and $30\mu m$. As mentioned before, the most common donors are phosphorus (P), arsenic (As), and antimony (Sb). The most commonly used acceptor dopant is boron (B). When the temperature is raised, the number of vacancies in the material for the placement of the dopant atoms is increased. These atoms penetrate first in the vacancies generated at the surface and then diffuse slowly into the interior.

The usual dopant sources are gaseous compounds, for example, BCl_3 (in boron doping) or $POCl_3$ (in phosphorus doping). The doping process is carried out (Fig. 8.4) in a quartz-tube furnace. The wafers to be processed are stacked up vertically into slots in a quartz carrier ("sled" or "boat") and placed in the "hot zone" with constant temperature over the length of the tube. When the gases flow along the tube, the surface of each wafer becomes saturated with the dopant, whose concentration reaches the order of $1 \times 10^{21} cm^{-3}$ (note that 100 percent saturation corresponds approximately to $1 \times 10^{22} cm^{-3}$). This is a predeposition step. The following diffusion can be performed in one of two ways:

1. The wafers remain in the furnace tube at the same temperature as the circulating gases. The diffusion continues at a constant dopant concentration at the surface.

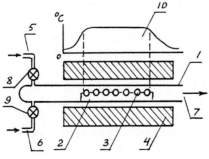

Figure 8.4 Diffusion furnace. 1 = cylindrical silica tube, 2 = silica "sled" or "boat," 3 = batch of wafers, 4 = heater, 5 and 6 = inlet of reactant gases, 7 = outlet of gases, 8 and 9 = valves, 10 = zone of constant temperature.

2. The wafers are exposed to a high temperature in an inert atmosphere provided by nitrogen or argon. In this case, the diffusion is carried out at a constant total amount of the dopant.

An alternative technique to thermal diffusion is *ion implantation*. This process allows doping to be controlled in terms of the concentration and penetration depth, and allows the placing of impurities into selected regions of the wafer. A high-vacuum system (Fig. 8.5) is used for developing and accelerating high-energy dopant ions that form a beam scanning the wafer. In order to create a positive ion source, a gaseous compound of dopant is bombarded with electrons. The ions are accelerated by a high-voltage potential difference (20 to 250kV) and enter an electromagnetic system where they are deflected by 90° and made more uniform, in terms of particles and energy (high-energy ions are separated). The beam follows through the acceleration electrodes, then through the x- and y-axis deflection plates, reaching the energy of several hundred keV. This beam is focused and scans the wafer surface by the variation of voltages applied to the horizontal and vertical plates. A masking layer over the wafer (of SiO_2, Si_3N_4, photoresist, films of gold, platinum, tungsten, for example) permits implantation only in the predetermined areas. A wafer-handling mechanism manipulates the position of wafers inside the vacuum chambers and makes it possible to treat a large number of wafers simultaneously.

The depth of the penetration of the ions into the surface of the wafer is typically between several tenths of a micrometer and one micrometer. With the increase of the accelerating voltage to a megavolt, a depth of penetration of several micrometers can be reached. If it is necessary to drive dopant deeper into the wafer, a postimplantation, high-temperature diffusion can be implemented.

During the implantation process, the ions of a dopant collide with the host crystal atoms at a high speed. As a result, a disordered imperfection in the crystalline structure is generated by forming an amorphous layer. To restore the structure to a well-ordered crystalline state, the wafer is subjected to an *annealing process* which is conducted at a temperature of about 1000°C and time length of 30 min. Laser-beam and electron-beam techniques are also used for annealing, when only the surface region of the wafer must be treated.

The ion deposition process is a more expensive process than an ordinary high-temperature deposition; however, it offers precise control over the density of dopants and the cleanliness of the process.

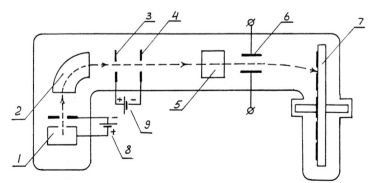

Figure 8.5 Ion implantation system. 1 = positive ion source, 2 = mass separation electromagnet, 3 and 4 = acceleration electrodes, 5 and 6 = y-axis and x-axis deflection plates, respectively, 7 = wafer-handling mechanism, 8 and 9 = sources of high voltage.

Oxidation

The oxidation of silicon can be a temporary measure for making the wafer surface in a fabrication process, or it can be a step in forming a protective or passivating layer on the crystal containing functional elements. The thickness of the layer is usually between 0.5 and 1 micrometer, and the layer is formed by the reaction Si + O$_2$ → SiO$_2$ in a high-temperature (950 to 1200°C) quartz-tube furnace similar to that for diffusion. The wafers are placed in the tube, heated, and exposed to gases containing O$_2$ and/or H$_2$O with N$_2$ used as a carrier gas. The rate of *thermal oxidation* depends on the temperature. For example, an oxide thickness of 0.5μm can be reached in 10 hours (using O$_2$-oxidation) if the furnace's temperature is about 1200°C and in 40 hours if the temperature is 1000°C. The oxide growth rate with H$_2$O is much faster than with O$_2$, but the oxide produced by O$_2$ is denser.

Oxide masking is used to mask an underlying silicon surface against the penetration of dopants in the diffusion or ion implantation process. An oxide layer can be patterned by the photolithographic process (see ahead) to produce openings where the oxide is removed and where the dopants can be deposited into the silicon. This is a method to make the pattern that originated in the photolithographic process. A cross-sectional view of a p-n junction obtained by diffusion through an oxide window is presented in Figure 8.6.

The diffusion is an isotropic process and dopants are spread through the window not only in the downward direction but also sideways, underneath the oxide. The oxide layer is called a *passivated junction* because it is chemically resistive and provides good protection of the structures in silicon against various environmental effects. The semiconductor junction obtained by diffusion is planar by nature; hence, it is called a *planar junction.*

The process of oxidation is accelerated by exposing the wafer to a high pressure. The time of the process is inversely proportional to the applied pressure. The speeding of the process allows the temperature to be decreased, which is always desirable since the probability of the formation of crystalline defects is decreased.

Photolithographic Processes

Microscopically small device patterns on silicon wafers are produced by means of a *photolithographic process* or *photoengraving*. The dimensions of a device as small as 2 micrometers can be obtained when an ultraviolet light is used for exposure, whereas electron-beam or X-ray exposure reduces the size to a small fraction of a micrometer (Fig. 8.7 [106]).

A brief description of the processing steps will be given in the following paragraphs.

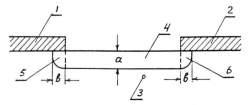

Figure 8.6 Diffusion masked by oxide layer. 1 and 2 = SiO$_2$ masking layers, 3 = n-type substrate, 4 = p$^+$ diffused layer, 5 and 6 = lateral diffusion under oxide layer ($b \approx 0.8\ a$).

Figure 8.7 Scanning electron micrograph (SEM) of 100nm-period grating lines obtained by ion etching in oxygen plasma (adopted from [106]).

1. The first step (Fig. 8.8) is *the preparation of the substrate* for application of a photoresist. The photoresist is a photosensitive material used for creating photographic images on the substrates. The photoresist should adhere well to the surface of the wafers. The surface should be free of contaminants and water. Thermally oxidized silicon is hydrophilic. The most effective pretreatment of the surface for removing moisture is dehydratization by heating. The wafers are baked in an infrared oven at a temperature of at least 100°C. This step is omitted if the wafers are taken to the photoresist application directly from the high-temperature furnace. A priming solution sometimes is applied to the surface before the photoresist in order to remove any residual moisture and provide better adhesion of coatings.

   A thin and uniform-thickness film of the primer is created by applying several drops of the primer to the surface and spinning the wafer at 3000 to 5000 revolutions per minute for 15 to 30 seconds.

2. *A photoresist coating* is made over the clean and dry surface (with or without the primer) by spinning a small amount of liquid photoresist applied to the surface similarly to the way the primer was applied. The thickness of the formed film is controlled by monitoring the spinning's speed. This thickness is in the range of 0.5 to 1.0μm.

3. In *a prebaking (soft baking)* cycle, the solvent from the photoresist coating is driven off by keeping the wafers in an oven at about 80°C for about 30 to 60 min. At the end of this cycle, the photoresist is hardened and turns into a semisolid film.

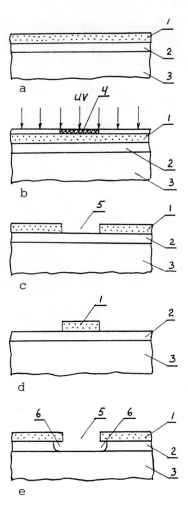

Figure 8.8 Steps of photolithographic process. a = primer and/or photoresist spinning, b = exposure of wafer to UV light, c = negative-photoresist development, d = positive-photoresist development, e = oxide etching; UV = ultraviolet light, 1 = primer and/or photoresist, 2 = silicon dioxide, 3 = silicon wafer, 4 = photomask, 5 = window with removed photoresist, 6 = overetched oxide.

4. *Mask alignment and exposure* are carried out with a photomask, which is obtained from the large-size drawing of a pattern by a photographical reduction of the original size. An intermediate step in the preparation of the mask is fabrication of a master photomask, which then can be used for making the large number of replications to be introduced to the wafer. The photomask is usually a piece of a thin glass (about 2mm thick) with a photographic emulsion or thin-film metal (usually chromium) pattern on one side. This pattern contains clear or opaque areas. The alignment of the photomask to the wafer is a very accurate operation. In some cases, the accuracy is within 0.5μm or better. The operation is carried out in a *mask aligner* with the wafer placed very close to the mask whose relative position is adjusted by alignment with reference marks on the wafer. After the alignment is achieved, the wafer and mask are brought into direct contact and exposed to a highly collimated ultraviolet light which penetrates through the transparent areas of the mask and illuminates (3 to 30 seconds) the photosensitive film on the wafer's surface. After this, the wafer is removed from the aligner and becomes ready for developing. A special alignment technique is necessary for the electron-beam and X-ray exposure systems. In this case, at least 0.1-μm positioning accuracy is required, and the photolithography process is performed with an electron image projector.

5. *Photoresist development* is provided by using proprietary solutions from photoresist manufacturers. The photoresists can be of two types: *negative and positive*. The negative photoresist is polymerized under ultraviolet radiation that makes it tougher and insoluble in the developer solution. The unexposed areas of the photoresist are readily dissolved by the solution; therefore, the areas on the wafer free from the photoresist correspond to the opaque pattern on the mask.

    The opposite process occurs with a positive photoresist: the areas exposed to ultraviolet radiation undergo depolymerization and become soluble in the developer solution. Finally, the replication of the photomask pattern contains areas with the removed photoresist in the places corresponding to the clear areas of the photomask. After development, the wafers are thoroughly rinsed.

6. In order to make the resist adhere better to the wafer, it is heat treated (*postbaked*) at a temperature of about 150°C for about 30 min.

7. The next step is *oxide etching*. The purpose of this operation is to remove silicon dioxide from the places defined in the developing process. Most commonly it is carried out by immersing the wafers in etching solution or by spraying this solution over the wafers' surfaces (this is called *wet etching*). The solution is usually composed of water (H_2O) and hydrofluoric acid (HF) : 10 : 1 H_2O : HF or ammonium fluoride (NH_4F) and HF : 10 : 1 NH_4F : HF. The solutions etch SiO_2 but do not attack silicon and photoresist to an appreciable extent. For the latter solution, the etch rate is about $0.1 \mu m$/min. at 25°C. The etching time should be controlled to remove SiO_2 from the windows. A prolonged etching can cause an undesirable undercutting of the oxide under the photoresist (Fig. 8.8e) with a widening of the window.

    In a *plasma etching* (also called a *dry etching*), reagent gases rather than liquids are used for removing the SiO_2. These gases are usually CF_4 (Freon 14) or C_2F_6. Wafers are placed inside a vacuum chamber that is filled with the gases. A radio-frequency field radiated inside the chamber ionizes the gas molecules and breaks up some molecules into radicals, forming a highly reactive plasma. The plasma's radicals react with silicon dioxide, forming oxygen and different gaseous silicon compounds that can be removed by venting. The rate of the destruction of the photoresist and silicon by plasma can be made smaller by removing the SiO_2. For example, when C_2F_6 is used, the ratio of the rates ("etch ratio") for SiO_2 and Si is 15 to 1. Dry etching allows the windows to form with dimensions smaller than one micrometer. This is not achievable in the wet-etching process because of some surface-tension effects. Wet etching is isotropic, which means that SiO_2 dissolves equally in all directions (Fig. 8.9). Undercutting the resist layer restricts the minimum pattern (line) width as can be understood from Figure 8.9. The dry-etching process allows anisotropy in etching, which helps to obtain the lines of submicron dimensions. A high degree of anisotropy gives a *planar plasma etching* that is performed with an electric field oriented perpendicular to the wafer's surface. This field directs the ions of plasma toward the wafer and makes the etching rate in the downward direction greater than in the sideways direction (Fig. 8.10).

    Another way of removing SiO_2 is mechanical rather than chemical. It is called *ion milling,* which is virtually a sputtering process. The surface of a wafer is bombarded by inert-gas ions [for example, argon (Ar^+)], knocking atoms of the oxide from the area opened for the ion beam. The beam is generated in a vacuum chamber and accelerated by a high-voltage

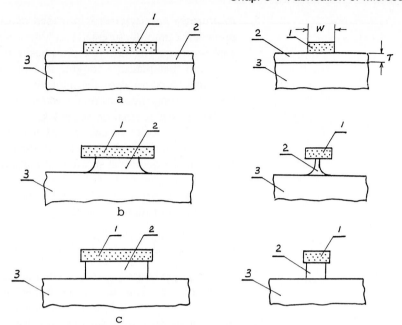

Figure 8.9 Isotropic and anisotropic etching. a = wafer with mask before etching, b = isotropic (wet) etching ($W = 2T$), c = anisotropic etching ($W \approx T$); 1 = etching mask (photoresist), 2 = film (SiO$_2$, Si$_3$N$_4$, Al, etc.), 3 = substrate (silicon).

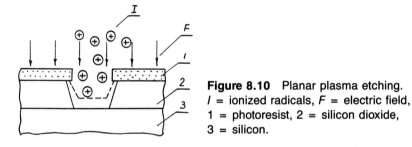

Figure 8.10 Planar plasma etching. I = ionized radicals, F = electric field, 1 = photoresist, 2 = silicon dioxide, 3 = silicon.

potential difference. This process is characterized by a great anisotropy of etching and an etching rate even higher than that for plasma etching.

8. If the resist is positive, the removal of the photoresist, called *photoresist stripping,* is made easier with organic solvents such as acetone. A negative photoresist is removed by mechanical scrubbing along with flushing in a hot sulfuric acid. Another way of stripping is to use a process similar to plasma etching with oxygen as a reagent gas. The resist is an organic compound containing carbon and hydrogen atoms. Oxidation forms gaseous CO, CO$_2$, and H$_2$O. When the gases are vented from the wafer's surface, they leave it free of the resist. This operation completes the photolithographic process.

Chemical-Vapor Deposition

The *chemical-vapor deposition* (CVD) process creates films from the chemical reaction of gases at the wafer surface. Chemical-vapor deposition reactions have been developed for many types of films. We will consider several of them: silicon dioxide deposition, silicon nitride deposition, and formation of silicon homo- and heteroepitaxial layers.

1. A commonly used reaction for *silicon dioxide deposition* is the interaction of oxygen with silane (SiH$_4$):

$$SiH_4 + O_2 \rightarrow SiO_2 + 2H_2 \qquad (8.1)$$

Two gases are brought into the reaction chamber where the substrates are heated to a temperature of about 450 to 600°C. The resultant passivating oxide layer, called *silox,* can cover any surface (for example, silicon, metal). Unlike the production of thermally grown oxide, silox's process requires a much lower temperature that does not affect the prior processes. However, the thermally grown oxide is more dense and is characterized by a higher dielectric constant.

2. *Silicon nitride* (Si$_3$N$_4$) *deposition* is also used for protection and passivation of the device. It is an effective mask for diffusion or ion implantation. The diffusion rate of oxidants (O$_2$, H$_2$O) is very small in Si$_3$N$_4$, and the oxidation of masked silicon is easily prevented. In the *local oxidation of silicon* (LOCOS) process, the wafer (Fig. 8.11) is masked by Si$_3$N$_4$, which is patterned by a photolithographic process. When the wafer is subjected to the thermal oxidation process, the oxide grows only in the areas free of nitride.

The reactions (at 600 to 800°C) between silane and ammonia (NH$_4$) or dichlorosilane (SiCl$_2$H$_2$) and ammonia producing nitride are as follows:

$$3\,SiH_4 + 4\,NH_3 \rightarrow Si_3N_4 + 12\,H_2$$

$$3\,SiCl_2H_2 + 4\,NH_3 \rightarrow Si_3N_4 + 6\,HCl + 6\,H_2 \qquad (8.2)$$

Nitride is etched with a hot (about 180°C) phosphoric acid (H$_3$PO$_4$). Photoresist can be also dissolved by the acid, and, therefore, a CVD oxide deposition is made on the top of the nitride; then, the photoresist is applied to the oxide and patterned for the following HF-etching. The steps of this process are illustrated in Figure 8.12.

Plasma etching of silicon nitride with an SiF$_4$/O$_2$ mixture is successfully used. Etch ratios for silicon and silicon dioxide are 5:1 and 50:1, respectively.

SiO$_2$ and Si$_3$N$_4$ have similar insulating but dissimilar mechanical properties. After growing, SiO$_2$ is in compression, but Si$_3$N$_4$ is in tension. Si$_3$N$_4$ is impervious to most silicon anisotropic etchants, whereas SiO$_2$ is not. Si$_3$N$_4$ effectively protects the underlying silicon against oxidation at high temperatures.

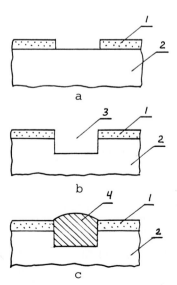

Figure 8.11 Process of local oxidation of silicon (LOCOS). a = silicon nitride film deposition and patterning, b = etching slots in silicon (Si$_3$N$_4$ is a mask), c = silicon oxide growth (Si$_3$N$_4$ is a mask); 1 = silicon nitride, 2 = silicon, 3 = etched slot, 4 = thermal oxide layer.

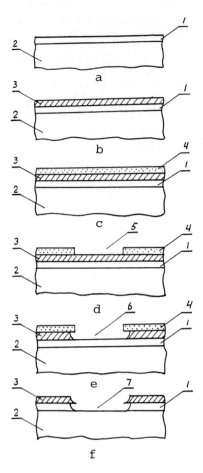

Figure 8.12 Obtaining pattern on silicon nitride layer by using CVD oxide as etching mask. a = CVD Si_3N_4 deposition, b = CVD SiO_2 deposition, c = spinning of photoresist film, d = photolithographic forming of window in photoresist, e = etching of CVD oxide using HF solution, f = etching of silicon oxide with H_3PO_4 and removal of photoresist; 1 = Si_3N_4, 2 = Si, 3 = SiO_2, 4 = photoresist, 5 = window in photoresist, 6 = window in SiO_2 after etching with HF, 7 = window in Si_3N_4 after etching with H_3PO_4.

Epitaxy or *epitaxial layer deposition* makes it possible to grow an upper layer on the top of the wafer and to have this layer in a single-crystal form as a crystallographic extension of the substrate's lattice. The layer can have a degree of purity much higher than that of the underlaying wafer. The *epitaxial growth* is possible only for a limited combination of the substrate and layer materials. However, the materials can be arbitrarily doped, and the combination of p- and n-materials with different degrees of impurities for the substrate and layer is attainable.

When the materials are the same (e.g., silicon-on-silicon), the process is called *homoepitaxy*; for different materials (e.g., silicon-on-sapphire), it is called *heteroepitaxy*. The thickness range of the epitaxial layers is 3 to $30\mu m$.

There are two basic types of chemical reactions that are used for epitaxial deposition [102–104]. One of them is reduction of chlorine compounds [silicon tetrachloride ($SiCl_4$), trichlorosilane ($SiHCl_3$)]; another one is pyrolytic (temperature-aided) decomposition [dichlorosilane (SiH_2Cl_2), silane (SiH_4)]. Table 8.1 gives some of the characteristics of the reactions.

Three types of apparatus are used for epitaxial growth: horizontal, vertical, and cylindrical reactors (Fig. 8.13). The substrates are placed in a fused silica reactor

TABLE 8.1

Reactions for Epitaxial Deposition of Silicon

No	Reaction	Temperature °C	Deposition rate µm/min.
1	$SiCl_4 + 2\,H_2 \rightarrow Si + 4\,HCl$	1150 to 1250	0.4 to 1.5
2	$SiHCl_3 + H_2 \rightarrow Si + 3\,HCl$	1100 to 1200	0.4 to 2.0
3	$SiH_2Cl_2 \rightarrow Si + 2\,HCl$	1050 to 1150	0.4 to 3.0
4	$SiH_4 \rightarrow Si + 2\,H_2$	950 to 1050	0.2 to 0.3

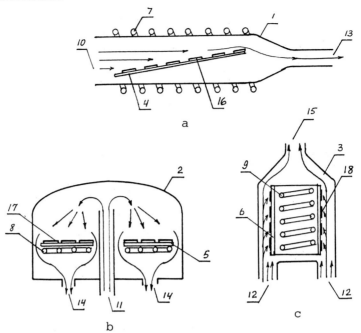

Figure 8.13 Epitaxial reactors. a = horizontal reactor, b = vertical ("pancake") reactor, c = cylindrical reactor; 1, 2, and 3 = chambers, 4, 5, and 6 = susceptors, 7, 8, and 9 = RF induction heating coils, 10, 11, and 12 = gas inlets, 13, 14, and 15 = gas outlets, 16, 17, and 18 = wafers.

chamber on susceptors, which are silicon-carbide-coated carbon plates. Radio-frequency (RF) eddy currents generated in the susceptors, as in the secondary coil of a transformer, produce the heat necessary for the reaction. RF energy is delivered from the coil (transformer's primary) mounted close to the susceptor. Epitaxial deposition takes place as the gases flow along the substrates. The carrier gas is ultrahigh-purity hydrogen. Doping can be provided along with the epitaxial deposition. Diopropan (B_2H_6) and phosphine (PH_3) are used for boron and phosphorus doping, respectively. Immediately before the deposition process, the surface of the wafer is conditioned. The natural film of oxide is removed by reduction with hydrogen at 1250°C for 10 to 20 min. This operation is followed by a dilute gaseous HCl or SF_6 light etch of silicon for 5 min. at 1250°C.

One of the requirements for the *heteroepitaxial process* is a close matching of the crystalline structure of the substrate to the deposited layer. Otherwise, the lattice will not be formed, and the layer will be polycrystalline or amorphous rather than crystalline. A thin epitaxial layer of silicon ($\sim 1\mu$m) on sapphire can have high isolation and low capacitance, which is important for devices with a low time constant.

Etching of Silicon

Isotropic and anisotropic (orientation-dependent) wet chemical etching [96, 102, 107–14] has been extensively developed for sensor technology. It is and will continue to be one of the most popular tools for shaping mechanical microstructures. Plasma techniques are also used for etching silicon.

In isotropic etching, silicon is uniformly removed in all directions accessible to the etchant. Usually, the etches are combinations of nitric (HNO_3), hydrofluoric (HF), and acetic (CH_3COOH) acids and water (HNA). HNO_3 oxidizes silicon; HF reacts with silicon oxide and forms complex soluble ions. The CH_3COOH acts as a diluent. For the solution with 7 parts of HF and 3 parts of HNO_3, a maximum etch rate of 800μm/min. at room temperature can be obtained. The etch rate can be varied and controlled by a change in the proportions of the active acid and diluent.

The geometry of an etched hole (Fig. 8.14) in silicon is changed with agitation during etching; it becomes deeper and rounder.

Anisotropic etchants attack silicon very rapidly in one direction but very slowly in another (Fig. 8.15). For example, if KOH at 80 to 100°C is used as an etchant for a (100)-plane wafer, the material will be removed quickly along the [100] direction and slowly in the [111] direction. As a result, V-shaped grooves can be fabricated. The angle between the (111) groove sidewalls and (100) silicon surface will be 54.74°, and it is defined by the relationships between the principal crystallographic planes (Fig. 7.23). It is obvious that for this etching, the maximum depth of the groove is defined by the width of the opening in the mask. The etching of vertical-walled slots is possible on (110) silicon substrate.

A heavily doped layer of silicon (e.g., by boron) makes etching slower and can stop it.

There are several factors that should be taken into account when the etching process for the shaping of silicon is designed:

1. The type of etching (isotropic or anisotropic) is defined by the etchant.
2. The rate of etching depends on the etching solution composition, temperature, and agitation.
3. The shape of the hole developed in silicon is controlled by the mask's pattern and the plane of the wafer from which the etching starts.
4. Self-stop in etching can occur because of reluctance to dissolving in a specific plane [e.g., (111)] or at the layer containing a high level of impurities (for EDP and KOH).
5. In the HNA system (1 : 3 : 8), the etch rate decreases with the decrease in the p-n dopant concentration.

Table 8.2 adopted from [96] gives a list of basic etchants and their characteristics. Three etchant systems—ethylene diamine, pyrocatechol, and water (EDP)—

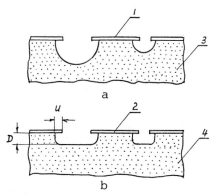

Figure 8.14 Geometries of isotropically etched silicon. a = isotropic etching with agitation, b = isotropic etching without agitation; D = depth of etching, U = undercutting ($D \approx U$), 1 and 2 = etching masks (SiO_2 or Si_3N_4), 3 and 4 = silicon substrates.

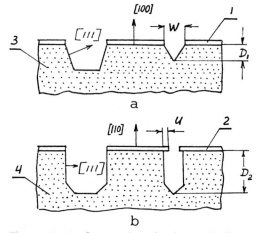

Figure 8.15 Geometries of anisotropically etched silicon. a = etching on (100) silicon wafer, b = etching on (110) silicon wafer; [100], [110], and [111] = crystallographic directions, D_1 and D_2 = depths of etching, U = undercutting (D_2/U can be as high as 400), W = width of window ($D_1 = \frac{1}{2}W$ tan 54.7°), 1 and 2 = SiO_2 or Si_3N_4 etching masks, 3 and 4 = silicon substrates.

TABLE 8.2
Etching Solutions for Silicon (Adopted from [96])

Etchant (Diluent)	Typical compositions	Temperature °C	Etch rate (μm/min.)	Anisotropic (100)/(111) etch-rate ratio	Dopant dependence	Masking films (etch rate of mask)
1	2	3	4	5	6	7
HF HNO_3 (water, CH_3COOH)	10ml 30ml 80ml	22	0.7 – 3.0	1:1	$<1 \times 10^{17}$ cm^{-3} n or p reduces etch rate by about 150	SiO_2 (300 Å/min.)
	25ml 50ml 25ml	22	40	1:1	no dependence	Si_3N_4
	9ml 75ml 30ml	22	7.0	1:1		SiO_2 (700 Å/min.)
Ethylene diamine Pyrocatechol (water)	750ml 120g 100ml	115	0.75	35:1	$>7 \times 10^{19}$ cm^{-3} boron reduces etch rate by about 50	SiO_2 (2 Å/min.) Si_3N_4 (1 Å/min.) Au, Cr, Ag, Cu, Ta
	750ml 120g 240ml	115	1.25	35:1		
KOH (water, isopropyl)	44g 100ml	85	1.4	400:1	$>1 \times 10^{20}$ cm^{-3} boron reduces etch rate by about 20	Si_3N_4 SiO_2 (14 Å/min.)
	50g 100ml	50	1.0	400:1		
H_2N_4 (water, isopropyl)	100ml 100ml	100	2.0	—	no dependence	SiO_2 Al
NaOH (water)	10g 100ml	65	0.25 – 1.0	—	$>3 \times 10^{20}$ cm^{-3} boron reduces etch rate by about 10	Si_3N_4 SiO_2 (7 Å/min.)

and two systems mentioned above—(HNA), KOH, and water—are of particular interest. EDP is anisotropic and highly selective, and it stops etching at highly boron-doped silicon. Suitable masks for EDP are SiO_2, Si_3N_4, Cr, and Au.

KOH and water is also an orientation-dependent solution. Its (110)-to-(111) plane etch-rate ratio is higher than that of EDP. Etching on a (110) wafer gives deep grooves with a minimum undercutting of the mask. Si_3N_4 is preferable for masking since the SiO_2 mask is etched at a high rate. Several examples of anisotropic etches of (100)-plane wafer are shown in Fig. 8.16.

In the *electrochemical etching* (ECE) [115] of silicon (Fig. 8.17), a wafer provided with a platinum electrode is immersed into HF/H_2O solution. A positive potential is applied to the wafer and a negative to the electrode. Under the difference in the potentials, holes in the silicon are accumulated at the Si/solution interface, and this causes a concentration of OH^- at the interface and promotes the oxidation of silicon and rapid dissolving of dioxide by HF. Holes, being positive hydrogen ions, flow to the platinum cathode where they form molecules that are released in the form of hydrogen bubbles. It is clear that the etching characteristics depend on doping, and it is possible to control the shape of the members that undergo etching by the variation of dopants.

One effective way to stop etching at the predetermined area is to deposit an n-type epitaxial layer and provide a voltage bias on the layer. P-type silicon's dissolution stops when the etching reaches the n-type layer. In this case, there is no need to have a buried p^+ layer. The highly doped p^+ film can be a source of mechanical stress in thin mechanical structures.

ECE *jet etching* generates small holes in silicon at a high rate of etching. A difference in potential is created between a narrow stream of etchant incident on one side of the wafer. Due to the agitation and delivery of fresh portions of etchant to the place of reaction, the reacted products are formed and removed rapidly.

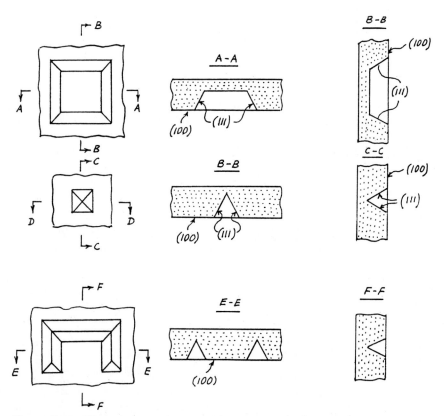

Figure 8.16 Examples of anisotropic etches of (100)-plane silicon wafer.

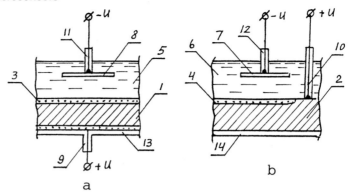

Figure 8.17 Electrochemical etching of wafers. a and b = arrangements with back and front contacts, respectively; $+U$ and $-U$ = positive and negative potentials at wafer and cathode, respectively, 1 and 2 = silicon, 3 and 4 = etched regions, 5 and 6 = HF solutions, 7 and 8 = Pt electrodes, 9 and 10 = back and front electrodes, 11 through 14 = electrical insulation.

ECE at a low current density and/or at the deficiency in OH^- (concentrated HF) leads to the forming of a brownish, not fully oxidized layer of silicon. It is a single-crystal silicon permeated with many fine holes and channels. Actually, it is a porous, spongelike layer, whose thickness can be of an order from 1 to $100\mu m$. The layer can be rapidly etched off because of the large surface of the silicon at pores exposed to the reaction.

Laser etching in a gaseous atmosphere can be used for generating practically any shape. A highly collimated laser beam can etch a groove of several micrometers in width and depth at a very high rate. Since the etching area is a small spot, a scanning process is required to treat the surface whose size is several orders of magnitude larger than the spot. As a result, the time of etching becomes large. For example [96], it takes more than 100 hours of scanning a 100-mm wafer in order to remove 1-μm layer of Si by a beam of a 20-W laser at a power density of $1 \times 10^7 W/cm^2$. Both chemical and local thermal effects are present in the laser etching, which takes place in atmospheres of HCl and Cl_2. The typical reaction is

$$4\,HCl + Si \rightarrow 2\,H_2 + SiCl_4 \tag{8.3}$$

The etching process is used whenever pits of different shapes are formed. The combination of etching and epitaxial depositing makes it possible to shape more complicated structures where the material can be not only removed but added. For example, a step-structure (Fig. 8.18) with planar surfaces is realized by etching material down to a B-doped p^+-Si layer (etch-stop layer) or to (111) surfaces and then by epitaxial growth of elevated hills.

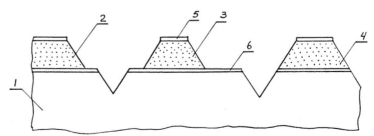

Figure 8.18 Silicon structure formed by etching and epitaxial deposition. 1 = etched substrate, 2, 3, and 4 = deposited silicon, 5 = silicon dioxide, 6 = B-doped p^+-Si layer.

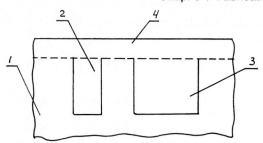

Figure 8.19 Structure with vertical-wall grooves filled with doped material. 1 = silicon substrate, 2 and 3 = grooves filled with p and n silicon, 4 = ordinary epitaxial layer.

Another example (Fig. 8.19) is a structure containing vertical-walled grooves that were etched first, then filled by the epitaxial deposition with p or n silicon, and finally buried beneath another epitaxial layer.

Metallization

The deposition of *a thin-film metal layer* on the surface of a crystal serves three purposes: it forms sensing or transduction parts of a structure, it provides the interconnection of various elements formed in the crystal, and it produces metallized areas called *bonding pads* for the bonding of wire leads for the outer connections. The typical sizes for the wire and pads are diameters of 25μm and square sides of 100μm by 100μm. The film thickness for the conductors and pads and the width of the conductors vary depending on the device; however, they commonly exceed 2μm.

Among the different materials used for metallization, aluminum is used most frequently due to several of its properties. It is a good conductor and forms low-resistance, nonrectifying contacts with p-type silicon and with heavily doped n-type silicon. It is easily deposited by vacuum evaporation. A good bond of aluminum with silicon is obtained by sintering or alloying at temperatures of 500 and 577°C, respectively.

The metallization process begins with the deposition of a thin metal film on the substrate which is placed in a vacuum evaporation chamber (Fig. 8.20). The pressure in the chamber is pumped down to about 1×10^{-6} to 1×10^{-7} torr (traditionally it is measured in torrs; 1 torr $= 133.32$Pa, 1 torr $= 1$mmHg, 1atm $= 760$torr). The material to be deposited is evaporated. The source of heat for the vaporization can be a resistively heated filament, a graphite crucible heated by induction from a

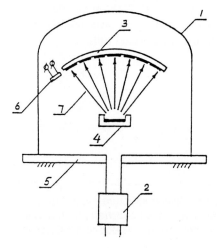

Figure 8.20 Vacuum evaporation system. 1 = vacuum chamber, 2 = vacuum pumps, 3 = substrate, 4 = evaporation source, 5 = base, 6 = quartz crystal, 7 = evaporant flux.

radio-frequency coil, an electron beam, or a laser beam focused on a target made from the depositing metal. When the material is heated up to a very high temperature, it starts to vaporize (see Table 8.3). The molecules of the metal travel in straight-line paths, hitting the substrates and condensing there to form a thin film. The high vacuum in the chamber precludes undesirable chemical reactions between the evaporant molecules and gases in the chamber; it also permits the molecules to reach the substrate without losing energy from collisions with gas molecules. The mean free path of gas molecules (mfp) depends on pressure: mfp $= 4.5 \times 10^{-3}$cm/torr. Therefore, for a pressure of 1×10^{-6}torr, the free path of molecules in the chamber is 45m. This means that for the typical 50-cm distance between the source and substrate, all the vaporized molecules directed to the substrates can reach them unimpeded.

Sputtering is also widely used for depositing not only metal but insulator films. The wafer to be coated is placed in a vacuum chamber.

The source of the material for depositing, called a *target,* is bombarded by neutral gas ions such as argon. The target has a several-hundred-volt negative potential with respect to the wafers. The positive ions dislodge the atoms of the target material, and atoms migrate toward the wafer, where they adhere to the wafer surface. The energy needed for the vaporization of the target material is obtained from the electric field, and it is not necessary to heat the material up to a high temperature. Keeping the material at a low temperature is important for depositing low-vapor-pressure refractory materials and making micromachined parts of them [116, 117].

During the evaporation process, the thickness of the film is monitored by an instrument provided with a quartz crystal. The crystal is exposed to the evaporant flux and changes the frequency of oscillation because of the change in the effective mass of the vibrating member when the film is condensed on its surface. A shift in the oscillation frequency is a measure of the monitored frequency.

A photolithographic process, similar to that for the treatment of Si and SiO_2, is used for producing patterns on the metal layer. Aluminum is etched by base

TABLE 8.3

Temperatures of Fusing and Evaporation for Some Elements (Adopted from [677])

Element	Fusing temperature °C	Evaporation temperature °C	Materials for evaporation sources (crucibles, boats, etc.)
Ag	961	1047	Ta, Mo, W, Al_2O_3
Al	660	1150	W, Ta, Mo, Nb
Au	1063	1465	W, Mo
Bi	271	698	Ta, W, Al_2O_3
C	3700	2681	+
Cd	321	264	W, Ta, Al_2O_3
Co	1490	1650	Nb
Cr	1800	1205	W, Nb, Mo, Ta
Cu	1083	1273	W, Al_2O_3, Nb, Mo, Ta
Ge	959	1251	Ta, Mo
Mg	651	443	W, Ta, Mo
Mo	2622	2533	+
Ni	1455	1510	W, BeO, Al_2O_3
Pt	1774	2090	W
Sb	630	678	Ta, W, Al_2O_3
Ta	2996	3070	+
Ti	1725	1546	W, Ta

Note:

1. A number of materials have an evaporation temperature lower than the fusing temperature, because of the ability of these materials to be transformed to the vapor state without passing through the liquid state (sublimation).

2. The mark + means that heating by an ion or electron beam is recommended.

solutions (KOH, NaOH) and acids (HCl, combination of H_3PO_4, HNO_3, and CH_3COOH). The typical etch rate is 1μm/min. at 50°C.

Metallization patterning is also provided using *a lift-off process*. In this process, a positive photoresist is applied to the wafer first. The resist is patterned by forming openings intended for coating with aluminum. After the deposition of aluminum, the wafers are immersed in an organic solvent (e.g., acetone) and subjected to ultrasonic agitation. The subsequent swelling and dissolution of the photoresist causes a lifting off of the metal film that is on the top of the photoresist. In this process, the metallization film should be thinner than the photoresist film. A fine line width of about 1μm can be obtained with this process.

Highly anisotropic aluminum etching (vertical-to-horizontal etch ratio) can be obtained using a CCl_4/He plasma. In this process, an aluminum line width of 1μm is produced, whereas the thickness of the line can be much greater.

A thin-film resistor fabrication is the metallization process for forming long-term stable and precision resistors which are frequently integrated with diffused or ion-implantated silicon resistors to form a sensor's network (e.g., temperature-compensated circuit). The commonly used materials for the deposition of the resistive films are nickel-chrome (80 parts of nickel and 20 parts of chromium), tantalum nitride, silicon-chrome, and "cermets" (ceramic metals).

The resistive elements in microelectronics are characterized by the sheet resistance expressed in ohms per square ($\Omega/\square$). The following consideration explains the meaning of this unit. If we have a film resistor, its resistance is

$$R = \frac{\rho l}{tw} = r\frac{l}{w} \tag{8.4}$$

where
$\quad\quad\quad R$ = resistance, Ω;
$\quad\quad\quad \rho$ = resistivity, $\Omega \cdot$ cm;
$\quad\quad l, t,$ and w = length, thickness, and width of the film, respectively, cm (it is common to express them in cm rather in m); and
$\quad\quad\quad r = \rho/t$ = sheet resistance, $\Omega/\square$.

Note that if $l = w$, $R = r$. The ratio l/w is called the *aspect ratio*. It determines the number of squares in the surface geometry.

The film is deposited on the silicon dioxide to be insulated from the silicon surface which can be electroconductive. In hybrid constructions, glass, alumina, glazed alumina, and sapphire are also used as a substrate for resistors.

Tantalum nitride films are highly reliable and stable. Tantalum nitride layers are often combined with titanium or chromium films (for good adhesion) and golden films (for higher conductivity). The resistivity of tantalum nitride films [102] is approximately 250$\mu\Omega \cdot$ cm. The thickness of the film is usually in the range between 100 and 500Å, which gives a sheet resistance range between 250 and 50$\Omega/\square$. The temperature coefficient of resistance (TCR) is -75.0×10^{-6}1/°C. The resistivity of nickel-chromium films is lower than that of tantalum nitride films. For example, for a 100-Å layer, the sheet resistance is 150$\Omega/\square$. This film is attacked by moisture and must be protected by a layer of glass or it must be hermetically sealed. Alloys of gold and aluminum are compatible with these films. Their TCR is typically approximately $+30.0 \times 10^{-6}$1/°C but can be made between 0 and $+50.0 \times 10^{-6}$1/°C by a change in composition.

The cermet films are constituted of mixtures of metals and oxides that are frequently deposited by flash evaporation in which the powder of an alloy is dropped onto a heated surface and evaporates immediately. A popular material for cermet films is the combination of chromium (Cr) and silicon monoxide (SiO). The sheet resistance varies depending on the ratio of the components. For example, 65% Cr and 35% SiO give 600Ω/$\square$; 70% Cr and 30% SiO give 300Ω/$\square$. There are limits in the Cr-to-SiO ratio. Beyond the limits, the resistance cannot be changed. The TCR for cermets is positive and does not exceed $100 \times 10^{-6}1/°C$. The coating of the film by silicon monoxide with the subsequent annealing increases the stability of the film.

Film deposition is also used for the fabrication of *thin-film capacitors*. The electrodes for the capacitors can be tantalum nitride and double layers of titanium and gold. The dielectric layer can be made by anodizing the tantalum nitride layer or by evaporating silicon monoxide. The obtainable capacitance density and working voltage is between 0.01 and 0.1μF/cm^2 and 30V, respectively.

Thermomigration

Thermomigration is utilized for the electrical connection of elements on two sides of a wafer and for the connections between wafers. Liquid eutectic Al/Si alloy droplets can migrate through a wafer if the silicon slice is subjected to a temperature gradient ($\sim 5°C$/mm) across the wafer (Fig. 8.21). Aluminum and silicon form a molten alloy that migrates toward the hotter side of the wafer due to the dissolution of the silicon atoms on the hot side of the molten zone. The thermomigration rate is about 3μm/min. at 1100°C. It was found [118] that wires of 30 to 160μm in width traversing a (100) wafer along the [110] direction are most stable. The wires can be close-spaced ($\sim 100\mu$m).

Thermomigration can be driven by a laser. This technique is important for a practical implementation.

Migration regions are sources for local mechanical stresses, which can be reduced by a postmigration thermal anneal.

Wafer Probing

Some functional tests of wafers are made before the wafers are separated into individual chips. These tests are performed on probe stations containing micromanipulators positioning the probes on the different parts of the electrical or electromechanical structure formed in the preceding steps. Dc and ac characteristics can be obtained and evaluated (often automatically). If the structure does not meet specifications, it is marked for sorting.

Die Separation

There are several techniques for the separation of a wafer into individual dies.

In the *scribe-and-break method,* the corner of an industrial diamond is drawn across the wafer with a light pressure, producing scribing streets that do not penetrate the entire wafer thickness. The scribe lines can be formed by sawing with diamond-impregnated saw blades. A pulsed laser creating lines by the localized melting of silicon is also used to accomplish the lines. After the lines are formed, the wafer is broken into dies by applying concentrated forces (e.g., rolling) along the lines. It should be noted that a (100) oriented wafer is fractured well because the natural

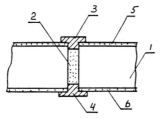

Figure 8.21 Structure with electrical connection of two sides of wafer by means of thermomigrated Al/Si alloy. 1 = n-Si, 2 = Al-doped p-Si migrated wire, 3 and 4 = deposited aluminum, 5 and 6 = electroconductive layers on wafers.

cleavage planes in this wafer exist along and perpendicular to the wafer's surface. In the (111) oriented wafer, the natural cleavage planes are parallel to the orientation flat and perpendicular to the wafer's surface. However, the planes perpendicular to the flat's and wafer's surfaces are not the natural cleavage planes, and fracturing along these planes can result in chipping at the edges and an unpredictable change in the die's geometry. Etching of the separation lines creates the minimum damage and stress of the dies. In this case, the wafers are photolithographically patterned and then etched over the lines by one of the etchants (e.g., potassium hydroxide or hydrozine).

Die Attachment

The packaging process usually begins with the mounting of the dies on a preform or in the device's housing. Traditional methods developed for microelectronic devices along with the techniques specifically designed for sensors utilize polyamide adhesives, epoxies, low-temperature-melt glasses (frits), and eutectics as bonding materials.

Bonding the parts by *epoxy* is inexpensive and easily automated. If an electrical contact and/or increased thermal conductivity between the attaching parts is required, epoxy is filled with a gold or silver powder. Curing of an epoxy is provided at temperatures from 125 to 175°C. Polyamide polymers can withstand higher processing temperatures than epoxies. The polymer bonding is soft and flexible, which is desirable for a number of applications. However, the organic bond can cause hysteresis and creep in the performance of the precision devices, if this bond transfers deformation, displacement, or force. The surfaces subject to the joint must be metallized, to enable them to be wet more easily by epoxy.

Frits are not as flexible as epoxy but allow a higher temperature for the parts and provide a stronger bond with a lower level of elastic imperfections. For bonding, a water suspension of frits is applied to the parts; then, they are combined, dried, and heated up to the melting point of frit. A good attachment of silicon and pyroceram (crystalline glass) parts is obtained with the frits having the following major components [119, 120]:

1. 58% PbO, 12% B_2O_3, 20% SiO_2, and 8% ZnO (fusing temperature 440°C).
2. 60% PbO, 15% B_2O_3, 6% SiO_2, 8% Fe_2O_3, 9% Cr_2O_3, and 2% Co_2O_3 (fusing temperature 530°C).

A *gold-silicon-eutectic-alloy* die bonding is one of the most effective techniques. This bonding is a good conductor of heat and electricity. It is strong and makes a reliable connection of a die to the substrate (usually to a housing or lead frame) fabricated from copper, Fe-Ni alloy, or a ceramic material. The Au-eutectic composition with 2 to 4 percent Si is most popular. Several other compositions are given in Table 8.4.

TABLE 8.4

Compositions and Melting Points for Die Attach Preforms (Source: [103])

Composition	Temperature (°C)	
	Liquidus	Solidus
80% Au 20% Sn	280	280
92.5% Pb 2.5% Ag 5% In	300	—
97.5% Pb 1.5% Ag 1% Sn	309	309
95% Pb 5% Sn	314	310
88% Au 12% Ge	356	356
98% Au 2% Si	800	370
100% Au	1063	1063

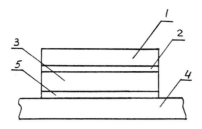

Figure 8.22 Silicon dice bonded with melting preform. 1 = silicon dice, 2 = dice metallization, 3 = melting preform, 4 = metallic or ceramic substrate, 5 = substrate metallization.

The die is usually bonded with melting preform placed between the die and substrate (Fig. 8.22), whose contacting surfaces are metallized in advance.

An expedient selection of materials having necessary properties for packaging (e.g., expansion coefficients) is important. In addition to the data in Table 7.1, Table 8.5 gives basic characteristics [103] of some materials, primarily adhesives and solders, employed in packaging. Thermal conductivity is one of the crucial parameters in defining the sink of heat from the dies and the maximum power that can be dissipated in the dies.

Field-assisted thermal bonding [96], originally developed for glass-metal thermal sealing [121], is good for mounting micromechanical structures, especially for those working in hostile environments. The essence of the method is in attaching two parts intended for bonding, heating them, and applying a voltage across the surfaces [122]. This technique is simple and forms a hermetic, strong, and reliable seal. Figure 8.23 illustrates three cases: (*a*) bonding bare or oxidized silicon to Corning borosilicate glasses (7070 or 7740)—the approximate composition [123] of 7740 glass is as follows: 80% SiO_2, 14% B_2O_3, 4% Na_2O, 1% Al_2O_3; (*b*) silicon-to-silicon bonding with sputtered 7740 glass between the parts; and (*c*) silicon-to-silicon bonding with thermal oxide passivating films and an electrostatic protecting metal shield on one of the parts. This shield is grounded and avoids damaging the structure when the high voltage for bonding is applied to the parts. It is remarkable that the bonding surface can contain aluminized lines without sacrificing the integrity and hermeticity of the seal.

It is believed that the mechanism of this kind of bonding can be explained [96] by a large electrostatic attraction between the parts under the applied voltage, which is accompanied by a high-energy pulse developing local high-temperature diffusion and fusing. In the further advances of the method [123–28], the mechanism of the bonding was studied in greater depth; it was found that it is possible to perform a room-temperature silicon-to-silicon, low-voltage (50V) bonding by using sputtered, low-melting-point glass [126] as an intermediate layer. The softening point of this

TABLE 8.5

Basic Characteristics of the Materials Used in Packaging (Source: [103])

Material	Thermal expansion $(1 \times 10^{-6})1/°C$	Young's modulus $(1 \times 10^9)N/m^2$	Thermal conductivity $W/cm \cdot °C$	Application
90-99% Al_2O_3	6.5	262.0	0.17	substrate
beryllia (BeO)	8.5	345.0	2.18	substrate
common Cu alloys	16.3–18.3	119.0	2.64	leadframe
Ni-Fe alloys (42 alloy)	4.1	147.0	0.15	leadframe
Au - 20% Sn	15.9	59.2	0.57	die bond adhesive and lid sealant
Au - 3% Si	12.3	83.0	0.27	die bond adhesive
Pb - 5% Sn	29.0	7.4	0.63	flip-chip
Au	14.3	78.0	3.45	wire metallurgy
Ag-loaded epoxy	53.0	3.5	0.008	die bond adhesive
epoxy (fused silica filler)	22.0	13.8	0.007	molding compound

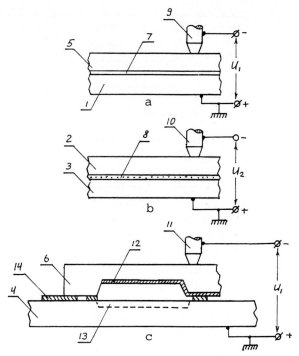

Figure 8.23 Field-assisted thermal bonding. a = glass-to-silicon bonding, b = silicon-to-silicon bonding, c = glass-to-silicon bonding with electrostatic shield; U_1 and U_2 = voltages of 1200 and 50V, respectively, 1 through 4 = silicon parts, 5 and 6 = Corning borosilicate glass, 7 = SiO_2, 8 = sputtered borosilicate glass (Corning 7740), 9, 10, and 11 = negative electrodes, 12 = metal electrostatic shield in etched well, 13 = circuitry, 14 = aluminized line.

glass (Iwaki Glass Co., #7570) is 440°C. The glass was deposited on a wafer by magnetron RF sputtering at a total pressure of 6mtorr and 30 percent oxygen concentration in argon ambient. The thickness of the layer was between 0.5 and 4μm, and the minimum voltage was 30V for 10 min. The pressure creating a contact between the bonding surfaces was 160kPa (the larger the pressure applied, the lower the bonding voltage required). This method can be used for bonding the glass-coated silicon to an aluminum film or an indium tin oxide film on a glass substrate.

If the bond is good initially, its interface is stronger than the constituent parts. The condition of the glass surface for bonding is critical. The surface must be smooth and clean. The glass should have the right composition.

The sputtering rate of 7740 glass should not be very high; otherwise, the Na^+ ions do not deposit on the substrate and the main bond mechanism can be missed. The deposit of 4μm film can take hours.

GaAs-to-glass electrostatic bonding [129] employs the hydrogen plasma treatment of GaAs and glass prior to the bonding. This treatment reduces the native oxide on GaAs and oxygen-contained residues in the surface of the glass. The glass has a thermal expansion coefficient nearly equal to that of GaAs, and the wafers are cut from bulk GaAs and mechanochemically polished. Chemical cleaning includes (1) degreasing in a boiling solution of ethanol-trichlorethylene-acetone; (2) deoxidation in HCl/ethanol (1:10); (3) etching in a $H_2SO_4 : H_2O_2 : H_2O$ (3:1:1 by volume) solution for 2 min; (4) rinsing in flowing $15\Omega \cdot$ cm deionized water; and (5) drying by blowing with N_2 and subsequent transferring to the plasma treatment chamber. The hydrogen plasma treatment (300 to 500°C) is carried out at an RF power of 500W,

10MHz and pressure in the reactor of 0.5torr. The treated surfaces of specimens of glass (DM-308) and GaAs are placed against each other, and a cathode electrode is held against the outer surface of the glass. The assembly is heated in a chamber in a nitrogen atmosphere at 180 to 500°C and at voltages of 20 to 1000V between the surfaces.

Wire bonding is needed to connect the metal pads on a chip to the various metal terminals that are parts of the sensor package. The most common methods of lead attachment are thermal compression bonding (Fig. 8.24) and ultrasonic bonding (Fig. 8.25). Typically, 20-μm gold or aluminum with 1 percent silicon wires are used for the attachment. Figure 8.24 illustrates the sequence of steps [103] of the so-called

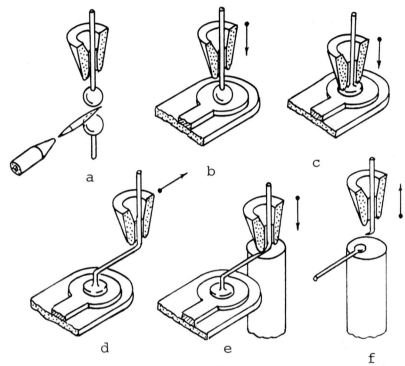

Figure 8.24   Sequence of steps in thermocompression bonding of fine wire. a = hydrogen flame or electronic spark forms ball at wire in glass or tungsten-carbide capillary, b = chip is heated (between 150 and 300°C), capillary descends, c = pressure is developed on ball and bonding is performed, d = capillary rises, forms wire loop, and now wedge bond can be performed anywhere on 360° area around ball bond, e = wedge-shaped bond is produced when capillary deforms wire against lead, f = capillary rises off lead, leaving bond. Now wire is prepared for forming new ball on its tail.

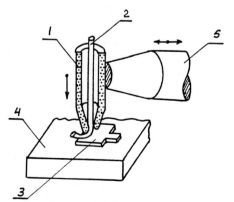

Figure 8.25   Ultrasonic bonding of fine wire. 1 = capillary tip vibrating at 20 to 60kHz, 2 = Au or Al with 1% Si wire, 3 = bonding pad, 4 = substrate, 5 = ultrasonic head.

tailless ball-and-wedge bonding cycle. A thermosonic bonding process is accompanied by ultrasonic vibrations that break through any metal oxide.

Other Technologies

Deposition of noncrystalline silicon films is one of the present trends in the fabrication of sensors [130–33]. Wide application of polycrystalline (poly-Si), microcrystalline (μc-Si), and amorphous (a-Si) silicon films is quite typical.

Poly-Si technologies are well developed for IC's. Their application simplifies the integration of mechanical elements (diaphragm, beams) with electronic elements (e.g., p-n junctions). A number of polycrystalline structures exploit porosity and grain-boundary properties in gas and temperature sensing. Amorphous Si has a wide photovoltaic application. In some cases (e.g., gate electrode of MOS transistor), the properties of poly-Si are not critical to the functioning of the device. However, if the device's feature size is decreased, nonuniformity in deposition can significantly affect the sensor's behavior.

The difference between single-crystal silicon and polysilicon is the presence of grain boundaries in the latter. A single-crystal silicon's mechanical properties depend on the crystallographic orientation. In the polysilicon, these properties are averaged over the domain of grain orientation.

Poly- and μc-Si films can be created by evaporation, HF-sputtering, chemical vapor deposition (CVD), or by CVD-fabricated a-Si followed by laser annealing [130]. A CVD process is commonly used for these films. It allows high deposition rates at relatively high pressure (0.1–1 bar). Evaporation is a high-energy process accompanied by a high temperature and inconvenient scattering of the Si evaporant. As a rule, sputtering and evaporation provide an a-Si layer if substrates have a temperature not greater than 600°C. Molecular beam deposition is an expensive procedure.

Poly-Si for sensors is usually deposited on a dielectric layer, typically on silicon dioxide and also on quartz glass, soda lime glass (softening temperature 400°C), Al ceramics, Si wafers, and metallic springs such as CuBe/AgPd/CuBe. The metallic springs are covered by an insulating layer of thermally grown or plasma-oxidized (Plasmox) SiO_2.

The formation of the silicon film structure can be explained by the following mechanism. In the initial stages of deposition, the structure of the film is mainly determined by the surface diffusion of the depositing material. Factors stimulating the surface diffusion help to form an ordered structure. Arriving silicon atoms combine with adsorbed ones, and form clusters and stable nuclei, which grow together and create a continuous film. The surface migration of atoms is increased with the increase in temperature, and the atoms are more likely to join an existing nucleus than to form a new cluster. The main factors determining surface migration and the growing of film are the deposition temperature, its rate, and its pressure. Higher temperature provides more energy for the migration process, the rate of deposition affects the time of migration, and the pressure, correlated with the presence of foreign atoms, opens or blocks the path for surface migration.

The amount of surface migration of the adsorbed atoms determines the structure of the continuous film.

An amorphous rather than polycrystalline film is formed when a limited surface migration is possible. At lower temperatures and pressures, the transition between the deposited amorphous and polycrystalline silicon occurs. At atmospheric pressure, the transition takes place at 680°C; in a low-pressure (0.2torr or 27Pa) CVD system, it occurs at 580°C.

At the lower deposition rate, the adsorbed impurities (especially O_2) decrease the surface migration and can also be adversely incorporated into the structure.

As mentioned above, a chemical vapor deposition (CVD), based on the pyrolysis (thermal decomposition) of silane (SiH_4), is usually used for forming a thin layer of Si. For the thicker-layer and higher-deposition rates, chlorinated compounds are employed (dichlorsilane, SiH_2CL_2; and silicon tetrachloride, $SiCl_4$).

Films deposited at 580°C in an amorphous form are crystalline after annealing at 800°C; they can crystallize during long heat treatment at 600°C.

Annealing reduces grain-boundary effects and provides a more uniform structure in films heavily doped with phosphorus (P). Excess P can degrade the properties of poly-Si film.

A typical LPCVD process for poly-Si [130] is characterized by the following: temperature of 630°C, pumping down to 5mbar, and cleaning in a flow of N_2, insertion of 100% SiH_4 at 360mbar with mass flow of 50SCCM. The deposition rate is 100Å/min, and the inhomogeneity of thickness on a 75-mm wafer is 2%. For this process, carrier gas is not required because of relatively high silane pressure.

A PECVD process allows obtaining poly-Si at temperatures above 450°C. At temperatures smaller than 350°C, a-Si is generated. This layer can be turned into poly-Si by laser annealing (wavelength of 530nm) or with halogen lamps. For a low-temperature process, a light-assisted recrystallization is preferable. RF power density in PECVD is 0.05W/cm³, silane flow is 20SCCM, Ar flow is 180SCCM, silane partial pressure is 0.25mbar, and deposition rate is 180Å/min.

It is possible to deposit a-Si and μc-Si at temperatures as low as 80°C using chemical transport in a low-pressure plasma; however, this process needs precision control since it is very sensitive to the fluctuations of parameters.

Several characteristics are of particular importance for the boron-doped LPCVD polycrystalline film used in pressure and temperature sensors; they are resistivity, temperature coefficient of resistance, gage factor, long-term stability, along with purely mechanical characteristics.

A typical layer for the sensors [131] has 0.5μm thickness; it is deposited onto an oxidized silicon wafer with an oxide thickness of 0.1μm. The layers are boron-doped using ion implantation with a subsequent annealing in N_2 at 950°C for 30 min. The layers are patterned by wet chemical etching using negative photoresist as an etch mask. Metallization is accomplished with aluminum vapor deposition and subsequent photolithographic patterning. Finally, the wafers are annealed at 470°C for 20 min. in an N_2.

The resistivity of the layers is always higher than that of a single crystal and changes from about 0.3×10^{-3} to $1 \times 10^3 \Omega \cdot$ cm as doping concentration varies from 1×10^{20} to $1 \times 10^{18} cm^{-3}$. Over a temperature range from -60°C to 160°C, depending on the implantation dose, the resistance change is not linear. Through selective doping, the temperature coefficient of the resistance α_R may be selected over a wide range, both positive and negative (from $+1.7 \times 10^{-3}$ to $-25.0 \times 10^{-3} K^{-1}$ for doping concentrations from 1×10^{20} to $1 \times 10^{18} cm^{-3}$, respectively). The temperature dependence increases with the decreasing doping concentration.

Under tension and compression, the resistance of the film increases and decreases, respectively. The resistance change is inversely proportional to the doping concentration. It is remarkable that, at the doping concentration of about $1.1 \times 10^{19} cm^{-3}$, $\alpha_R = 0$, the gage factor is of about 30, which is roughly 15 times as large as for metal gages and is 1/3 of that for diffused piezoresistors. This factor reduces with temperature, and its temperature coefficient lies between -2×10^{-3} and $-1 \times 10^{-3} K^{-1}$.

Surface effects play only a secondary role in the film characteristics; this is why long-term stability is high even for nonpassivated resistors. The drift of the resistance at a temperature of 125°C over a time period of 1000 hours is less than 5×10^{-3}. A plasma deposition of Si_3N_4 will result in an improvement of stability.

Temperature characteristics can be easily varied by a change in the boron concentration and by the connection of a trimmed resistor parallel or in series with the temperature-sensitive one.

Surface diffusion of the absorbed atom during deposition influences the mechanical characteristics of the film (e.g., response to stress). This diffusion is determined by the deposition temperature, rate and pressure, and the presence of impurities (O_2) retarding the motion of the absorbed atoms.

Poly-Si characteristics can be significantly affected by the grain boundaries, as well as by the differently oriented crystallites in the film.

The elastic constants of this material are mainly defined by the grain nature rather than the grain boundaries' properties. For example, Young's modulus is 150 to $170G(N/m^2)$ for polysilicon and $190G(N/m^2)$ for single-crystal silicon. By contrast, deformation and fracture are greatly affected by grain boundaries. The grain boundaries block the propagation of dislocations in grains, but, on the other hand, they are weak locations in the structure. Poly-Si grain sizes range from 0.02 to $20\mu m$; for μc-Si, they are below $0.02\mu m$. In a μc-Si the small grains are embedded in a matrix of a-Si.

The fracture strength of polysilicon is 2 to $3G(N/m^2)$. Grain boundaries in poly-Si significantly decrease thermal conductivity, which was found to be $0.3W/cm \cdot K$ in a LPCVD polysilicon film (compare with $1.4W/cm \cdot K$ for a bulky material).

One of the most notable features of poly-Si films is their ability to sustain temperatures above 200°C, while still retaining some features of crystalline Si. The following characteristics of a-Si are essential (some of them are not obtainable in crystalline materials): (1) a high optical absorption coefficient in the visual range; (2) ability to obtain crystallike properties by mixing microcrystalline and amorphous phases; (3) low deposition temperature (200 to 300°C) and possibility of using organic films for substrates; (4) uniformity of the films deposited over a large area; (5) stability and reliability of films over years; (6) good quality of films deposited on curved surfaces; and (7) possibility of using photolithographic and etching [134] processes.

The other *amorphous* material becomes popular in sensor structures. It is *magnetic material* [132] which is characterized by (1) the lack of the magnetic anisotropy of a crystal, high permeability, and low loss; (2) high resistivity (much higher than in permalloy); (3) great mechanical strength due to the lack of dislocations, grain boundaries, etc.; and (4) the Curie temperature of 200 to 500K.

Among the various technologies, *thick-film deposition* offers flexibility in the choice of materials and design and easy integration with electronic circuits and in packaging [135, 136].

In the production of passive components, specially formulated pastes or inks are applied onto a ceramic or insulating substrate in a definite pattern and sequence. The pastes are applied by *screen printing* and then fired in an atmosphere of air or nitrogen, forming films of 10 to $20\mu m$ thick. The material of the pastes includes a primary functional component (metal, metal-oxide, semiconductor, or dielectric) and bonding agent (glass frit or metal oxide). These components are prepared in particle form and suspended in an organic vehicle.

In order to define a pattern, a screen made of stainless-steel woven mesh with open areas is positioned at a precise distance above the substrate. The ink is spread on the screen by a spreading bar and is forced to penetrate through the open areas in the screen, thus forming the desired pattern on the surface of the substrate. The screen is then lifted away from the substrate, leaving the ink in place. The viscous properties of the ink are such that it is prevented from flowing away from the predetermined contour; however, it can fill in the gaps left by the wire mesh. After setting, the ink is dried (at about 150°C), and then the patterns are fired to be converted into a usable film. There is a time-temperature plateau (at about 350°C) in the early part of the firing cycle when the organic vehicle is burnt off. Depending on the type of ink, the firing temperature and firing period vary to provide proper chemical reaction and glass flow. After quenching at room temperature, the ink's material becomes part of the substrate.

One interesting alternative to screen printing is forming a thick-film pattern by attaching an adhesive film to the substrate's surface with the subsequent firing cycle similar to that just described. This film, similar to Scotch tape and *transfers,* carries printed patterns of the material to be formed onto a substrate. By a simple pressing of the tape against the substrate's surface, the pattern is transferred to the substrate which becomes ready for firing [137, 138].

The functional characteristics of the films depend on the nature of the primary components and on their size, volume fraction, and chemical environment. The screen printing process and choice of substrate material along with variations of the materials give a great flexibility to the technology. Other features of thick-film technology are (1) the possibility of tracking the parameters of the fabricated components by means of an inexpensive automatic technique (trimming operations such as sandblasting and laser trimming), (2) ease of integration of the transduction element with IC's, and (3) convenience in automation and mass production.

Some remarks should be made on the change in some characteristics of the sensitive layers as compared to corresponding sensitive elements obtained through different technologies. For example, the reduced sensitivity to the change of the physical values (e.g., film resistivity versus temperature) can be explained by the scattering of the electrons in the layer having some microdefects (grain boundaries, porosities), or by the influence of the impurities added to the film for improving sintering at low temperature, and by other factors. The following is a brief description of several thick films whose technologies and applications are successful or potential.

Piezoresistive thick films based on RuO_2-Ruthenates [136] exhibit high gage factors (10 to 18) and a low temperature coefficient of both resistance ($+20$ to $50ppm/°C$) and gage factor (200 to $300ppm/°C$). The thick-film gages have a good long-term stability for the unstrained and strained states; they are immune to contamination and mechanical roughness and can operate in a wide temperature range.

The application of thick films in the constructions of *temperature sensors* (Table 8.6) has a number of advantages. Among them are ease of integration with IC's, a wide choice of shape and size, and the possibility of achieving tight tolerance by a simple trimming of resistance. Resistances of these films are functions of

TABLE 8.6

Characteristics of Thick-film Materials for Temperature Sensors (Adopted from [135])

Sensor	Main ink component	Sensitivity	Range		Notes
Thermocouple	PdPtAu alloy AuPd alloy	—	—		—
Thermopile	Au + PtAu alloy	$\sim 20\mu V/°C$	-200 to $+500°C$		—
Thermistors	Mn-Co-Ru Spinel type	$B = 2500$–$5000K$	-50	$+200°C$	$\rho = 10^4$–$10^6\Omega \cdot cm$
Thermistors	Mn-Ru Spinel type	$B = 2000K$	-50	$+200°C$	$r = 10^3\Omega/\square$
RTD	ZrO_2-RuO_2-RhO_2	$\alpha = -3000ppm/°C$	20	120°C	NTC
RTD	ZrO_2-RuO_2	$\alpha = 1270ppm/°C$	20	120°C	PTC
RTD	Au	$\alpha = 2700$–$3400ppm/°C$	0	700°C	$r = 10m\Omega/\square$
RTD	Pt	$\alpha = 3700$–$3900ppm/°C$	0	200°C	$r = 80$–$90m\Omega/\square$
RTD	Au	$\alpha = 2840ppm/°C$	0	700°C	$r = 2$–$3m\Omega/\square$
RTD	Pd, Ag	$\alpha = 275ppm/°C$	0	700°C	$r = 25$–$35m\Omega/\square$
RTD	Ni	$\alpha = 5290ppm/°C$	0	200°C	$r = 90$–$120m\Omega/\square$
Capacitor	$BaTiO_3$ + $SrTiO_3$	$\Delta C/C = 65\%$	5	100°C	reactance *vs.* T
Capacitor	V_2O_5 + Ag	$R(75°C)/R(60°C) = 0.5$–3	—		sharp ΔR at 68°C

temperature and are characterized by positive or negative temperature coefficients (PTC and NTC respectively).

CdS and CdS/CdTe *screen-printed films* are used for photovoltaic (with a 12.8 percent efficiency) and photoconductive conversion [139–42].

Nickel-based, screen-printed, thick-film conductors reveal magnetoresistive effects with a maximum relative change in resistance of about 0.6 percent [143]. Sensitivities of the film to the magnetic field and temperature are affected by the peak-firing temperature and firing atmosphere.

Various *semiconductive-oxide inks* have been developed for sensing humidity (Table 8.7). The resistivity of the thick films for measuring humidity is a decreasing logarithmic function of H_2O partial pressure. The resistance layers are sensitive to relative humidity (RH) over the range 5 to 95 percent RH. Due to the specific advantages of thick-film technology, the integration of the humidity-sensing film with an on-chip heater and temperature-sensing element is easily provided. The heater is used for a rapid regeneration of the film after exposure to heavy humidity, and the temperature sensor is employed for temperature compensation.

Thick films sensitive to gas concentration are generally semiconductor oxides (Table 8.8). They are often associated with a catalytic metal whose role is to improve the sensor's sensitivity to one or several reactions.

A *combination of several thick films* based on different materials can be used for measuring the gas concentration of a mixture of different gases. For example [144], films made of SnO_2, WO_3, and $LaNiO_3$ with a Pt-based thick-film heater are applicable for detecting the amount of CO and C_2H_5OH (heating gas).

Some of the *thick-film* semiconducting layers (e.g., TiO_2) have a selective *sensitivity to* the *oxygen* partial pressure [145]. The resistance of such a film changes (see Chapter 2, Electrode Elements) as a function of the oxygen content. It is remarkable that, during the sensor's operation, this film is exposed to a very severe environment (combustion engine exhaust gas) and can withstand it.

Thick films for polarographic oxygen sensors are fabricated [146] by depositing gold lines, silver electrodes, dioxide glazes, and electrochemically formed silver-chloride layers on an alumina substrate.

Advances in the application of thick films can be envisaged in devices employing ion-sensitive doped photoresists, polymeric membranes entrapping solutions, or enzymes for the measurements of substances such as urea, glucose, or penicillin [147].

TABLE 8.7

Characteristics of Thick-film Materials for Humidity Sensors (Adopted from [135])

Main ink component	Type	Sensitivity $\Delta R/R\%$ or $(\Delta C/C)\%$/RH%	Temperature (°C)	Notes
$ZrCr_2O_4$ + glass	R	9.3	−10 to +40	$\rho = 6.2 \times 10^7 \Omega \cdot cm$
$MnWO_4$ + V_2O_5	R	8.8	−10 +40	$\rho = 6.2 \times 10^7 \Omega \cdot cm$
$NiWO_4$	R	8.2	−10 +40	$\rho = 8.0 \times 10^4 \Omega \cdot cm$
$MgCr_2O_4$ + glass	R	8.3	−10 +40	$\rho = 8.3 \times 10^7 \Omega \cdot cm$
$CoAl_2O_4$	R	3.1	−10 +40	—
glass ceramic + Al_2O_3	C	2.5 (50Hz)	+5 +35	$\Delta C/C\Delta T < 1\%/°C$
standard dielectrics	C	<1	+5 +60	60 · 100% RH
$BaTiO_3$ + glass	dew point	2500$^+$	+5 +60	$\rho = 9.0 \times 10^6 \Omega \cdot cm$
RuO_2 + glass	dew point	3500$^+$	+5 +60	$\rho = 9.0 \times 10^6 \Omega \cdot cm$

R = resistive sensor, *i.e.*, resistance R is a function of relative humidity RH.

C = capacitive sensor, *i.e.*, capacitance is a function of RH.

+ = R(75%RH)/R(at dew point).

TABLE 8.8

Characteristics of Thick-film Materials for Sensing Gas Concentration (Adopted from [135])

Main ink component	Test gas	Sensitivity $\Delta R/R$ (%) at 1000 ppm	Operating temperature (°C)	Notes
$SrFe^{3+}_{0.7}Fe^{4+}_{0.3}O_{2.65}$	ethanol	0.1	470	—
SnO_2 + Pd	$CH_4/CO/C_2H_5OH$	2/1.1/8.3	400	—
WO_3 + Pd	$CH_4/CO/C_2H_5OH$	1.5/9/25	400	—
ZnO + Pd	$CH_4/CO/C_2H_5OH$	1/2.2/17	400	—
SnO_2ThO_2 + hydrophobic silica	CO	90(500ppm)	200	$\Delta R/R\Delta T = 3.3\%°C^{-1}$ $\Delta R/R$ = 2.2% per % RH, selective with respect to H_2
SnO_2 + Pd, Pt	$C_2H_5OH/CO/H_2$	−60/−2/+45	250	—
SnO_2 + Pd, Pt	$C_2H_5OH/CO/H_2$	−90/−5/−50	400	—

A *maskless ion implantation* is found effective [148] for the fabrication of miniature Hall sensors. In this process, a very sharp ($\sim 0.3\mu m$ in diameter) ion beam of Si^+ is focused on the surface of a (100)-oriented semi-insulating GaAs wafer. The beam is extracted from a liquid-metal ion source with Au-Si eutectic alloy. During the implantation, two orthogonal lines are drawn with the ion beam. The overlapped region of the two lines forms the sensing region. The line dose used is from 8.0×10^7 to 1.5×10^9 ions/cm². The contact region is also implanted with the same Si beam to a dose of 5×10^{13} ions/cm². Then the wafers are annealed in an N_2 ambient at 850°C or 900°C for 1 minute. Finally, the electrodes of Au-Ge defined with a lift-off technique are metallized and alloyed.

Si_3N_4 *films of a thickness reaching* $300\mu m$ can be formed [149–51] *by nitradation of porous silicon.*

Through selection of the appropriate starting density and morphology of porous silicon, the process can produce either encapsulated pores, porous nitride, or solid silicon nitride. The starting material is porous, and the volume expansion during the conversion to silicon nitride is into the voids, thereby reducing the residual stress. Porous silicon is monocrystalline p- or n-type silicon which contains a very high density of micropores or channels etched into silicon during anodization in hydrofluoric acid. To obtain a solid nitride film, a silicon density of 75 to 80 percent is required. If the density is much less than 75 percent, for example, 50 percent, a silicon nitride layer becomes porous. Good results are obtained on 10 to $20\Omega \cdot cm$, p-type (100) Si anodized in a 49-weight-percent HF at the formation current density of $80mA/cm^2$ for 20 min. Nitridation is performed in a furnace with ammonia at 0.1atm for 1 hour at a formation temperature of 725°C.

Besides the sacrificial layers in micromachining technology, *porous silicon* is used [152, 153] in humidity, liquid, and gas sensors and also in microchromatography packing support. Computer-generated simulations of porous propagation during the anodization in HF solutions give a good correlation with the data obtained by Transmission Electron Microscopy. The precis of a model for computer simulation is that pore growth is rate limited by the diffusion of a reactant, holes, or an electrochemical equivalent from the bulk silicon to the silicon/solution interface.

Three-dimensional electron-beam lithography [154] is realized by a fine focusing of an electron beam and a delineation micron-order resist pattern on the three-dimensional substrates.

A *local oxidation of silicon* (LOCOS) *process* [155] *for micromotors* is distinct from the process described above due to the employment of a LOCOS step prior to the standard surface-micromachining deposition and patterning of the structural and sacrificial layer. Local oxidation of silicon is accomplished by depositing and patterning a thin Si_3N_4 layer. This layer resists the diffusion of oxygen, preventing the thermal oxidation of the substrate below Si_3N_4, thus selectively masking the growth of the thermal SiO_2. In general, the process consists of alternatively depositing and patterning structural/conducting, sacrificial, and insulating layers. By immersion in a highly selective wet etchant (HF), the sacrificial layers (SiO_2) are removed, releasing the structure for motion.

Laterally *mobile structures* allowing rotation and sliding motion are used in sensors and actuators. The fabrication process of these structures [156] uses *lamination of two oxidized wafers* to form a sacrificial layer that will attach the mobile structure to the substrate. The wafers have an oxide layer approximately $2\mu m$ thick in between. During the process, the upper wafer is thinned and oxidized. After lithography, a remaining masking film defines the lateral geometry. The side walls of the mobile structure are determined by etching. The revealed oxide is removed and an overlay is laminated to the substrate surface. In a final step, the sacrificial layer is dissolved in a HF bath, and the mobile part is released for motion.

A technique named *ion shower doping* is proposed for forming a highly doped amorphous-Si n^+ layer in the fabrication of pin-type photodiodes [157]. The hydrogenated amorphous silicon (a-Si:H) films have been widely employed in photosensitive devices. In these applications, highly doped p- or n-type thin layers and low defect density are required. The conventional technique for this process is gas phase doping into a-Si:H films by plasma chemical vapor deposition (PCVD). The ion shower doping enables doping impurities selectively by low energy ions. A very simple apparatus develops a highly excited plasma that is generated in the ion source using both RF electric field and magnetic field. A mixture of phosphine (PH_3) and hydrogen (H_2) is used as the impurity gas source, which is introduced into the ion source and discharged at a pressure of about 5×10^{-4}torr. Impurity ions are accelerated by the potential difference of a few kilovolts. The ions form a shower of ions into a-Si:H films without mass separation. Selective doping can easily be achieved by this technique.

A *scanning tunneling microscope* (STM) can be used for depositing *nanometer-dimensioned wires* on a substrate [158]. The wires are created by atoms of gold flying off the tip of the probe, propelled by an applied electric field. When the distance between the emitting tip and the substrate is of macroscopic order, the difference in potentials required for deposition reaches thousands of volts. STM's can get within a few nanometers of the substrate. At those distances, the electric fields bonding atoms to the tip overlap with those of the substrate, and only a short 3-volt pulse is required to knock the atoms off. This process is fast and reproducible.

An argon-ion laser can be used [159] for *deposition of doped silicon lines from* mixtures of *silane and diborane*. Piezoresistive properties of the lines can be of interest for the construction of mechanical sensors. The lines are deposited on sapphire, using a gas mixture that contains 1 percent diborane in silane. These lines exhibit polycrystalline characteristics with the gage factor less than about 20.

The lines deposited on SOS substrates show anisotropic characteristics in their piezoresistivity properties with a maximum gage factor of 44. A higher gage factor of 77 can be obtained by using a gas mixture of 0.1 percent diborane in silane. The deposition of silicon strips is carried out by the pyrolytic deposition of silane, SiH_4, in a reaction chamber using the visible radiation (488 and 514.5nm) of an argon laser. The laser's beam is focused onto the sample through a window of the chamber. The diameter of the focused spot is around 8 μm. With this method, a very localized deposition of different materials is possible.

A *capacitive pressure sensor* can be *fabricated* [160] by a batch process using a *heat-decomposition* material *(poly-α-methylstyrene)*. The sensor's structure is composed of a photosensitive glass substrate, a lower electrode (Au/Ti), an insulator (SiO_2), and a metal diaphragm (Ni). The photosensitive glass is used in order to prepare a pore on the back side of the decomposition material. The fabrication process starts with depositing the lower electrode and then insulator on the substrate. Next, poly-α-methylstyrene is coated on the insulator and is patterned in a particular shape with the usual photolithographic technique. After this, Ni film is deposited by electroplating at 75°C. The pore in the glass is formed by immersing it in a dilute HF solution. Finally, poly-α-methylstyrene is decomposed in a vacuum at 180°C with generation of a monomer gas that is removed from the cavity through the pore. The metal diaphragm becomes movable and works as a second electrode of the capacitor.

Tactile imagers [161] for precision robotic applications are fabricated by a *dissolved-wafer process* using diffused bulk-silicon row lines and metal-on-glass columns which offer the simplest steps, fast response, and high-force resolution.

Bulk and surface *micromachining of GaAs* structures [162] has potential for further micromechanical development of III-V-compound sensors—for instance, in applications where mobile microstructures are combined with lasers, photodetectors, or piezoelectric devices. Among the advantages of GaAs in comparison with silicon for sensor application are its nonzero piezoelectric coefficient, its higher Peltier coefficient, its direct band gap facilitating efficient two-way conversion between electrical energy and light, its higher radiation hardness, its lower stiffness, and its wider operative temperature range. However, GaAs is more expensive than silicon, has a lower degree of crystalline perfection, is more difficult to process, and has lower mechanical strength. It is believed that with the rapidly progressing technology of GaAs, its mechanical strength can be rather high if micromachining is properly performed. There are two basic schemes of GaAs micromachining: anisotropic etching of bulk GaAs and selective etching using $Al_xGa_{1-x}As$ as an etch-stop [163, 164]. Surface micromachined mobile structures can be produced in GaAs by a sacrificial layer technique similar to that used for polysilicon. In bulk micromachining, ordinary photolithography is used. Masks of negative photoresist can be applied at one or at two sides of a wafer and aligned. The etchants are (1) Br_2 (2 vol. % and $\geq 99.8\%$ purity) mixed with CH_3OH ($\geq 99.5\%$ purity); (2) the mixture of H_3PO_4, CH_3OH, and H_2O. The first etchant gives well-defined [111] side surfaces on (001) wafers. The second etchant produces a flat and well-defined etch-pit bottom but not-so-well-defined sides. A possible fracture strength of carefully micromachined Si bars can be of the order of $4-8G(N/m^2)$.

An epilayer etch stop with GaAs and $Al_xGa_{1-x}As$ layers is used for an isotropic, nonselective etching using the 5 $H_2SO_4 : H_2O_2 : H_2O$ etchant.

For surface micromachining, hydrofluoric acid (HF) is known to etch $Al_xGa_{1-x}As$ selectively with selectivity relative to GaAs of 1×10^9 or higher. The process scheme is basically the same as for surface micromachining of polysilicon structures but in a different material system.

PECVD techniques are used to *deposit polycrystalline diamond films* [165] that have properties identical to those of natural diamond (hardness, high thermal conductivity, electrical insulation). The films are produced using low-pressure, plasma-enhanced CVD techniques, which typically employ a methane and hydrogen-feed gas stream at temperatures between 600°C and 1000°C in a special reactor system.

A plasma-enhanced chemical vapor deposition (*PECVD*) can be used [166] for the *fabrication of a rigidly suspended diamond plate,* which is made in a self-aligned manner. The basic steps of the process are preparation of a wafer with diamond/SiO_2/Si_3N_4 and Si_3N_4 layers on the top and back, respectively; patterning and etching of the back; depositing of a photoresist on the top of the wafer; illuminating it from the back with the subsequent silicon etch from the top (self-aligning).

Zinc-oxide thin film exhibits good piezoelectric and pyroelectric properties [167, 168]. Highly oriented films can be deposited on SiO_2/Si, $SiO_2/Poly-Si/Si$, and Si_3N_4 substrates. For 1-μm-thick films, the piezoelectric coefficient $d_{33} = 14.4 \times 10^{-12}C/N$ and the pyroelectric coefficient $p^\sigma = 1.4 \times 10^{-9}C/cm^2 \cdot K$. In sensor designs, voltage responses are 5.2mV/g (piezoelectric effect) and 150mV/K (pyroelectric effect) at 300K. Planar magnetron sputtering is used for the deposition because of its low film damage due to electron bombardment and high deposition rate. It is found that the best thin-film crystallinity corresponds to the deposition conditions carried out at a forward sputtering power of 200W with a 10mtorr ambient gas mixture consisting of 50 percent oxygen and 50 percent argon, a substrate-to-target distance of 4cm, and a substrate temperature of 230°C during the deposition.

When the film is deposited on Si_3N_4, the basic steps are the following. Si_3N_4 deposition is carried out at 835°C over thermally grown oxide using low-pressure CVD with dichlorosilane to ammonia at a gas ratio of 5 to 1. An anisotropic etchant (EPW) is used to form the silicon-nitride-2μm diaphragm by etching the wafer from the backside.

Next, polysilicon electrodes are formed on the diaphragm over CVD SiO_2. After this, another layer of CVD SiO_2 is deposited, and, over this, a 0.3-μm layer of ZnO is sputter-deposited with the c-axis oriented perpendicularly to the plane of the diaphragm. Triode, dc and RF magnetron sputtering is used [169]. The process temperature is between 150 and 450°C and sputter rates are up to 20μm/h.

Aluminum is sputter-deposited after CVD SiO_2 is laid down; then, aluminum is patterned and sintered. Zinc oxide is a very reactive material so that further processing of wafers with ZnO must avoid the risk of contamination and of damaging the layer.

The after-the-deposition cleaning cycle of ZnO is composed [169] of (1) ultrasonic cleaning in acetone for 5 min., rinsing in demi water, and drying; (2) ultrasonic cleaning in toluene for 5 min., rinsing in demi water, and drying; (3) ultrasonic cleaning in dimethylsulfoxide for 5 min., rinsing in demi water, and drying. Rinsing in demi water must be brief since ZnO is soluble in water.

ZnO is attacked by all common acids and bases and also by photoresist developer. For quick removal of ZnO, any acid can be used—for example, concentrated HCL or fuming HNO_3. For pattern definition, several etchants are recommended: (1) 1 percent HCL (etch rate in μm/min., e.r., is 10); (2) 6g of NH_4CL + 4ml of NH_4OH + 30ml of H_2O (e.r. is 0.5); (3) 1ml of H_3PO_4 + 1ml HAc + 10ml of H_2O (e.r. is 1.5). Aluminum is not attacked by these etchants.

For patterning aluminum on top of zinc oxide without its etching at the same time, the following solution is used: 1g of KOH + 10g of $K_3Fe(CH)_6$ + 100ml of H_2O (its e.r. is 0.5). For the same combination of chemicals and 600ml of H_2O, e.r. is 0.2. Agitation by a magnetic stirrer for the ZnO and Al etching is required.

Silicon nitride layers on zinc oxide for its passivation is successfully performed by plasma-enhanced chemical vapor deposition (PECVD) in a parallel-plate reactor at 350°C. A gas mixture of 1000SCCM N_2, 10SCCM SiH_4, and 3SCCM NH_3 is fed into the reactor (SCCM stands for *standard cubic centimeter per minute*). The pressure in the reactor is kept at 1torr, the electrode spacing is 20mm, and RF power is 70W at 100kHz. The thickness of Si_3N_4 on ZnO layers reaches 200 to 500nm. Due to its easy application through spin coating, polyimide layers can also be applied on zinc oxide-covered wafers [170, 171]; however, it is less useful for application in sensors because of its large absorption of acoustic energy.

It is noteworthy that the film encapsulated by layers of SiO_2 has resistivity of $3 \times 10^7 \Omega \cdot cm$, relative permittivity of 10.3, and static charge decay time of longer than 32 days. These qualities make it possible to fabricate sensors with a near dc response (down to 0.15Hz) to the measurands. Since the film is sensitive to deformation and heat radiation, multifunction sensing devices can easily be built.

It is possible to use planar technology for fabricating an actuation mechanism [172] of a cantilevered bimorph consisting of alternating layers of metal and piezoelectric zinc oxide. Each metal layer is divided into two electrodes that can be individually addressed, thereby enabling the cantilever to move in three orthogonal directions. The bimorph consists of metal electrodes (0.5μm thick), insulating dielectric film (0.2μm thick), and piezoelectric ZnO (3.0μm thick). The ZnO films are prepared by reactive sputtering of pure zinc in O_2/Ar (80%/20%) plasma in a dc magnetron sputtering chamber.

Recent studies of ferroelectrics [173] have demonstrated that *ferroelectric film* of the lead-based pervoskite structure ferroelectrics can be deposited upon both silicon and GaAs substrates and annealed to the proper ferroelectric crystalline form. They have excellent piezoelectric characteristics and electrical strength, and they offer new possibilities for piezoelectric sensors.

Tin oxide films whose resistance is sensitive to organic vapors (such as ethyl alcohol and ethyl ether) can be integrated with polysilicon heaters [174] and fabricated on thin ($\sim 2\mu$m) silicon membranes generated by anisotropic wet chemical etching using ethylene diamine-pyrocatechol (EDP) as the etchant. This etchant has a fast etch rate ($\sim 80\mu$m/h) in the Si [100] direction, a slower rate ($< 10\mu$m/h) in the Si [111] direction, a very slow rate (< 20nm/h) for SiO_2, and a close-to-zero rate for heavily boron-doped silicon (doping concentration $\geq 5 \times 10^{19}$cm^{-3} generated by ion implantation). A boron-doped layer (implantation at dosage of $\sim 5 \times 10^{16}$cm^{-2} and implant voltage of ~ 200keV) is used as the etch-stop. Tin oxide thin film is prepared either by sputter-deposition [175] or by metalloorganic deposition (MOD) [676]. For the MOD technique, an ink is prepared by dissolving tin 2-ethylhexanoate in xylene. The ink is spun onto a silicon wafer, fired to form 100nm to 200nm tin oxide film, and then patterned either by reactive ion etching or by wet chemical etching.

An aluminum/chromium double layer with a thickness of 1μm/50nm is used to form metal interconnects for the tin oxide as well as the polysilicon heater.

Iridium oxide (IrO_2) *film sputtered* on alumina substrate is being considered [176, 177] as a potential pH electrode for high-temperature (above 100°C) solutions. Iridium is sputtered from a pure (99.99%) iridium, 5cm-diameter target with a 13.5MHz RF planar magnetron source at a 10cm target-to-substrate distance and approximately 0.40Pa total pressure. Argon and oxygen are mixed in a 1:1 ratio to oxidize the growing film. A 1μm-thick film can be obtained on an alumina substrate at temperatures of 30 to 40°C and 240°C.

There is a trend [177] to use *indium tin oxide* (ITO) rather than the currently used polycrystalline silicon as the electrode material in *solid-state image sensors*. Silicon as the electrode material is easy to control and process but has the disadvantage of absorbing light, which reduces the sensitivity of the sensor. ITO is both transparent and conductive, but the technology and theory have to be developed if it is to be used successfully. ITO electrodes are applied in a magnetron sputter-deposition system forcing indium and tin atoms out of an indium tin cathode by ion bombardment at low pressure in a plasma of argon and oxygen. The indium and tin atoms combine with the oxygen to form ITO, which is deposited on an oxidized silicon wafer. The pattern is produced using the same methods as in normal chip production: the ITO layer is etched away locally using photoresist and exposure in conjunction with masks. Good electrode contacts are obtained by developing a double layer of aluminum on titanium tungsten.

Oxide coatings for acoustic wave sensors [178] provide resistance to thermal or chemical degradation, large sorption capacities for chemical sensing, controlled microstructure, easy modification of the chemical nature of the surfaces, and minimal changes in the film's viscoelastic properties during sorption. *Sol-gel chemistry* associated with the coating process involves the hydrolysis and condensation of metal alkoxides to form inorganic polymers. By varying the reaction conditions, the structure of these polymers can be changed. Films are prepared by spin- or dip-coating with the subsequent forming procedure. Dense films are formed from weakly branched polymers, whereas high surface area porous films are fabricated from highly ramified polymers or dense colloidal particles.

The principles of *micromachining* are applicable to *crystalline quartz* to shape fine transduction elements for different sensors of mechanical values. Chemical etching [179–81] is preferable although other means are available (e.g., dry etching). The etchants used for quartz are warm solutions of hydrofluoric acid, HF, and ammonium fluoride, NH_4F [182]. The effect of the etching solution on a mechanically lapped surface differs from the effect observed on a chemically polished surface [183, 184]. The quality of the surface finish prior to etching is important for micromachining. The variations of etching rates (μm per hour) for different bath compositions are given in Table 8.9 [182]. It is noteworthy that the etching rate is always much faster along the z-axis and that both the etching and the anisotropy increase with the proportion of HF. When the amounts of NH_4F and HF are varied in etching solutions, the profiles of the etched wafer vary greatly as well. The etch rate has an exponential variation with temperature. For example, the etch rate for the concentration of NH_4F to HF around 1 is increased by factor 1.6 for every ten degrees for temperatures around 50°C.

Mechanical parts obtained by chemical micromachining have a good surface quality and display a mechanical strength higher than that fabricated by traditional means [180].

Anodization is an oxidation process that is performed in an electrolytic cell where oxidation and reduction take place at an anode and cathode, respectively. This process is primarily used to form quality thin dielectric films on certain metals [102]. The oxidizing parts are connected to the positive terminal of the power source (anode). The cathode (usually made of a noble metal) is connected to the negative terminal. The primary oxidizing agent is H_2O. Anodization is effective for the materials that form coherent films of metal oxides—for example, aluminum, tantalum, and silicon. The process is practical for producing protecting films, trimming resistors, or forming a capacitor dielectric. Tantalum and tantalum nitride are materials widely used in the anodization process. During the process, the current through the electrodes is kept constant with the increase in the oxide film, and the potential difference between the electrodes is increased.

The thick film of a conductor can be built up by *electroplating* where metal is reduced from metal salts dissolved in solutions. Plating takes place on the cathode, whereas material at the anode is etched. A typical gold-plating process is described by two reactions:

1. At the cathode: $Au(CN)_2^- + e^- \rightarrow Au + 2CN^-$
2. At the anode: $Au + 2CN^- \rightarrow Au(CN)_2^- + e^-$

The current density at an electrolytic cell is directly proportional to the film thickness, time, metal ionic charge, film density, and Faraday constant. It is inversely proportional to the molecular weight.

TABLE 8.9

Variations of Etching Rate (Etching Solution is at 25°C)

Bath composition (mole/L)	Etch rate in $\mu m/h$*		
	R_x	R_y	R_z
10.9 HF	0.020	< 0.005	9.60
7.2 HF + 4 NH_4F	0.025	0.005	2.55
5.4 NH_4NH_2	0.015	0.015	1.10
5.4 NH_4HF_2 + 1.8 NH_4F	0.015	0.015	0.75

*R_x, R_y, and R_z are etch rates along x, y, and z axes, respectively.

The solution composition, pH, current density, temperature, and agitation are the important process parameters.

Conducting polymers deposited on a gold layer have a gas-sensing application [185]. The substrates are fabricated by sublimation of gold onto alumina following baking in air at 200°C for 12 min. Electrochemical polymerization of N-methylpyrrole onto the substrates is followed by evaporation of a second metal electrode on top of the polymer film. The prepared constructions are gold/poly (N-methylpyrrole) perchlorate/gold- and gold/poly (N-methylpyrrole) perchlorate/indium-sandwiched structures.

Polyimide (PI) *films* are extensively used as structural and sensing components of various microsensors, especially as intermetal dielectrics, passivation layers, sacrificial layers, and *surface planarizers* [186]. One of the goals of microelectronic technology is to minimize the use of high-temperature processing and wet-etching chemistry. PI can be plasma-etched and fully treated at temperatures of 300 to 400°C. After curing, it is chemically inert, mechanically tough, flexible, and thermally stable up to 450°C. PI provides excellent planarizing of irregular surfaces. It has a low relative permittivity (3.2 to 3.4) at audio frequencies and low loss over a wide frequency range. The hygroscopic nature of PI allows its use in relative humidity sensors, but this property turns into a disadvantage since it leads to increased insulator conductivity, loss of adhesion, and corrosion. Sensitivity to variation in preparation procedures of PI's and ionic contamination can also be problems. The investigation performed with the polyamic acid precursor of pyromellitic dianhydride (PMDA) and oxydianaline (ODA) combined N-methyl pyrrolidone (NMP) solvent shows that the PI's interface with silicon is stable, whereas the PI's interface with air changes significantly.

The study of creep and recovery of PI thin film is essential for understanding the viscoelastic properties of the material. The use of circular membranes in measuring the properties has some advantages. The behavior of one commercially available PI was studied in detail [187]. This PI is the imidized form of the poly-amic acid precursor that results from the reaction of benzophenone-tetracarboxylic dianhydride (BTDA) with a blend of metaphenylenediamine (MPDA-40%) and oxydianiline (ODA-60%). It was determined that the creep compliance of the material is a very nonlinear function of the stress and also requires a threshold strain before creep begins.

A *square PI diaphragm fabrication* process [188, 189] was developed for the verification of the load-deflection behavior of elastic membranes and the adhesion of thin films. The process begins with forming a p^+ etch stop in a silicon wafer by boron deposition at 1175°C for 120 min. in an environment of 90% N_2 and 10% O_2. Then, a thermal oxide is grown at 990°C for 75 min. (15 min. dry O_2, 45 min. steam, 15 min. dry O_2). This step gives a resulting oxide thickness of 3200Å. A standard photolithographic technique and 50/50 hydrazine/water solution is used to form a thin (4.7μm) diaphragm. The polyimide (BTDA/ODA/MPDA) is spin cast on the wafer in multiple coats with a prebake in air at 135°C for 14 min. after each coat. Finally, the film is cured in nitrogen at 400°C for 45 min., and the silicon diaphragm supporting the film is etched away in SF_6 plasma, thus forming a free-standing membrane of 5 to 11μm thickness.

High-temperature (up to 900°C) *ceramic sensors* of mechanical quantities are *fabricated* [190] from a fine-grained, high-purity alumina (>99% Al_2O_3), which is characterized by a high strength, nearly constant Young's modulus, and high volume resistivity between room temperature and 1000°C. Molding and machining of semi-products before sintering are the typical operations in the preparation of plates, rods, tubes, and other shapes. Grinding and boring parts with high-speed tools as well as microsandblasting are rather time-consuming and difficult to apply for fragile structures' operations. Laser machining offers essential advantages and is characterized by a reasonable speed in machining, low reaction forces, flexibility in operation, and automatic control. Tolerances of about $10\mu m$ can be achieved by this technique.

Thin *sputtered-platinum metallization* ($1\mu m$) is a common technique for creating electroconductive layers on ceramics for sensors operating at high temperature. These layers are resistive to corrosion, have small differences in coefficients of thermal expansion with respect to alumina, and have high ductility.

Reaction soldering is the operation of joining ceramic to ceramic by an alloy of copper and 27 percent of titanium. Two surfaces with foil of this material ($\sim 50\mu m$-thick) at a contact pressure of about $0.1N/mm^2$ are heated for 30 min. at $1050°C$. Foils of platinum can easily be joined to alumina by a similar process.

An electrical connection for a sputtered *Pt* layer can be provided by a *thermosonic wedge bonding* the platinum wire. A more rugged connection is *spot welding* of relatively thick wire ($> 0.1mm$) to a platinum pin that has been pressed into a hole (Fig. 8.26) of substrate before or after sputtering.

Microholes ($15\mu m$-diameter) and *microparts* (a turbine of 0.4mm diameter, a shaft of $4\mu m$ diameter) can be *machined* [191] using a micro *electro-discharge machining* technique. The source of discharging pulses is a relaxation generator. A rotating mandrel with a capillary holds a very fine electrode ($0.5\mu m$ or less) which is in close ($1\mu m$) contact with the part under treatment. Voltage (short pulses of 60 to 110V) is applied between the electrode and part, thus creating the discharge. A wide variety of materials (tungsten carbide, stainless steel, silicon) can be treated by this method with submicron accuracy.

A thin-film *shape memory alloy* of nickel and titanium (TiNi) for microactuators is fabricated [192] by being sputtered onto deposited and cured polyimide film, which is used as a spacer and sacrificial layer. Sputtering is provided in an argon plasma at a current of 50mA and pressure of 0.05mtorr for about 4 hours at a deposition rate of $0.5\mu m/h$. After the spinning, baking, and patterning of photoresist, films are etched in hydrofluoric- and nitric-based solutions. Finally, the structure is released through a dry-etch process consisting of reactive ion etching of the sacrificial polyimide layer in 1:1 $SF_6:O_2$ plasma at high pressure (100mtorr) and moderate power (150W). Gold can be used as a sacrificial layer. This layer is etched in a potassium iodide-gold etch.

Submicron features in film deposition are formed due to a recently developed technique [193] called *shadow deposition*. This technique uses an electron beam gun to disperse materials in a vacuum. The materials are sprayed onto a tilted substrate containing microscopic hills and valleys. Due to the linear trajectory of the dispersing particles, they will not be deposited in the shadows of hills. This process can eliminate the etch step that typically follows the deposition.

Shaping of microstructures is one of the most common processes in the fabrication of microsensors. Different types of etching, laser treatment, and other processes described in the preceding paragraphs are used to sculpt simple and sometimes complicated parts. The creativity and ingenuity of designers in the development of the shape forming processes are quite remarkable. The photomicrographs given below are expressive and indicative of the progress in this field. One of the first successful works in the shaping of a resonator [194] is shown in Figure 8.27. The following photographs (Figs. 8.28–8.32) illustrate representative elements of sensor constructions with short explanations in the figure captions. Details of technologies can be found in the cited references placed in the brackets.

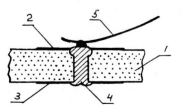

Figure 8.26 Spot-welded connection for sputtered Pt layer. 1 = Al_2O_3, 2 and 3 = Pt layers, 4 = Pt pin, 5 = Ni wire.

Figure 8.27 Scanning electron microscope (SEM) photograph of 150μm-long, 135μm-thick polysilicon resonant cantilever with 150nm-thick sensor polymer [194].

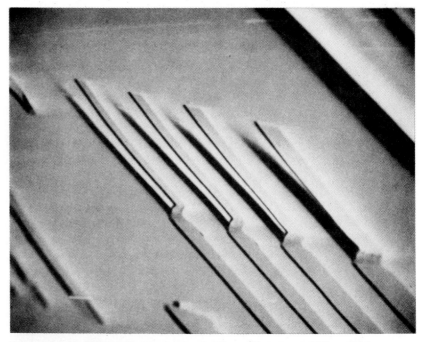

Figure 8.28 SEM photograph of polysilicon 60μm-long, 800nm-thick, 3.5μm-gap cantilevers [195].

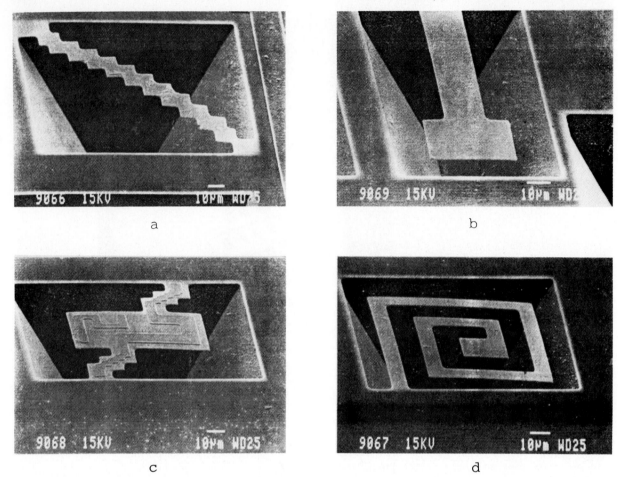

Figure 8.29 SEM photographs of micromechanical structures [196]. a = microbridge, b = paddle, c = bridge with metal layer, d = coil.

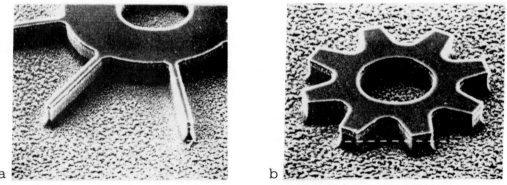

Figure 8.30 SEM photographs of free-standing rotating parts [197]. a = 40μm-thick, 600μm-diameter turbine with 20μm-wide by 100μm-long blades, b = 300μm-diameter gear.

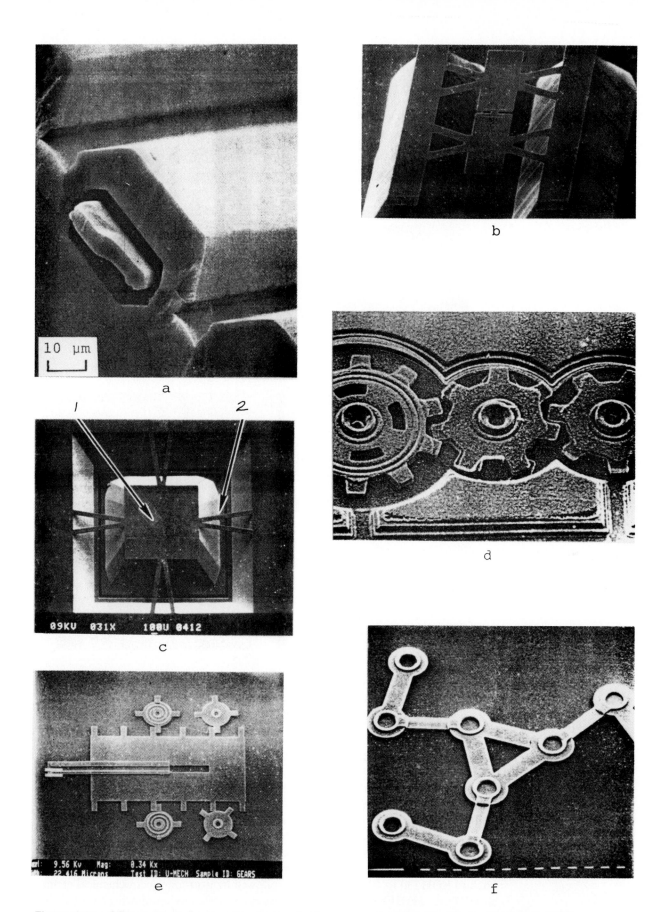

Figure 8.31 SEM photographs of motional microstructures. a = resonator vibrating in silicon cavity [198], b = silicon resonant structure for pressure sensor [199], c = proof mass (1) and eight constraining bridges (2) of microaccelerometer [200], d = gear train (smallest gear measures approximately 120μm) [201], e = gear-slide combination (movable slide measures 210 by 110μm) [202], f = link mechanism (each white dash is 10μm-long) [203].

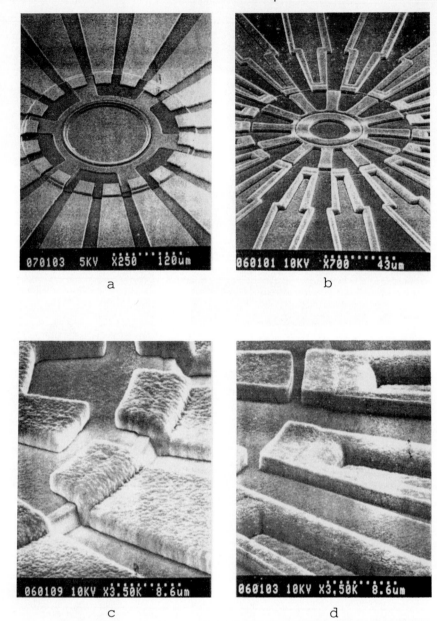

Figure 8.32 SEM photographs of microfabricated motors and their elements
[204]. a = three-phase, variable-capacitance, top-drive micromotor, b =
side-drive motor, c and d = close-up views of elements of rotor and stator.

FABRICATION OF PRESSURE-SENSITIVE MICROSTRUCTURES

Diaphragms and membranes are key elements of different sensors, especially
pressure sensors. Various technologies developed for the fabrication of these elements include shaping thin layers of silicon and other materials in order to create
flexible microstructures, form transduction and signal-processing elements, joint
parts, and accomplish some other operations.

The silicon diaphragms are normally etched [205–207] in anisotropic etchants
such as KOH (30 to 40% KOH/H_2O) followed by isotropic electrochemical etching
in HF (e.g., 5%HF) or EDP. The process is controlled by monitoring the etch time
and using etch-stop layers of high boron concentration. In electromechanical etching, the etch stop occurs at a passivating potential that can be obtained in the vicinity

of a p-n junction between an epitaxial layer and substrate of opposite doping. Electrochemical etching is the superior technique where low stress and doping concentration in the etch-stop layer are of importance. The shapes of the anisotropically etched membranes are bordered by slow-etching crystal planes. The electrochemical etching stops are usually formed by an epitaxial layer of n-type silicon grown on a p-type wafer. The doping concentration is low enough to keep internal stresses at a minimum and to make feasible the integration of electrical components on the membrane.

An ion-implantated or diffused p-n junction can be used for an etch stop instead of an epitaxial layer. Junctions of 3 to 12μm thicknesses are fabricated [205] using a standard $POCL_3$ diffusion process at $1000°C$ or phosphorous ion implantation for predeposition followed by a drive-in diffusion at $1100°C$. (100)-oriented, 75mm, boron-doped wafers, for a resistivity of 11 to $16\Omega \cdot$ cm, are subjected to a standard IC processing to provide a p-n junction, patterns, and masking layers. Then they are etched in KOH solution at $50°C$.

In a piezoresistive structure, strain-sensitive areas are realized by ion implantation.

A fabrication of high-quality structures [206] can involve as many as seven masks to obtain precision and reliable micromachining.

Using NH_4F solution for electrochemical isotropic etching of the circular diaphragm has some advantages [208]. The solution has little toxicity. It has almost no etching action on the silicon dioxide or silicon nitride when an electric field is applied. Therefore, these materials can be used for selective etching masking.

Three main technologies have been used to fabricate piezoresistive silicon strain gages: diffusion or ion implantation, polysilicon deposition on SiO_2, and forming of an epitaxial layer of silicon-on-sapphire substrate.

The first one is commonly used but limited by a $125°C$ operational temperature because of the leakage current in the rear junction. The sensitivity and reproducibility of the second one are limited. The third one does not allow batch fabrication.

A number of processes have been developed for forming piezoresistive pressure-sensitive structures (Fig. 8.33a, Schemes 83, 96). *The simplest process* [209] (Fig. 8.33b to g) starts with an n-type (100) wafer that is polished on both sides; then, SiO_2 or Si_3N_4 is grown or deposited on both surfaces. A photolithography step next defines oxide cuts in the top surface. The process continues with the introducing of boron into silicon through the cuts using either ion implantation or diffusion techniques. Another photolithography step is used to open small contact windows at the ends of the doped regions. Then metal is deposited onto the front surface of the wafer. Photolithography is again used to define the metal into paths leading from the resistors to larger pads. A thin silicon diaphragm is created next. Photolithography and front-to-back alignment are performed in order to open a window in the insulator on the back of the wafer. After this, the wafer is immersed in an anisotropic etchant, and a cavity is formed. Most commonly, the diaphragm thickness is controlled by timing the etch. For a very thin diaphragm, etch-stop techniques must be employed (EDP technique, voltage across p-n junction stopping etch process).

A *silicon-on-insulator* "SIMOX" (Separation by IMplanted OXygen) *technology* developed for VLSI application gathers all the needed qualities [210] of piezoresistive structures. They include high sensitivity, freedom from hysteresis and creep, high temperature range (up to $400°C$), and micromachining capabilities.

The SIMOX structure is formed by two main steps: a deep implantation of oxygen and high-temperature annealing. The dose of oxygen is $1.8 \times 10^{18}cm^{-2}$. It is implanted at a temperature higher than $500°C$. The very-high-temperature annealing ($>1300°C$) allows the elimination of the defects created during the implantation. As a result, a silicon top layer (~ 200nm) on the buried oxide (~ 400nm) is formed. The top layer is uniformly p-type doped by ion implantation and dry-etched. 75-mm, (100)-orientation silicon wafer and conventional photolithographic techniques are used in this process. The doped resistors are [110] oriented (maximum sensitivity) and interconnected by a metallization film.

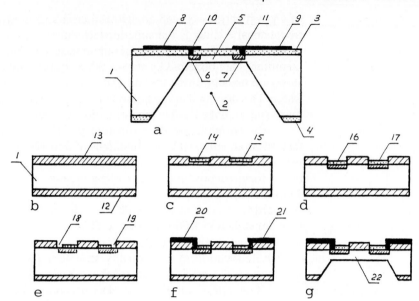

Figure 8.33 Simple silicon piezoresistive pressure-sensitive structure (adopted from [209]). a = structure diagram, b to g = fabrication process; 1 = Si, 2 = anisotropically etched cavity, 3 = insulator (SiO_2 or Si_3N_4), 4 = etch mask, 5 = thin sensing diaphragm, 6 and 7 = piezoresistors, 8 and 9 = metallizations, 10 and 11 = contact holes, 12 and 13 = SiO_2 or Si_3N_4, 14 and 15 = insulator cuts using photolithography, 16 and 17 = layers with implanted or diffused boron, 18 and 19 = opened window for contacts, 20 and 21 = metallizations, 22 = thin diaphragm.

Silicon-on-sapphire (SOS) technology [211] allows obtaining a structure that can operate at temperatures up to 425°C. Sapphire substrates are hard and chemically stable. They have excellent elastic properties and provide good electrical insulation.

Monocrystalline silicon films of about $1\mu m$ thick are grown on sapphire in epitaxial reactors and contain a given amount of p-type boron impurities, which define the electrical and mechanical properties of the piezoresistors. These resistors are formed by means of standard photolithographic processing. Molybdenum is used for metallization, and contact between Si and Mo is achieved by alloying them at $\sim 550°C$. The contacts are plated by gold. A diaphragm (of about $30\mu m$ thick) is produced by mechanical grinding of the substrate on the reverse side. Mounting of the sapphire part can be provided by metallization and subsequent anodic bonding to Corning 211 glass. It should be noted that SOS has the disadvantage of difficulty in fabricating small diaphragms because of the mechanical grinding of sapphire substrates. Epitaxially stacked structures of Si(100)/Al_2O_3(100)/Si (100) [212] are free of this drawback. In order to prepare the structure, first, single-crystal Al_2O_3 (100) films are grown epitaxially on Si(100) substrates by low-pressure chemical vapor deposition (LPCVD) or gas source molecular beam epitaxy (MBE), and, second, by growing the Si(100) epitaxial films on Al_2O_3(100)/Si(100) substrates using a method similar to SOS epitaxial growth. Pyrolysis of N_2-bubbled $Al(CH_3)_3$ (TMA) and N_2O at a pressure of 30torr in an RF-heated vertical reaction tube is conducted at substrate temperatures ranging from 950 to 1050°C. The growth rate depends on the TMA flow rate and is between 50 and 120 Å/min. [213]. Si epitaxial growth is carried out on the Al_2O_3/Si substrate using SiH_4 gas by the CVD method (at 1040°C) or gas source MBE method (at 750°C). The growth rates are $1.0\mu m/min$. and 120 Å/min. for the CVD and MBE methods, respectively. Al_2O_3 films cannot be etched by KOH (7mol/l) at 70°C or diluted by HF at 20°C for 60 min. Therefore, the Al_2O_3 film can be used as an etch-stop layer. For a 4mol/l KOH solution and 72°C, a maximum anisotropic etch rate is $55\mu m/h$. Passivation films of SiO_2 are prepared on both sides of the SOI wafer at 1000°C by wet oxidation for 240 min. The ability to obtain very flat surfaces of Al_2O_3 films is one of the remarkable features of this process. These surfaces are much better than those in etched silicon. A cross-sectional view and image of the diaphragm fabricated by this method are shown in Figure 8.34 [dimensions in (a) are in μm].

A microfabrication of silicon capacitive *tactile sensor arrays* is based on using SOI material produced by electric field-assisted silicon wafer bonding technology [214]. In this structure, an etch stop is formed by the Si-SiO_2 boundary in SOI material when etching is provided in a KOH solution having a great difference between the etch rate of Si and SiO_2.

A force/pressure-variable capacitor (Fig. 8.35) is formed between a selectively metallized glass and a metal layer on the oxide-nitride layer over one side of the diaphragm.

The fabrication process begins with thermal oxidizing and H_2 plasma treatment (activating) two mirror-polished p-type (100) silicon wafers with the subsequent production of the SOI structure by bonding the wafers at 850 to 900°C and 50Vdc for 30 min. (*b* and *c* in Figure 8.35). An optical mirror-polishing is then performed to thin the wafer's sides. The wafer is first oxidized (1.2μm) on both sides (*d*) and then patterned (6μm) in the capacitor area (*e*) with anisotropic etch solution: $KOH:H_2O$ = 50g:100ml. This solution etches Si, SiO_2, and Si_3N_4 at the etch rates of 5000, 30, and 1 Å/min., respectively (the etch-rate ratios of 170 and 5000 are achieved). After this, the wafer is oxidized again (*f*) and a 1200Å LPCVD Si_3N_4 is deposited (g). The SiO_2 and Si_3N_4 layers are patterned on the thin side of the wafer, and Al electrodes are formed by sputtering (*h*). Finally, the wafer is patterned on the thick side; then, silicon and a small portion of SiO_2 is etched off with the same solution (*i*).

The remaining part of SiO_2 (less than 1μm) is removed by a solution of $HF:NH_4F:H_2O$ = 3:6:10 (*j*). This process greatly improves the flatness and uniformity of the thin (20 ± 0.1μm) diaphragms in the sensor arrays. The roughness of the diaphragm surface is only 50Å.

A polycrystalline germanium film prepared by a *plasma-assisted chemical vapor deposition* can be used as a piezoresistor on a stainless-steel diaphragm cov-

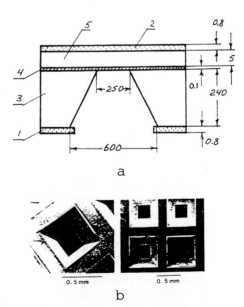

Figure 8.34 Si/Al_2O_3/Si-structure diaphragm (adopted from [212]).
a = schematic cross-sectional view,
b = SEM image; 1 and 2 = SiO_2,
3 = p-Si(100), 4 = Al_2O_3, 5 = Si.

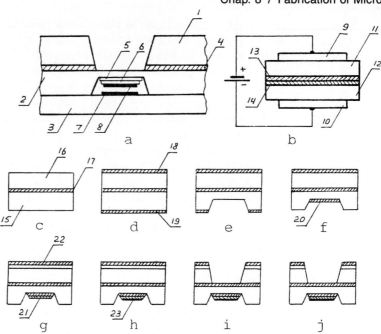

Figure 8.35 Capacitive tactile sensor (adopted from [214]).
a = structure diagram, b = field-assisted bonding system, c to
j = fabrication process; 1 and 2 = Si, 3 = glass, 4 and 5 = SiO$_2$,
6 = Si$_3$N$_4$, 7 and 8 = Al, 9 and 10 = electrodes, 11 and 12 = Si, 13
and 14 = SiO$_2$, 15 and 16 = Si, 17, 18, 19, and 20 = SiO$_2$, 21 and
22 = Si$_3$N$_4$, 23 = Al.

ered with an insulating layer [215]. The plasma CVD method uses the decomposition of germane (GeH$_4$) and phosphine (PH$_3$) in a fused-quartz reactor with an RF glow discharge. Germane gas is diluted to 2.2 percent by hydrogen, and phosphine gas is diluted to 81ppm by hydrogen. In order to obtain n-type material with a high carrier concentration, phosphorous is doped into film (undoped samples show p-type conduction). The carrier concentration is 5×10^{18} to 1×10^{20}cm^{-3} and is almost proportional to the PH$_3$/GeH$_4$ ratio. The Hall mobility is 5 to 10cm^2/V·s and is decreased as the ratio is decreased. The gage factor of the doped film is 30 to 40. The temperature coefficient of the gage factor for the temperature range 20 to 100°C is 1100ppm/°C in films prepared at 400°C and with PH$_3$/GeH$_4$ ratio of 1.85×10^{-2}. The temperature coefficient of resistance is about 2000ppm/°C for temperatures from 20°C to 70°C. The sensitivity of the films is more than 1.5 times higher than that in polycrystalline silicon thin films. The sensing layers can be perfectly separated from the substrates, and devices can work at higher temperatures than those using diffused layers. However, the temperature coefficients of the gage factor and of resistance are larger than those for polycrystalline silicon.

Corrugated diaphragms with high linearity and travel can be produced in silicon microstructures by etching grooves in the top surface of a wafer, by diffusing an etch stop, and by a back-side etching to form the corrugations [70]. For a diaphragm of thickness between 0.6 and 8μm, the deepness of the corrugations of 3 to 20μm can be made. Depending on the desired final diaphragm contour, the grooves can be isotropically or anisotropically etched, forming any pattern: circular, spiral, serpentine, or radial (Fig. 8.36). A conventional plasma-etching technique and boron doping for an etch stop are used in this process. Nearly the same procedure can be used to produce corrugated diaphragms from materials other than silicon, for example, silicon dioxide or nitride. Any substance that can be formed on the silicon surface and withstand the final silicon etch can be accepted in this process and offer more design flexibility.

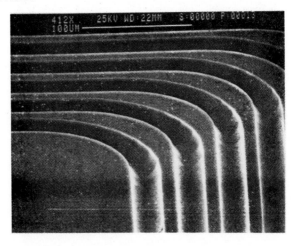

a

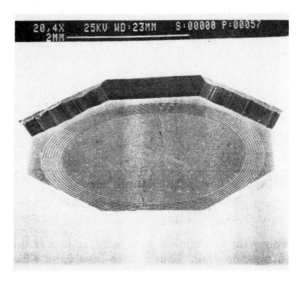

b

Figure 8.36 Corrugated microdiaphragm (adopted from [70]). a = plasma-etched grooves for corrugation, b = back-side view of bossed, corrugated diaphragm.

A general trend in shaping diaphragms is the wider use of dry etching.

More details on the fabrication of fine diaphragms can be found in literature [216-19].

At this point, it would be instructive to retrace the evolution over the years [220] (beginning from the middle of the fifties) of the construction of the piezoresistive silicon pressure-sensor structure, since it has become the most popular component of microsensors.

The first silicon sensors were made of a thin bar cut out of a wafer. The bar had two electrodes attached to the ends. Two or four bars were cemented to a metal diaphragm (Fig. 8.37) and exposed to pressure, which deflected the diaphragm and caused strain in the bars. The change in the resistance of the material was proportional to the pressure. This was a model made originally of foil or wire strain gages but with much higher sensitivity to pressure, since the gage factor in semiconductors can be two orders of magnitude greater than that in metals. Unfortunately, this

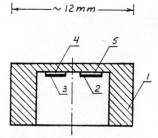

Figure 8.37 Pressure sensor with metal diaphragm and bonded silicon strain gages. 1 = metal constraint with machined diaphragm, 2 and 3 = silicon strain gages, 4 and 5 = epoxy bond.

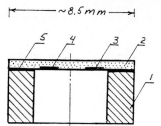

Figure 8.38 Pressure sensor with silicon diaphragm and diffused strain gages. 1 = metal constraint, 2 = silicon diaphragm, 3 and 4 = diffused silicon strain gages, 5 = epoxy bond.

sensor had a number of drawbacks, the most significant of which were temperature errors and poor stability because of a mismatch of silicon-glue-metal interface, high price defined by low yield, and other factors.

A conceptional breakthrough in technology (Fig. 8.38) was the diffusion of strain gages directly into the diaphragm. Glue was eliminated, and gages became an integral part of the deforming diaphragm. Due to this arrangement, the drifts determined by glue (hysteresis, creep, etc.) were avoided.

In the next stage of improvement (Fig. 8.39), the diaphragm was shaped as a cap. The metal constraint was replaced by a silicon one, and a gold-silicon eutectic bond was used instead of epoxy. These measures helped improve performance and lowered cost.

The further evolution of the technology (Fig. 8.40) was related to the precise alignment of crystallographic planes in silicon for better strain gage positioning, using isotropic and then anisotropic etching on an entire wafer simultaneously and also developing a masking process identical to other IC masking steps. The thickness of the diaphragms was controlled by etch time. Glass frit was introduced as a bonding material for joining the diaphragm and constraint wafers. The entire process was performed in a batch mode, which led to reduced cost.

Along with a further decrease in price, new methods of forming the strain gages by ion implantation simplified the control of the electrical parameters of conducting ele-

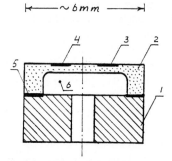

Figure 8.39 Pressure sensor with silicon-cup, Si-Au-bonded diaphragm. 1 = silicon constraint with drilled hole, 2 = silicon-cup diaphragm, 3 and 4 = diffused strain gages, 5 = Si-Au bond, 6 = mechanically milled and/or chemically isotropically etched cavity.

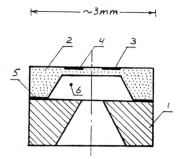

Figure 8.40 Pressure sensor with anisotropically etched, glass-frit-bonded diaphragm. 1 = silicon constraint with anisotropically etched hole, 2 = silicon-cup diaphragm, 3 and 4 = diffused strain gages, 5 = glass-frit bond, 6 = anisotropically etched cavity.

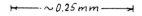

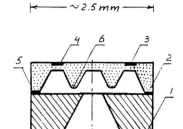

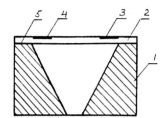

Figure 8.41   Pressure sensor with ion-implanted strain gages. 1 = pyrex constraint, 2 = silicon diaphragm, 3 and 4 = ion-implanted strain gages, 5 = anodic bond, 6 = overpressure stop.

Figure 8.42   Pressure sensor with silicon-to-silicon bond diaphragm. 1 = silicon constraint, 2 = silicon diaphragm, 3 and 4 = ion-implanted strain gages, 5 = silicon-to-silicon bond.

ments. Electrostatic bonding (Fig. 8.41) was used for the connection of a diaphragm to a pyrex constraint having thermal expansion characteristics that matched those of silicon. A special shaping of the diaphragm provided the sensor with overrange protection.

Among the latest advantages (Fig. 8.42) is silicon-to-silicon wafer lamination [221–23]. This process makes it possible to perform a molecular bond between two or more silicon wafers, reducing the size of structures and increasing the level of mechanical integration on silicon chips. The process starts with polishing the wafers' surfaces and then boiling the wafers in nitric acid to make the surfaces hydrophilic with a high density of OH groups attached to the surface of silicon atoms. When the two bonding surfaces are placed in contact and heat treated (300 to 800°C) in oxidizing and nonoxidizing ambients, OH groups convert to water. Additional oxygen becomes free to bond to silicon, while hydrogen diffuses through the silicon lattice. At higher temperatures (800–1400°C) oxygen tends to diffuse into the crystal lattice. The addition of a small amount of oxygen from the gap between wafers has little effect on a wafer's electrical properties. Silicon atoms fill in microscopic surface-roughness voids by diffusion.

There are some variations in the process [221] (see Die Attachment), which are adopted for fabrications of different sensors.

A dimensional control by a programmable etch stop improved performance of the pressure-sensitive elements. Some results in research on space charge control in CMOS devices was implemented in the improved stability of sensors. With the progress in technology, the dimensions of sensors were significantly reduced (from approximately 12 to 0.25mm). It should be noted that the implementation of new advances in the applied sciences is keeping pace with further improvements in technology.

FABRICATION OF SURFACE-SENSITIVE STRUCTURES

Among the surface-sensitive devices, FET and MOS transistors have become extremely popular for the measurement of charges in chemicals, microdisplacements, and other quantities. The fabrication of the sensing FET's and MOSFET's does not differ in principle from that of conventional transistors. The basic feature of FET and MOS sensors lies in the arrangement of a charge-controlling gate that is furnished with a movable electrode or with a substance collecting (or modifying) charges at the gate's region. Figure 8.43 indicates a simplified standard fabrication process [102, 224, 225] of a p-channel enhancement MOSFET with involvement of the typical operations described previously. This process requires four masking steps and one diffusion. The deposition of the sensitive layer is usually based on general thin-film techniques [226–29] such as sputtering, evaporation, CVD, or plasma deposition.

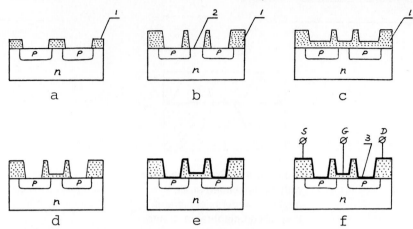

Figure 8.43 Fabrication of p-channel enhancement MOSFET. a = SiO$_2$
layer is grown on n-type substrate; p-type impurities are diffused by standard
masking and etching of windows; two formed p-regions are source and drain;
b = thicker layer of SiO$_2$ is grown over surface; three openings in oxide
layer are provided by masking and etching; c = thin layer (~1000Å) of
SiO$_2$ is formed at gate area; d = source and drain areas are etched away
after masking and etching; e = aluminum is deposited over surface; f =
aluminum connecting areas of source, gate, and drain are removed by etching
following masking; D, G, and S = drain, gate, and source, respectively;
1 = SiO$_2$; 2 = gate region; 3 = aluminum (adopted from [225]).

Another example is processing of p-Si/Al thermopile [230], which consists of
several cantilevered thin (10μm) beams surrounded by a thick, heat-sinking rim. Besides
the p-silicon and aluminum elements, each cantilever is partially covered by a layer
absorbing radiation. The main steps of the fabrication process are shown in Figure 8.44.

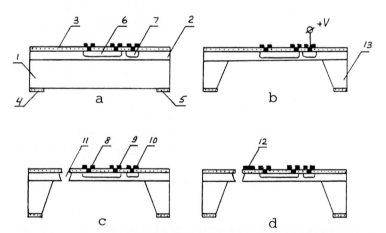

Figure 8.44 Simplified process-flow chart for p-Si/Al thermopile
(adopted from [230]). a = n-type epilayer is grown on p-type
(100)-silicon wafer; after oxidation and LPCVD Si$_3$N$_4$-layer growing
on back side, p- and n-type diffusion is carried out and aluminum
contacts are formed [n$^+$-islands are created to realize good contact
during electrochemically controlled etching (ECE)]; b = ECE is
used to etch Si up to etch stop at epilayer; c = plasma etch
process and ion etching in CF$_4$ + 6% O$_2$ (see details in [231]) are
used to open "channels" and form cantilevers; d = black coating
with absorbent material; +V = voltage applied to junction for etch
stop, 1 = p-Si, 2 = n-epi, 3 = SiO$_2$, 4 and 5 = Si$_3$N$_4$,
6 = p$^+$-diffusion, 7 = n$^+$-diffusion, 8, 9, and 10 = deposition of
aluminum, 11 = etched "channel", 12 = black coating, 13 = rim.

chapter 9

Micromechanical Sensing and Actuating Structures

Similar to the traditional transducers, microsensors that treat mechanical quantities are dominant. In this chapter we combine devices that form a group of the most successfully produced or developing microsensors. Micromotors are also given attention because they are "the closest relatives" to microsensors. Integration and interaction of these two types of devices in one system become obvious.

ELECTROMECHANICAL MICROSENSORS

Several microstructures converting mechanical quantities into electrical signals are discussed in this section. Strictly speaking, miniature strain gages should also be among these devices. However, strain gages are given special attention in a separate section because of their wide application and popularity. Like conventional transducers, electromechanical microsensors can be categorized either in terms of a measurand (e.g., acceleration, pressure, etc.) or with respect to the physical concept governing a quantity conversion. With these two categories in mind, we will use two approaches in the descriptions to keep the presentation of the material concise.

Surface Acoustical Wave (SAW) Microsensors

SAW microsensors [232] have great potential for sensing nearly all forms of physical and chemical phenomena. Sensors based on SAW technology have been researched and developed since the 1970's; however, their technology is less mature than that based on resistance, capacitance, or inductance modulation. The recent intensive work being conducted to advance the sensors' design looks promising.

Rayleigh surface acoustic waves are electromechanical waves (Fig. 9.1) that are usually generated on surfaces containing piezoelectric material and lithographically patterned electrodes ("interdigitated electrodes").

In general, surface acoustic waveguides are formed by confining a channel (or core) region on a substrate with a surrounding region (or cladding) having a faster wave velocity [233]. The material is excited by electrical signals in the range of frequencies between 10MHz and 1GHz. The velocity at which the waves travel along the surface is about 5 orders of magnitude lower than the velocity of electromagnetic

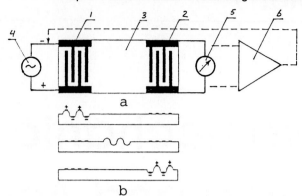

Figure 9.1 Schematic diagram of patterned SAW
element (a) with outline of wave propagation (b).
1 = electrodes which launch waves, 2 = electrodes
that receive waves, 3 = substrate, 4 = ac generator,
5 = signal indicator, 6 = amplifier.

waves. The waves have longitudinal and vertical shear components. The latter can
interact with the medium, responding to the properties of the medium by changing
the amplitude, frequency, or phase of the propagating oscillations. The vibratory
deformation of the piezoelectric material is associated with the electrical field generated in the material, which can be sensed and used for the quantitative evaluation
of the propagation characteristics.

Coating the device with materials sensitive to the particular physical or chemical stimulus expands the list of the values that can be sensed. Rayleigh waves are
chiefly generated in SAW devices; however, other types of waves (Fig. 9.2), distinguished by their velocities and displacement directions, may have advantages in
new designs [234, 235]. Some of these waves are Lamb waves, Love waves, Bluestein-
Gulyaev waves, Stonely waves, surface-skimming bulk waves (SSBW), and deep bulk
acoustic waves (DBAW). Typically, the amplitude of the waves is about 10 Å, and the
wavelengths are in the range of 1 to 100 μm. The particular type of waves can be selectively created by choosing the type of the crystalline material and the frequency of
the excitation voltage, and by patterning the electrodes. The energy-loss mechanism
must also be considered when choosing the type of waves. The use of Lamb waves
results in a high sensitivity of thin-membrane devices. The application of bulk acoustic waves is useful for sensing physical and chemical properties in liquids and gases.

One of the attractive features of SAW sensors is their direct response to the
change in the inertial mass, elasticity, and viscosity of the materials contacting with
the device surface [236]. The waves generated on a piezoelectric substrate have an
associated electric field that can be used to probe the field outside the sensor. This
field can also detect the dielectric and conducting properties of the medium. If the
waves are generated on a magnetostrictive substrate, they will be affected by magnetic fields and the permeability of the material surrounding the sensor. The changes
in the dimensions and in the thermoelastic properties of the substrate are reflected in
the characteristics of wave propagation. If a displacement, force, or acceleration can

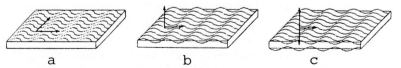

Figure 9.2 Illustration of three types of acoustical waves in solids.
a = horizontally polarized shear wave, b = Rayleigh surface wave,
c = symmetric Lamb wave (adopted from [232]).

be converted into the strain of the substrate, the wave propagation characteristics become sensitive to these mechanical quantities.

A SAW structure is relatively simple in the fabrication of a delay-line oscillator or resonator. Recently, progress has been made in integrating high-frequency electronics for the excitation of the structure and for signal conditioning. Monolithic structures combining sensing and signal-treatment functions have been created.

In order to generate and detect acoustic waves and to build the associated electronics on silicon, piezoelectric overlays (e.g., ZnO) on silicon are used. The sandwiched structure, $ZnO-SiO_2-Si$, [237–239] contains metal patterns placed on top of the ZnO layer or at the interface between ZnO and the insulating SiO_2 layer. Admittedly, a SAW device is more expensive than a conventional component providing a similar conversion of a measurand into an electrical signal. However, superior linearity and dynamic range can make the SAW sensor suitable for some applications.

In the design of SAW devices, the sensing of a physical quantity can be performed by detection of the amplitudes, phases, or frequencies of the waves. The reading of frequencies is preferable, since it is most precise due to the well-developed digital technique. The delay-line oscillator is usually formed by a piezoelectric structure combined with a feedback amplifying circuit. The frequency of oscillation ω [rad/s] is stable if the following relationship is satisfied:

$$\omega = (2\pi n - \varphi_e)/\tau, \qquad \tau = L/v \tag{9.1}$$

where n = integer;
φ_e = electrical phase shift of the amplifier, rad;
τ = delay time of the SAW device, s;
v = surface wave velocity, m/s; and
L = acoustic path length of the delay line, m.

n is usually small, because of some restrictions in the bandwidth of the associated circuit. The RF amplifier gain must be sufficient to cover the losses in the delay line (10–20dB). Typically, the short-term stability of the resonant frequency (e.g., over 10 seconds) is between 0.01 and 0.1ppm. The signal-to-noise ratio is in the range of 100 to 1000.

The temperature sensitivity of *SAW sensors* [240, 241] depends on the orientation and type of crystalline material used for the fabrication of the device. Lithium niobate employed in temperature sensors has a high temperature coefficient of delay time: 94 ppm/°C; ST-cuts of quartz have this coefficient near to zero at 25°C. SAW temperature sensors have a resolution of about 1×10^{-3}°C, good linearity, and low hysteresis.

The temperature dependences of distance L and velocity v are described by

$$L = L_0[1 + \epsilon(T - T_0)] \qquad \text{and} \qquad v = v_0[1 + \delta(T - T_0)] \tag{9.2}$$

where ϵ = temperature expansion coefficient of the material in the direction of propagation, 1/K;
δ = temperature coefficient of propagation velocity, 1/K;
T = temperature, K; and
T_0 = reference temperature, K.
L_0 and v_0 represent L and v at T_0.

Considering that $\epsilon(T - T_0) \ll 1$ and $\delta(T - T_0) \ll 1$, we obtain the expressions for the temperature dependence of the delay time (see equation 9.1):

$$\frac{1}{\tau} \cdot \frac{d\tau}{dT} = \frac{1}{L_0} \cdot \frac{dL}{dT} - \frac{1}{v_0} \cdot \frac{dv}{dT} \tag{9.3}$$

Since

$$\epsilon = \frac{1}{L_0} \cdot \frac{dL}{dt} \quad \text{and} \quad \delta = \frac{1}{v_0} \cdot \frac{dv}{dt}, \quad \frac{1}{\tau} \cdot \frac{d\tau}{dT} = \epsilon - \delta. \quad (9.4)$$

Similarly, the temperature coefficient of ω is obtained as follows:

$$\frac{1}{\omega_0} \cdot \frac{d\omega}{dT} = \delta - \epsilon \quad (9.5)$$

where ω_0 is ω at T_0.

The sensitivity to temperature can influence the performance of the device adversely if the measurand is not temperature. In this case, commonly used measures, such as arranging differential measuring structures and using references, are applied to alleviate the problem.

In *SAW pressure sensors* [242, 243], wave velocities are affected by stress developed in the crystal on which the wave is propagating. A typical construction (Fig. 9.3a) contains a quartz-made diaphragm with two SAW delay lines on the diaphragm surface. One of the lines is part of an oscillator and forms a frequency signal proportional to the pressure; another line serves as a reference and mainly responds to temperature. Even compensated, many pressure sensor designs suffer from the lack of thermal stability. There was an attempt to use a SAW resonator as a transduction element combined with a bellows similarly to the construction developed for quartz transducers (see the section on resonant elements in Chapter 4.) However, a construction with a freely deformed crystal that does not have any linkage with an additional sensing element is regarded as a more appropriate approach for a SAW-type device.

Force and *acceleration sensors* [244] are conveniently arranged using a quartz cantilever (*b*) in which signal and reference delay lines are formed at two sides of a beam. One end of the beam is fixed and the other is subject to force. The bending deformation of the cantilever causes tension stress at one side of the beam and compression stress at the other. Respectively, the frequencies of oscillation of the delay-line oscillators are increased and decreased. The difference between the two frequencies is the output proportional to the force. Due to the differential circuit, close location of the two lines, and uniformity of the quartz crystal properties, the temperature drifts are minimized. A seismic mass added to the free end of the cantilever makes the *device sensitive to acceleration* [245, 246]. Such an accelerometer is rugged, and has low hysteresis and a fast response to acceleration.

A *voltage* applied to two electrodes across a substrate, in which a surface acoustic wave is propagating, can be *measured* [247, 248]. An electric field directed normal to the surface stimulates a change of crystal stiffness that gives a shift in the wave velocity (*c*). The major advantages of this sensor are in its extremely high input impedance and good dielectric isolation.

The *measurement of displacement* can be accomplished by the use of two separate lithium niobate plates carrying SAW delay lines sliding on each other and having acoustic coupling of the surface wave from one plate to the other through a thin layer of silicon oil. The coupling through the air gap occurs when the plate separation is less than 0.4nm. According to a report on the design [249], a frequency drift of 300Hz/μm can be obtained when the central frequency is 8.39MHz.

The temperature sensitivity of lithium niobate can be used for building a *gas flow sensor* [250]. A lithographically patterned meander-line heater incorporated onto the SAW delay-line chip raises the temperature of the device to about 50°C. The anemometer-like performance of the sensor gives a reading of gas flow with a sensitivity of 11Hz/SCCM in the range of 2200–4000SCCM.

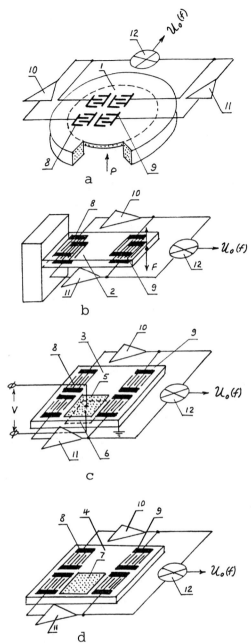

Figure 9.3 Schematic diagrams of SAW sensors (source: [232]). a = pressure sensor, b = force sensor, c = voltage sensor, d = chemical sensor; P = pressure, F = force, V = voltage to be measured, $U_0(f)$ = output signal, 1 = diaphragm, 2 = cantilever, 3 and 4 = substrates, 5 and 6 = electrodes, 7 = chemically selective coating, 8 and 9 delay lines, 10 and 11 = amplifiers, 12 = RF signal mixer.

A *film thickness sensor* [251] employs the effect of a decrease in the frequency of a SAW delay-line oscillator when a metal film is deposited in a vacuum on a surface carrying acoustic waves. The effective mass of the vibrating layer is increased due to the deposition that gives the shift in the frequency. The sensitivity is $106Hz/\overset{\circ}{A}$ at the central frequency of 52MHz.

Normally, when a thin fluid layer comes in contact with a SAW delay path, the waves are severely attenuated, and no appreciable output signal is observed. However, if the fluid layer is sufficiently thick, it is possible for the SAW path to excite acoustic waves in the fluid. It has been found [252] that the thickness and physical properties of the fluid layer modify the acoustic wave properties that can be exploited in the development of microsensor concepts.

The *mobility of carriers* in a weakly conducting semiconductor can be *measured* [253] if the film is placed in close proximity to the SAW device and affects the oscillations due to the acoustoelectric coupling.

The *transition of a polymeric film* from a firm to a soft state can also be *measured* [254] by a SAW device. Softening of the polymer film, which is in contact with the device surface, causes a significant attenuation of the surface wave amplitude.

In a recent design (Fig. 9.4) [255], multiple-frequency acoustic waves are excited on a ST-cut quartz substrate using 25 pairs of interdigitated electrodes that allow the differentiation of pressure and temperature measurement from one another and from changes in surface mass.

The application of *SAW devices in chemical sensors* has been intensively investigated since the end of the seventies. The typical construction (Fig. 9.3d) of a SAW chemical sensor is a plate with delay lines and a chemically selective coating that absorbs the species of interest, resulting in an increased mass loading of the vibrating structure. The chemically induced change in the film's electrical conductivity can also cause a change in the wave propagation [256].

Coating of the surface of a quartz resonator with a hygroscopic polymer, such as sodium salt of poly (styrene) sulfonate, gives approximately 3.6kHz per 1% relative humidity change when the central frequency is 158MHz [232]. The response is rapid and reversible (~ 5s on uptake and ~ 25s on drying). A number of coatings have been designed for different vapors [257–262] including water, hydrogen sulfide, nitrogen dioxide, organic solvents, organophosphorus compounds, and so on. The materials employed for coating the bulk wave piezoelectric resonators are, as a rule, acceptable for use in SAW devices. The improvement of selectivity of these materials is an area currently being actively investigated [263].

Along with their application in chemical sensors, the use of *SAW devices in biosensors* has become more appreciable. For example, the sensor can be used for measuring the immunochemical binding reactions in a liquid phase [264], for gelation detection [265], and so on.

In the development of SAW sensors, new opportunities [232] are provided by a wider application of different types of waves, by development devices based on the generation of SAW in magnetostrictive materials, and by using the structures in which the combination of piezoelectric and magnetostrictive effects is exploited. Pulse-echo techniques can be used for obtaining information using a single SAW device that can replace an array of sensors. Of special interest are microperistaltic pumps, which are capable of displacing fluid contacting the vibrating surface. Such machines can generate and control extremely low flow rates of fluid.

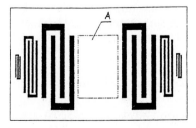

Figure 9.4 Schematic diagram of multifrequency acoustical wave device (source: [255]). A = interaction region; 25 interdigitated electrodes for central frequencies of 16, 40, 100, and 250MHz are used in construction.

Resonant Microsensors

Microresonators that respond to various physical and chemical quantities are being developed intensively at the present time, and the number of quantities that can be measured is becoming greater. The attractive features of conventional resonant-type transducers are advanced not only by the extremely small sizes of microresonators but also by some of the characteristics that are not attainable from regular-size devices, such as the possibility of optical excitation of vibration, the extremely high quality factor of the mechanical system, and others.

Many general issues of the theory of vibrating elements discussed in the Resonant Elements of Chapter 4 are applicable to the analysis of resonating microstructures; however, there are some specific aspects of theory that are primarily defined by the semiconductor nature of the microresonator materials when they are in use, and by the small size of the device.

The approximate approaches in calculations have found their application in the analysis of the vibratory characteristic of microresonators, and one such technique, Rayleigh's method, is used most commonly to determine the resonant frequency of microstructures [266, 267]. The essence of the calculation is based on the premise that for a conservative system (a system without damping) the maximum kinetic energy is equal to the maximum potential energy. This method is quite universal and can be used for the analysis of systems having more than one degree of freedom, distributed or lumped parameters, a plane or corrugated shape, and so on. We will consider several examples [10, 44] illustrating the application of the method in solving representative problems.

Suppose there is a system (Fig. 9.5) consisting of an elastic element, such as a beam, with several lumped masses that can include the mass of the beam itself. Assuming that the system vibrates at the fundamental mode and there is no phase shift in the vibrations of each mass, the displacement of i-th mass y [m] and its velocity $\dot{y}$ [m/s] are written as

$$y_i = Y_i \sin(\omega_0 t + \varphi) \quad \text{and} \quad \dot{y}_i = Y_i \omega_0 \cos(\omega_0 t + \varphi) \tag{9.6}$$

where Y_i = amplitude, m;
φ = phase shift, rad; and
ω_0 = angular frequency, rad/s.

The masses pass the equilibrium position (at the x-axis) simultaneously. At this moment, $y_i = 0$, $\dot{y}_i$ is at the maximum, and the kinetic energy of the system T [N · m] is also at the maximum:

$$T = \sum_{i=1}^{n} \frac{m_i \dot{y}_i^2}{2} = \frac{\omega_0^2}{2} \sum_{i=1}^{n} m_i Y_i^2. \tag{9.7}$$

The potential energy U [N · m], being proportional to the displacement of the system from the equilibrium, is zero at this moment since $y_i = 0$, but U attains the maximum at the time of maximum deflection of the masses. The condition of conservation of energy requires $T = U$; therefore,

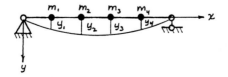

Figure 9.5 Model of vibrating system. m_i = lumped mass, y_i = displacement of mass.

$$\frac{\omega_0^2}{2} \sum_{i=1}^{n} m_i Y_i^2 = U \quad \text{and} \tag{9.8}$$

$$\omega_0^2 = \frac{2U}{\displaystyle\sum_{i=1}^{n} m_i Y_i^2}. \tag{9.9}$$

In order to calculate ω_0, the shape of the elastic curve of the beam must be known. Approximations are usually used for the presentation of this shape to eliminate the lengthy calculations associated with differential equations for determining the curve.

In the first example, we will calculate the fundamental frequency of the longitudinal vibrations of a bar (Fig. 9.6) having mass m_0 [kg], length l [m], cross-sectional area A [m^2], and Young's modulus of material E [N/m^2] . An added mass m [kg] is attached to the end of the bar. Assuming that the displacement of the bar's end is Δl [m] and the displacements of various sections of the bar are linear functions of y, we first will find the potential energy U from the standard relations of the theory of elasticity:

$$dU = \frac{1}{2}Nd(\Delta l), \quad U = \int_0^{\Delta l} \frac{1}{2}Nd(\Delta l) = \frac{1}{2}N \cdot \Delta l = \frac{1}{2}\frac{(\Delta l)^2}{l}EA \tag{9.10}$$

since

$$N = \sigma A = \frac{\Delta l}{l}EA \quad \text{and} \quad \sigma = \epsilon E = \frac{\Delta l}{l}E$$

where N = force along the y-axis, N;
   σ = normal stress in the bar, N/m^2; and
   ϵ = strain in the bar, dimensionless.

Now we will find $\displaystyle\sum_{i=1}^{n} m_i Y_i^2$.

$$\sum_{i=1}^{n} m_i Y_i^2 = \int_0^l \frac{m_0}{l}\left(\Delta l \frac{y}{l}\right)^2 dy + m(\Delta l)^2 = \frac{1}{3}m_0(\Delta l)^2 + m(\Delta l)^2 \tag{9.11}$$

Finally, the relevant substitutions into formula 9.8 give

$$\omega_0^2 = \frac{EA}{l(m + \frac{1}{3}m_0)}. \tag{9.12}$$

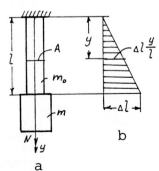

a   b

Figure 9.6 Model of bar with added mass at end (a) and diagram of displacements (b) along y-axis.

As shown in this formula, only one third of the bar's mass is added into the total effective mass of the vibrating body (compare with the effective mass of the diaphragm given in formula 3.72).

One of the typical applications of Rayleigh's method is the calculation of the fundamental resonant frequency for beams vibrating at a flexural mode. The displacement of any section of the beam is a function of coordinate x and time t:

$$y = y(x, t). \tag{9.13}$$

As before, function y can be introduced as a product of the function of coordinate $f(x)$, and the function of the time $q(t)$ [it is common to assume $q(t)$ is sinusoidal]: $y = q(t) \cdot f(x)$; therefore, the speed of the transverse motion dy/dt of an infinitesimal element of length dx and its kinetic energy dT will be

$$\frac{dy}{dt} = \dot{q}(t) \cdot f(x), \qquad dT = \frac{1}{2}m(x) \cdot dx \left(\frac{dy}{dt}\right)^2 \tag{9.14}$$

where $m(x)$ = mass per unit of length at the point with coordinate x, kg/m, and
   $m(x) \cdot dx$ = mass of the element, kg.

The total potential energy of the beam is the integral taken over the beam's length l [m]:

$$T = \frac{\dot{q}^2(t)}{2} \int_0^l m(x) \cdot f^2(x) dx. \tag{9.15}$$

If $M(x)$ and $\theta(x)$ are the bending moment and slope of the elastic curve at point x, the work done on the beam's element dx and stored as elastic energy is an elementary potential energy:

$$dU = \frac{1}{2}M(x)d\theta. \tag{9.16}$$

The potential energy for the total beam's length is

$$U = \frac{1}{2}\int_0^l M(x)d\theta. \tag{9.17}$$

From the theory of beams,

$$\frac{1}{R} = \frac{M(x)}{EJ} \tag{9.18}$$

where R = radius of curvature, m, and
   J = moment of inertia of beam's section, m^4.

Since the deflection of the beam is small (see Fig. 9.7),

$$\theta = \frac{dy}{dx} \quad \text{and} \quad \frac{1}{R} = \frac{d\theta}{dx} = \frac{d^2y}{dx^2}. \tag{9.19}$$

The substitutions into formula 9.17 give

$$U = \frac{1}{2}\int_0^l \frac{M^2(x)}{EJ}dx = \frac{1}{2}\int_0^l EJ\left(\frac{d^2y}{dx^2}\right)^2 dx. \tag{9.20}$$

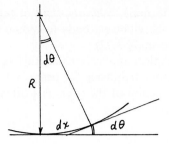

Figure 9.7 Slope and radius of curvature of elastic curve.

Inserting in this formula

$$\frac{d^2y}{dx^2} = q(t) \cdot \ddot{f}(x) \tag{9.21}$$

will determine the potential energy:

$$U = \frac{q^2(t)}{2} \int_0^l EJ[\ddot{f}(x)]^2 dx. \tag{9.22}$$

Assigning as in formula 9.6 a sinusoidal function for $q(t)$ and equating each to the other the maximum values of T and U, we obtain

$$q(t) = Y \sin \omega_0 t, \qquad \dot{q}(t) = \omega_0 Y \cos \omega_0 t$$

$$\omega_0^2 = \frac{\displaystyle\int_0^l EJ(x)\,[\ddot{f}(x)]^2 dx}{\displaystyle\int_0^l m(x) \cdot f^2(x) dx}. \tag{9.23}$$

Note that the maximums of T and U are calculated by substituting the maximum values for q and $\dot{q}$ ($q_{\max} = Y$ and $\dot{q}_{\max} = \omega_0 Y$).

When the resonant frequency is calculated for a simply supported beam with a uniform cross-sectional area [$m(x) = \text{const} = m$], the representation of the elastic curve by a sine function is convenient. For example, the deflection of the beam in Figure 9.8a can be approximated as $y_0 \sin (\pi x/l)$.

$$\ddot{f}(x) = -\left(\frac{\pi}{l}\right)^2 y_0 \sin \frac{\pi x}{l} \tag{9.24}$$

The resonant frequency is readily calculated by substituting the $\ddot{f}(x)$ into equation 9.23:

$$\omega_0^2 = \frac{\displaystyle\int_0^l EJ \left(\frac{\pi}{l}\right)^4 y_0^2 \sin^2 \frac{\pi x}{l} \cdot dx}{m \displaystyle\int_0^l y_0^2 \sin^2 \frac{\pi x}{l} \cdot dx} = \frac{\pi^4 EJ}{ml^4}. \tag{9.25}$$

Finally,

$$\omega_0 = \pi^2 \sqrt{EJ/ml^4}. \tag{9.26}$$

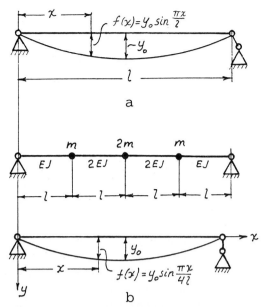

Figure 9.8 Flexural deformation of beams with elastic line approximated by sine functions. a = beam with uniform stiffness and mass along length, b = beam with lumped masses and nonuniform stiffness along length.

It can be shown [268] that a clamped-clamped beam, affected by an axial load causing total strain ϵ, has a fundamental natural frequency of flexural vibrations calculated from

$$\omega_0^2 = \frac{Eh^2}{\rho l^4}\left(1 + \frac{2l^2}{7h^2}\epsilon\right) \tag{9.27}$$

where ρ = density of material, kg/m³, and
h = beam's thickness, m.

The total strain ϵ is the sum of the built-in strain and the applied strain.

Calculations for beams with a nonuniform flexural rigidity EJ along the beam's length and with lumped masses are performed similarly. The following is an example of such a calculation for a beam shown in Figure 9.8b. The elastic line is approximated by the function $y_0 \sin(\pi x/4l)$. The maximum potential energy of the deflected beam is

$$U = \frac{1}{2}\int_0^{4l} EJ[\ddot{f}(x)]^2 \cdot dx = \frac{1}{2}\left[2EJ\int_0^l y_0^2\left(\frac{\pi}{4l}\right)^4 \sin^2\frac{\pi x}{4l}\cdot dx\right.$$
$$\left. + 2\cdot 2EJ\int_l^{2l} y_0^2\left(\frac{\pi}{4l}\right)^4 \sin^2\frac{\pi x}{4l}\cdot dx\right]. \tag{9.28}$$

After integration,

$$U = EJy_0^2 l\left(\frac{\pi}{4l}\right)^4\left(\frac{3}{2} + \frac{1}{\pi}\right). \tag{9.29}$$

Next is the calculation of

$$\sum_{i=1}^{n} m_i Y_i^2 :$$

$$\sum_{i=1}^{n} m_i Y_i^2 = 2m\left(y_0 \frac{\sqrt{2}}{2}\right)^2 + 2my_0^2 = 3my_0^2 . \tag{9.30}$$

Using formula (9.9) gives

$$\omega_0^2 = \frac{EJ\pi^4}{384l^3 m} \left(\frac{3}{2} + \frac{1}{\pi}\right) . \tag{9.31}$$

As mentioned in the beginning of this description, this method of calculation is applicable to idealized systems in which the dissipation of energy due to the effects of (1) internal thermoelastic friction in the material; (2) damping by the elements of the vibrator supporting construction; and (3) damping by the gas surrounding the vibrator are not considered. However, the performance of a number of sensors depends on and frequently is defined by these characteristics. In many designs, special attention is paid to creating structures and conditions of operation that will provide a high level of quality factor Q [269–271]. Normally, the higher the quality factor of the system, the better the measurement accuracy of the resonant frequency and, thus, higher the sensitivity of the sensor. Experiments show that a high-Q vibrator encapsulated in a vacuum enables detection of a force that corresponds to a mass of $100\mu g$. Monocrystalline silicon has potentially low damping because of low internal friction. It is reported that an extraordinarily high Q of 600 000 has been obtained [269] on a resonator oscillated in a vacuum at a frequency of 24kHz. Having a value of about 15 000 at atmospheric pressure, this vibrator can be used for a high precision sensor application, such as a pressure or absorption sensor. The vibrator (Fig. 9.9) is a miniature ($2 \times 4 \times 0.37$mm^3) rectangular vane with two bars attaching the vane to a supporting vane. These bars are each 2mm long and have a hexagonal cross section. The vane is electrostatically excited to vibrate torsionally around the long axis of the bar. In order to decouple the resonator from the clamping, a special frame surrounding the resonator is included into the construction. This frame itself can vibrate in a torsional mode but with about one-tenth of the vane's frequency. This kind of suspension is essential for achieving such a high Q. The structure is etched in (100) silicon (p-type, $5\Omega \cdot$cm) from both sides of a wafer simultaneously in 40%KOH at 60°C with SiO$_2$ mask.

The precision of measurement possible with resonant microsensors is inherently dependent upon the frequency stability of the sensor's output. This in turn is

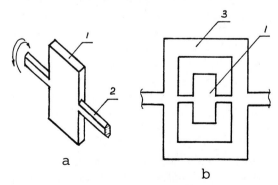

Figure 9.9 Torsion vibrator (a) and vibrator in frame (b). 1 = vane, 2 = torsion bar, 3 = frame.

defined by damping in the sensor's resonator. It is shown in one thorough work on the subject [271] that thermoelastic internal friction becomes a major part of resonant damping as the dimensions of a resonator become small. Several relationships are offered for the calculation of the damping effects:

$$\delta = \frac{1}{2Q} = \Gamma(T) \cdot \Omega(f), \tag{9.32}$$

$$\Gamma(T) = \frac{\alpha^2 TE}{4\rho C_p}, \tag{9.33}$$

$$\Omega(f) = \frac{2f_d f}{f_d^2 + f^2}, \quad \text{and} \tag{9.34}$$

$$f_d = \frac{\pi K}{2\rho C_p t^2} \tag{9.35}$$

where δ = thermoelastic critical damping ratio, dimensionless;
   Q = quality factor, dimensionless;
   $\Gamma(T)$ = peak magnitude of damping, dimensionless;
   $\Omega(T)$ = frequency response of the damping in relation to f_d, dimensionless;
   α = coefficient of thermal expansion of the resonator material, 1/K;
   T = resonator's absolute temperature, K;
   E = Young's modulus, N/m^2;
   ρ = vibrator's material density, kg/m^3;
   C_p = heat capacity at a constant pressure, J/kg $\cdot$ K or m^2/s^2 $\cdot$ K;
   f_d = characteristic damping frequency, Hz;
   f = operating frequency, Hz;
   K = thermal conductivity, W/m $\cdot$ K or kg $\cdot$ m/s^3 $\cdot$ K; and
   t = resonator's thickness, m.

The relevant material properties of single-crystal silicon and quartz given in Table 9.1 have been used for determining δ and Q [271]. It was found that the calculated and experimentally obtained data are in good agreement (Fig. 9.10).

As can be seen from the analysis, the resonator should be designed in such a way that its operating frequency is arranged away from the characteristic damping frequency. Note that thermoelastic friction in quartz is more than one order of magnitude higher than in silicon. In order to obtain a more explicit calculation of internal friction, we should take into account the dependence of thermal expansion and the elastic modulus on crystallographic direction.

Free-standing vibrating structures can be fabricated by lateral etching of a sacrificial silicon dioxide layer that has been sandwiched between the substrate and silicon film [268, 272, 273]. The devices containing such fabricated beams have a length between 100 and 500μm; are driven into resonance via electro-thermo-mechanically,

TABLE 9.1

Data on Material Properties for Calculation of Q (Source: [271])

Property	Units (in CGS)	Silicon	Quartz
α	ppm/K	2.60	13.70
E	1 × 10^{12} dyne/cm^2	1.70	0.78
ρ	g/cm^3	2.33	2.60
C_p	J/g $\cdot$ K	0.70	0.75
K	dyne/K $\cdot$ s	1.57	0.10
$\Gamma(T)$ at 300K	1 × 10^{-4}	1.06	11.34

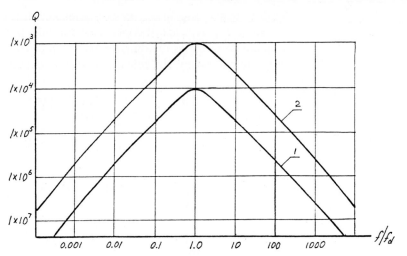

Figure 9.10 Q versus f/f_0 for silicon (1) and quartz (2).

optically, or electrostatically induced forces; and shift their frequency in response to the force creating strain in the beams. These beams can serve not only as resonating spring elements in force-measuring sensors, but also as components in displacement, attitude, acceleration, and other mechanical-value sensors. The resonant frequencies of such vibrators are quite high, between 100 and 1000kHz. The equations describing the excitation forces in electrostatically driven macro- and micro-devices are the same (see Chapter 3, Electrostatic Forces and Nonlinear Dynamics). Experimental verification shows that calculations quite accurately describe the real forces acting in the structures.

The electro-thermo-mechanical excitation is based on the thermal expansion of the beam material [268, 274], which results in a periodic thermal gradient producing a driving mechanical moment (Fig. 9.11). The resistive heating is proportional to the square of the applied harmonic voltage $U_{ac} \cos \omega t$; therefore, the frequency of the ac driver signal must be at half the resonant frequency of the structure if the ac voltage is the only excitation agent. However, if the ac signal is superimposed on a dc bias U_{dc}, the ac signal can be at the resonant frequency, which is more convenient in terms of the arrangement of a driving circuit. In this case, the electro-thermally generated heat $P(t)$ [W] is given by [275]:

$$P(t) = \frac{U_{dc}^2 + 0.5U_{ac}^2}{R} + \frac{2U_{dc}U_{ac} \cos \omega t}{R} + \frac{0.5U_{ac}^2 \cos 2\omega t}{R} \qquad (9.36)$$

where R = electrical resistance of the resistor generating heat, Ω.

The generated heat consists of three components: static, dynamic with frequency ω, and dynamic with frequency 2ω, the first, second, and third terms, respectively, in formula 9.36. The static component does not stimulate the vibratory

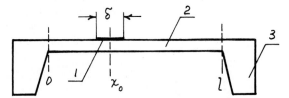

Figure 9.11   Model of electro-thermally excited resonating beam (source: [275]). 1 = resistor generating heat, 2 = beam, 3 = support.

motion, but creates a steady heat flux along the beam. The temperature-induced effects can be determined if we find the average temperature elevation ΔT_{av} [K]. The calculation of ΔT_{av} is simplified if we assume that (1) heat is uniformly generated over the resistor, (2) all heat flows through the beam (heat transfer by conduction) towards the clamped edges, and (3) the heat-sink conditions are ideal at the edges. If the heat generation is concentrated at the center of the resistor (the coordinate x_0), temperature ΔT_{av} is found as a function of x_0/l:

$$\Delta T_{av} = \frac{U_{dc}^2 + 0.5 U_{ac}^2}{R} \cdot \frac{l[x_0/l - (x_0/l)^2]}{2bhK} \tag{9.37}$$

where b, h, l = beam's width, thickness, and length, respectively, m, and
  K = thermal conductivity, which is 157W/m·K for Si.

A compressive axial stress σ [N/m^2] created by temperature change ΔT_{av} is calculated as follows:

$$\sigma = -E\alpha \cdot \Delta T_{av} \tag{9.38}$$

where E = Young's modulus, N/m^2, and
  α = coefficient of thermal expansion, 1/K.

The resonant frequency ω_0 can be determined using formula (9.27) or the following:

$$\omega_0 = \omega_r \sqrt{1 + \sigma/\sigma_{cr}} \tag{9.39}$$

where ω_r = resonant frequency with no stress in the beam ($\sigma = 0$), rad/s, and
  σ_{cr} = critical Euler buckling load, N/m^2.

$$\omega_r = 6.46(E/\rho)^{1/2} \cdot h/l^2, \qquad \sigma_{cr} = 4\pi^2 \, EJ/Al^2 \tag{9.40}$$

where A = cross-sectional area, m^2 (the other notations as before).
  The critical static power P_{cr} at which the beam will buckle is found from the equation

$$P_{cr} = \frac{2\pi^2}{3} \cdot \frac{Kb}{\alpha[x_0/l - (x_0/l)^2]} \cdot \left(\frac{h}{l}\right)^3 \cdot \tag{9.41}$$

As is evident from the preceding discussion, the thermal excitation causes a rise in the resonator temperature, which induces an axial stress leading to the shift in frequency. This influence is more significant than the effects of the temperature dependency of the material parameters. A high sensitivity to stress indicates that the mechanical packaging should be made considering the issues of heat transfer in the construction.

The thermal time constant of the system τ [s] is the other important characteristic of the thermal excitation. It is calculated as [268]

$$\tau = l(L - l)/D \tag{9.42}$$

where D = thermal diffusivity, cm^2/s (for silicon it is 0.87).
  Simple calculations indicate that τ is of the order of a tenth of a millisecond for a typical beam's length of 200μm, whereas the resonant frequency of such a beam is about 650kHz. In other words, the period of resonant oscillation is much less than the thermal time constant, and the efficiency of excitation is significantly decreased.

As a result, a large drive signal is required for the system excitation, the dc component is correspondingly increased, and the resonant frequency shifts. In addition to this, the quality factor also becomes a function of driving level. The shift of frequency interferes with the intended application of the device; consequently, driving structures via thermal excitation heaters is not very advisable.

The mechanism of optical excitation is close to that of electrothermal excitation with one difference: the driving force stimulating the periodic deformations of a vibrator is created not only by the optical-thermal-mechanical conversion, but also due to the photoelectrical processes in the material leading to the vibratory displacements of its elementary particles. A resonator can be excited by the absorption of pulsed laser or LED light transmitted through a fiber. An example of a typical construction of a vibrator [276] is a silicon microfabricated cantilever with dimensions $1000 \times 80 \times 5\mu m^3$ and with a paddle of $300 \times 300\mu m^3$ at the free end. The cantilever deflection can be measured by a polysilicon piezoresistor located at the cantilever's base, by a reflective multimode fiber-optic pick-up system, or by other means. The cantilever is excited by light at 780 or 830nm, and it oscillates at a frequency of about 2kHz. The vibratory system has a quality factor of about 40. Vibrations can be induced by a source generating a light noise [277] rather than a signal of single frequency. This arrangement is convenient when a number of vibrators having different resonant frequencies are excited by one source of energy. An interesting design is a self-excited resonator maintaining its resonance by optical feedback [278], which is provided by the formation of a Fabry-Perot cavity between the microresonator surface and the end of the optical fiber. The resonators in this construction are metal-coated silicon dioxide strips having a length of the order of 200μm.

Recently, there have been several developments [279–281] of structures resonating parallel to the plane of the substrate by means of one or more interdigitated capacitors (Fig. 9.12) called electrostatic combs. Besides resonators, these structures with laterally moving parts have taken a variety of shapes including microscopic rotating disks restrained by springs [202, 282, 283], gripping devices, and tweezers [284]. Some advantages of the lateral motion resonators are (1) less damping effect of air, leading to higher quality factors, (2) linearity of the electrostatic-comb drive, and (3) flexibility in the design of the suspension for the resonator. The quality factor for such resonators approaches 50 000.

In several novel designs, the principle of a resonator's amplitude or frequency modulation by physical quantities is successfully realized for building microsensors. A solid-state gyroscopic sensor [285] exploits the piezoelectric effect in quartz both to excite a reference vibration in the plane of a tuning fork and to sense the vibration normal to this plane due to the angular velocity of applied rotation. The resonance frequency is quite high, 38.5kHz, and, thus, enables the sensor to operate in

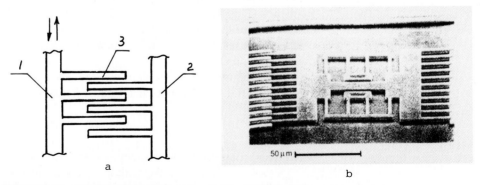

Figure 9.12　Outline of comb structure (a) and SEM of linear resonant plate (b) with two electrostatic-comb drives (source: [280]). 1 = movable plate, 2 = stationary plate, 3 = electrode.

a vibration-rich environment. The fork contains sense electrodes, and the amplitude of the signal picked up from these electrodes is proportional to the measured speed of the device's rotation around one of the axes.

Tiny piezoelectric crystals attached to a pendulum are used for the measurement of acceleration [286]. The nominal oscillation frequency of one resonator is 35kHz, and a full-scale frequency change is 3500Hz. The structure is differential; with an increase of one frequency, the other one is decreased, and a digital counter takes the difference of the two frequencies. Another construction is a two-axis silicon microaccelerometer [200], which consists of two pairs of polysilicon resonant bridges orthogonally attached to a silicon proof mass. Acceleration in the plane of the substrate causes differential axial loads on the opposing microbridges, giving a shift of their resonant frequencies. The acceleration is measured as a difference in the resonant frequencies. The device has a 1.45-mg mass; the dimensions of one bridge are $250 \times 100 \times 1.6\mu m^3$. The bridges are driven electrostatically, and their sensitivity to the acceleration is 160Hz/g.

Several micromachined silicon resonant pressure sensors show promising characteristics. In one of them [287], a resonator fabricated on the surface of a diaphragm (Fig. 8.31b and Fig. 9.13) consists of a coupled pair of 8-μm-thick suspended rectangular plates. The device is batch fabricated from single-crystal silicon. Pressure deforming the diaphragm distorts and stresses the ribs that support the plates, modifying the resonant frequency of the system. The vibrations can be excited optically or by a piezoelectric crystal incorporated into the construction. The resonator can vibrate at a minimum of fifteen vibrational flexural modes. Each of them can be regarded as a resonance of the first-order antisymmetrical Lamb acoustic wave that propagates in the resonator's plate. In one model used for measuring pressure, the point of the maximum deflection (antinodal point) was at the center of the vibrator (Fig. 9.13). At this mode, both plates are driven up and down simultaneously by the supporting strips, and the frequency of oscillation changes from 98 to 128kHz with a differential pressure change from 0 to 100kPa. The quality factor of the device is between 13 000 and 21 000. It is typical for such a structure that some modes exhibit very close frequencies of resonance within a certain range of differential pressure. This closeness leads to coupling between the modes and limits the use some of them in a practical system because the electronics could easily be misled in the coupling region (there would be a "jump" of frequency at the electrical output).

Another construction [288] is a sensor fabricated from a single silicon crystal containing two resonant strain gages held in vacuum cavities on the surface of the

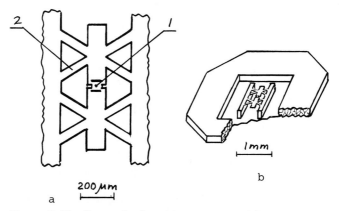

Figure 9.13 Resonator for pressure sensor (a), pressure-sensitive diaphragm (b) with resonator on it (source: [287]).
1 = untinodal point for one of modes of vibration,
2 = support.

diaphragm. Due to this arrangement, the gages are isolated from the surrounding fluid (Fig. 9.14 and Fig. 8.31a). The strain gages are fabricated by a three-dimensional semiconductor process: a silicon interface chip is bonded to the diaphragm chip by thermal oxidation bonding, and an Ni-Fe alloy pressure port is bonded by Au-Si eutectic. A strain gage is constituted of a flexure resonator with four ends fixed to the surface of the diaphragm. The strain caused by the pressure-deformed diaphragm alternates the frequency of the resonator. With the pressure increase in the cavity, the frequency of the central gage is increased and the frequency of the corner gage is decreased. A permanent magnet creates the field whose interaction with the side strips of the H-shaped pattern creates a driving force and amplifier-input voltage. With the pressure change from 0 to about 100kPa, the frequency shift of each gage is about 6kHz (the central frequency is ~46kHz). The Q factor is about 50,000, which gives an oscillation stability of better than 1ppm. The accuracy is as high as 0.01% FS. The temperature coefficient of the frequency is -40ppm/K.

The pressure-sensitive sensor shown in Figure 9.15 [289] consists of a square dual-diaphragm cavity structure that vibrates in a balanced torsional mode. This mode, in combination with a suspension along the two crossing node lines of the electrostatically driven parts, yields a high Q factor of 2,400 in air and 80,000 in a vacuum. The dependence of the frequency of vibration on pressure is a result of pressure-induced changes in the torsional stiffness of the vibrating body. The fabrication process includes anisotropic silicon etching and thermal bonding. The excitation and displacement-detection method is realized by arranging the electrodes

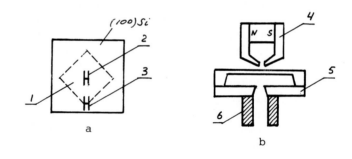

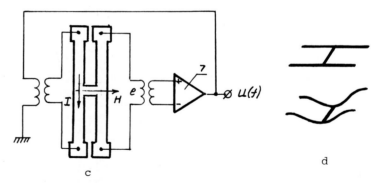

Figure 9.14 Silicon resonant-pressure sensor with electromagnetic excitation (source: [288]). a = top view of diaphragm chip, b = outline of structure, c = schematic diagram of gage and excitation, d = resonant mode; e = pickup voltgage, *I* = excitation current, *H* = magnetic field, *U(f)* = frequency output, 1 = diaphragm, 2 and 3 = resonant strain gages, 4 = permanent magnet, 5 = interface chip, 6 = pressure port, 7 = amplifier.

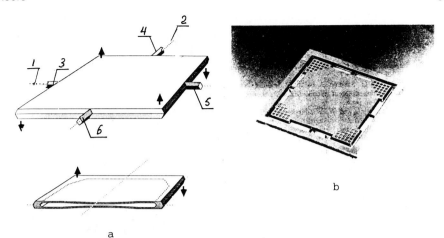

a

b

Figure 9.15 Torsionally vibrating pressure-sensitive diaphragm structure (a) and photograph of diaphragm with extended perforated corners (b) (source: [289]). 1 and 2 = node lines and supports, 3–6 = supports. Differential-pressure sensor has cavity-connecting channel in one of supports.

of the capacitors at the moving corners of the diaphragm and at the stationary portion of the structure underneath the corners. The Q factor can be enhanced by increasing the masses at the corners. The oscillation is accompanied by viscous energy losses, especially at the corners where the amplitude of vibration is large. These losses are reduced by making holes at the corners through which the air is free to move from the upward to the downward side (Fig 9.15b). The typical central frequency of the device is 18kHz; the frequency's change is between 14 and 19% per bar. The inherent uncompensated temperature sensitivity of the frequency is − 16ppm/K.

A miniature cantilever oscillator with a paddle tip of single Si crystal fabricated by lithography and an etching technique can be used as a vacuum sensor, covering the range from 1×10^{-2}Pa to atmospheric pressure, 1×10^{5}Pa [290]. The paddle tip (Fig. 9.16) is driven electrostatically by applying dc-biased ac voltage between the cantilever and stationary electrode. Depending on the cantilever and gap dimensions, the natural frequency of oscillation is between about 76 and 3 660Hz. With the pressure change over the given range, the amplitude of the paddle vibration experiences changes by approximately three orders of magnitude. The effects of the damping force from the ambient gases are explained by the free molecule flow and viscous flow models.

Several techniques have been developed for measuring mass flow by means of vibrating microstructures. A thermally excited microbridge with thin-film resistors embedded into it [291] changes its resonant frequency in response to the variation in the flow rate. The bridge is made across a V-shaped groove (Fig. 9.17) by KOH

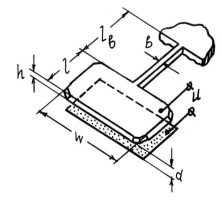

Figure 9.16 Schematic structure of cantilever vibratory vacuum sensor (source: [290]). $0.01 \leq b \leq 0.70$mm, $0.04 \leq h \leq 0.35$mm, $1.48 \leq l_b \leq 6.00$mm, $1.80 \leq w \leq 9.00$mm, $1.20 \leq l \leq 3.80$mm, $0.06 \leq d \leq 0.56$mm, U = driving voltage.

Figure 9.17 SEM photograph of silicon nitride microbridge ($600 \times 200 \times 2.1 \mu m^3$) with phosphorus-doped polysilicon resistors and chromium-gold interconnection, suspended across V-shaped groove in (110)-silicon substrate (source: [291]).

front-side anisotropic etching of (110)-silicon substrate. The resistors are used as strain gages for detecting and as heat generators for forcing the vibrations. The ac and dc voltages are superimposed and applied to the excitation resistor. The generated dynamic heat induces a bending mode vibration by thermal expansion of the bridge's upper layers. A feedback loop amplifier fed from the Wheatstone-bridge-connected strain gages completes the excitation circuit. The static temperature of the microbridge, caused by the dissipation in the resistors, is dependent on the heat transfer by conduction and forced convection. This temperature influences the frequency of oscillation. The central frequency of 85kHz at a 20°C temperature elevation of the microbridge is shifted by 0.8kHz when the flow changes from 0 to 10SCCM.

It has been shown [292] that the drag forces developed by the moving fluid on a vibrating object placed into the flow damps the resonance oscillations of the vibrator proportionally to the mass flow. One construction of a tiny sensor based on this concept is composed of a small cantilever blade driven at resonance (1–5kHz range) by piezoelectric actuators. A typical blade is made of steel or silicon with sizes $7 \times 2 \times 0.05 mm^3$. The blade is attached to the piezoelectric bimorph. The tip of a typical blade vibrates with an amplitude of 0.5mm. The effect of flow on frequency is very small; however, the amplitude of vibration for constant excitation varies appreciably with a flow velocity change. For example, it is reduced approximately by factor seven when air flow velocity increases from 0 to 24m/s (the vibrator's frequency is 1 125Hz, and the Q factor is about 500).

Similarly to conventional vibratory transducers, micromachined resonators can be used for the measurement of the properties of some substances when these properties can affect the mass, stiffness, or viscous friction of the vibratory member. The substances can be applied to the vibrating surface either directly or via an absorber that coats the surface of the member. Examples of such devices can be found in [293–295].

Microaccelerometers

The basic principles of operation of a microaccelerometer and a traditional accelerometer (see Chapter 5, section Inertial-mass Elements) are the same: an elastically suspended and damped proof mass experiences a force proportional to the acceleration; this force, being translated into the displacement or strain of a transduction element member, develops an electrical signal proportional to the acceleration. In terms of strain- or displacement-to-electrical signal conversion, the transduction elements used in accelerometers can be divided into several groups:

1. *Switch-type* [296]. This is the most simple element. It is fabricated of a silicon, silicon dioxide, or metal fine beam deflecting under acceleration and providing closure of a single-point contact.

2. *Piezoresistive-type* [209, 297–300]. As a rule, these elements are silicon semiconductor strain gages formed in the areas of maximum stress caused by the acceleration-induced force. Possible applications of ink-made thick-film strain gages for microaccelerometers have been reported [301].

3. *Piezojunction transistor-type* [302, 303]. Similarly to the strain-gage type, the strain-sensitive transistor is formed in the silicon monolithic structure at those spots that experience the maximum stress due to acceleration.

4. *Capacitive-type* [304, 305]. Commonly, these elements are formed by an anisotropic etching technique of silicon with the depositing of a thin metal film, such as gold, on the moving and stationary surfaces forming the capacitance.

5. *Piezoelectric-type* [306, 307]. These elements have a basic configuration similar to the conventional piezoelectric element (including the bilaminar one), but the shaping of the charge-forming beam is provided at the fraction-of-a-millimeter scale. The surface-acoustic-wave devices can be related to this group.

6. *Servo-type* [308–311]. In servo microaccelerometers, the displacement- or strain-sensitive elements are variable capacitors or strain gages. The restoring force or torque is developed by the electrostatic or electromagnetic systems.

7. *Micromachined resonator* [312, 313]. The excitation is normally electrostatic or either photothermal (using incident light from an optical fiber) or electrothermal (using joule heating in a diffused resistor).

Figure 9.18 illustrates different suspensions of the proof mass in the etched-silicon structures. The strain gages are normally formed on the bridges where flexural stresses are concentrated. The surface of a moving plate is used to form one of the electrodes of the capacitance-type sensor. A cantilever structure provides the highest sensitivity of the transduction element to the acceleration (*a* and *g*). The suspension of the proof mass at four points (*b, c,* and *f*) increases the stiffness of the system for the angular displacements around the axis perpendicular to the plane of the chip. In some constructions (*e*) mass is added to one side of the moving plate. Under acceleration, the plate experiences angular displacement. In the construction shown in (*f*), the outer frame plays a role of proof mass, whereas the central part is used for support. This arrangement decreases the sizes of the device. The cantilever with three bridges supporting the mass at one side is used in the resonant-type sensor. The upper bridge is excited to vibrate at the flexural mode; and the frequency of vibration, proportional to the stress in the bridge due to the acceleration-induced force, is proportional to the acceleration.

In the design of microaccelerometers, a number of factors should be taken into account:

1. In order to obtain the maximum performance from an piezoresistive-type accelerometer, the product of the two parameters of primary interest has to be maximized. These two parameters are sensitivity and resonant fre-

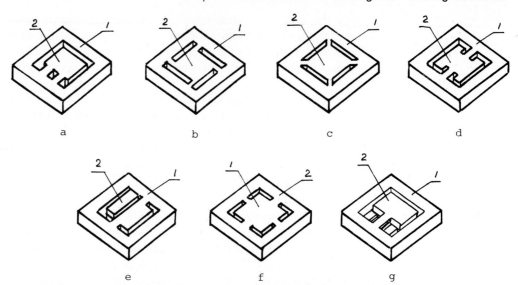

Figure 9.18 Suspension of proof masses in microaccelerometers. a = two bridges at one side of proof mass, b and c = four bridges at corners, d = two bridges at central line (translational deflection of mass), e = two bridges at central line (angular deflection of mass), f = four bridges at central lines (support is at center, and surrounding frame is proof mass), g = three bridges at two planes; 1 = support, 2 = proof mass.

quency. The sensitivity is defined by the strain at the places that are chosen for the concentration of stress and location of strain gages. The frequency can be increased if the deforming structure is made stiffer. Good results can be obtained if the areas with gages are thinned, while the other parts of the moving elements are made thicker. However, the increase in thickness leads to an increase in mass and ultimately to a decrease in frequency. Details of the geometry of these elements must be thoroughly analyzed to find the optimum for frequency and sensitivity.

2. Overload protection is another item to be mentioned. Usually a flexible element of the sensor continues to bend under overload accelerations in the same way that it bends under normal accelerations. When the stress exceeds a tensile stress, the beam breaks. An effective measure to prevent the device from breaking is adding stops in the designs (Fig. 9.19). A precision placement of stops is important since the overall travel of the moving parts is usually within several micrometers. In the other directions perpendicular to the direction of deformation, protection is provided by the strength of the beams themselves.

3. There are three types of damping that are used in microaccelerometers: air, liquid, and electrical. The latter is provided in the force-balance sensors by means of a feedback loop. Damping is necessary for the depression of the unwanted high-amplitude vibration of the moving system at its mechanical resonance. It is also desirable for obtaining a more uniform amplitude characteristic and for creating a critical damping. The latter allows a high-sensitivity conversion in a relatively flexible system, yet with a good range of the frequencies of the measuring acceleration. Liquid damping is quite effective but it suffers from several problems [209]: reduction in the sensitivity of the device because of the density of the liquid (which tends to buoy the proof mass), viscosity variation with temperature, and cost associated primarily with the liquid filling procedure. Gas damping [314, 315] has minimal problems related to density and viscosity. Air damping is

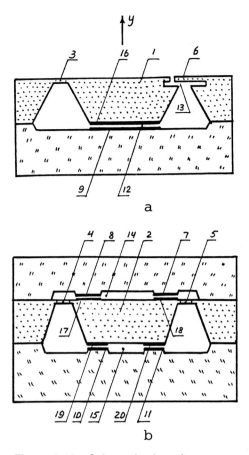

Figure 9.19 Schematic view of cantilever beam (a) and beam with bridges at end points (b). Y = direction of acceleration; 1 and 2 = proof masses; 3, 4, and 5 = bridges; 6, 7, and 8 = stops for upward travel of proof mass; 9, 10, and 11 = stops for downward travel of proof mass; 12, 13, 14, and 15 = spaces for damping; 7–11 and 16–20 = electrodes for capacitive sensing and/or generating force. Note that in piezoresistive sensors, strain gages are formed at bridges; in capacitive sensors, electrodes at gaps are used for sensing displacement and inducing restoring force.

usually the method of choice for correctly designed systems. Air damping is achieved by choosing the proper gap in the system between the moving and stationary parts. Since the gap is defined also by the prescribed travel of the proof mass and by the condition of the maximum allowed deformation, there should be an engineering tradeoff between the parameters of interest.

4. An accelerometer designed to measure the acceleration along one of the axes should not sense the acceleration in the other directions. This is achieved by the design of the mechanical structure in such a way that its compliance along the axis of interest is much greater than along the other axes. If piezoresistive elements are used, the orientation of the elements

and their interaction [300] can be optimally chosen and any undesirable responses of the system will be depressed.

The typical processes described in the preceding sections are applicable to the fabrication of the elements of microaccelerometers and for the assembly of their parts. Such processes as anisotropic etching, diffusion, passivation, and metallization are typical in the fabrication of the devices.

The concept of self-test and self-calibration has found its development in accelerometers [316–318]. The typical self-test is conducted by applying a normalized voltage to the static-force generating electrodes and by recording the output of the sensor. When the test and acceleration signals are superimposed, the check of performance is made by the subtraction of the memorized signal, generated by the acceleration only, from the combined signal.

For the calculation of basic characteristics of microaccelerometers, the methods and formulas introduced in the preceding sections can be used. The most common calculation is conducted by applying principles of the theory of elasticity for determining strains, displacements, and natural resonant frequencies of elastic structures. We will illustrate a solution of a two-dimensional problem [300] that can be easily modified for the analysis of those structures in Figure 9.18 that can be reduced to a beam (Fig. 9.20) having two fixed supports at the ends, elastic sections next to the ends (bridges), and a stiff central section representing the proof mass. We are interested in finding the two major relationships, assuming that the device in question utilizes strain gages formed in the flexible sections of the beam. The first relation is the strain $\epsilon(a)$ along the x-axis versus acceleration, and the second one is the resonant frequency $f(a)$ versus acceleration.

For the system shown in Figure 9.20, the differential equation governing elastic lines is [319]:

$$\frac{Eb_1d_1^3}{12} \cdot \frac{d^2w}{dx^2} = M_0 - a\left(m_1 + \frac{m_2}{2}\right)x + \frac{m_1a}{2l_1}x^2 \qquad (9.43)$$

where
E = Young's modulus of the elastic section material, N/m^2;
b_1 = width of beam's elastic section, m;
d_1 = elastic section's thickness, m;

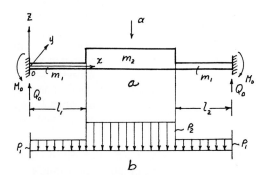

Figure 9.20 Model of two-dimensional structure with two elastic sections and central proof mass. a = beam, b = diagram of distributed load emerged due to acceleration a; l_1 and l_2 = lengths of elastic sections, M_0 and Q_0 = moment and transverse force in supporting points, m_1 and m_2 = masses of elastic and stiff sections, respectively, p_1 and p_2 = distributed loads induced by acceleration.

w = beam's deflection along the z-axis, m;

M_0 = supporting point moment, N · m;

a = acceleration along z-axis, m/s^2;

x = coordinate along x-axis, m;

m_1 and m_2 = masses of elastic and stiff sections, respectively, kg; and

l_1 = length of elastic section, m (it is assumed that $l_1 = l_2$).

The solution of this equation allows deriving $\epsilon(a)$ for the obvious conditions: $(m_1/m_2) \ll 1$ and $\dot{w} = 0$ at $x = 0$ and $x = l_1$. The strain for the fiber at the surface of the flexible section is

$$\epsilon(a) \approx \frac{3l_1 m_2}{2Eb_1 d_1^2}\left(2\frac{x}{l_1} - 1\right)a. \tag{9.44}$$

The dynamic response [320] is found from the following equation:

$$\frac{Eb_1 d_1^3}{12} \cdot \frac{d^4 w}{dx^4} = -\frac{m_1}{l_1} \cdot \frac{d^2 w}{dt^2}. \tag{9.45}$$

In this equation, w is a function of two variables: coordinate x and time t. The separation of variables

$$w = X(x) \cdot T(t) \tag{9.46}$$

results in the time-dependent equation

$$T(t) = C_0 \cos(2\pi f t + \alpha) \tag{9.47}$$

and in the position-dependent equation

$$X(x) = C_1 \cos\left(\frac{\lambda}{l_1}x\right) + C_2 \sin\left(\frac{\lambda}{l_1}x\right) + C_3 \cosh\left(\frac{\lambda}{l_1}x\right) + C_4 \sinh\left(\frac{\lambda}{l_1}x\right) \tag{9.48}$$

where C_0, C_1, C_2, C_3, C_4, and α are constants, and

$$\lambda = \left(\frac{48\pi^2}{E} \cdot \frac{m_1 l_1^3}{b_1 d_1^3}f^2\right)^{1/4} \tag{9.49}$$

The constants of integration are calculated when the following conditions are taken into account: the deflection is allowed only in the z-direction, the flexing section is rigidly fixed at the support and at the proof mass section, and the transverse force is continuous. Therefore,

$$X\bigg|_{x=0} = \frac{dX}{dx}\bigg|_{x=0} = \frac{dX}{dx}\bigg|_{x=l_1} = 0 \quad \text{and} \tag{9.50}$$

$$E\frac{b_1 d_1^3}{12} \cdot \frac{d^3 w}{dx^3}\bigg|_{x=l_1} = \frac{m_2}{2} \cdot \frac{d^2 w}{dt^2}\bigg|_{x=l_1}. \tag{9.51}$$

The eigenvalue equation results from the boundary conditions

$$\frac{m_1}{m_2} = \frac{\lambda}{2} \cdot \frac{1 - \cos\lambda \cdot \cosh\lambda}{\sin\lambda \cdot \cosh\lambda + \cos\lambda \cdot \sinh\lambda}. \tag{9.52}$$

There exists an infinite number of λ for the given ratio m_1/m_2. Using equation 9.49 and series expansion of equation 9.52 enables one to calculate the resonant frequencies of the structure. The fundamental frequency f_0 [Hz] is approximately defined as

$$f_0 = \frac{1}{2\pi} \sqrt{\frac{2Eb_1d_1^3}{l_1^3 m_2}}.$$ (9.53)

Concluding this section, we will note that the market for microaccelerometers is expanding quite rapidly. It is anticipated [312] that a huge demand for micromachined sensors will arise from automotive industries.

Pressure Microsensors

Single-crystal piezoresistive pressure sensors, which are the most successful microelectronic sensing devices, have found their applications in biomedical instruments, automotive equipment, process-control systems, and so on. The outstanding mechanical and electrical properties of single-crystal silicon make them ideal for different applications [321].

A number of issues related to the structure, fabrication, and function of pressure microsensors have been considered in the preceding sections. In this section, we will briefly review only the material that has not been discussed, especially the most recent advances in sensor designs.

Due to the convenience of their fabrication, the absolute majority of pressure-sensitive structures are built with a rectangular diaphragm. Each strain gage is a two-terminal element whose longitudinal and transverse axes, X and Y, respectively, are oriented in such a way with respect to the crystallographic axes that the sensor's response is optimized. As a rule, the condition of optimization corresponds to the maximum sensitivity. The strain gage, in general, is sensitive to stress in three principal axes, and to shear stress in three planes. Besides stresses, voltages and currents related to three directions must also be specified in order to define the functions of the element. A simple-geometry strain gage formed in the diaphragm is subjected to planar stress, and one's attention can be limited to the voltage and current in the X direction and stress in X direction. However, if a strain gage is integrated into a planar force collector, the principle stresses are in the X and the Y directions. A number of orientations are used in sensors. Some of them are given below as suggested in [209]:

1. (110) plane, $<X> = <111>$,* and $<Y> = <111>$.

    This orientation is commonly used in sensors. The [111] is the axis of densest packing, and the maximum piezoresistive effect is observed along this axis.

2. (110) plane, $<X> = <110>$, and $<Y> = <001>$.

    In certain geometrical situations, cross-axis sensitivity is not desirable. The net change in the resistance of the gage is the sum of the on-axis and cross-axis responses to stress. If the signs of these sensitivities are different, the net change in resistance is reduced. This situation takes place when the gage is placed at the center of a square or round diaphragm where stresses are equal and orthogonal. For such a case, the total sensitivity can be increased if the cross-axis effect is minimized. For the given orientation, the cross-axis sensitivity is close to zero.

3. (100) plane, $<X> = <011>$, and $<Y> = <011>$.

    This is the most widely used orientation for silicon integrated sen-

---

*Angle-bracket numbers define the strain-gage X- or Y-axis vectorial orientation within (100) and (110) planes as shown in Fig. 7.23.

sors. (100) material is commonly used in conventional integrated circuits and is widely available. This orientation provides equal- and opposite-sign sensitivities for the on- and cross-axis stresses. In a clamped silicon diaphragm, the maximum stress is normal to the edge of the diaphragm and relatively small parallel to the edge. Therefore, it is desirable to have a high cross-axis sensitivity in the sensor placed at the edge. Radial and transverse gage combinations allow equal and opposite resistance changes when the gages are placed at the high-stress edge of a diaphragm. This arrangement is convenient for connections in a Wheatstone bridge.

4. (100) plane, $<X> = <010>$, and $<Y> = <001>$.

The elements having this orientation show a small piezoresistive effect and they can be used for sensing temperature if fabricated simultaneously with the piezoresistive elements. The temperature coefficients of resistance for both strain- and temperature-sensitive elements are equal, which simplifies interfacing with signal conditioning electronics, such as a microprocessor.

Another strain-sensitive gage used in pressure microsensors is a transverse voltage shear strain gage described later in this chapter in the section about semiconductor strain gages. This gage is placed at one of the diaphragm's edges and its arms make a 45° angle with the edges to sense the maximum shear stress.

In the calculation of sensors' performance characteristics, we usually are interested in finding the diaphragm's stress-pressure and deformation-pressure characteristics. In addition to the simplified relationships given in Chapter 5 in the section Elastic Elements, we will reproduce the calculation [322] of a diaphragm's deformation and stress based on a finite element analysis. A schematic diagram of a pressure sensor chip micromachined in (100)-Si (the most common orientation) is shown in Figure 9.21. The deflection of the diaphragm's surface is described by the following relationships:

$$w(\xi, \eta) = w(0, 0) \cdot F(\xi, \eta), \tag{9.54}$$

$$F(\xi, \eta) = (\eta^2 - 1)^2(\xi^2 - 1)^2(1 + C_1\eta^2 + C_2\xi^2 + C_3\eta^2\xi^2), \tag{9.55}$$

$$w(0, 0) = C_0 w^4 P/D, \tag{9.56}$$

$$D = Eh^3/(1 - \nu^2), \quad \xi = 2x/l, \quad \eta = 2y/w \tag{9.57}$$

where
    $w(\xi, \eta)$ = deflection of a point on the diaphragm's surface having coordinates ξ and η [m];

    ξ and η = normalized coordinates of points located at the distances x and y from the diaphragm axes, respectively, dimensionless;

    $w(0, 0)$ = deflection at the diaphragm center, m;

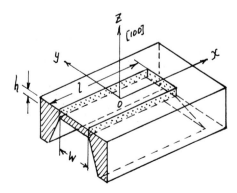

Figure 9.21 Schematic diagram of pressure-sensor chip with rectangular diaphragm (source: [322]).

$C_0, C_1, C_2,$ and C_3 = constant coefficients, dimensionless;

P = pressure difference across the diaphragm, Pa;

D = flexural rigidity, N·m;

h, l, w = thickness, length, and width of the diaphragm, respectively, m; and

ν = Poisson's ratio, dimensionless.

Stress components σ_x and σ_y [N/m^2] are calculated from the given set of equations:

$$\sigma_x(\xi, \eta) = A(F_{xx} + \nu F_{yy}), \tag{9.58}$$

$$\sigma_y(\xi, \eta) = A(\nu F_{xx} + F_{yy}), \tag{9.59}$$

$$A = 48C_0 f(z) P \left(\frac{w}{h}\right)^2, \tag{9.60}$$

$$F_{xx} = a^{-2}(\eta^2 - 1)^2 \{C_2 - 2 + \eta^2(C_3 - 2C_1) \\ + 6\xi^2[1 - 2C_2 + \eta^2(C_1 - 2C_3)] + 15\xi^4(C_2 + \eta^2 C_3)\}, \tag{9.61}$$

$$F_{yy} = (\xi^2 - 1)^2 \{C_1 - 2 + \xi^2(C_3 - 2C_2) \\ + 6\eta^2[1 - 2C_1 + \xi^2(C_2 - 2C_3)] + 15\eta^4(C_1 + \xi^2 C_3)\} \tag{9.62}$$

where $f(z) = 1 - (2z/h)$ describes the linear decrease from the surface ($z = 0$), and $a = l/w$ is the aspect ratio of the plate. The following table gives the values of coefficients C_0 to C_3 for square ($a = 1$) and rectangular ($a = 2$) plates.

a	C_0	C_1	C_2	C_3
1	1.38×10^{-3}	0.219	0.219	0.320
2	2.64×10^{-3}	0.0223	1.444	0.338

The relatively new silicon *micromachined pressure switch* [323, 324] contains an etched, square, 1-mm^2 diaphragm fabricated by the same technique used in the fabrication of pressure sensors, but, instead of pressure-sensitive strain gages being used, a metal pattern is deposited in an etched shallow well to form contacts in the inner surface of the diaphragm (Fig. 9.22). Metal electrodes opposing those on the diaphragm are made on a glass substrate. Under pressure, the diaphragm deforms, and the contacts on the silicon touch the electrode on the glass, making the contacts close. The resistance of one closed contact is about 1kΩ. The pressure set points range from 0.5 to 13bar. The switch can operate at temperatures from $-150°C$ to $+200°C$. The repeatability of the set point and the pressure hysteresis are better than 0.015bar.

As we know, a conventional piezoresistive, micromachined design employs a flat, square, rectangular, or round diaphragm with four piezoresistive elements forming a Wheatstone bridge. This construction has proved to be successful for the measurement of full-scale pressure ranges of 35kPa and higher. For the measurement of low pressures between 0.1×10^3 and 7×10^3kPa, the thickness of the flat diaphragm must be significantly decreased in order to obtain an appreciable deflection

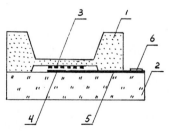

Figure 9.22 Cross-sectional view of pressure switch (source: [324]). 1 = silicon, 2 = glass, 3 = contact, 4 = electrode, 5 = diffused feedthrough, 6 = bond pad.

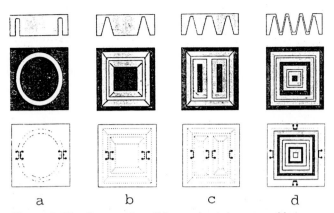

a   b   c   d

Figure 9.23 Geometries of bossed pressure-sensitive highly-linear diaphragms (source: [325]). a = round bossed, b = square bossed, c = double bossed, d = convoluted; C-shaped elements are strain gages.

or the strain needed for an adequate sensitivity of the instrument. This decrease in the thickness leads to an undesirable pressure-deflection or pressure-strain characteristic. The solution of this problem has been found in the creation of diaphragms with local stiffening and with the placing of strain-sensitive gages in the thinned areas [325]. The rigid bosses allow the diaphragm to sustain relatively high curvatures, resulting in high bending moments and strains in the thin regions where the strain gages are located. Various geometries of bossed and convoluted diaphragms having enhanced-linearity characteristics are illustrated in Figure 9.23. The design of such diaphragms is based on a fully integrated mechanical, electrical, and process analysis. Finite element modeling and analytic models including stress control layers, doping levels, junction depth, etc. [70, 322, and 325–327] are the elements of this analysis. A typical finite-element model that is used to calculate the diaphragm's properties is shown in Figure 9.24.

During the design and fabrication of such a sensitive diaphragm, special attention must be given to a number of factors [325]:

1. *Precise dimensional control of boss geometry.* Rather deep bosses are etched using a variety of KOH, EDP, electrochemical and high concentration etch-stop techniques. This process should be economical as well as accurate.

2. *Precision control of diaphragm thickness.* The diaphragm's thickness is between 2 and 12μm, and the trend is to decrease this size. To achieve precision in etching such diaphragms, high-concentration-layer and p-n-junction-isolation etch-stop techniques are used.

3. *Implant depth.* Since the typical junction depth of 1 to 5μm is comparable with the diaphragm's thickness, the junction depth should be shallow and

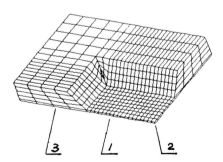

Figure 9.24 Finite-element mesh shown for one quadrant of diaphragm chip (source: [327]). 1 = diaphragm, 2 = chip center, 3 = rigid rim.

well-controlled to avoid a reduction in sensitivity (the active gage layers should not be spread over the excessive depth).

4. *Resistor geometries.* Due to the presence of bosses, a small area is available for the placement of resistors, and the resistor's geometries are accordingly small.

5. *Stress control.* Stresses of the diaphragm, which can be caused by imperfections in mounting and differences in the thermal expansion of parts, must be avoided to prevent the diaphragm from buckling and bimetallic-effect stress. The materials used to form the structure are borosilicate glass, silicon, epitaxially deposited silicon for the basic flexure, and silicon dioxide and silicon nitride for dielectrics.

A search for optimal diaphragm shapes and technologies is in progress and gives new solutions (e.g., a twin-isle diaphragm in [328]). Another advance in the development of pressure sensors has been reached in the building of structures having both a high sensitivity to pressure and a tolerance of very high pressure. The typical sensor used for a 35-70kPa full-scale application can survive overpressures of up to 35 000kPa [321], which means that the sensor has a 500× overpressure protection. This level of overpressure tolerance is not possible using conventional sensor structures in which the diaphragm is not restricted to movement under pressure. Advanced processes in the fabrication of pressure-sensitive structures have made possible constructions with integral stopping surfaces (Fig. 9.25). With a pressure increase, the stopping surface beneath the diaphragm limits the diaphragm's displacement. It is notable that sensors show no degradation in performance characteristics after repetitive cycles of overpressure. For example, a sensor having a sealed cavity and polysilicon diaphragm [329] with full-scale pressure ranging from 0.24 to 2.0MPa, after repetitive cycles of overpressure to 20MPa, still has a zero repeatability of better than 0.1% of span.

It is proved that fusion bonding of these, only silicon-made, parts has a significant impact on the characteristics of sensors because the thermal mismatch of parts is practically eliminated.

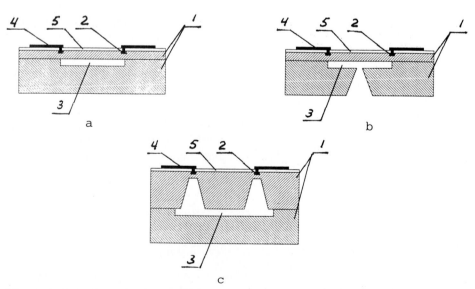

Figure 9.25 Silicon fusion bonded pressure sensors with integral overpressure protection (source: [321]). a = absolute-pressure sensor, b = gage-pressure sensor, c = sensor with bossed diaphragm; 1 = silicon, 2 = piezoresistor, 3 = etched cavity, 4 = metal, 5 = insulator.

Polysilicon as a material for piezoresistive diaphragms has been investigated in several works, such as [330 and 331]. The electrical and mechanical characteristics of polysilicon are extremely sensitive to deposition conditions [332]. Polysilicon, when LPCVD-deposited from silane, has many forms [330]. For deposition near 580°C, a very fine-grained silicon film is obtained, whereas deposition near 640°C produces rather coarse-grained or large-crystallite films. As they are deposited, films produce compressive strain that can be reduced by anneal procedures (3 hours at 1150°C). It should be noted that the piezoresistive coefficients of polysilicon are substantially lower than those in single-crystal silicon, and transverse stress sensitivity is only about 20% of the longitudinal coefficient. This creates a problem in the use of the conventional resistor-bridge arrangement of the diaphragm. The long-term stability of a polysilicon spring element working under high-stress and temperature-excursion conditions has not been fully studied. The temperature characteristics of microcrystalline and polycrystalline silicons greatly depend on doping concentration, grain size, external stress, non-ohmic contact resistance, and so forth [333].

As we know, a conventional pressure sensor with diffused piezoresistors usually contains a p-type region in an n-type substrate. Being reverse-biased as diodes, the resistors are isolated from each other and from the substrate. When the temperature of the substrate exceeds approximately 120°C, the diodes' leakage currents increase substantially, disturbing the isolation between the resistors and the normal mode of sensor operation. Several works have been devoted to the design of a high-temperature pressure sensor in which the resistors would have something other than a diode structure for the insulation. In one of the designs [334], sapphire is used as a substrate, on which a silicon-on-sapphire (SOS) process deposited layer allows the formation of isolated piezoresistors operating in the range of temperatures between -40 and $+200$°C. In this construction, a [100]-oriented SOS wafer is used, the epitaxial thickness is 2.0 to 2.5μm, the diameter of diaphragm is 5mm, and its thickness varies between 100 and 350μm for sensors having pressure ranges of 0.5×10^6 and 5×10^6 Pa, respectively. A comparatively high performance has been reached without any zero and temperature compensation: zero shift and sensitivity shift are within 0.1%/°C each for the temperature range of -40 to $+200$°C. In the most advanced construction [332], a basic concept of silicon fusion bonding for building high-temperature pressure sensors is used. The essence of the fabrication process is in preparing dielectrically-isolated, single-crystal resistors on one silicon wafer and then covalently bonding them to the oxidized surface of another single-crystal silicon wafer. In this simple process, no incompatible intermediate layers are required; all resistors are ideally isolated. The silicon fusion bond is formed by annealing the wafer assembly in a furnace at temperatures above 900°C. The sensors are capable of operating over the temperature range from -40 to $+250$°C.

The stability of output is better than $\pm 0.1\%$ of FS during a two-month period. Short-term stability is better than 0.01% of FS.

The wide application of piezoresistive pressure sensors has stimulated the development of devices with integrated signal conditioning for amplifying signals and providing thermal compensation [335, 336].

Figure 9.26 depicts several structures of capacitive pressure microsensors whose composition and operation are easy to understand. They all contain a diaphragm, moving and stationary electrodes, and elements of mounting. The electrodes can be made by deposition of metal or by heavy doping of silicon. Fusion bonding of silicon parts is the most attractive operation for joining the parts. In some designs, Pyrex glass substrates and intermediate layers are used to form the construction. A study of side effects related to the sensitivities and stabilities of ultrasensitive sensors [341] shows that temperature offset and pressure-sensitivity coefficients can increase by more than an order of magnitude due to the mismatch in the thermal expansion coefficients of silicon and dielectrics, and also because of the dependence of stress on temperature.

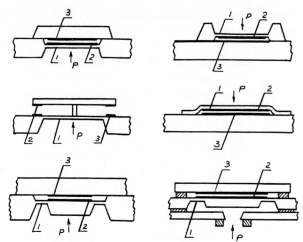

Figure 9.26 Structures of capacitive pressure microsensors (sources: [71, 72, 337-340]).
P = pressure, 1 = diaphragm, 2 = moving electrode, 3 = stationary electrode.

Since the nominal capacitance of the sensors' electrode couples is extremely small, sometimes units of pF, the CMOS signal-processing circuits are designed to anticipate the intrinsic drawbacks of these structures; for example, sensitivity to environmental noise, effects of parasitic capacitances, and so on [340, 342]. In some designs, the capacitive sensors are superior to piezoresistive sensors in several respects. They have higher sensitivity, greater long-term stability, lower temperature effect, and lower power consumption [339]. With recent advances in etch-stop techniques, fusion bonding, and other methods of fabrication, a thin silicon diaphragm having a large area can be fabricated. This makes it possible to obtain highly sensitive and accurate diaphragms. Considering the many different applications of pressure sensors, one can assume that piezoresistive and capacitive solid-state sensors will coexist, and each will be used in the area where the characteristics of a given type provide the best match to the application.

A micromachined piezoelectric pressure sensor for the detection of low-level sound (Fig. 9.27) has been fabricated using a deformable silicon membrane with ZnO film deposited by RF planar magnetron sputtering [343]. The total process (see p. 340) is completely compatible with CMOS technology because it uses conventional fabrication technologies, and materials. As seen from the diagram in Figure 9.27, the upper movable electrode is made of aluminum film. A lower stationary electrode is

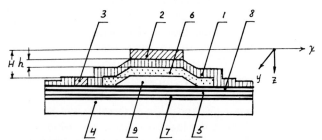

Figure 9.27 Structure of micromachined piezoelectric sensor (source: [343]). H = total thickness of structure, h = thickness of ZnO layer, 1 = ZnO, 2 and 3 = Al, 4 = Si substrate, 5 and 6 = poly-Si, 7 = SiO_2, 8 = Si_3N_4, 9 = sealed air cavity.

formed by depositing $0.1\mu m$ LPCVD polysilicon and is electrically encapsulated by $0.1\mu m$ LPCVD Si_3N_4.

The crystal structure of ZnO depends on the process of deposition. It can be cubic or dihexagonally polar as in this device. The matrix of piezoelectric constants is given by formula 9.109 in Piezoelectric Elements, and the values of the coefficients are $d_{15} = -8.3 \times 10^{-12}$C/N, $d_{31} = -5 \times 10^{-12}$C/N, and $d_{33} = 12.4 \times 10^{-12}$C/N.

The electrical response of the sensor can be obtained if we calculate an average polarization P_{av} [C/m^2] using the average values of stresses σ'_{xav} and σ'_{yav}, which are found from formulas 9.58 and 9.59.

$$P_{av} = d_{31}(\sigma'_{xav} + \sigma'_{yav}), \tag{9.63}$$

$$\sigma'_{xav} = \frac{1}{h}\int_{-H/2}^{-(H/2-h)}\sigma_{xav}dz, \qquad \sigma'_{yav} = \frac{1}{h}\int_{-H/2}^{-(H/2-h)}\sigma_{yav}dz, \tag{9.64}$$

$$\sigma_{xav} = \frac{4}{lw}\int_0^{l/2}\int_0^{w/2}\sigma_x dxdy, \qquad \sigma_{yav} = \frac{4}{lw}\int_0^{l/2}\int_0^{w/2}\sigma_y dxdy \tag{9.65}$$

where $H, h, l,$ and w = thickness of structure, thickness of ZnO layer, and length and width of the membrane, respectively.

The fabricated sensors have dimensions ranging from $50 \times 50\mu m^2$ to $250 \times 250\mu m^2$; they make use of a 0.95-μm-thick sputtered layer of ZnO on a polysilicon membrane $2.0\mu m$ thick. The sensors exhibit approximately 0.36mV/μbar sensitivity and 3.4dB variation over a frequency range of 200Hz to 40kHz.

In addition to the junction-type pressure-sensitive structures described in the section in this chapter on semiconductor strain gages, several designs based on effects in FET and MOS structures should be mentioned in our review. One of them [344] is a pressure-sensitive insulated-gate field-effect transistor consisting of a p-type silicon substrate in which two n^+ regions, the source and drain are formed. The structure also includes an elevated-gate electrode made of aluminum film, sandwiched in a PECVD Si_3N_4 microdiaphragm resembling a pillbox, and the gate oxide. The 0.5-μm space between the gate electrode and the oxide is a vacuum or air. Deflected by pressure, the diaphragm varies the gate capacitance, modulating the conductance of the channel (see Field-effect Transistors, Chapter 7). The drain current of one of the specimens varies by approximately $200\mu A$ when the applied pressure changes from 0 to 300mmHg.

In another design [345], a CMOS ring oscillator is integrated on a square silicon diaphragm with a center boss and makes use of the piezoresistive effect in the MOS field-effect transistor. The ring oscillator [346] consists of an odd number of inverting gates connected in a ring. They are commonly used to determine the gate delays of integrated circuits. Their frequency is inversely proportional to the gate's average delay time. The mechanical stress generated in the diaphragm by the applied pressure changes the mobility of the charge carriers in the MOSFET's of the oscillators due to the piezoresistive effect; consequently, the drain current, the delay time per inverting gate; and the frequency become functions of pressure. For the device in question, the oscillator frequency ranges between 20 and 30MHz. With the pressure change from 0 to 3bars, the frequency's relative change is approximately 3×10^{-2}.

It was shown in [347] that a MOS structure containing a permanently charged dielectric layer (electret) between the gate and the bulk of the MOS structure has a sensitivity of drain current to pressure change reaching $35\mu A/100$mmHg. The structure consists of a thin conductive diaphragm, air gap, electret, silicon dioxide layer, and a silicon substrate. This structure is typical for the MOS device and differs from that described in the section on metal-oxide-silicon FET's in Chapter 7 by the addition of the flexible diaphragm and electret layer.

Due to the intensive development of robotics, the sensing of touch has become an indispensable operation for robotic manipulation and assembly. Highly sensitive tactile elements responding to forces perpendicular and tangential to surfaces are in demand. The piezoresistive and pyroresistive nature of silicon is exploited to create tactile sensors [348–351]. Typically, these sensors are arranged in an array containing silicon islands in a flexible polyimide substrate. The islands are interconnected by leads embedded in the substrate. Each island contains one or more touch-sensitive elements. The element is a multilayered structure composed of polyurethane and polyimide coatings forming a touch-exposed surface. Under this coating, several consecutive layers, including silicon, metal interconnect, polyimide, polyurethane, and rigid support form the structure. Applied pressures are detected by silicon strain gages, which are the cores of the devices. The alternatives of these sensors are capacitive tactile sensors [352].

The measurement of blood pressure is one of the most frequently performed processes in medical practice. Catheters furnished with a pressure sensor at the tip, or other techniques in which a tiny sensor is in direct contact with blood, are quite often used in clinics. In these cases, reusable pressure transducers have been traditionally used. After each use the transducer needed service for the replacement of at least one part, and also needed sterilization, recalibration, and so on. Typically, these transducers could not survive more than 10 to 100 applications. The new technology of sensors, which has taken advantage of batch processing techniques to produce silicon chips with a low cost, has allowed the creation of a new generation of disposable blood pressure sensors [353]. These sensors are quite remarkable products. Along with their low cost, they are reliable, and their metrological characteristics are consistent with the existing standards.

The structure of the sensor is typical: a small chip ($2.5 \times 2.5 \times 0.4 \text{mm}^3$) with a thin diaphragm (5 to $250\mu\text{m}$ thick). A fully active Wheatstone bridge is arranged on the diaphragm surface. The electrical circuit also includes trimming elements for calibration and temperature compensation. The essential characteristics of these sensors are as follows: pressure range, -30 to $+300\text{mmHg}$; operating temperature range, 15 to 40°C; overpressure, -400 to 4000mmHg; excitation, 4 to 8V dc or rms ac with a frequency within 5kHz; output impedance, $<3000\Omega$; sensitivity, $5\mu\text{V/V/mmHg}$; linearity and hysteresis, $< \pm 2\%$ of pressure or $\pm 1\text{mmHg}$, whichever is greater. The concept of disposable pressure sensors can be expected to spread to other industries.

Another biomedical application of piezoresistive pressure sensors is related to arterial tonometry. This is a technique for blood pressure measurement that uses an array of pressure transducers pressed against the skin overlying an artery. In one of the new constructions [354] for tonometer transducer arrays, several transducers share a single long diaphragm. The sensors provide the information necessary for continuous measurement throughout the heart's pumping cycle.

An intrauterine pressure transducer [355] also contains a silicon pressure-sensitive diaphragm with piezoresistors in it. When this transducer is applied to the patient externally on the abdomen over a fluid-filled part of the uterus, it will measure the intrauterine pressure through the uterine wall and the skin.

An ultraminiature capacitive pressure sensor [356] has been designed to be mounted inside a 0.5-mm diameter catheter that is suitable for multipoint pressure measurements from within the coronary artery of the heart. The sensor contains a silicon micro-diaphragm of $290 \times 550 \times 2\mu\text{m}^3$ surrounded by a supporting rim $12\mu\text{m}$ thick, both defined by the boron etch-stop technique. A hybrid interface circuit chip is also a part of the sensor. It provides a high-level output signal with minimal distortion.

A capacitive pressure microsensor [357] is suitable for implantation in the human heart. Low power consumption and size are the defining factors for choosing

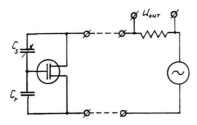

Figure 9.28 Schematic of two-wire capacitance-to-voltage converter with signal and reference capacitors (source: [357]).

a capacitance-type sensor. To protect the patient against an accidental dc leakage current, which can cause fibrillation of the heart by electrocution, the device is fed by alternating current. For the ac mode of operation, a capacitive voltage divider and an impedance converter (Fig. 9.28) have been designed and fabricated in a standard CMOS process. The divider includes sensing and reference capacitors connected to a MOSFET that, being controlled by the pressure signal, modulates the current in the power line. This results in a two-wire approach free of dc components.

A small fiber-optic sensor having an outer diameter of 0.5mm [358] is attractive for intravascular and on-heart application due to its small size and the non-electrical functions of parts contacting the living body. The sensor (Fig. 9.29) consists of a transparent, wave-guiding cantilever beam, which is mounted on a silicon bar structure containing a recessed area. Over this area, the cantilever beam is allowed to bend when exposed to an applied force. The light passing through the optical cantilever beam is reflected, and the intensity of the reflected light becomes a function of the bending force or the measuring pressure applied to the beam through a thin silicon rubber membrane.

Microactuators and Micromotors

Some of the more recent growth pertaining to silicon micromachining is in the development of tiny devices that can generate mechanical energy. Micromotors can be fabricated in sizes smaller than ever imagined (5μm thick and 75μm in diameter), and they can be applied in building microrobotic, micropositioner, microteleoperator, and other systems. In measuring devices, they can play an auxiliary role in cases where a small displacement, force, or vibration should be generated. We will briefly review the basic concepts of micromachines and abstain from detailed considerations of the subject, since it does not directly relate to sensors.

In terms of the physical principle of the conversion of electricity into mechanical action, there are several concepts that can be realized in practical constructions of micromachines.

In spite of their diversity, the operation of all *electrostatic drivers* is based on the same principle: inducing a force between two electrodes carrying electrical charges. All these devices form several groups: motors [204, 359–366], linear actuators [367–372], three-degree-of-freedom actuators [373], microvalves [374, 375], microgrippers [376], and levitating microstructures [377, 378]. The basic calculation of electrostatic force developing in microdrivers can be found in Chapter 3, Electrostatic Forces and Nonlinear Dynamics.

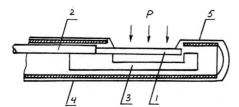

Figure 9.29 Miniature fiber-optic biomedical pressure sensor (source: [358]). P = pressure, 1 = cantilever glass beam, 2 = optical fiber, 3 = silicon bar, 4 = cannula, 5 = silicon rubber shell.

Piezoelectric motors and actuators make use of the reverse piezoelectric effect: voltage applied to a crystal causes its deformation. In a piezoelectric motor [379], a longitudinally oscillating rod, which is positioned asymmetrically on a roller-shaped rotor, drives the rotor. The driving force is developed due to the lateral displacement of the resonator tip. A lateral motion of a single crystal and stacked piezoelements are used in actuators [380], precise positioning mechanisms [381], micropumps [382], and valves [383, 384].

Piezoelectric bimorph plates provide a relatively large displacement of the driving members. The applications of these structures are similar to those just described in actuators [385–387], micropumps [388], and microvalves [389]. The deformation-voltage relationships for piezoelectrics are given in the last section of this chapter.

Devices with miniature *magnetic and electromagnetic mechanisms* are used in microvalve structures [390], experimental micromachines for cataract surgery [391], and micro-manipulators with magnetic levitation [392].

Heated *bimetallic structures* that provide the operating force are exploited in various actuators [393–396]. The calculation of bimetal plates is introduced in Chapter 5, Elastic Elements.

The direct *optical control* of a *microactuator* described in [397] is based on a direct optical-to-mechanical power conversion due to a purely quantum-mechanical process involving photons and matter on the atomic level. The device is a hybrid. Potential energy from an electrical power supply adds a charge to the capacitor formed by the cantilever-ground plane assembly. The charge is removed, and potential energy is converted to kinetic energy or to work due to the optical signals that cause a photoelectric leakage current to migrate through the capacitor air gap.

An *electro-pneumatic driving mechanism* [398–400] typically consists of a sealed cavity filled with liquid or gas. One wall of the cavity is flexible. When the fluid is heated by an electrical heater or light radiation, the pressure in the cavity is increased due to thermal expansion of the fluid. This expansion causes the flexible wall to move, providing mechanical action.

Shape-memory-alloy (NiTi) *actuators* are based on effects arising from a change from the martensitic to the austenitic phase of the material at a conversion temperature of about 80°C. During this change, reversible forces and deformations can be created [401–404].

In an *electrodynamic micropump,* fluid forces are generated by the interaction of the electric field and charges in the fluid. The hydrodynamic effects allow fluid transport and developing pressure to drive hydraulic valves and switches [405, 406].

An *ultrasonic micromotor* [407] uses traveling ultrasonic flexural waves to create motion. These waves, originated in the stator, cause the movement of surfaces that rub against the rotor and activate its rotation.

An electrically controlled *interfacial tension* between a liquid and solid ("electrowetting") can be the driving mechanism of a *micropump* for fluid systems, such as microchemical dispensers, injectors, and reactors, integrated cooling systems for high-density microelectronics, and so forth [408].

Direct fluid pumping with no moving mechanical parts is the most notable feature of this pump.

A micro-*superconducting actuator* [409] provides levitaton and driving achieved by the Meissner effect and the Lorentz force. The Meissner effect consists of the expulsion of magnetic flux from the interior of a piece of superconducting material as the material undergoes the transition to the superconducting phase. Because of its diamagnetism, the material tends to oppose the applied magnetic field, providing levitation of magnets. The system, composed of a stator (a superconducting YBaCuO film and etched copper film on polyimide) and a slider (a Nd-Fe-B permanent magnet), is able to levitate an 8-mg slider and keep it at about 1mm above the stator.

Bismuth and graphite are materials that can be used for *diamagnetic levitation of micromechanical bearings* [410]. It has been found that modern magnets on the order of 100 to 300μm will self-levitate over diamagnetic materials without supporting external fields. This suggests a number of noncontact micromechanical bearing applications.

Precise etching processes allow the creation of microminiature fluidic amplifiers [411] using nitrogen as the working fluid. The gain of the amplifier is greater than unity when the supply pressure is between 10 and 80psi. The amplifier can be used as a circuit element in integrated fluidic circuits or, combined with electronics, as a part of a transducer system.

Advances in the development of micromechanical moving structures have stimulated the creation of a family of micromechanical etched-from-silicon constructional components: microgears and turbines [412, 413], microbridges, cantilever beams, coils, suspended membranes [414], anchored pin joints, self-constraining joints, multi-joint cranks, sliders, gears combined with sliders, and many other elements [415]. Some of these parts are shown in the figures of Chapter 8, Other Technologies.

SEMICONDUCTOR STRAIN GAGES

Similar to a metal strain gage, a semiconductor strain gage changes its resistance with applied stress. The change of resistivity of semiconductor strain gages is appreciable and can be two orders of magnitude larger than the resistance change due to mechanical deformation. Unlike for metal strain gages, the term $(\Delta\rho/\rho)/(\Delta l/l)$ in formula 2.51 dominates and determines the high sensitivity of semiconductor gages.

It was found that a number of semiconductors are sensitive to stress, the best-known being Si, Ge, InSb, InP, GaAs, and GaSb. Silicon prevails over others in application because it is chemically inert and its technology is extensively developed in industry.

The strain gages of silicon are fabricated of ingots that are sliced to form small bars or patterns. It is typical in the construction of strain gage transducers to have the gages integrated with a silicon-made spring element. As described in the section on fabrication of pressure-sensitive microstructures in Chapter 8, the strain-sensitive layers are fabricated by diffusion of impurities into low-conductive material; growing crystal needles during vaporization and condensation of the material from a gaseous phase; epitaxial growth; polycrystal thin-film deposition; and other methods. The most popular gage constructions are a thin strip cut from an ingot and an elastic transduction element (e.g., a diaphragm) having strain-sensing resistors integrated with the element. The gages made as strips are thin patterns of different shapes (Fig. 9.30). Like

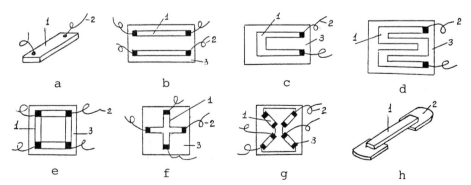

Figure 9.30 Different patterns of strip-type semiconductor strain gage.
a = simplest structure, b = combination of two elements for thermal compensation,
c = C-shaped pattern, d = zigzag pattern, e = square-shaped electrical bridge,
f = cross-shaped electrical bridge, g = element with diagonal orientation of bars,
connected in bridge, h = element with two heavy tabs for mounting on unbonded
structure; 1 = strain-sensitive material, 2 = electrical leads, 3 = backing.

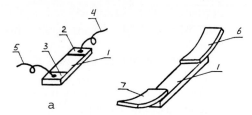

Figure 9.31 Semiconductor strip strain gage. a = element with wire leads, b = element with ribbon leads; 1 = silicon strip, 2 and 3 = metal pads, 4 and 5 = wire leads, 6 and 7 = ribbon leads.

metal gages, they are primarily used in constructions where they are bonded on a member whose deformation is measured. The elements are fabricated in such a way as to have a maximum sensitivity when they are deformed along the longest axis. Variations in direction of two or four strips combined together are employed for thermal compensation and/or for measurements of strains along two or three axes. A zigzag- or C-shaped pattern provides larger resistance and a greater output signal. Sometimes these patterns are more convenient for the orientation of leads at one side of the device. Figure 9.31 illustrates the basic construction of a strip element. The silicon strip has two areas at the ends with deposited metal pads. Two electrical leads are made of wire or fine ribbon attached to them using methods described in Chapter 8, Die Attachment.

In a diffused element (Fig. 9.32), a n-Si base has a p-Si diffused layer with dimensions $l \times w \times h$. The layer forms a junction with the base. The conductivity along the y'-axis is small if the layer is back-biased, but, along the x'-axis, the layer works as a stress-sensitive conductor when its resistance is measured between the leads.

An epitaxial element is formed on a substrate, which can be made of silicon or some other crystalline material, such as sapphire. A layer of an electroconductive semiconductor, as a rule, silicon, is grown on the surface. This layer with the metallizations and electrodes forms a strain gage that has an atomic bond to the substrate. The construction of a gage grown from a gaseous phase does not differ significantly from that of the strip gage. A thin-film semiconductor element is similar to one made of metal. An application of SOI technology for strain gages is given in Chapter 8,

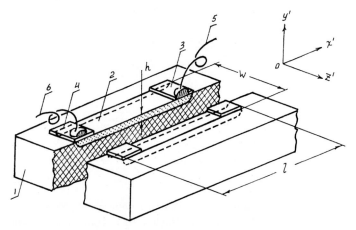

Figure 9.32 Diffused strain gage shown in cut. 1 = n-Si base, 2 = p-Si diffused layer, 3 and 4 = metallizations, 5 and 6 = leads.

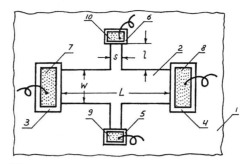

Figure 9.33 Transverse voltage shear strain gage. 1 = n-Si spring element, 2 = p-Si diffused strain-sensitive resistor, 3–6 = p$^+$-Si pads for contacts, 7 and 8 = power-supplying electrodes, 9 and 10 = output electrodes.

Basic Operations in Fabrication of Microsensors; and Fabrication of Pressure-sensitive Microstructures. The SOI process has a number of advantages in forming the elements. One of the original constructions of a strain-sensitive structure is a transverse-voltage shear strain gage [416, 417], schematically shown in Figure 9.33. In this element, the cross-patterned layer of p-Si is diffused in an n-Si spring member. When the resistor is stressed, a transverse electric voltage can be developed by a longitudinal current. It was found that the shear stress developed in the element plays a major role in inducing the transverse voltage. This voltage is picked up across the width of the element through the side arms. The ends of the arms have heavily doped p$^+$ pads that are used for better ohmic contact of electrodes. It should be noted that contemporary spring elements usually contain two or four active gages that form a half or full electrical bridge. Two examples illustrate the configuration of this structure. A cantilever (Fig. 9.34) with four C-shaped diffused gages is stretched and compressed at its upper and lower surfaces, respectively, when the cantilever undergoes bending deformation under force F. All the gages are identical since they are made on the same die and during the same technological cycle. Another example is a pressure-sensitive diaphragm (Fig. 9.35), which is formed by etching down silicon and diffusing impurities to form gages (see Chapter 8, Diffusion; and Etching of Silicon for details). When pressure P deforms the diaphragm, the gages experience different-sign stresses, causing the change of their resistances.

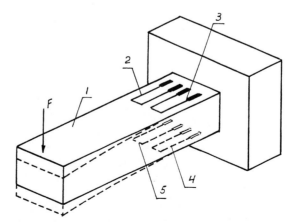

Figure 9.34 Cantilever integrated strain gage element. F = force, 1 = cantilever, 2–5 = C-shaped strain gages.

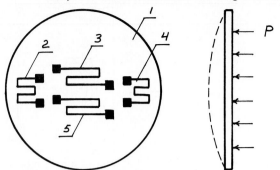

Figure 9.35 Pressure-sensitive integrated diaphragm with four gages. P = pressure, 1 = diaphragm, 2–5 = strain gages formed on diaphragm surface.

All major characteristics of strain gages depend on the type of material and on the impurities in the material. They are also defined by the orientation of the gages' axes with respect to the crystallographic axes if the element is fabricated of a crystalline material. The direction of axes, strain, and current in the element are mutually dependent parameters. Strain-gage effect in silicon and in other crystalline semiconductors is anisotropic. Similarly to the calculations given in the examples of Appendix 2 and in the last section of Chapter 7, the calculations of the piezoresistive characteristics of semiconductors are conducted using the principles of tensor algebra [417–422].

If the corners of a cubic crystal are oriented along the x'-, y'-, and z'-axes of an arbitrary coordinate system, the electric field components E_i' [V/m] and current density components J_j' [A/m^2] are related by a symmetric resistivity matrix [417]:

$$\begin{bmatrix} E_x' \\ E_y' \\ E_z' \end{bmatrix} = \begin{bmatrix} \rho_1' & \rho_6' & \rho_5' \\ \rho_6' & \rho_2' & \rho_4' \\ \rho_5' & \rho_4' & \rho_3' \end{bmatrix} \cdot \begin{bmatrix} J_x' \\ J_y' \\ J_z' \end{bmatrix} \qquad (9.66)$$

where ρ_1' to ρ_6' = components of resistivity tensor, [$\Omega \cdot$ m].

The relationship between the resistivities and stresses is determined by using piezoresistive coefficients π_{ij} [m^2/N], which are given in the square matrix of 36 components:

$$\begin{bmatrix} \rho_1' \\ \rho_2' \\ \rho_3' \\ \rho_4' \\ \rho_5' \\ \rho_6' \end{bmatrix} = \begin{bmatrix} \rho_0 \\ \rho_0 \\ \rho_0 \\ 0 \\ 0 \\ 0 \end{bmatrix} + \begin{bmatrix} \pi_{11}' & \pi_{12}' & \pi_{13}' & \pi_{14}' & \pi_{15}' & \pi_{16}' \\ \pi_{21}' & \pi_{22}' & \pi_{23}' & \pi_{24}' & \pi_{25}' & \pi_{26}' \\ \cdot & \cdot & \cdot & \cdot & \cdot & \cdot \\ \cdot & \cdot & \cdot & \cdot & \cdot & \cdot \\ \cdot & \cdot & \cdot & \cdot & \cdot & \cdot \\ \pi_{61}' & \pi_{62}' & \pi_{63}' & \pi_{64}' & \pi_{65}' & \pi_{66}' \end{bmatrix} \cdot \begin{bmatrix} \sigma_{xx}' \\ \sigma_{yy}' \\ \sigma_{zz}' \\ \sigma_{yz}' \\ \sigma_{zx}' \\ \sigma_{xy}' \end{bmatrix} \qquad (9.67)$$

where σ_{xx}', σ_{yy}', and σ_{zz}' = components of normal stress, [N/m^2];

σ_{yz}', σ_{zx}', and σ_{xy}' = components of shear stress, [N/m^2];

ρ_0 = isotropic resistivity of unstressed crystal, [$\Omega \cdot$ m].

If axes x', y', and z' coincide with the crystallographic axes X, Y, and Z (Fig. 7.23) of cubic crystal, the only nonzero components of π are

$$\pi_{11} = \pi_{22} = \pi_{33}; \qquad \pi_{12} = \pi_{23} = \pi_{31} = \pi_{21} = \pi_{32} = \pi_{13}, \qquad \text{and}$$

$$\pi_{44} = \pi_{55} = \pi_{66}. \qquad (9.68)$$

The matrix for π's is reduced to the following.

$$\begin{matrix} \pi_{11} & \pi_{12} & \pi_{12} & 0 & 0 & 0 \\ \pi_{12} & \pi_{11} & \pi_{12} & 0 & 0 & 0 \\ \pi_{12} & \pi_{12} & \pi_{11} & 0 & 0 & 0 \\ 0 & 0 & 0 & \pi_{44} & 0 & 0 \\ 0 & 0 & 0 & 0 & \pi_{44} & 0 \\ 0 & 0 & 0 & 0 & 0 & \pi_{44} \end{matrix} \tag{9.69}$$

Coefficients π_{11}, π_{12}, and π_{44} are called longitudinal, transverse, and shear piezoresistive coefficients, respectively. They depend on the type of material, its conductivity (n or p), dopand concentration (resistivity), and temperature. Table 9.2 illustrates the values of the coefficients for silicon of different conductivities.

Note that in formulas 9.66 and 9.67, the directions of E's, J's, and σ's with the same subscripts coincide (e.g., E_x, J_x, and σ_{xx} are directed along the X-axis).

For the main axes, tensor formula 9.67 is developed:

$$\rho_1 = \frac{E_x}{J_x} = \rho_0[1 + \pi_{11}\sigma_{xx} + \pi_{12}(\sigma_{yy} + \sigma_{zz})] + \pi_{44}(J_y\sigma_{xy} + J_z\sigma_{xz})$$

$$\rho_2 = \frac{E_y}{J_y} = \rho_0[1 + \pi_{11}\sigma_{yy} + \pi_{12}(\sigma_{xx} + \sigma_{zz})] + \pi_{44}(J_x\sigma_{xy} + J_z\sigma_{yz})$$

$$\rho_3 = \frac{E_z}{J_z} = \rho_0[1 + \pi_{11}\sigma_{zz} + \pi_{12}(\sigma_{xx} + \sigma_{yy})] + \pi_{44}(J_x\sigma_{xz} + J_y\sigma_{yz}). \tag{9.70}$$

If field E_i, current density J_i, and strain σ_{ii} are in the same direction, the relation between them is simplified. For example, if a strain gage is oriented along the x-axis, voltage is applied to the two end contacts, as in Figure 9.32, current flows along the resistive path, and the only normal stress σ_{xx} acts along the x-axis ($\sigma_{yy} = \sigma_{zz} = 0$), the relation between E_x, J_x, and σ_{xx} will be

$$E_x/\rho_0 = J_x(1 + \pi_{11}\sigma_{xx}). \tag{9.71}$$

For any other orientation of the element, the coefficient π'_{11} can be found using the direction cosines:

$$\pi'_{11} = \pi_{11} + 2(\pi_{44} + \pi_{12} - \pi_{11})(l^2m^2 + l^2n^2 + m^2n^2) \tag{9.72}$$

where l, m, and n = direction cosines (see Appendix 2) of the angles between the direction of deformation and axes X, Y, and Z, respectively.

Equation 9.71 is usually written in a general form with π'_{11} denoted as π_l (longitudinal coefficient) and E, J, and σ without subscripts:

$$E/J = \rho = \rho_0(1 + \pi_l\sigma). \tag{9.73}$$

TABLE 9.2

Coefficients π for Silicon

Material	ρ $\Omega \cdot cm$	π_{11}	π_{12} $1 \times 10^{-11} m^2/N$	π_{44}
p-Si	7.8	+6.6	−1.1	+138.1
n-Si	11.7	−102.2	+53.4	−13.6
p-Si	0.02	+4.53	−0.75	+94.5
n-Si	0.02	−72.6	+38.0	−9.5

Cofactor $(l^2m^2 + l^2n^2 + m^2n^2)$ is at maximum when an element is oriented along the [111]-axis of the cubic crystal and when π_{11} and $(\pi_{44} + \pi_{12} - \pi_{11})$ have the same signs (e.g., as in p-Si). The minimum will occur if they have different signs. For example, the cosines for [111]-direction are $l = m = n = 1/\sqrt{3}$ and $\pi'_{11} = \pi_{11}/3 + 2(\pi_{12} + \pi_{44})/3$.

The resistivity of a metal strain gage is independent of shear stress and defined only by normal stresses:

$$\frac{E_x}{J_x} = \rho_1 = \rho_0(1 + \pi_{11}\sigma_{xx} + \pi_{12}\sigma_{yy} + \pi_{13}\sigma_{zz}). \tag{9.74}$$

In metal, $\pi_{12} = \pi_{13}$. For constantan, π_{11} and π_{12} are 1.50×10^{-12} and $2.25 \times 10^{-12} m^2/N$, respectively.

The expressions for the gage factor K of a semiconductor and metal strain gages are the same (see explanations to formula 2.51 in the section on strain-gage elements in Chapter 2):

$$K = 1 + 2\nu + \frac{\Delta\rho}{\rho} \cdot \frac{1}{\epsilon} \tag{9.75}$$

where $\Delta\rho$ = change in the resistivity of the material due to strain ϵ, $\Omega \cdot m$; and

$$\rho = \rho_0 + \Delta\rho \quad \text{and} \quad \epsilon = \frac{\sigma}{B} \tag{9.76}$$

where B = Young's modulus, N/m^2.

Therefore, from formulas 9.73 and 9.76,

$$\Delta\rho/\rho = \pi_l\epsilon B. \tag{9.77}$$

Substituting this ratio in formula 9.75 gives

$$K = 1 + 2\nu + \pi_l B. \tag{9.78}$$

Since $\pi_l B > (1 + 2\nu)$ for semiconductor gages, it is usually assumed that $K \approx \pi_l B$.

By using formulas 9.72 and 9.78 for calculating K, the gage factor can be found for different crystallographic directions. Figure 9.36 illustrates the relation between the resistivity, impurity content, and temperature for n- and p-Si. Figure 9.37 shows the dependence of the silicon gage factor on resistivity for several crystallographic directions. As shown in the graph in Figure 9.37, the gage factor is positive for p-Si and negative for n-Si. The factor increases with the increase in the resistivity of the material. For the semiconductor strain gages, the fractional change in resistance $\Delta R/R$ is a nonlinear function of strain and temperature. For the given orientation of the gages with respect to the crystallographic axes, the presentation of $\Delta R/R$ as a polynomial is practical:

$$\frac{\Delta R}{R} = \frac{298}{T}C_1(\rho_0)\epsilon + \left(\frac{298}{T}\right)^2 C_2(\rho_0)\epsilon^2 + \left(\frac{298}{T}\right)^3 C_3(\rho_0)\epsilon^3 + \ldots \tag{9.79}$$

where T = absolute temperature, K;

C_1, C_2, and C_3 = factors defined by resistivity of material, K, K^2, and K^3, respectively; and

ϵ = strain, dimensionless.

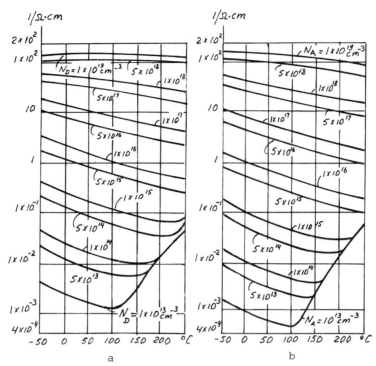

Figure 9.36 Resistivities of n-Si (a) and p-Si (b) as functions of temperature and densities of donor atoms (N_D) and acceptor atoms (N_A).

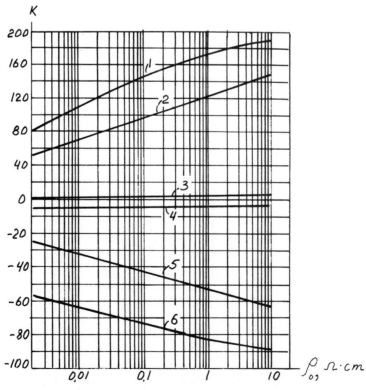

Figure 9.37 Gage factor K versus resistivity ρ_0 of Si. 1 = for p-Si[111], 2 = for p-Si[110], 3 = for p-Si[100], 4 = for n-Si[111], 5 = for n-Si[110], 6 = for n-Si[100].

The third term of this expansion contributes not more than 1% of the total amount of $\Delta R/R$ and can be neglected; therefore, for room temperature, function $\Delta R/R = f(\epsilon)$ is usually given as

$$\frac{\Delta R}{R} = C_1(\rho_0)\epsilon + C_2(\rho_0)\epsilon^2. \tag{9.80}$$

The gage factor can be calculated as

$$K = d(\Delta R/R)/d\epsilon = C_1(\rho_0) + 2C_2(\rho_0)\epsilon. \tag{9.81}$$

Figure 9.38 gives the values of C_1 and C_2 for n- and p-Si in two orientations and resistivities of the material in the range between 0.01 and 0.04$\Omega \cdot$ cm. For example, for the gage fabricated of high-resistance [111]-oriented p-silicon [423], the resistance change at room temperature (25°C) will be

$$\frac{\Delta R}{R} = 175\epsilon + 72625\epsilon^2. \tag{9.82}$$

For the high-resistive n-Si [100],

$$\frac{\Delta R}{R} = -125\epsilon + 26000\epsilon^2. \tag{9.83}$$

For the p-Si [111] having $\rho_0 = 0.02\Omega \cdot$ cm,

$$\frac{\Delta R}{R} = 119.5\epsilon + 4000\epsilon^2. \tag{9.84}$$

For the n-Si [100] having $\rho_0 = 0.03\Omega \cdot$ cm,

$$\frac{\Delta R}{R} = -100\epsilon + 10000\epsilon^2. \tag{9.85}$$

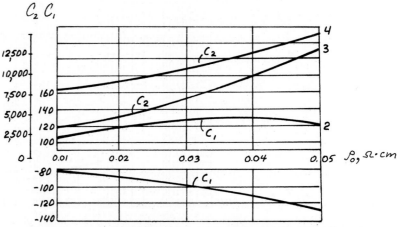

Figure 9.38 Coefficients C_1 and C_2 of formula (9.81) as function of resistivity ρ_0. 1 = C_1 for n-Si[100], 2 = C_1 for p-Si[111], 3 = C_2 for p-Si[111], 4 = C_2 for n-Si[100].

The gage factor is greater for gages made of high-resistive materials. However, they are characterized by higher nonlinearity of ϵ to $\Delta R/R$ conversion and a higher temperature coefficient of resistivity. For some semiconductor strain gages, the sensitivity to strain is 50 to 70 times greater than that for metal elements; but high sensitivities of semiconductors to temperature require special measures for compensation for the resistance and gage-factor drifts. It is a common practice to use four strain gages formed in one silicon die. These gages are connected in a bridge and, due to the ratiometering nature of the bridge circuit, the temperature errors are significantly diminished. In some applications, the bridges have, integrated with them, two or three additional resistors that are connected in a series with the bridge and parallel with one of its arms. These resistors are laser-trimmed during the transducer calibration. The trimming is provided for the normalization of the output characteristic and for additional compensation (if needed) for temperature drifts. Using the additional resistors decreases the sensitivity of the circuit to the change of the gage resistance, and a compromise is required to find the optimum in the signal-to-noise ratio.

The transverse-voltage element has less susceptibility to thermal disturbance than the conventional bridge-type construction. This advantage can be explained, first of all, by a smaller active area of the element and inherently symmetrical configuration of the signal-producing arms.

The expression for the transverse voltage V'_y can be obtained from equation 9.66:

$$E'_y = (\rho'_6/\rho'_1)E'_x, \qquad V'_y = V_s(W\rho'_6/L\rho'_1) \tag{9.86}$$

where W and L = width and length of the resistor, m, and
V_s = voltage supply, V.

The values of ρ'_6 and ρ'_1 can be found from equation 9.67:

$$\rho'_1 = \rho_0 + \pi'_{11} + \sigma'_{xx} + \pi'_{12}\sigma'_{yy} + \pi'_{16}\sigma'_{xy}, \tag{9.87}$$

$$\rho'_6 = \pi'_{61}\sigma'_{xx} + \pi'_{62}\sigma'_{yy} + \pi'_{66}\sigma'_{xy}. \tag{9.88}$$

The location and orientation of the strip within the spring element can be chosen such that, to the first approximation, $\rho'_1 = \rho_0$, and ρ'_6 is proportional to the tension to be measured. When the strain gage is integrated with a pressure-sensitive diaphragm, a maximum pressure sensitivity is obtained [416] for the gage that is located at the middle edge of a (001)-plane square diaphragm, and the resistor strip has a 45° inclination angle against the edge. In another structure [417], the resistor strip is diffused in the center of the same-orientation diaphragm, and the resistor strip also has a 45° inclination angle against the edge. If this angle is other than 45°, ρ'_6 will decrease, and the sensitivity of ρ'_1 to pressure becomes appreciable. The dependence of ρ'_1 on the tension can be used to compensate ρ'_1 for its nonlinearity. For example, if ρ'_6 is a quadratic function of a tension T [N/m^2] to be measured and ρ'_1 is a linear function of T, the ratio of ρ'_6 to ρ'_1 can be linearized:

$$\rho'_6 = a(1 + bT)T \qquad \text{and} \qquad \rho'_1 = \rho_0(1 + dT), \tag{9.89}$$

$$\frac{\rho'_6}{\rho'_1} = \frac{a}{\rho_0}T\frac{1 + bT}{1 + dT} \tag{9.90}$$

where a, b, and d are constants.

By finding a specific inclination angle giving $b = d$, the last cofactor in formula 9.90 becomes equal to unity, and ratio ρ'_6/ρ'_1 is linearized.

The voltage induced across the resistive layer creates a current whose path is inside the layer; therefore, the voltage at the output terminals is reduced. Coefficient f inserted in formula 9.86 takes this effect into account:

$$V'_y = V_s f \frac{W}{L} \cdot \frac{\rho'_6}{\rho'_1}. \qquad (9.91)$$

f increases with the ratios L/W and l/s (Fig. 9.33); for $(L/W) > 3$ and $(l/s) > 0.7$, practically, $f \approx 1$.

Different technologies related to strain gages have been considered in Chapter 8, Other Technologies, and the choice of the method of fabrication of the elements depends on the concrete requirements of the characteristics of the transducer. It should be noted that the metrological characteristics of the device are defined not only by the nominal intrinsic properties of the material, but also by the process of fabrication of the strain-sensitive layers, reliability of contacts, protection of a structure against a chemically aggressive environment, and so on.

Significant successes have been achieved in the application of gages in the structures used for measuring mechanical values, especially pressure. Some more attention will be given to these structures in the following paragraphs.

Strain-sensitive Junction Structures

Advances in semiconductor technology stimulated a search for high-efficient junction-based structures for conversion of force or displacement into electrical signals. In one of the earliest works [424], it was reported that the breakdown and conducting voltages of p-n junction semiconductor diodes vary with hydrostatic pressure. The pressure-sensitivity coefficient of the Zener diode is negative irrespective of the breakdown mechanism and direction of current. It was also noted that the current in a germanium diode fluctuates in response to a localized stress at the junction area. This effect, called "anisotropic stress effect," was exploited for the design and manufacture of a family of pressure-sensitive transistors called "Pitran" [425]. It is assumed that the effect of stress on p-n junctions is the result of changes in the energy band gap and of a generation-recombination mechanism. The most remarkable feature of this transducer is its ability to produce large changes in the electrical characteristics of a p-n junction for very small mechanical forces, allowing the construction of transducers that combine small size, high sensitivity, mechanical resonance frequency, and electrical output (as much as 90% of the applied voltage). Being a transistor, the device can have its output modulated electrically as well as mechanically. Besides the dc output, the frequency and pulse-width modulating circuits are easily constructed with the use of the transistor. Scheme 56 shows the construction of the device, and Figure 9.39 displays several essential characteristics of the Pitran. In spite of a number of valuable characteristics, this device could not compete with the semiconductor strain gages, which have superior metrological characteristics and are more economical in production. The stress-concentrating needle in the transistor creates a high-level local stress in the surface of crystal at the micrometer-level of deformation. Problems associated with the stability of the mechanical joints of such a construction are not simple. The thermal mechanical drift combined with the thermal electrical drift is typical for the transistor. There is a need for special measures for the stabilization of transistor transfer characteristics.

Several more structures of strain-sensitive transistors have been suggested to eliminate the needle in the construction and to create the concentration of stress by different means [426–428]. Typically, the junction is formed at the area of maximum stress of a beam that undergoes bending deformation; for example, at the rigidly

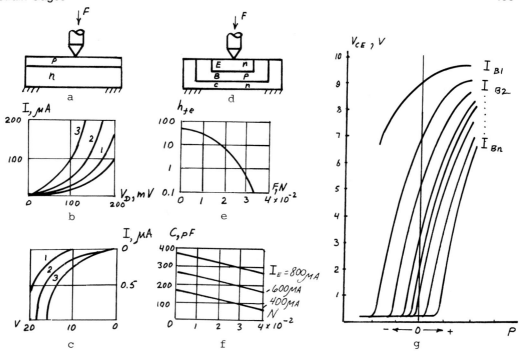

Figure 9.39 Characteristics of pressure-sensitive diode and transistor. a = diode's structure, b = germanium diode's forward current versus voltage for loads of 1×10^{-2}, 2×10^{-2}, and 3×10^{-2}N (curves 1, 2, and 3, respectively), c = diode's reverse current versus voltage, d = transistor's structure, e = silicon transistor's forward-current transfer ratio versus force F, f = transistor's output capacitance C versus force, g = transistor's collector-emitter voltage versus pressure for different base currents.

fixed end of a cantilever. The highest concentration can be achieved by thinning the sensitive area as shown in Figure 9.40. This figure shows a MOS transistor [see Metal-oxide-silicon FET's (MOSFET's), Chapter 7] that is integrated with an elastic cantilever having a neck concentrating stress at the area of the transistor. The transistor includes the standard elements: source, drain, channel, and gate. However, the channel is made of cadmium sulfide film, which is a piezoelectric material (see Piezoelectric Elements, Chapter 9). Under deformation, the stress in the channel produces charges that control the current through the transistor, as described in the section on field-effect transistors (FET's) in Chapter 7. Thus, the current becomes a function of the applied force or displacement. Compensation for the temperature of these elements must be thoroughly provided.

There is a probability that new developments in microelectronic technology will make it possible to improve and advance the construction of strain-sensitive transistors and make them widely acceptable for contemporary instruments.

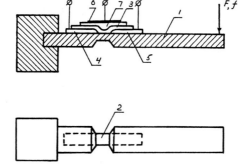

Figure 9.40 Schematic diagram of strain-sensitive MOS-transistor combined with cantilever. F and f = force and displacement, 1 = cantilever, 2 = neck, 3 = channel of CdS, 4 = source, 5 = drain, 6 = gate, 7 = insulator.

PIEZOELECTRIC ELEMENTS

Piezoelectric devices produce mechanical work when they are excited electrically, or they generate electrical energy when they are driven mechanically. The most extensively used materials in these devices are crystals—quartz, tourmaline, Rochelle salt (sodium potassium tartrate), lithium niobate, lithium tantalate, cadmium sulfide, zinc oxide and ceramics—barium titanate, ammonia dihydrogen phosphate (ADP), lead zirconate—lead titanate (PZT), and so forth.

The crystalline materials are inherently piezoelectric: they exhibit electrical charges with the deformation of the crystal lattice. The synthetic piezoceramic materials are made by baking small crystallites under pressure and placing the specimens in a strong dc electric field where they are polarized along the field direction and acquire piezoelectric properties. In piezoelectric ceramics, the piezoeffect is particularly strong. From a design point of view, they are very universal because the versatile shapes and sizes of parts can be obtained by simple molding. However, when a precision electromechanical operation is required, the monocrystalline rather the polycrystalline structure is preferable. Recent developments in new transduction materials have led to the invention of piezoelectric materials such as polymers and flexible piezoelectric ceramics [429]. The elements fabricated of these materials have practically no constraints in size, and they are light and flexible. Piezoelectric polymer films are presently made of polyvinylidene fluoride and are often referred to as PVF_2 or PVDF.

Because of the specific properties of piezoelectrics, three areas of their application are most noteworthy:

1. Conversion of displacement, acceleration, force, pressure, stress, and strain into an electrical signal (direct effect);
2. Production of displacement, and force, or radiation of vibratory energy (e.g., acoustic wave) in response to the applied voltage (reverse effect); and
3. Operation in resonating structures for the most efficient radiation of energy, sensing of physical values, filtration of electrical signals, and generation of stable pulses (direct and reverse effects).

Due to its mechanical perfections, crystalline quartz (SiO_2) is one of the most attractive materials for the construction of transduction elements. Quartz belongs to the triogonal trapesohedral class (32) of the rhombohedral subsystem [306] (Fig. 9.41). The lattice type is hexagonal and characterized by one axis of two-fold perpendicular symmetry and separation by angles of 120°C. There are no centers of symmetry and no planes of symmetry. The reference axes X, Y, and Z are called electrical, mechanical, and optical, respectively. There are three X-axes perpendicular to the Z-axis and passing through the crystal edges (axes of two-fold symmetry). Three Y-axes perpendicular to the Z-axis pass through the centers of the crystal faces. The Z-axis is the axis of three-fold symmetry.

The piezoeffect in quartz is illustrated by Figure 9.42. The silicon atoms are in four-coordination with oxygen and form a SiO_4 tetrahedron, which is the basic unit of the structure. Each oxygen is shared with two silicon atoms. Each ion of Si with charge $+4e$ is linked with four ions of O having charge $-2e$ each. In a still condition, all the charges are compensated and the cell is neutral. If the cell is deformed along the X- or Y-axis, the O-ion is displaced, and positive or negative charges are formed, as shown in Figure 9.42. The same model can be used to explain the reverse effect. An element will vibrate at a high amplitude if it is cut from the crystal and excited by an ac voltage having a frequency close to the frequency of the natural mechanical resonance. These vibrations are intense, and the element establishes a frequency of oscillation in the electrical circuit providing excitation. Since the

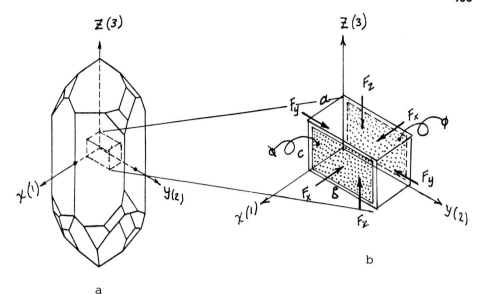

Figure 9.41   Quartz crystal. a = crystal, b = element cut from crystal under stress.

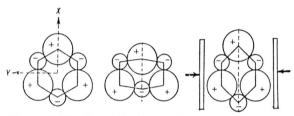

Figure 9.42   Piezoeffect in quartz.

mechanical properties (Table 9.3) of quartz are very stable, the frequency of the electrical signal is also stable. The mechanical state of the element can be changed by applying external force, varying its temperature, or adding a mass to the vibrating body. These factors alternate the frequency of vibration, making the element sensitive to the specific physical values. The range of the quartz resonator frequencies is fairly wide: from units of kilohertz to hundreds of megahertz. The electrical and mechanical properties of the elements greatly depend on their shape, size, and

TABLE 9.3

Some Properties of Quartz (Adopted from [306])

Property	Value $\parallel Z$	Value $\perp Z$	Temperature (T) dependency
Thermal conductivity [cal/(cm · s · °C)]	29×10^{-3}	16×10^{-3}	decreases with T
Dielectric constant	4.6	4.5	decreases with T
Thermal expansion coefficient [1/°C]	7.1×10^{-6}	13.2×10^{-6}	increases with T
Electrical resistivity [$\Omega \cdot$ cm]	0.1×10^{15} (ionic)	20×10^{15} (electronic)	decreases with T
Young's modulus [N/m^2]	9.7×10^{10}	7.6×10^{10}	decreases with T
Density [kg/m^3]	2.65×10^3		
Curie temperature [°C]	573		

direction of cut. The simplest form of the element is a circular or rectangular flat plate with metallization on the surfaces. Metal films are used to pick up or to deliver electricity to the crystal. Table 9.4 illustrates [430] the basic piezoelectric actions of elements. The deformation of the element can be quite complex. For example, if the element is composed of two bonded plates, a shear deformation can be accompanied by twisting.

The analysis [431] of crystalline elements is based on a three-dimensional presentation of the electromechanical characteristics of piezoelectrics.

TABLE 9.4

Basic Actions of Piezoelectric Elements (Source: [430])

Basic deformations	*Typical applications*
Thickness shear	Delay-line transducers; ultrasonic transducers; accelerometers; high-frequency resonators
Face shear	Headphones; microphones; twister "Bimorph" phonograph cartridges
Thickness expansion	Delay-line transducers; ultrasonic testing and cleaning transducers; high-frequency resonators
Transverse length expansion	Sonar transducers and hydrophones; "Bimorph" phonograph cartridges; headphones; microphones; tweeters; clock drivers; heart pacers
Parallel length expansion	Sonar radiating transducers; ultrasonic bonders and welders; ultrasonic solder cleaners
Poled along this axis Planar expansion	"Bimorphs" and "Unimorphs" for labs; tweeters for cameras, radios, and simple alarms; transmitters and receivers for intrusion alarms

As described previously, the charges developed on the surfaces of the element are proportional to the applied force. For the analysis, polarization P [C/m^2] is usually taken as a measure of charge amount. P can be regarded as a charge density (by definition, P is a vector quantity equal to the electric dipole moment per unit volume of material). The measure of force is stress σ [N/m^2]. The relation between these two values is established by using the proportionality coefficient d [C/N], which is called the *piezoelectric constant:*

$$P = d\sigma. \tag{9.92}$$

The reverse effect is described by another equation:

$$x = gP \tag{9.93}$$

where x = strain, dimensionless, and
g = constant coefficient, m^2/C.

In this formula, the magnitude of P is defined by the source supplying charges. It is known that

$$P = \epsilon_0(\epsilon_r - 1)E \tag{9.94}$$

where ϵ_0 = permittivity of free space, C^2/N $\cdot$ m^2;
ϵ_r = dielectric constant, dimensionless; and
E = electric field strength, N/C or V/m.

It is proved in thermodynamics that

$$\epsilon_0 g(\epsilon_r - 1) = d. \tag{9.95}$$

Therefore, the same constant is used for the description of the direct and reverse effects:

$$x = d \cdot E. \tag{9.96}$$

Another essential characteristic of the element is electromechanical coupling coefficient k. By definition, its square is equal to the ratio of the electrical energy generated by the element per unit volume W'_e [N/m^2] to the total energy per unit volume needed for its deformation W' [J/m^3].

$$k^2 = W'_e/W', \qquad W' = W'_e + W'_m \qquad \text{and} \qquad W'_m = Bx^2/2 \tag{9.97}$$

where W'_m = mechanical component of W', N/m^2, and
B = Young's modulus, N/m^2.

The effective modulus of elasticity B_{ef} found from the equation $W' = B_{ef}x^2/2$ is larger than B:

$$B_{ef} = B/(1 - k^2). \tag{9.98}$$

When the reverse piezoeffect is used, the energy of the power supply is spent for the charge of the element as a capacitor: $W_e = CU^2/2$ [J] (C [F] is capacitance, and U [V] is voltage), and for the deformation of the structure: W_m [J].

The total energy W [J] is a sum: $W = W_e + W_m$. Since $W > W_e$, the effective value of the capacitance C_{ef} is determined from the set of relationships [431]:

$$W = C_{\text{ef}}U^2/2, \qquad W_m = W - W_e, \qquad k^2 = W_m/W, \qquad (9.99)$$

$$\frac{C_{\text{ef}} - C}{C_{\text{ef}}} = \frac{\epsilon_{\text{ef}} - \epsilon}{\epsilon_{\text{ef}}} = k^2 \qquad (9.100)$$

where ϵ_{ef} = effective dielectric constant determining C_{ef}.

The values of k are measured in experiments, and they are the same for the direct and reverse effects. The coupling coefficient is a true measure of the piezoelectric effect because it is independent of the level of elastic stiffness or permittivity. This coefficient can be specific and defined not only by the properties of the material but also by the features of the construction. For ceramic materials, k is defined as a planar electromechanical coupling coefficient k_p ("planar coupling"). This coefficient characterizes a radial vibration of a thin disk with electrodes on its surfaces.

The mechanical quality factor Q_M is the characteristic that is given by the ratio of the equivalent reactance to the equivalent resistance of losses in a vibrating element. The magnitude of Q_M is determined from experiments using the following formula:

$$Q_M = \frac{f_n^2}{2\pi f_m Z_m C(f_n^2 - f_m^2)} \qquad (9.101)$$

where f_n and f_m = frequencies at maximum and minimum impedances of element, respectively, Hz;

$\qquad Z_m$ = minimum impedance of element, Ω; and

$\qquad C$ = element capacitance (usually measured at 1000Hz), F.

Curie temperature T_c given in °C is also an important characteristic determining the natural limit in the application of materials.

As it follows from the theory in Mechanical Characteristics of Semiconductors in Chapter 7, the stress can be introduced by a tensor of rank two having nine components σ_{ij}. Due to the symmetry, six components are left: σ_1, σ_2, and σ_3 for the compression or tension (taken negative); σ_5, σ_6, and σ_7 for the shear.

From equations 9.92 and 9.96,

$$P_i = \sum_{j=1}^{6} d_{ij}\sigma_j = d_{ij}\sigma_j \qquad (9.102)$$

$$x_j = \sum_{i=1}^{3} d_{ij}E_i = d_{ij}E_i \qquad (9.103)$$

where i = 1, 2, and 3 are the numbers of the components of polarization, and

$\qquad j$ = 1, 2, . . . , 6 are the numbers of the components of mechanical stresses or strains.

The dielectric constant is a third-rank tensor containing $3 \times 6 = 18$ components d_{ij}. Depending on the type of its lattice, the crystal can be characterized by a number of terms different from zero. For an isotropic material free of piezoeffect, all the terms are zero. The piezoelectric behavior of a crystal can be described in detail if the set of data related to the piezoelectric constants d_{ij}, elastic compliances S_{jl} [m²/N], and permittivity ϵ_{ik} are given. This set can be introduced by a 9×9 matrix [431] in which columns are associated with mechanical stresses and field strengths, and rows refer to strains and polarizations.

	σ_1	σ_2	σ_3	σ_4	σ_5	σ_6	E_1	E_2	E_3
x_1	S_{11}	S_{12}	S_{13}	S_{14}	S_{15}	S_{16}	d_{11}	d_{21}	d_{31}
x_2	S_{21}	S_{22}	S_{23}	S_{24}	S_{25}	S_{26}	d_{12}	d_{22}	d_{32}
x_3	S_{31}	S_{32}	S_{33}	S_{34}	S_{35}	S_{36}	d_{13}	d_{23}	d_{33}
x_4	S_{41}	S_{42}	S_{43}	S_{44}	S_{45}	S_{46}	d_{14}	d_{24}	d_{34}
x_5	S_{51}	S_{52}	S_{53}	S_{54}	S_{55}	S_{56}	d_{15}	d_{25}	d_{35}
x_6	S_{61}	S_{62}	S_{63}	S_{64}	S_{65}	S_{66}	d_{16}	d_{26}	d_{36}
P_1	d_{11}	d_{12}	d_{13}	d_{14}	d_{15}	d_{16}	ϵ_{11}	ϵ_{12}	ϵ_{13}
P_2	d_{21}	d_{22}	d_{23}	d_{24}	d_{25}	d_{26}	ϵ_{21}	ϵ_{22}	ϵ_{23}
P_3	d_{31}	d_{32}	d_{33}	d_{34}	d_{35}	d_{36}	ϵ_{31}	ϵ_{32}	ϵ_{33}

$$(9.104)$$

This matrix is symmetrical ($S_{jl} = S_{lj}$, $\epsilon_{ik} = \epsilon_{ki}$). Generally, it is composed of 45 terms including 21 compliances, 6 permittivities, and 18 piezoelectric constants. The following is an example of how the expansion can be written for P's, E's, and ϵ's:

$$P_1 = \epsilon_{11}E_1 + \epsilon_{12}E_2 + \epsilon_{13}E_3$$

$$P_2 = \epsilon_{21}E_1 + \epsilon_{22}E_2 + \epsilon_{23}E_3$$

$$P_3 = \epsilon_{31}E_1 + \epsilon_{32}E_2 + \epsilon_{33}E_3. \tag{9.105}$$

For a crystalline quartz (see the arrangement of the components in the left lower corner of relations 9.104) the matrix for d_{ij} is

$$\begin{vmatrix} d_{11} & -d_{11} & 0 & d_{14} & 0 & 0 \\ 0 & 0 & 0 & 0 & -d_{14} & -2d_{11} \\ 0 & 0 & 0 & 0 & 0 & 0 \end{vmatrix} \tag{9.106}$$

For lithium niobate ($LiNbO_3$) and lithium tantalate ($LiTaO_3$),

$$\begin{vmatrix} 0 & 0 & 0 & 0 & d_{15} & -d_{22} \\ -d_{22} & d_{22} & 0 & d_{15} & 0 & 0 \\ d_{31} & d_{31} & d_{33} & 0 & 0 & 0 \end{vmatrix} \tag{9.107}$$

For the cubic crystals of CdS, ZnS, and ZnO,

$$\begin{vmatrix} 0 & 0 & 0 & d_{14} & 0 & 0 \\ 0 & 0 & 0 & 0 & d_{14} & 0 \\ 0 & 0 & 0 & 0 & 0 & d_{14} \end{vmatrix} \tag{9.108}$$

For the piezoceramic material with residual polarization along the Z-axis,

$$\begin{vmatrix} 0 & 0 & 0 & 0 & d_{15} & 0 \\ 0 & 0 & 0 & d_{24} = d_{15} & 0 & 0 \\ d_{31} & d_{32} = d_{31} & d_{33} & 0 & 0 & 0 \end{vmatrix} \tag{9.109}$$

The subscripts 1, 2, and 3 indicate the x-, y-, and z-axes, respectively.

In the piezoelectric constants, the first subscript refers to the direction of the field; the second subscript refers to the direction of the strain. For example, in $d_{ij} = d_{33}$, $i = 3$ indicates that electrodes are perpendicular to the 3-axis, and the piezoelectric-induced strain or the applied stress is in the 3-direction.

A simplified notation with one subscript is usually taken for the dielectric constants: $\epsilon_{11} = \epsilon_1$, $\epsilon_{22} = \epsilon_2$, and $\epsilon_{33} = \epsilon_3$. For the defined orientation of the crystallographic axes, $\epsilon_{ik} = 0$ for $i \neq k$.

Tables 9.5 and 9.6 give the most essential characteristics of several piezoelectric monocrystalline and ceramic materials.

Following is a list of a dozen more materials [432] with two characteristics in the parentheses: k and ϵ for the elements transducing longitudinal waves (k's are the first figures):

$LiTaO_2$ (0.34, 10.8); $Ba_2NaNb_5O_{15}$ (0.57, 32); $Na_{0.5}K_{0.5}Nb_3$ (0.46, 306); $LiGaO_2$ (0.30, 8.5); $LiIO_3$ (0.49, 6.6); $PbTiO_3$ (0.78, 126); $PbNb_2O_6$ (0.63, 240); $Bi_{12}GeO_{20}$ (0.19, 38.6); $BaSi_2TiO_8$ (0.11, 11); CdSe (0.12, 10.2); GaAs (0.02, 12.5); and plastic film, PVF_2 (0.12, 4).

The composition of many materials and their technologies are property of the manufacturers and are continuously under development.

In terms of their operational characteristics, all the materials can be divided into several classes:

1. In the first class are materials for the conversion of low-level mechanical or acoustical signals. In this case, the important characteristic is the piezoelectric constant. The dielectric and mechanical losses are not essential.

2. Materials belonging to the second class work under the conditions of high electrical and mechanical strain, such as in power radiators of ultrasonic energy. Low losses, high magnitude of the quality factor, and high resistance to depolarization are important for this application.

3. Materials in the third class work in the elements for which the stability of frequency of the generated signals is determined by the oscillatory characteristics of the element (filters, clocks, etc.). A high quality factor, low temperature drifts, and high stability in time are required for this application.

4. Materials in the fourth class work in extreme conditions; for example, at elevated temperatures. There are several materials that can operate at relatively high temperatures (above 300°C). Exceeding the Curie temperature destroys the spontaneous polarization and, as a result, piezoelectrical properties of the material.

When an element works in the mode of conversion of mechanical deformation into an electrical signal or reverse, its performance is described by the basic relationships given in the beginning of this section.

TABLE 9.5

Characteristics of Piezoelectric Monocrystals (Adopted from [431])

Crystal	Piezoelectric constant d_{ij}, 1×10^{-12} [C/N]	ε_{ef}	k_{max}
Quartz (SiO_2)	$d_{11} = 2.31$; $d_{14} = 0.7$	$\varepsilon_1 = 4.52$; $\varepsilon_3 = 4.63$	0.1
ZnS	$d_{14} = 3.18$	$\varepsilon = 8.37$	<0.1
CdS	$d_{15} = -14$; $d_{33} = 10.3$; $d_{31} = -5.2$	$\varepsilon_1 = 9.35$; $\varepsilon_3 = 10.3$	0.2
ZnO	$d_{15} = -12$; $d_{33} = 12$; $d_{31} = -4.7$	$\varepsilon_3 = 8.2$	0.3
KDP, KH_2PO_4	$d_{14} = 1.3$; $d_{36} = 21$	$\varepsilon_1 = 42$; $\varepsilon_3 = 21$	0.07
ADP, $NH_4H_2PO_4$	$d_{14} = -1.5$; $d_{36} = 48$	$\varepsilon_1 = 56$; $\varepsilon_3 = 15.4$	0.1
Rochelle salt at 34°C	$d_{14} = 345$; $d_{25} = 54$; $d_{36} = 12$	$\varepsilon_1 = 205$; $\varepsilon_2 = 9.6$; $\varepsilon_3 = 9.5$	0.97
$BaTiO_3$	$d_{15} = 400$; $d_{33} = 100$; $d_{31} = -35$	$\varepsilon_1 = 3000$; $\varepsilon_3 = 180$	0.6
$LiNbO_3$	$d_{31} = -1.3$; $d_{33} = 18$; $d_{22} = 20$; $d_{15} = 70$	$\varepsilon_1 = 84$; $\varepsilon_3 = 29$	0.68
$LiTaO_3$	$d_{31} = -3$; $d_{33} = 7$; $d_{22} = 7.5$; $d_{15} = 26$	$\varepsilon_1 = 53$; $\varepsilon_3 = 44$	0.47

TABLE 9.6

Characteristics of Piezoelectric Ceramics (Adopted from [431])

Material	d_{ij}, 1×10^{-12}[C/N]			ε	k_p	Q_M	T_c, [°C]
	$-d_{31}$	d_{33}	d_{15}				
$BaTiO_3$	45-78	100-190	260	1400	0.2-0.36	100-300	120
$BaTiO_3 + \sim 5\%$ $CaTiO_3$	43	77	240	1200	0.25	≥ 300	105
$BaTiO_2 + 8\%$ $CaTiO_3$ and 12% $PbTiO_3$	27	77	112	450	0.25	≥ 350	160
$Pb(Zr_{0.53}Ti_{0.47})O_3 + (0.5-3)\%$ La_2O_2 or Bi_2O_3, or Ta_2O_5, or Nb_2O_5	119	282	380	1400	0.47	70	290
Acceptors to $Pb(Zr_{0.53}Ti_{0.47})O_3$, K^+ replacing Pb^{2+} or Fe^{3+}, Sc^{3+}, Co^{3+}, In^{3+} replacing Ti^{4+}	100	200	—	1050	0.43-0.45	≥ 200	280
$(Pb_{0.6}Ba_{0.4})Nb_2O_6$	40-67	100-167	—	1600-1800	0.20-0.28	150-300	265-330
$(K_{0.5}Na_{0.5})NbO_3$	49	160	—	420	0.45	240	420

To be more specific, we will consider a right parallelepiped (Fig. 9.41) cut from the crystal. This element's faces are parallel to crystal planes XY, XZ, and YZ. The metallizations are on two faces perpendicular to the X-axis. Under force F_x [N] parallel to the X-axis and causing compressing stress σ_1, the polarization P_1 (see expressions 9.102 and 9.104) and charge corresponding to P_1 on the faces q_1 [C] will be

$$q_1 = A_1 P_1 = A_1 \sigma_1 d_{11} = A_1 \frac{F_x}{A_1} d_{11} = F_x d_{11} = 2.31 \times 10^{-12} F_x \quad (9.110)$$

where A_1 = area of the face perpendicular to X-axis, m^2.

If the element is deformed by force F_y along the Y-axis, the charge will be different:

$$q_1 = A_1 P_1 = A_1 \sigma_2 d_{12} = A_1 \frac{F_y}{A_2} d_{12} = F_y \frac{cb}{ac} d_{12} = -F_y d_{11} \frac{b}{a}$$

$$= -2.31 \times 10^{-12} F_y \frac{b}{a} \quad (9.111)$$

where $\quad A_2$ = area at the face parallel to X-axis, m^2;

σ_2 = stress normal to plane XZ, N/m^2; and

a, b, and c = parallelepiped's dimensions, m.

The relation $d_{12} = -d_{11}$ follows from matrix 9.106. Note that (1) the polarity of output signals is different for these two cases; (2) the magnitude of the signal for F_x is independent of the elements' dimensions; (3) the output for F_y can be increased by the increase in ratio b/a; and (4) force F_z acting along the Z-axis will not produce charges since $d_{13} = 0$ for quartz.

If the element is bent as a cantilever by force F_s and shear stress σ_4 is induced,

$$P_1 = d_{14} \sigma_4. \quad (9.112)$$

If the element experiences a uniform compression, the following expansion obtained from matrix 9.104 is used:

$$P_1 = d_{11} \sigma_1 + d_{12} \sigma_2 + d_{13} \sigma_3. \quad (9.113)$$

Since $d_{13} = 0$, P_1 is reduced to

$$P_1 = d_{11}\sigma_1 + d_{12}\sigma_2 = d_{11}(\sigma_1 - \sigma_2). \tag{9.114}$$

For cuts having faces not conforming to the major planes, the calculations should be provided using principles of tensor algebra (see Appendix 2; Chapter 7, Mechanical Characteristics of Semiconductors).

The dynamic behavior of the element as a force-to-voltage converter should be analyzed from two points of view: electrical and mechanical. When a force (e.g., along X-axis in Fig. 9.41) is applied to the element, its deformation and emergence of charges are not instantaneous. Because of the inertia of moving parts, friction in the structure, and elastic properties of the material, the deformation is a time-dependent process. After the charges are formed, the current of electrons flows into the signal conditioning circuit. This current is also a function of time since the element is capacitive by nature and the circuit coupled with the element usually contains resistors and capacitors.

We will analyze the dynamic responses of the system using the example of Figure 9.41 as a model. Assuming that the deformation is linear, the balance of forces is introduced by a second-order differential equation:

$$m \frac{d^2(\Delta a)}{dt^2} + \eta \frac{d(\Delta a)}{dt} + \beta(\Delta a) = F_x \tag{9.115}$$

or, in operator notation,

$$(mD^2 + \eta D + \beta)\Delta a = F_a \tag{9.116}$$

where D = operator;
  m = effective mass of deforming structure, kg;
  η = friction coefficient, $N \cdot s/m$;
  β = spring factor, N/m;
  Δa = incremental deformation in the X-direction, m; and
  F_a = acting force, N.

According to Hooke's law, the stress developed in the crystal is

$$\sigma_1 = B_1 \frac{\Delta a}{a} \tag{9.117}$$

where B_1 = Young's modulus of the material along the X-axis, N/m^2.
  From formula 9.117, the force F_x causing deformation is

$$F_x = \sigma_1 A_1 = B_1 A_1 \frac{\Delta a}{a}. \tag{9.118}$$

Substituting Δa from formula 9.116 into formula 9.118 gives

$$G_m(D) = \frac{F_x}{F_a} = \frac{B_1 A_1}{a} \cdot \frac{1}{mD^2 + \eta D + \beta}. \tag{9.119}$$

The ratio $F_x/F_a = G_m(D)$ is a mechanical transfer function giving the relationship between the mechanical input F_a and mechanical output F_x. The total transfer function $G(D)$ should reflect the ratio of the electrical output to the mechanical input. A convenient electrical output can be voltage U (Fig. 9.43)

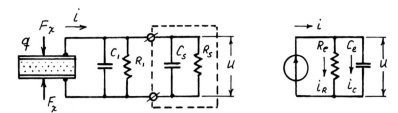

Figure 9.43 Equivalent electrical circuits of piezoelectric element converting force into electrical signal.

proportional to the input force F_a. We can determine the mechano-electrical transfer function as $G_e(D) = U/F_x$ and find that

$$G(D) = G_m(D) \cdot G_e(D) = \frac{F_x}{F_a} \cdot \frac{U}{F_x} = \frac{U}{F_a}. \qquad (9.120)$$

An equivalent electrical circuit for calculating the force-to-voltage transducing (Fig. 9.43) includes the charge generator, leakage resistance of the crystal R_1, element's capacitance C_1, equivalent resistance R_s, and capacitance C_s of a signal conditioner input. R_s and C_s model leakages in cable, some additional capacitors and resistors connected parallel to the input, and so on. The reduced equivalent circuit is composed of a current generator producing current i [A], and equivalent resistance and capacitance in parallel connection:

$$i = \frac{dq_1}{dt} = d_{11} \frac{dF_x}{dt} \text{ (see formula 9.110)}, \qquad R_e = \frac{R_1 R_s}{R_1 + R_s},$$

$$\text{and} \qquad C_e = C_1 + C_s. \qquad (9.121)$$

The equation for currents is

$$i = i_R + i_c = \frac{U}{R_e} + C_e DU = d_{11} DF_x = \frac{U}{R_e} + C_e DU \qquad (9.122)$$

where i, i_R, and i_c = currents in the element, R_e, and C_e, respectively, A.

From formula 9.122, the electrical transfer function will be

$$G_e(D) = \frac{U}{F_x} = \frac{d_{11}}{C_e} \cdot \frac{\tau D}{1 + \tau D}; \qquad G_e(j\omega) = \frac{d_{11}}{C_e} \cdot \frac{j\omega\tau}{1 + j\omega\tau} \qquad (9.123)$$

where $\tau = R_e C_e$ = time constant of circuit, s.

The magnitude $|G_e|$ of G_e (D) is frequency-dependent:

$$|G_e| = \frac{d_{11}}{C_e} \cdot \frac{\omega\tau}{\sqrt{1 + (\omega\tau)^2}} \qquad (9.124)$$

where ω = angular frequency of F_x, rad/s.

If τ is high enough, the increase in frequency leads to $(\omega\tau)^2 \gg 1$ and $|G_e| \approx d_{11}/C_e$. For $\omega = 0$, $|G_e| = 0$. It means that the electromechanical "fraction" of the element operates as a differentiating cell (see Chapter 3, Pressure-sensitive Capacitive Elements; Chapter 4, Variable-reluctance Elements) and it is suitable for measurements of alternating rather than static forces. Using a charge amplifier with high-resistance input having negligible leaks enables quasi-static measurements. Note that connecting a capacitor parallel to the element (increase in C_e) gives an increase in τ leading to a smaller attenuation of low frequency signals. At the same time, the general level of $|G_e|$ is decreased since C_e is in the denominator of formula 9.124.

Making substitutions of $G_m(D)$ and $G_e(D)$ from equations 9.119 and 9.123 into equation 9.120, we will find that

$$G(D) = \frac{d_{11}A_1B_1}{aC_e} \cdot \frac{\tau D}{1 + \tau D} \cdot \frac{1}{mD^2 + \eta D + \beta}, \tag{9.125}$$

$$G(j\omega) = \frac{d_{11}A_1B_1}{aC_em} \cdot \frac{j\omega\tau}{(1 + j\omega\tau)(\omega_0^2 - \omega^2 + jh\omega)}, \quad \text{and} \tag{9.126}$$

$$|G| = \frac{d_{11}A_1B_1}{aC_em} \cdot \frac{\omega\tau}{\sqrt{(\omega_0^2 - \omega^2 - h\omega^2\tau)^2 + \omega^2(h + \omega_0^2\tau - \omega^2\tau)^2}} \tag{9.127}$$

where $\omega_0 = \sqrt{\beta/m}$ and $h = \eta/m$. $\tag{9.128}$

At the resonance frequency w_0, the output has a sharp increase (Fig. 9.44). For many applications, the plateau in the characteristic $|G| = f(\omega)$ between frequencies ω_1 and ω_2 is the most convenient operational range, since the conversion is uniform for the signals of different frequencies. If the element is a part of a microphone, an acoustic correction of its frequency characteristic can be performed as described in Chapter 5, Acoustical Elements.

The operation of the piezoelements at resonance has a specific interest and is considered in Chapter 4, Resonant Elements; and Chapter 9, Electromechanical Microsensors.

The reverse piezoelectric effect has various applications especially at the resonance mode when the oscillations are extremely intensive.

It follows from equation 9.96, that the strains along the X- and Y-axis, $\Delta a/a$ and $\Delta b/b$, respectively, can be expressed as functions of the field strength E_1, along the X-axis:

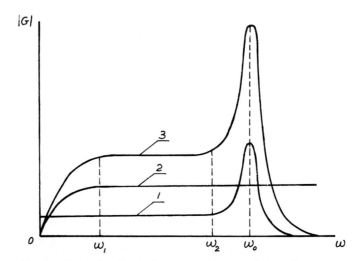

Figure 9.44 Amplitude-frequency characteristics of piezoelectric element. 1 = response of mechanical structure, 2 = response of electrical circuit, 3 = combination of characteristics 1 and 2 (transfer characteristic of element as function of frequency of input force).

$$\frac{\Delta a}{a} = d_{11}E_1, \quad \frac{\Delta b}{b} = -d_{11}E_1. \tag{9.129}$$

The voltage U_1 [V] applied across the electrodes is

$$U_1 = E_1 \cdot a; \text{ therefore,}$$

$$\Delta a = d_{11}U_1 \quad \text{and} \quad \Delta b = -d_{11}\frac{b}{a}U_1. \tag{9.130}$$

In order to enhance the level of electrical or mechanical outputs, the elements are stacked in columns (Scheme 27) or form a flexible bimorph structure. Multilayer constructions made of thin or thick films are also applicable. The shapes of elements vary. They can be disks, squares, solid and hollow cylinders, hemispheres, rings, and so forth. Depending on the orientation of the electrodes and polarization, the direction of electrical or mechanical action can be longitudinal, tangential, radial, torsional, etc. The characters of static or vibratory deformations can be tension-compression, flexing, twisting, or a combination of these.

Temperature characteristics of piezoelectric elements are quite complicated and greatly depend on composition, treatment of material, and features of design. All the major characteristics of the materials are temperature-dependent. But there is a common observation that these characteristics follow the nature of the dielectric constant-temperature characteristic. The experimental evaluation of temperature responses for concrete conditions and materials is the most practical approach. Some compositions based on Pb and Zr are extremely stable. Compounds with $BaTiO_3$ show variations in characteristics defined by the additions. Rochelle salt is temperature-unstable. Coating the elements by films for protection against moisture is always desirable and often necessary. The condensation of moisture on surfaces and its penetration in pores change the insulation of parts and quality of materials. By nature, a piezoelectric element has a high-resistance and capacitive output. The random charges induced due to the friction of insulators, (e.g., in cables or due to the changes in their capacitance) can deteriorate measurements. There are specially designed cables for operation with the elements. However, the integration of the element with a signal conditioner and elimination of the cable is desirable.

chapter 10

Temperature- and Light-sensitive Microstructures and Fiber-optic Sensors

Many transduction mechanisms of temperature- and light-sensitive sensors are similar. In various applications light sensors and fiber-optic sensors are united in one system. This situation makes it expedient to combine and describe the mentioned sensors in one chapter.

SOLID-STATE TEMPERATURE SENSORS

In spite of the diversity in solid-state sensors of temperature, the vast majority of them use the pronounced phenomena in semiconductors: dependence of their electroconducting properties on thermal agitation. This is the reason why we will pay primary attention to the semiconductor devices in this chapter. It is advisable, however, to review Electrical Characteristics of Semiconductors in Chapter 7 and the descriptions in this book of a number of solid-state structures that are characterized in terms of temperature-to-electrical value conversion, (e.g., piezoelectric materials, ferrites, etc.) since these structures can be alternatives to the semiconductors in sensor design.

Thermistors

Thermistors are semiconductor temperature-to-resistance transduction elements. They cannot approach the laboratory accuracy of platinum-wire and film devices (see Temperature-sensitive Elements, Chapter 6), but they are inexpensive, smaller in size, and provide a high-output signal. The resistance versus temperature characteristic of thermistors is exponential (Fig. 10.1). Standard resistances are from 10Ω to $20M\Omega$ at 25°C. The maximum continuous temperature is 450°C. The temperature coefficient of resistance can be positive or negative. For most thermistors, it is negative. Typically, this coefficient lies in the range 1 to 6% per °C. The semiconductor material used for thermistor fabrication is usually a combination of metallic oxides or sulfides. Several examples of these materials are Fe_3O_4 ($FeO\ Fe_2O_3$) with the addition of $MgO\ Cr_2O_3$ or $2ZnO\ TiO_2$; Fe_2O_3 with a small addition of TiO_2; NiO or CoO with the addition of Li_2O; Mn_3O_4; compositions $CuO\text{-}Mn_3O_4$, $NiO\text{-}Mn_3O_4$,

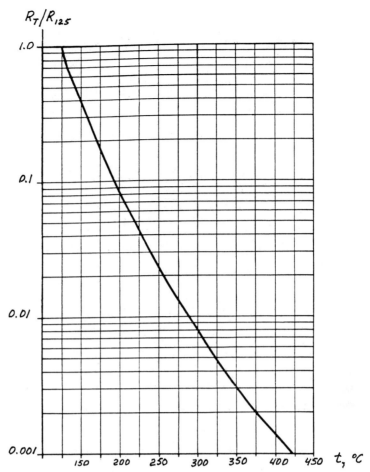

Figure 10.1 Typical relation between relative change of resistance R_T and temperature t for thermistor (source: [433]). R_{125} = resistance at 125°C.

and $NiO-Co_3O_4-Mn_3O_4$; and $MnCo_2O_4$ and $MnCr_2O_4$ combined with iron oxides. Control of the resistivity of the material in a wide range is possible due to the additions that change the lattice of the crystals, and, therefore, alter the passage of electrons. Some other oxides are usually added to the major component in order to improve the reproduction of the characteristics. Polycrystal ceramic materials are used for thermistors with a positive temperature coefficient. The resistivity of these materials is chiefly defined by the conditions at the borders of cells, which in turn are controlled by the dielectric constant.

The oxides used in the thermistors have electronic conductivity. According to the principles of electroconductivity considered in Chapter 7, Electroconductivity of Solids, the number of electrons passing the energy gap in a semiconductor is increased with temperature due to the energy that electrons obtain from the heat source. The increase in the electron flow is reflected in the resistance-temperature characteristic by a decrease in the resistance.

The most common and simplest presentation of the thermistor characteristic is

$$R_T = A \exp (B/T) \tag{10.1}$$

where R_T = zero power resistance (no self-heat) at absolute temperature T, Ω;
   A = constant defined by the thermistor resistance, Ω; and
   B = constant that depends on thermistor resistance and material, K.

For different materials, B varies from 2000 to 6000. If a thermistor is at reference temperature T_0, its resistance R_{T_0} will be

$$R_{T_0} = A \exp (B/T_0). \tag{10.2}$$

By taking the ratio of formulas 10.1 and 10.2, we will find that

$$R_T = R_{T_0} \exp B\left(\frac{1}{T} - \frac{1}{T_0} \right). \tag{10.3}$$

From this equation,

$$B = \frac{TT_0}{T_0 - T} \ln \frac{R_T}{R_{T_0}} \quad \text{and} \quad T = BT_0 / \left(B + T_0 \ln \frac{R_T}{R_{T_0}} \right). \tag{10.4}$$

The temperature coefficient of resistance α_R can be calculated as

$$\alpha_R = \frac{1}{R_T} \cdot \frac{dR_T}{dT} = -\frac{B}{T^2}. \tag{10.5}$$

Equation 10.2 is valid only for a narrow temperature range [433] since the material constant B increases with temperature. For the typical thermistor, an error of $\pm 0.3°C$ in the range of 0 to 50°C can affect the calculations when formula 10.2 is used. More exact reproduction of the real transducer performance can be obtained if the following equations are used:

$$R_T = \exp (A_0 + A_1/T + A_2/T^2 + A_3/T^3) \tag{10.6}$$

$$1/T = a_0 + a_1 \ln R_T + a_2(\ln R_T)^2 + a_3(\ln R_T)^3 \tag{10.7}$$

where A_0, A_1, A_2, and A_3 = unique constants for equation 10.6—the units are K^0, K^1, K^2, and K^3, respectively, and

a_0, a_1, a_2, and a_3 = unique constants for equation 10.7, 1/K.

Equations 10.6 and 10.7 require a minimum of four calibration points in order to determine the constants. These equations can be used over wider temperature spans with one improved curve fit [433]. A maximum error of 0.0015°C is encountered for a typical thermistor operated over the range of 0 to 100°C.

Thermistors are fabricated in four basic shapes: rod, disc, flake, and bead. In order to obtain the final shape of a semiconductor body, the metallic oxides are intimately mixed and ground to a homogeneous powder. This powder is then bound by an organic binder, shaped to the required form, sintered by firing, and then dried. After this process, the units are aged to achieve electrical stability. The proportion of the ingredients is controlled carefully by chemical analysis since the ingredients' composition mainly determines the resistivity of the material. Minor variations in the resistivities of the final product are corrected by a slight change in the firing temperature, which lies between 1000 and 1350°C.

Rod-type thermistors are usually made by combining the powdered oxides with an organic binder to form a plastic body; extruding the material through a die, usually of a circular cross section; and cutting to length. The bodies are dried and fired slowly to avoid cracking and distortion. The process steps are analogous to those used in the fabrication of ceramic parts. The sintered rods are provided with metallized contacts at the ends that are fabricated by metal spraying or by applying and sintering metallic pastes (see Chapter 8, Other Technologies). The units are completed by pressing on wired end-caps or by soldering on lead wires.

Disc- and washer-shaped bodies are made in machines similar to those used for making medicinal tablets. Powder provided with a binder is dried, recrumbled, and fed into a die where it is formed at high pressure into a relatively hard disc. The other fabrication steps are similar to those for rod-shaped elements. The units may have lead wires or attached base plates. The resistance of a unit can be precisely adjusted by removing small areas of metal from one of the faces of the disc.

Thermoflakes are fabricated as thick-film thermistors that have no substrate backings. Their lead wires are fired directly into their electrodes. The high surface-to-mass ratio associated with flakes results in a low heat capacity and fast response time. The elements are ideally suited for infrared detection. In some designs, the surfaces of the flakes are coated by a special infrared energy absorber. This coating provides more uniform absorption over the spectral wavelength band.

Thermistor beads are characterized by a wide range of bead-body and lead-diameter sizes. They vary from 0.13 to 1.4mm for the body and from 0.018 to 0.1mm for the lead's diameter. The typical configuration of a bead is small, olive-like, and hermetically sealed in a glass body with two leads at one side or at opposite sides of the body. The lead material must (1) have a high melting point to withstand sintering temperature, (2) have a coefficient of thermal expansion equal or close to that of the thermistor material, and (3) be chemically resistive, especially to oxidation during sintering. These requirements lead to the use of platinum or one of its alloys. In the fabrication of beads, a slurry made of powdered oxides mixed with a dilute organic binder is applied to parallel-held lead-wires, forming drops of the material resting on string. After the bead is dried and sintered, and the string is cut to form individual elements, the bead is encapsulated in the glass seal. Bare beads (without glass coating) also have some application; however, their characteristics are not stable.

When a thermistor is a part of a special-purpose sensor, it can be placed inside a housing, such as a hypodermic needle, catheter, or sheath. The construction of a typical thermistor is shown in Scheme 156.

The reader can find interesting information on the linear conversion of a thermistor's resistance into frequency given in [434].

Silicon Resistive Temperature Sensors

As we have shown in Electroconductivity of Solids in Chapter 7, silicon resistivity varies with temperature, which results from the variation of the mobility of the free electrons and holes with temperature. In lightly doped silicon ($\rho > 10\Omega \cdot cm$), the mobility μ varies with temperature T [K] approximately as $\mu \propto 1/T^{5/2}$. Since the resistivity ρ is reciprocal to the conductivity σ (see Chapter 7, Electroconductivity of Solids), and σ is directly proportional to μ, the following relation takes place:

$$\rho = (1/\sigma) \propto T^{5/2}. \tag{10.8}$$

From this expression, the temperature coefficients of resistivity α_ρ and resistance α_R [1/K] can be found:

$$\alpha_\rho = \frac{1}{\rho} \cdot \frac{d\rho}{dT} = \frac{5}{2} \cdot \frac{1}{T} \quad \text{and} \quad \alpha_R = \frac{1}{R} \cdot \frac{dR}{dT} = \frac{5}{2} \cdot \frac{1}{T}. \tag{10.9}$$

For 25°C (298K), $\alpha_R = 8.4\%$, which is small relative to α_R for the typical thermistor. With the increase in the doping level above $1 \times 10^{15} cm^{-3}$, α_R decreases and becomes negative when the doping level exceeds about $3 \times 10^{18} cm^{-3}$. It is convenient in a design to have the silicon temperature-sensitive layer integrated with other functional components in one die, especially for temperature compensation purposes.

A standard polysilicon process is implemented for fabrication of the temperature-sensitive resistors (see Chapter 8, Other Technologies). Some details of the process and description of the resistor characteristics can be found in [435]. The authors used a 50-mm, [111], 5Ω · cm, p-Si wafer as a starting material. The steps of fabrication were (1) wet oxidation; (2) atmospheric plasma chemical vapor deposition of undoped, 600nm-thick Si at 750°C; (3) patterning by photolithography and etching; (4) boron implantation at the energy of 100keV; (5) second oxidation and annealing; (6) opening holes for the contacts; (7) forming the contacts and leads by metallization; and (8) separation of the chips.

An intermediate step in the process can be a second implantation of boron into the temperature-independent resistor, which is used in parallel connection with the temperature-sensitive resistor for the linearization of the transfer characteristic.

Thermal Sensors Based on Transistors

Transistors lend themselves to use as temperature sensors [436] in the temperature range of − 55 to + 150°C by exploiting a base-emitter voltage V_{BE} dependence on temperature T. Figure 10.2 illustrates three basic methods in reading the temperature.

1. Voltage V_{BE} of a single transistor is used as a measure of T.
2. The difference ΔV_{BE} between two voltages V_{BE1} and V_{BE2} of two transistors gives a reading of T. This voltage is proportional to the absolute temperature (PTAT).
3. The combination of two voltages V_{BE} and ΔV_{BE} is utilized for determining the T (intrinsic reference device).

In the following calculation, we will derive the transfer characteristic V_{BE} (T) assuming in the discussion that the collector-base voltage of the sensor transistors is biased at zero volts. In this case, the collector current I_c [A] is defined similarly to the diode current (Chapter 7, Semiconductor Junctions; equation 7.23):

$$I_c = I_s \exp \left(\frac{qV_{BE}}{kT} \right) \tag{10.10}$$

where I_s = saturation current (reverse leakage current), A.

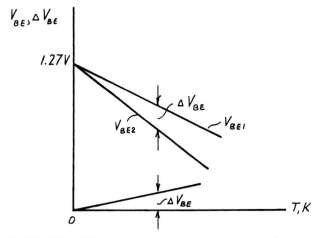

Figure 10.2 Plot illustrating modes of operation of transistor temperature sensors. Two identical transistors operate at different collector currents.

I_s strongly depends on temperature; this fact is reflected in the modified equation 10.10 [437]:

$$I_c = C'T^\eta \exp \frac{q(V_{BE} - V_{GO})}{\bar{k}T} \qquad (10.11)$$

where C' = constant, A/K^η;
   η = constant, which is slightly related to the doping level; and
   V_{GO} = extrapolated band-gap voltage at 0K, V.

To develop the equation for $V_{BE}(T)$, we suppose that $I_c \propto T^m (m = $ some power of T). The function will be written for an arbitrary temperature T and a specified reference temperature T_r by substituting T^m for I_c in equation 10.11 and applying the latter for each temperature:

$$V_{BE}(T) = V_{GO}\left(1 - \frac{T}{T_r}\right) + \left(\frac{T}{T_r}\right)V_{BE}(T_r) - (\eta - m)\left(\frac{\bar{k}T}{q}\right)\ln\left(\frac{T}{T_r}\right). \quad (10.12)$$

It is convenient to express $V_{BE}(T)$ as a sum of three terms: a constant term, a linear term, and a nonlinear term. In the following expression, they are separated by braces:

$$V_{BE}(T) = \left\{V_{GO} + (\eta - m)\frac{\bar{k}T_r}{q}\right\} - \left\{\lambda T\right\}$$
$$+ \left\{(\eta - m)\frac{\bar{k}}{q}\left[T - T_r - T\cdot\ln\left(\frac{T}{T_r}\right)\right]\right\}. \qquad (10.13)$$

In this expression, the first term is constant, the second one is a linear function of T (PTAT), and the third term is a nonlinear function of T:

$$\lambda = \frac{1}{T_r}[V_{GO} + (\bar{k}T_r/q)(\eta - m) - V_{BE}(T_r)]. \qquad (10.14)$$

The order of magnitude and weight of the different terms of the last two equations can be illustrated by the typical parameters of one of the sensors reported in [436]. For this example, current I_c is taken constant ($m = 0$), reference temperature $T_r = 323$K, and $V_{BE}(T_r) = 547$mV. The empirical values of V_{GO} and η are 1171mV and 3.54, respectively; and the relevant substitutions into the formulas give $V_{GO} + (\bar{k}T_r/q)\eta = 1269.5$mV, $\lambda = 2.24$mV/K, and the value of the last term in equation 10.13 for $T = 293$K is -0.44mV.

For relatively small temperature changes $\Delta T = T - T_r \ll T_r$, equation 10.13 can be approximated using the Taylor expansion:

$$V_{BE}(T) = V_{GO} + (\eta - m)\left(\frac{\bar{k}T_r}{q}\right) - \lambda T - \frac{1}{2}(\eta - m)\left(\frac{\bar{k}T_r}{q}\right)\left(\frac{\Delta T}{T_r}\right)^2. \quad (10.15)$$

In practical designs, the base and collector terminals of the sensor are short-circuited. This measure is undertaken to eliminate the Early effect (the dependence of I_c on the collector-base voltage), which can be a source of errors. In this case, the transistor is reduced to a two-terminal sensor. Special attention should be paid to the elimination of self-heating. The range of the bias current is in the range 10 to 100μA.

In measurements, equations 10.13 and 10.15 are usually approximated as $V_{BE} = 1.27 - CT$ ($C = $ const). Since $V_{BE}(0) = 1.27$V, a single measurement of $V_{BE}(T)$ will completely define the voltage-versus-temperature relationship for the transistor. It is also possible to offset differences in transistors by setting V_{BE} (T) equal to a specific value. This is accomplished by adjusting the bias current of the device.

As we have mentioned above, the PTAT transistor sensor produces an output voltage or current proportional to the absolute temperature. The base-emitter voltage for a single transistor is found from equation 10.10:

$$V_{BE} = \frac{\bar{k}T}{q} \ln \frac{I_c}{I_s}. \tag{10.16}$$

If two transistors are exposed to the same temperature, the difference ΔV_{BE} between two voltages V_{BE1} and V_{BE2} will be

$$\Delta V_{BE} = V_{BE1} - V_{BE2} = \left(\frac{\bar{k}T}{q}\right) \ln \left(\frac{I_{c1}}{I_{c2}} \cdot \frac{I_{s2}}{I_{s1}}\right). \tag{10.17}$$

Ratio $r = I_{s2}/I_{s1}$ is defined by the emitter-base junction areas of the two transistors. Ratio $\rho = I_{c1}/I_{c2}$ is a constant when the collector currents are maintained at a prescribed level. Therefore, equation 10.17 is reduced to a simple form:

$$\Delta V_{BE} = T \frac{\bar{k}}{q} \ln r\rho = C''T \tag{10.18}$$

where C'' [V/K] is a constant.

The ratiometric nature of the sensor and linear response to the measurand are attractive features of this device. The output signal from the sensor is greatly compensated for the effects of processing inequalities. Transistors with closely matched characteristics are usually chosen to form the transduction couples. However, the signal from the sensors is small and the influence of some disturbances can be appreciable.

A problem common to the single transistor and PTAT device is a large initial offset voltage. The temperature-dependent base-emitter voltage is superimposed on this value. This means, for example, that the detection of a 0.1K change in temperature at 300K would require a resolution of 3000:1. To decrease this ratio, it is necessary to remove the offset voltage from the output signal. This will require more complicated supporting electronics, which will increase the price of the device. The intrinsic reference sensor has an initial offset of about one-tenth of the other two devices. The sensor exploits the intrinsic band gap voltage as a reference [436]. It should be noted that the three basic approaches in the arrangement of the sensors are characterized by the same accuracy.

Integrated Thermocouples

Microelectronic technology makes it possible to fabricate miniature thermocouples, whose principle of operation is the same as those described in Chapter 6, but whose sizes and configurations can significantly differ. In silicon, several methods can be employed to form a thermocouple or thermopile. Diffusion, ion implantation, oxidation, epitaxial growth, and metallization are just a few processes that lend themselves to the fabrication of these devices. For the silicon/aluminum thermopile shown in Figure 10.3a, a thin layer of n-type silicon is deposited on a p-type substrate. Next, p-type silicon is either diffused or ion-implanted into the n-epilayer. The surface is passivated by growing a layer of silicon dioxide. Finally, aluminum metallization is performed to provide the second material in the thermopile's construction and to connect the thermocouples in series [438, 439].

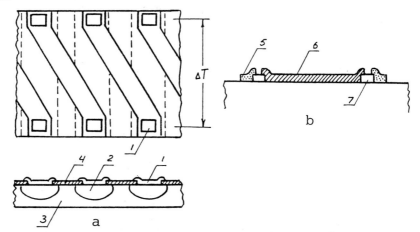

Figure 10.3 Si/Al thermopile (a) and thin-film thermocouple (b).
ΔT = difference in temperatures, 1 = Al, 2 = p-Si, 3 = n-Si epilayer,
4 = SiO_2, 5 = metal A, 6 = metal B, 7 = metal interconnect.

Figure 10.3b shows a thermocouple that was made using only metallization on the surface of Corning glass. A thermopile sensor using five copper/palladium thermocouples [440] shows a response to temperature of about several μV per K. The sensor is also sensitive to the hydrogen content in a gas mixture. The device is formed by a standard photolithographic technique on a Cr-Cu film (4000Å thick) deposited by thermal evaporation. The Pd film, 4000Å thick, is deposited by dc sputtering and defined by the lift-off process.

One problem that exists, when a thermocouple or thermopile is fabricated on silicon, is distinguishing between the sensing junction and the reference junction. Silicon is a good heat conductor. Since the sensor is only about 1% of the total thickness of the wafer, the remaining thickness acts as a thermal short circuit. In order for a portion of the sensor to be able to conduct heat, the material underneath it must be removed (Fig. 10.4). This not only makes the section thermally active, it also increases the thermal resistance, which in turn increases the sensitivity of the device. When needed, the thermal short circuit property of silicon is used for conduction of heat.

The fabrication techniques used in the thermo-sensing structures of thermocouples and thermopiles are similar to those used in the fabrication of integrated circuits. This allows small sensors with processing and control circuitry to be located on the same chip.

As is explained above, a thermopile is just a series of thermocouples. The three most widely used thermopile structures are the closed membrane, the cantilever

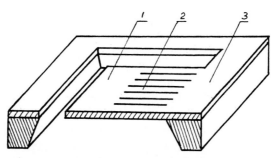

Figure 10.4 Thermal isolation of reference junction from sensing junction in cantilever beam structure. 1 = interaction area, 2 = thermopile, 3 = reference junction.

beam, and the floating membrane. Each of these structures utilizes an area for sensing the temperature changes by a thermopile arrangement. The closed or round membrane sensor has a circular interaction area located at its center. The thermopile structure radiates out from this center to the thermally short-circuited rim. The rim forms the reference junction while a heating resistor embedded in the interaction area forms the sensing junction. Measurements are made by setting the heating resistor such that the interaction between the sensor and the ambient provides a fixed temperature reference. A change in the ambient produces a proportional change in the overall temperature of the device. This is the most sensitive of the three structures mentioned above. The cantilever beam structure has a rectangular interaction area supported on one side by the thermopile array. The thermopile senses the temperature difference between the area and the rim. The larger interaction area along with the reduction in the number of thermocouples in the thermopile reduces the sensitivity of this device in relation to the closed membrane. Increasing the length of the interaction area will increase the sensitivity of the sensor (Fig. 10.4). The floating membrane has the largest interaction area. It is suspended on all four sides by only a few suspension beams. The thermopile uses these beams to place its sensing junctions on the interaction area. The small number of suspension beams leads to a low thermopile sensitivity, but the high thermal resistance of the interaction area increases the overall sensitivity of the device.

All three structures are better suited to thermopiles than those that are fabricated using metallization on the surface of a substrate. Considerations such as mechanical stress, ambient conditions, package size, and yield set limits on the size and selection of the appropriate structure. Some typical applications of thermopile elements are true rms converters, vacuum sensors, flow sensors, and infrared sensors [441].

Other Temperature Sensors

In a CMOS *temperature sensor* [442], the output current is presented as the difference between positive- and negative-temperature-coefficient currents (PTC and NTC, respectively). The PTC current is derived from a PTAT cell, while the NTC current is derived from a threshold voltage reference source. The method of subtraction of the currents in the design diminishes to a certain extent the effects of the thermal dependence of the resistors used in the circuit. For the temperature-to frequency conversion, the CMOS transistor is used in a weak-inversion operation mode. At this mode, the transistor appears to behave similarly to p-n junctions with respect to their thermal behavior. Therefore, a PTAT cell can also be made in CMOS technology.

The operation of the *lateral-field silicon temperature sensor with charge injection* [443] is based on the same effect applied in resistive thermal sensors; that is, on the temperature dependence of the mobility of carriers. The sensor is a temperature-dependent delay line that can be a part of an oscillator whose frequency is modulated by the temperature. In the sensor's structure, the electrons injected into a channel are transported along the channel with a velocity determined by the applied lateral electric field and the mobility of carriers. The temperature-dependent mobility results in a temperature-dependent traversing time. As a result, there will be a delay in the path between the electron-injecting and collecting electrodes. Temperature sensors and pseudo-thermopiles can be constructed using such a device with the advantage of simple A/D conversion and compatibility with bipolar processing.

A *high-resolution thin-film germanium temperature sensor* [444] is fabricated by a deposition of 250-nm-thick amorphous germanium film at the rate of 0.5nm/s and at a residual pressure during evaporation of better than 5×10^{-7}mbar. Polyimide, glass, and polished alumina substrates of thicknesses ranging from 0.1 to 0.2mm are used. The a-Ge films are tempered at 420K for 4h. Patterning of the films is provided by a plasma-enhanced etching with CH_4 and O_2 mixtures (ratio 4:1) with standard photoresist techniques. Conducting paths are deposited by the evaporation of 0.25-μm-thick titanium-gold-titanium layers. Titanium acts as an adhesion layer as well as a diffusion barrier to hinder Au migration into Ge. Passivating is provided by a 3-μm-thick silicon nitride layer. The conductivity of film σ is an exponential function of T: $\sigma = \sigma_0 \exp(-\gamma T^{-1})$, where σ_0 and γ are constants.

The characteristics of the sensors of a 0.14×0.1-mm^2 sensitive area typically have a nominal resistance in the range of $1 \times 10^5 \, \Omega$, at a temperature coefficient of about 2%K^{-1} for 25°C. The material has a free-carrier density of 8×10^{17}cm^{-3}. The time constant is about 3ms for the sensor suddenly inserted into liquid. Temperature resolution is 0.1mK with a bandwidth of 1kHz.

A hybrid *magneto-temperature sensor* using contacts on a $YBa_2Cu_3O_{7-\delta}$ (YBCO) *superconductor* shows [445] both the change of resistance with respect to the magnetic field intensity and the change of resistance as a function of temperature. The substantial parts of the sensor are contacts formed between the superconductor and silver wire. One of the contacts (type A) is prepared by the conventional ultrasonic bonding method; another contact (type B) is formed by the spot-welding method. The type A contact is affected by the magnetic field below the superconducting critical temperature T_c, which is 93K for this material. The type B contact is not sensitive to the magnetic field; however, it responds to temperature. A hybrid-temperature sensor is formed by fabricating both contacts on the surface of the YBCO ceramics. Note that the resistance of the YBCO ceramics below T_e is zero. Therefore, it is assumed that the detectable resistance changes are due to the electrical processes at the contacts. The resistance change is in the range of fractions of an ohm and units of ohms.

An *ultrasonic distributed temperature sensor* is a part of a measuring system [446] that is composed of a magnetostrictive transceiver, transmission line, and sensor (see the section in Chapter 3 on magnetostrictive elements). The transceiver is an electroacoustic converter. When electrically energized, it generates an ultrasonic pulse and, when acting as a receiver, it converts reflected ultrasonic pulses back to electrical signals. During operation, a short ultrasonic pulse is generated by the transceiver and sent via the transmission line to the sensor. The sensor can be in the form of a rod, wire, or ribbon made from materials such as tungsten, stainless steel, titanium, plastic, glass, and so forth. The sensor is divided into several sections similar to those in a magnetostrictive filter (Fig. 3.53c). A portion of the transmitted ultrasonic energy reflected from the sections is returned to the transceiver. The time difference between the pulses received from two adjacent sections depends on the sound velocity in that zone. Average temperature in the zone can be monitored from the arrival time, since sound velocity varies with temperature. Several materials have been investigated for operation up to 800°C: Inconel 600, nickel, stainless steel 304, Kanthal, and Ni-Cr. Their sensitivity was found sufficient for sensor application. For example, for Inconel 600, the slope of the propagation-time versus temperature curve varies between 1.5×10^{-4} and 4.0×10^{-4} per°C.

The inherent advantages of the acoustic techniques based on this sensor are that the measured temperature is the true average zone temperature rather than the temperature at a point; the sensor is rugged, it is not sensitive to electrical and magnetic fields; the system is acoustical, and hence better suited than a purely electrical system for use in explosive and hazardous areas.

Smart temperature sensors include in one self-contained package not only transduction elements, but also control and processing circuitry. Linearization, compensation for some adverse factors, and self-calibration are a few of the conditioning functions smart sensors may include. In one design [447], a CMOS-compatible lateral bipolar transistor was used as a sensing element (in a lateral transistor, the current from emitter to collector flows laterally). The temperature measurement was determined by the ratio of the temperature-dependent current I_t and a reference current I_r. This device is similar to the intrinsic-reference transistor sensor in that the ratio I_t/I_r is proportional to the ΔV_{BE} referred to a reference voltage. The lateral bipolar transistor was used as a sensing element because of its temperature sensitivity, linearity, and long-term stability, which are superior to those of a standard CMOS device.

PHOTODETECTORS

Various types of optical detectors will be covered in this section. With the rapid progress in optical instrumentation and communication systems, photodetectors have been intensively developed during the last two decades. Contemporary devices allow detection and measurement of light characteristics in a wide spectrum of wave lengths and intensity of radiation. The principles of operation of photodetectors and their designs are quite different. Within the room available for this topic, we will review only basic principles and features of photosensors and refer the reader interested in in-depth studies of the subject to informative books on optoelectronics [448–450], and to periodical reviews on the subject, such as [451].

By their principle of conversion, photodetectors can be divided into two classes: thermal and photon devices. In thermal detectors, light absorbed by a sensing element is first converted into heat and then into an electrical value. Depending on the absorption characteristics of the sensing element, the output of the detector will be proportional to the received energy in the wide or narrow range of the light's wavelength. Photon detection is accompanied by quantum events that affect the electrical state of the light-absorbing substance, inducing an electrical signal proportional to the rate of absorption of light quanta. The energy of photon E [J] is

$$E = h\nu = hc/\lambda \qquad (10.19)$$

where h = Planck's constant = 6.625×10^{-34} J · s;
$\quad \nu$ = frequency of electromagnetic radiation, Hz;
$\quad c$ = speed of light, m/s; and
$\quad \lambda$ = wavelength of electromagnetic radiation, m.

There is a certain minimum in E in order to initiate the photon processes; in other words, there is a maximum in the wavelength above which conversion is not possible (see equation 10.19).

Quantum events in the light-sensitive material may be generated also by thermal excitation creating a noise background for a light-induced signal. With the increase in λ (above 3μm), measures for reduction of the thermal effects are necessary. This is the reason why long-wave infrared detectors quite frequently require cooling for normal operation.

Thermal Detectors

The operation of thermal detectors is similar to that of temperature sensors (see Temperature-sensitive Elements, Chapter 6; and Solid-state Temperature Sensors, Chapter 10) with some specific features that can be described by using a simple model such as the one shown in Figure 10.5. In this model, the incoming radiation causes the change in the temperature of the sensing part, whose temperature is a measure of the radiation. Because of the thermal flux leakage toward the heat sink,

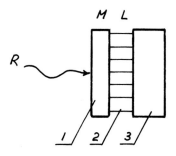

Figure 10.5 Model of thermal detector.
M = heat capacity, L = heat conductance,
R = incoming radiation, 1 = sensing part,
2 = thermal link, 3 = heat sink.

it is important to know the dynamics of the heat exchange. The relevant differential equation describing the process is similar to equation 6.18 with some modifications. We will introduce the heat capacity of the sensing part as $M = \rho_s V_s C_p$ [J/K], the rate of heat absorption as $B = \bar{h}_1 A \theta_m$ [J/s], and thermal conductance as $N = \bar{h}_1 A$ [J/K·s]. The equation describing the dynamics will be

$$M \frac{d\theta_s}{dt} + N\theta_s = B. \tag{10.20}$$

Note that the major difference between equations 6.18 and 10.20 is in the sense of term B. For this case, it represents the absorption of radiation energy.

Using operator notation for equation 10.20, the transfer function for the sensing part of the sensor will be given as

$$\frac{\theta_s}{B} = \frac{1}{MD + N} = \frac{1}{N} \cdot \frac{1}{\tau D + 1} \tag{10.21}$$

where $\tau = M/N = \rho_s V_s C_p / \bar{h}_1 A$ is the time constant.

It is obvious from expression 10.21 that the decrease in M and N magnitudes is desirable in terms of sensitivity and can be achieved if the absorbing area and mass of the sensing part are small. When M is chosen, N should not be taken too small, otherwise τ will be high and the dynamic response will be slow. The typical value for τ is 1×10^{-3}s. A random thermal fluctuation in the energy flow rate out of the element ΔB limits the value of the minimum detectable signal ΔB, as given by the following relation [452]:

$$\Delta B = \sqrt{4\bar{k}T^2 N \Delta f} \tag{10.22}$$

where $\bar{k}$ = Boltzmann constant = 1.38×10^{-23}J/K;
      T = absolute temperature, K; and
    Δf = frequency bandwidth of signals, Hz.

In the practice of measurements, the minimum detectable power is about 5×10^{-11}W for the sensors with an active area of about one square centimeter.

The sensing element is usually made of fine metal strips with coatings of materials providing a uniform absorption for a wide range of wavelengths. Encapsulating the elements in a vacuum increases the stability of sensors because of their isolation from the moving air unpredictably cooling the sensitive surface. Using the differential technique and reference sensors sensitive to temperature only is advantageous for the measurement of weak radiation.

Thermocouples, temperature-sensitive resistors, and junctions are in contact with the sensing element and provide a temperature-to-electrical-value conversion. For measurements of far-infrared radiation, deeply cooled (4.2K) carbon-resistive thermometers are used.

Pneumatic Detectors

A pneumatic detector for the measurement of radiation is quite similar to that described in the first section of Chapter 1 (Fig. 1.1) for the measurement of temperature. The pressure in an airtight chamber is proportional to the amount of radiation falling on the chamber's wall and causing a change of its temperature. Standard pressure-to-electrical-output techniques are employed to form the output. In one of the most sensitive devices, the Golay cell, a flexible silvered diaphragm operates as a mirror whose focal length depends on the pressure within the chamber. The change in the pressure deflects a light beam directed to the diaphragm surface. After passing through the system composed of lenses and grating, the reflected beam reaches a light detector whose photocurrent is a function of the deflection [448].

Pyroelectric Detectors

Pyroelectric detectors are not as sensitive as thermoelectric detectors, but they have a simple, robust construction and fast response to the measured radiation. The construction of a typical detector (Fig. 10.6) includes a thin slab of ferroelectric material with two electrodes on its surface. One of the electrodes is transparent. Molecules of the material develop a permanent dipole moment. Below the Curie temperature, the aligned dipoles give polarization to the material. The polarization is changed under the thermal agitation when the material is exposed to the radiation. The change in the polarization state is reflected in the current flowing through the load resistance. The voltage drop across this resistance can be calculated in the same manner as it is for piezoelectric materials sensitive to deformation. The equivalent circuit in Figure 9.43 is applicable for pyroelectric detectors if we assume that the current developed by the detector is

$$i = A \frac{dP}{dt} \tag{10.23}$$

where P = polarization proportional to the incident radiation, C/m^2, and
$\quad\quad A$ = detector's active area, m^2.

The dependence of P on temperature T [K] is given by a pyroelectric coefficient γ [C/m^2 · K]:

$$P = \gamma \cdot T. \tag{10.24}$$

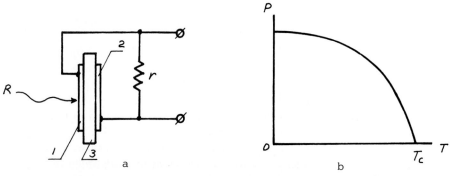

Figure 10.6   Schematic diagram of pyroelectric detector (a) and spontaneous electrical polarization for ferroelectric material (b). *P* = polarization, *R* = radiation, *r* = load resistance, *T* = temperature, T_c = Curie temperature, 1 = transparent electrode, 2 = regular electrode, 3 = ferroelectric material.

The choice of material for the sensor depends on the range of the radiation signals to be measured. Triglycine sulfate (TGS) has a high sensitivity, but a low Curie temperature (49°C). Piezoceramic materials are widely used for the sensors. They have Curie temperature ranges from about a hundred to several hundred degrees centigrade (see Table 9.6). The value of γ [C/m$^2 \cdot$ K] varies for different materials. For example, for TGS -3×10^{-4}; for BaTiO$_3$ -5×10^{-6}; and for LiTaO$_3$ -1.6×10^{-4}. Since the equivalent circuit of the sensor is characterized by a high impedance, the use of a JFET follower integrated with the transduction element is desirable. The minimum of the detected power is defined by the frequency bandwidth of signals. It is believed that this minimum can go down to 1×10^{-8}W for the $\Delta f \approx 1$Hz. The response time varies, but remains in the nanosecond region. Wavelength response extends to the wavelength of 100μm. There are reports such as those in [453, 454], on a successful use of PVDF film in the measurement of different radiations.

Photoemissive Devices

A photoemissive effect in metals takes place if a photon energy $h\nu$ from a source of radiation enters the metal, is absorbed in it, and gives up its energy to an electron. The electron that obtained the quantum of energy should be able to overcome the surface potential barrier defined by work function E_0 [J]. No electrons will be emitted if $h\nu < E_0$ or in terms of wavelength: $\lambda > hc/E_0$. Elastic collisions of photons with electrons lead to the emission. With the increase in the probability of inelastic collisions, the number of escaped electrons is decreased. The ratio of the number of the emitted electrons to the number of absorbed protons is called the quantum yield or quantum efficiency. In photoemissive devices, special materials with low work function and high efficiency are usually used. In pure metals, efficiency is low, about 0.1%, and work function is high. This is why metals are rarely used for photocathodes. There are two types of photoemissive surfaces utilized for practical designs: *classical* type and *negative electron affinity* (NEA) type. The former is made of an evaporated layer (e.g., NaKCsSb) composed of alkali metals and some elements from group V of the periodic table. NEA surfaces are fabricated by the evaporation of caesium or caesium oxide onto the surface of a semiconductor material, such as GaAs. These photocathodes have a high quantum efficiency in the range of waves of up to 0.9μm. Figure 10.7 illustrates the quantum efficiency η [percent] as a function of wavelength for several materials. The letter S is commonly accepted for denoting the classical type of materials; for example, "S 20" is NaKCsSb.

The simplest device that uses the photocathode is the vacuum and low-pressure gas-filled *photodiode* (see Scheme 175). A neutral gas like argon at low pressure (less than 1 torr) is used for filling. The sensitivity of the gas-filled photodiode is higher than that of the vacuum diode. Collisions of electrons with gas atoms ionize them and generate further electrons, thus overall current gain is of the order of 10. The voltage applied across the diode is a few hundred volts.

The current amplification in the *photomultipliers* (see Scheme 175) G is introduced by a simple relation:

$$G = N^n \tag{10.25}$$

where N = average number of secondary electrons emitted at each dynode surface
for each incident electron, and
n = number of dynodes.

In practical constructions, G exceeds a million. A number of contemporary photomultipliers have a semitransparent cathode evaporated onto the inner surface

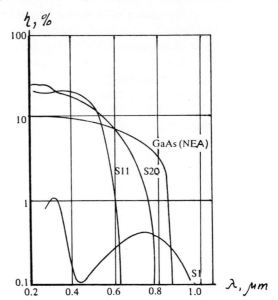

Figure 10.7 Quantum efficiency as function of wavelength for several materials (adopted from [448]).

of one end of the envelope (see Figure 10.8 and compare it with Scheme 175). The cathode layer should be thick enough to let the photons be absorbed in the material; at the same time, the layer should not be too thick, since the photons must penetrate to the inner surface in order to create the electrons.

N varies proportionally to the interdynode voltage and depends on the type of the material used for emission. For example, for Be-Cu dynodes, N changes between 0 and 6 proportionally to the voltage change from 0 to approximately 200V. Sets of slats and grids between the dynodes are used in some designs in order to focus the electron beam, which is advantageous in terms of the device's efficiency and response time. The electrons released from the cathode and dynodes have a spread in velocities; they traverse slightly different paths. These are two main reasons why they have a different transit time to traverse from the cathode to the anode and, as a result, some limits in a faithful reproduction of a fast optical signal. Using fewer dynodes and focusing the electron beam make the propagation of the electrons more uniform and improve the dynamic response. Some new materials for the cathode make it possible to have a higher value of N. These features allow the obtaining of devices with a 30-ns transit time.

There are several sources of noise in a photomultiplier that limit its ability to treat low-level signals. One of them is *dark current,* which is a main source of *noise.* The dark current i_T [A] constitutes a current due to thermionic emission at the

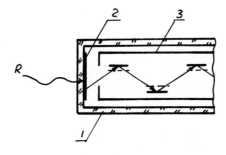

Figure 10.8 Schematic diagram of multiplier with semitransparent cathode. R = incident radiation, 1 = envelope, 2 = cathode, 3 = part carrying dynodes, grids, etc.

absence of the incident radiation activating the cathode. The Richardson-Dushman equation [455] gives the value of i_T:

$$i_T = aAT^2 \exp\left(-\frac{E_0}{kT}\right) \tag{10.26}$$

where a = constant = $1.2 \times 10^6 \text{A/m}^2 \cdot \text{K}^2$ for pure metals;
      A = cathode's area, m^2; and
      T = absolute temperature, K.

Reducing T is desirable for decreasing i_T, especially for low-level-work-function cathode materials. Because of the statistical nature of i_T, it varies randomly.

    Another source of noise is *shot noise,* which arises from the discrete nature of the electronic charge. The arrival rate of electrons to any point of an electrical circuit fluctuates slightly, giving a variation of the current at this point [456]:

$$\Delta i_f = \sqrt{2iq\Delta f} \tag{10.27}$$

where Δi_f = root mean square variation in current, A;
      i = current in circuit, A;
      q = electron's charge, C; and
      Δf = frequency band, Hz.

The dark current defines [456, 448] the minimum detectable signal W_m [W]:

$$W_m = \frac{W}{i}\sqrt{2i_T q \cdot \Delta f} \tag{10.28}$$

where W = optical power falling on the photocathode, W.

    A statistical spread in the secondary electron emission about the mean value of N is a source of so-called *multiplication noise,* which is appreciable when N is near to unity and gives an increase of the expected noise at the anode ($G\sqrt{2iq \cdot \Delta f}$).

    The *Johnson noise* is present in the anode load resistor. This noise is usually much smaller than the dark current or shot noise.

Photoconductive Detectors

An electron that absorbed a photon of frequency ν, provided that $h\nu \geq E_G$, may be raised from the valence band to the conduction band of a semiconductor. The last expression can be written in terms of wavelength,

$$\lambda \leq hc/E_G, \tag{10.29}$$

which defines the band-gap wavelength λ_g for the largest value of wavelength that can cause the transition of the electron:

$$\lambda_g = hc/E_G. \tag{10.30}$$

    The presence of the photon-induced electron in the conduction band gives an increase in the semiconductor's conductivity. This phenomenon, named *photoconductivity,* is employed in photoconductive detectors. The detector can be schematically introduced by a slab of a semiconductor material with metal electrodes on opposite ends and with a plane exposed to the incident radiation (Fig. 10.9). Suppose

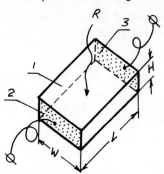

Figure 10.9 Schematic diagram of photoconductive slab. R = incident radiation, H, L, and W = dimensions, 1 = semiconductor, 2 and 3 = metallization.

we have a radiant flux of monochromatic radiation directed perpendicularly to the plane of the slab. In this case, the transmitted irradiance I [W/m^2] will be

$$I = I_0 \exp(-\alpha H) \tag{10.31}$$

where I_0 = incident irradiance, W/m^2;
  α = absorption coefficient, 1/m; and
  H = slab's thickness.

Figure 10.10 illustrates α as a function of λ for several materials used for the detectors. Note that for $\lambda > \lambda_g$, α is comparatively small and increases rapidly with the decrease of λ below λ_g. If the slab is thick enough ($H \gg 1/\alpha$) and all the falling radiation is absorbed, the total number of generated electron-hole pairs per second will be $\eta I_0 WL/h\nu$, where η is the quantum efficiency of the absorption and WL is the area of the slab's plane. If we take the ratio of the number of generated carriers per second to the slab's volume WLH, the generation rate per second, per unit of volume r_g [1/s · m^3] can be obtained:

$$r_g = \frac{\eta I_0}{h\nu H}. \tag{10.32}$$

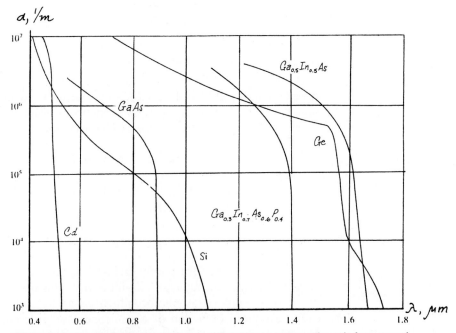

Figure 10.10 Optical absorption coefficient versus wavelength for several materials used for photoconductive detectors (source: [448]).

An effective quantum efficiency parameter is *photoconductive gain G*. It is a ratio of the electrons' flow rate from the detector under illumination Δi [A] to the rate of generation of charges in the device $\Delta i'$ [A]:

$$G = \frac{\Delta i}{\Delta i'}. \tag{10.33}$$

Current i [A] through the semiconductor slab expressed as a function of applied voltage U [V], dimensions, and conductivity σ will be

$$i = \frac{V}{R} = U\frac{WH}{L}\sigma \qquad \text{since} \qquad R = \frac{1}{\sigma}\cdot\frac{L}{WH} \tag{10.34}$$

where R = resistance of the slab between the electrodes, Ω.

From equation 10.34 and equation 7.15 in Chapter 7, Electrical Characteristics of Semiconductors,

$$\Delta i = U\frac{WH}{L}\Delta\sigma \qquad \text{and} \qquad \Delta\sigma = \Delta nq\mu_n + \Delta pq\mu_p \tag{10.35}$$

where Δn and Δp = excess in carrier concentration, $[1/m^3]$.

The recombination rate or carrier concentration r_r $[1/m^3 \cdot s]$ depends on Δn, Δp, and on the minority carrier lifetime τ_c [s]. For the equilibrium and charge neutrality conditions, $\Delta n = \Delta p$, and the recombination rate must be equal to the generation rate, $r_r = r_g$; therefore,

$$r_r = \frac{\Delta n}{\tau_c} = \frac{\Delta p}{\tau_c}, \qquad \Delta n = \Delta p = r_g\tau_c,$$

$$\Delta\sigma = r_g\tau_c q(\mu_n + \mu_p), \qquad \text{and} \qquad \Delta i = \frac{UWH}{L}r_g\tau_c q(\mu_n + \mu_p). \tag{10.36}$$

The rate of generation of charges $\Delta i'$ is readily found as a product:

$$\Delta i' = qr_g WHL. \tag{10.37}$$

Substitutions of Δi and $\Delta i'$ from formulas 10.36 and 10.37 into formula 10.33 gives

$$G = \frac{\tau_c U}{L^2}(\mu_n + \mu_p). \tag{10.38}$$

The increase in τ_c, U, and μ's and decrease in L are desirable for obtaining high efficiency G. However, the rise in τ_c gives a longer response time.

The detector output per unit energy input is increased proportionally to the wavelength of the incident light up to λ_g. Beyond λ_g, the output falls to zero since the photons' energy is not sufficient to excite carriers across the band gap. The quantum efficiency falls off with the increase in λ.

The fluctuations in the rates of generation and recombination of electron-hole pairs are the main source of noise in photoconductive detectors. Optical and thermal processes define this noise. The former process depends on the size of the band gap. Reduction of temperature reduces the noise. The common practical condition for the temperature is $T < E_G/25k$. If the thermally generated noise is negligibly small, the rms of the current resulting from generation-recombination Δi_{gr} [A] is

$$\Delta i_{gr} = \sqrt{\frac{4iqG\Delta f}{1 + 4\pi^2 f^2 \tau_c^2}}. \tag{10.39}$$

The designations of terms in this equation are given above. For $f \ll 1/(2\pi\tau_c)$, Δi_{gr} has a small dependency on f. For $f > 1/(2\pi\tau_c)$, Δi_{gr} gradually reduces with the increase of f. At the frequencies $f \gg 1/(2\pi\tau_c)$ ($\sim 1\text{MHz}$ and higher) the Johnson noise dominates. Flicker, or so-called $1/f$-noise, predominates in frequencies below approximately 1kHz. Rms of the flicker-noise current Δi_f [A] is

$$\Delta i_f = i\sqrt{B \cdot \Delta f/f} \tag{10.40}$$

where $B = \text{const} \approx 1 \times 10^{-11}$, dimensionless.

Cadmium sulfide (CdS) *and cadmium selenide* (CdSe) are two widely used materials for inexpensive detectors. They are characterized by high G (1×10^3 to 1×10^4) and comparably slow response ($\sim 50\text{ms}$). In the detector construction, a polycrystalline film is deposited on a ceramic substrate. The film is comb-like patterned, and metal (typically gold) contacts are formed along the contour of the sensitive layer. Conforming to the conditions following from equation 10.38, the configuration of the pattern provides a large sensitive area and a small interelectrode spacing.

Lead sulfide (PbS) is used for detection of the radiation with a wavelength ranging from 1 to 3.4μm and up to 4μm if the detector is cooled to $-30°\text{C}$. The typical response time is 0.2ms and dark resistance is about 1MΩ. The characteristics of the sensor (G and the time constant) can be varied by a change in the process of forming the PbS film.

Indium antimonide (InSb) in the form of a single crystal is known as a low-resistance material ($\sim 50\Omega$) that can detect radiation of a 7μm wavelength if the crystal is cooled to liquid-nitrogen temperature (77K). The response time is around 50ns. This material can operate at room temperature; however, the signal-to-noise ratio is worse than in the case of low-temperature application.

Mercury cadmium telluride ($\text{Hg}_x\text{Cd}_{1-x}\text{Te}$) is a compound of HgTe and CdTe. By a change in the relative content of the components, the band gap of the material can be between 0 and 1.6eV. The peak sensitivity can lie in the region of 5 to 14μm, which corresponds to the peak emission wavelength of bodies and good atmospheric transmission. For normal operation of the detector, cooling is necessary.

Zinc- and boron-doped germanium cooled to 4K (liquid helium temperature) is capable of responding to radiation of a 20- to 100-μm wavelength.

Photodiodes

If the depletion region of a junction is affected by photons, electron-hole pairs formed in the region can be separated by a high internal field at the depletion zone. A photodiode having the junction exposed to radiation can work as a detector. The generated charges are measured between the terminals directly (*photovoltaic* mode of operation) or by reading the reverse-bias current (*photoconductive* mode of operation). The electron-hole pairs generated away from the depletion zone can contribute to the current if they diffuse to the edge of the zone and then recombine. The equivalent circuit of the photodiode is shown in Figure 10.11. In this circuit, the current generator produces current i_λ proportional to the absorbed light. An ideal diode simulates a p-n junction with a forward current i_d. The internal impedance of the cell is modeled by the parallel and series terms: resistor R_{sh} carrying current i_{sh}, resistor R_s with current i_{ex}, and capacitor C_d with current i_c. In the photovoltaic mode, the load resistor R_L is connected parallel to the output. In the photoconductive mode,

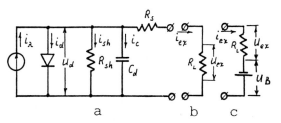

Figure 10.11 Photodiode equivalent circuit. Parts a and b are combined for photovoltaic mode, parts a and c are combined for photoconductive mode of operation.

a source of external bias voltage U_b is connected to the cell. This source is in series with the load resistor. Current i_λ is proportional to the light irradiance falling on cell I_0 and to the area of cell A:

$$i_\lambda = \frac{\eta I_0 A q \lambda}{hc}. \tag{10.41}$$

Capacitance C_d should be taken into account if the cell operates at a relatively high optical modulation frequency. It can be assumed for low frequencies, that $i_c = 0$.

The diode output can be introduced as a voltage drop U_{ex} [V] across the load resistance. In order to calculate the diode transfer characteristic, we will use the basic equation for a diode (formula 7.23) and assume that $i_c = 0$.

The relevant equations for the circuit operating in a photovoltaic mode are

$$i_\lambda = i_d + i_{sh} + i_{ex}, \qquad U_{ex} = U_d - i_{ex} R_s, \qquad U_d = i_{sh} R_{sh},$$

$$i_d = i_0 \left[\exp \frac{q U_d}{\bar{k} T} - 1 \right] \tag{10.42}$$

where i_0 = reverse-bias leakage current, A.

Assuming that R_L is much larger than the diode's internal impedance ($i_{ex} = 0$), two equations from the last set can be simplified:

$$i_\lambda = i_d + i_{sh} \qquad \text{and} \qquad U_{ex} = U_d. \tag{10.43}$$

Substitutions of i_d and i_{sh} give

$$i_\lambda = i_0 \left[\exp \left(\frac{q U_d}{\bar{k} T} \right) - 1 \right] + \frac{U_d}{R_{sh}} \qquad \text{or}$$

$$\exp \left(\frac{q U_d}{\bar{k} T} \right) = 1 + \frac{i_\lambda}{i_0} - \frac{U_d}{i_0 R_{sh}}. \tag{10.44}$$

Typically, $i_0 \approx 1 \times 10^{-8}$A, $R_{sh} \approx 1 \times 10^8 \Omega$, $U_d \approx 0.6$V, $U_d/(i_0 \cdot R_{sh}) \approx 1$, and $i_\lambda \gg i_0$. From these conditions and from the last equation, we have approximately

$$\exp \left(\frac{q U_d}{\bar{k} T} \right) = \frac{i_\lambda}{i_0} \qquad \text{and} \tag{10.45}$$

$$U_{ex} \approx U_d = \frac{\bar{k} T}{q} \ln \left(\frac{i_\lambda}{i_0} \right). \tag{10.46}$$

Substituting the value of i_λ from formula 10.41 into formula 10.46 gives

$$U_{\text{ex}} = \frac{\bar{k}T}{q} \ln \left(\frac{\eta I_0 q \lambda A}{hci_0} \right).$$

(10.47)

The current-voltage characteristics of a photodiode for different irradiances are shown in Figure 10.12. If the maximum power must be extracted from the diode (*solar cell*), it operates in the quadrant marked by A. At present, silicon cells have an efficiency of 15%; they produce 0.6V per cell with current density 80A/m².

When a diode operates in a *photoconductive mode,* its current saturates at i_0 for small values of reverse bias, which is about two orders of magnitude smaller than the bias voltage ($\approx$ 10V and higher). The equation for the currents can be written as

$$i_\lambda = i_0 + i_{\text{sh}} + i_{\text{ex}}.$$

(10.48)

The orders of magnitude for currents i_0 and i_{sh} are 10 and 100nA, respectively. Provided that for operating conditions, current i_λ is of the order of microamperes and higher, we can assume that $i_\lambda \approx i_{\text{ex}}$; therefore,

$$i_{\text{ex}} = \frac{\eta I_0 A q \lambda}{hc}.$$

(10.49)

Unlike in the photovoltaic mode, where output voltage is a logarithmic function of the input, in the photoconductive mode, the output current is a linear function of the input. Among other advantages of the photoconductive mode are faster response, better stability, and greater dynamic range. The presence of a dark current ($i_0 + i_{\text{sh}}$) is a limit defining the minimum of optical signals to be detected. Shot and generation noises are typical for the device; however, recombination noise does not affect the performance since the carriers are separated in the depletion zone before their recombination.

Silicon is widely used for diodes. The quantum efficiency for the silicon diodes reaches 80% for wavelengths between 0.8 and 0.9μm. A typical silicon photodiode structure is shown in Figure 10.13. A junction is formed by a heavily doped p layer (p$^+$) and a relatively lightly doped n material. The depletion region extends into the n material. A heavily doped n$^+$ layer is formed at the bottom of the die to obtain a good ohmic contact with the metallization. There exist some structures that are illuminated from the side parallel to the junction. In these diodes, a good sensitivity for wavelengths close to the band gap limit can be obtained. Some antireflection coatings on the front surfaces of the detectors are used for increasing the efficiency; for example, a ($\lambda/2$)-thick silicon dioxide coating. One of the main characteristics of

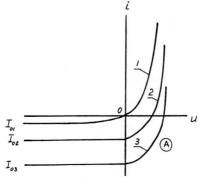

Figure 10.12 Current-voltage characteristics for photodiode. Curves 1, 2, and 3 correspond to irradiances: $I_{01} < I_{02} < I_{03}$.

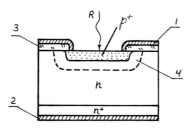

Figure 10.13 Structure of silicon photodiode. R = radiation, 1 and 2 = metallization, 3 = SiO_2, 4 = depletion region.

a diode is *responsivity,* which is the ratio of the sensor's output to the incident flux at the stated wavelength. This characteristic ($S = i_{ex}/I_0 A$) for a typical silicon diode is shown in Figure 10.14.

The response time of photodiodes is defined by several factors. One of them is the diffusion time of carriers to the depletion region. The time τ_d [s] taken for carriers to diffuse distance d [m] is found from a simple relation:

$$\tau_d = d^2/D_c \qquad (10.50)$$

where D_c = minority carrier diffusion constant, m^2/s.

The carriers generated within the depletion region respond rapidly, while those generated outside take time to diffuse toward the region, increasing the response time of the device. So-called pin-structures (PIN) have an extended depletion region and, therefore, faster response. These structures contain consecutive p^+-, intrinsic-, and n-Si-regions. Only a few volts of reverse bias are needed to extend the depletion region all the way through to the n-region, providing higher sensitivity and faster response.

Another cause of delay in the diode response is the drift time of carriers through the depletion region. The velocities of carriers tend to saturate with the increase of field across the depletion layer width W, acquiring constant velocities v_{sat}. If the carriers are generated at one edge of the depletion region, and one of the carriers must traverse the width W, the transit time τ_{dr} will be

$$\tau_{dr} = \frac{W}{v_{sat}}. \qquad (10.51)$$

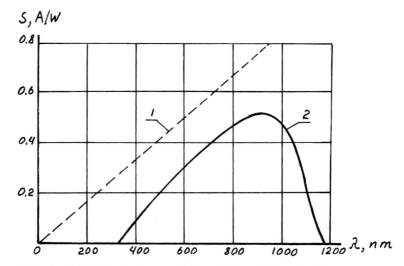

Figure 10.14 Typical responsivity of silicon diode. 1 = ideal photodetector, 2 = real photodetector.

The third factor influencing the response time is the junction capacitance C_j [F] effect. The value of this capacitance is given by

$$C_j = \frac{A}{2} \sqrt{\frac{2q\epsilon_r\epsilon_0}{U_0 - U} \cdot \frac{N_D N_A}{N_D + N_A}} \tag{10.52}$$

where
ϵ_r = dielectric constant of material, dimensionless;
ϵ_0 = permittivity of empty space, F/m;
U_0 = zero-bias junction potential, V;
U = external bias, V; and
N_D and N_A = donor and acceptor densities, $1/m^3$.

With two assumptions, $U \gg U_0$ and $N_A \gg N_D$ (for the p^+-n junction), expression 10.52 simplifies:

$$C_j = \frac{A}{2} \sqrt{2q\epsilon_r\epsilon_0 N_D/U}. \tag{10.53}$$

We have considered the relationships for an abrupt junction, whose capacitance is inversely proportional to $\sqrt{U}$. For a more realistic junction, such as for the linearly graded one, the mentioned dependence will be $\sqrt[3]{U}$. It is important to note that with the increase in the external bias, the value of the junction capacitance is decreased.

In the diode equivalent circuit (Fig. 10.11), $R_{sh} \gg R_L$ and $R_s \ll R_L$; this allows us to simplify the circuit and have it introduced by the sources i_λ, C_j, and R_L connected in parallel. The transfer function for the electrical conversion U_{ex}/i_λ can be found by the calculation of the simplified circuit as a current divider assuming $C_d \approx C_j$:

$$\frac{U_{ex}}{i_\lambda} = R_L \frac{1}{1 + TD} \tag{10.54}$$

where $T = R_L C_j$ = time constant, s.

This integrating network gives attenuation of the output with the increase in frequency.

Several factors should be considered when the possibility of the time constant reduction is discussed. According to formula 10.53, the decrease in A and N_D and the increase in U lead to the reduction in T. However, a reduction in A makes focusing the light beam more difficult. A reduction in N_D and an increase in U make the depletion region wider, giving an increase of the drift transit time (equation 10.50). With the reduction in N_D, the bulk resistance of the structure becomes greater, which can be reflected in an increased time constant. A compromise for these conditions should be reached when the device is designed. Contemporary technology allows fabrication of silicon diodes with a response time of less than 1ns. These diodes operate in the wavelength range of 0.4 to 1μm, and, besides a high quantum efficiency, provide quite linear conversion. They are known as all-purpose devices available at a low price. Photodiodes made of germanium can cover a wavelength range up to 1.8μm, but their quantum sensitivity is low and their noise level is comparably high. Compounds like GaInAs, HgCdTe, and GaInAsP also serve as sensitive materials for photodiodes.

Monolithic heteroepitaxial sensor arrays on Si for thermal imaging applications have been fabricated [457] of different IV-VI narrow-gap semiconductors that have been grown on (111)-Si using stacked intermediate fluoride buffer layers. All the spectral range from 3μm up to 14μm can be covered using different materials like PbS, PbEuSe, PbTe, and PbSnSe. With this technique, fully monolithic, low-noise, high-performance infrared sensor arrays can be produced on a Si substrate containing all the signal conditioning circuitry.

Avalanche Photodiodes

Internal amplification of current in a photodiode can be achieved if there is a need to match the diode to a low-resistance load that is good for obtaining a fast response to optical signals. Using the avalanche gain effect in a diode makes it possible to generate a huge number of secondary electron-hole pairs, giving the high gain. The avalanche conditions are created by increasing the reverse-biased voltage just slightly below the breakdown value. As a high field intensity (larger than 10^5V/cm) is established across the junction, carriers acquire sufficient energy to produce additional electron-hole pairs due to inelastic collision. The energy developed in this process is sufficient to move electrons from the valence band into the conduction band. This process is enhanced by a factor that is represented by an exponential function of the bias voltage. This voltage should not be too high; otherwise a self-sustaining avalanche current can flow in the absence of photoexcitation. Since the high voltage is applied to the diode, it is important to have a uniform field distribution across the device for ensuring the normal mode of operation. This condition can be achieved by several methods. One of them is the formation of a guard-ring structure that restricts the avalanche region to the central illuminated part of the cell. An example of such a structure can be a relatively low n-doped ring surrounding a heavily doped (n^+), exposed-to-light area. Because of the low doping, the depletion region extends an appreciable distance into the ring. Consequently, at the vicinity of the guard ring, the electric field strength is not as strong as it could be without the ring, and the field is distributed more uniformly over the sensitive area. Since the diode's current gain greatly depends on the applied voltage, the bias voltage must be well stabilized. A high sensitivity of the diode characteristics to temperature can be substantially overcome by using a differential connection of two diodes, of which one is masked from the incident light, but exposed to temperature and operates near the breakdown point. The rms value of the shot-noise current Δi_f [A] is

$$\Delta i_f = M \sqrt{2iq \cdot \Delta f} \qquad (10.55)$$

where M stands for the number of secondary carriers created by the individual photogenerated carrier. Due to the statistical nature of the carrier multiplication process, the overall rms noise current may be expressed as [458]:

$$\Delta i_f = M\sqrt{2iq \cdot \Delta f F}, \qquad F = M[1 - (1 - 1/r)(M - 1)/M^2] \qquad (10.56)$$

where F is called the excess noise factor. It is defined as the ratio of the actual noise to the noise of an ideal device if the multiplication process is noiseless. In the expression for F, r is a ratio of the electron-to-hole ionization rate probabilities. For Si, r is 10 to 100 in the spectral region 0.8 to 0.9μm. This means that silicon diodes have quite a low level of noise. The quantum efficiency of Si diodes reaches near 100%. Their response time is around 1ns; current gain—100; and excess noise factor—about 5. Silicon becomes transparent in the long-wavelength region (1 to 1.6μm); materials having narrower band gaps and pronounced light absorption have to be used for this region. The noise level in germanium diodes is higher than that in diodes made of silicon (dark current in Ge is about 1×10^{-7}A). In spite of this, the characteristics of recently developed Ge diodes are quite remarkable: quantum efficiency—about 80% at $\lambda = 1.3\mu$m; bandwidth—500MHz; gain—10; excess noise factor—7. Besides Si and Ge, other materials such as InGaAs or InGaAsP are also used to make diodes sensitive to long waves. Planar epitaxial avalanche photodiodes with enhanced blue sensitivity have been developed for scintillation detectors [459].

Schottky Photodiodes

Schottky photodiodes [460] employ the generation of carriers in metal-semiconductor junctions. The structure of the diode is composed of a thin, light-transparent metal film deposited on the surface of a semiconductor substrate (Fig. 10.15). Another side of the substrate is also metallized. As the surface is illuminated, electron-hole pairs are generated within the depletion region at the metal-semiconductor junction. The carriers are separated by the electric field at the junction, similarly to those in the p-n junction diode. The current established in the external circuit connected to the two metallized surfaces is proportional to the incident irradiance. Typically, gold is used for the metallization and n-Si for a substrate. One of the features of this device is its ability to enhance selectively the blue and near-ultraviolet radiation, which can penetrate through the thin metal layer and activate the junction. An integrated circuit with a GaAs diode can operate in a frequency range of 900MHz–1.3GHz [461].

Phototransistors

Like the avalanche photodiodes, phototransistors can provide amplification of the photocurrent inside the structure. These sensors are called charge-coupled devices [462–464]. The basic construction of the transistor is similar to that described in the section on bipolar transistors in Chapter 7 with two major differences: (1) the base region has access for the light rays, and (2) normally, the base does not have any external connection. Referring to the circuit shown in Figure 7.12 and to the designations given in Chapter 7, we can easily obtain the expression for the emitter current:

$$I_E = (I_B + I_R)(h_{\text{fe}} + 1). \tag{10.57}$$

In the calculation of this current we assume that the collector current includes two components: one is I_R and another one is found with relation 7.24. With no light excitation, $I_B = 0$ and the dark current will be $I_R(h_{\text{fe}} + 1)$. This current is much greater than I_R since typically h_{fe} is about 100. With illumination, the base current will be

$$i_\lambda = \eta(I_0 Aq\lambda/hc), \tag{10.58}$$

and the emitter current, which we can consider an output, becomes

$$I_E = (i_\lambda + I_R)(h_{\text{fe}} + 1) \approx i_\lambda h_{\text{fe}} \tag{10.59}$$

since $i_\lambda \gg I_R$ and $h_{\text{fe}} \gg 1$.

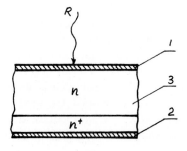

Figure 10.15 Schottky photodiode. R = incident radiation, 1 and 2 = metallizations, 3 = semiconductor substrate.

This linear and high-gain response is quite attractive for a design. Some advances in the development of new structures [465] have made it possible to create transistors that are extremely sensitive to light and have a high signal-to-noise ratio. However, two main drawbacks should be taken into account: (1) the range of optical signals' frequencies is limited (about 200kHz) because of a comparably high time constant defined by the base junction capacitance and resistance; (2) there is a decrease in the gain for low irradiance levels, associated with a purely electrical dependence of the forward-current transfer ratio h_{fe} on the base current $I_B = i_\lambda$.

ISFET [466] and metal-insulated n/p semiconductor thyristors [467] for photosensing are under development and represent a potential interest.

Charge-coupled Devices (CCD's)

One of the methods of optical image conversion into electrical signals is based on using an array of discrete detectors. Each detector produces an electrical equivalent of the incident irradiance that is read out by scanning. The array can contain many thousands of detectors that constitute the basic building blocks of the device. One of the most popular detectors in the array is the *metal-oxide-semiconductor* (MOS) capacitor (Fig. 10.16). The gate is biased positively with respect to the silicon. Under illumination, electron-hole pairs are generated in silicon and separated by the field across the material. Electrons are attracted to the surface of the silicon under the gate and remain trapped there while the gate potential is kept positive. The amount of charge formed by the electrons is proportional to the amount of radiation obtained from the light rays. The read-out technique is based on moving the charges from detector to detector along the chain of MOS cells by applying to the gates a repeated sequence of variable-polarity potentials. Fast development in the design and fabrication of CCD's has led to a variety of constructions [468, 469] whose essence, however, is as described above.

Thin-film Diffraction-grating Radiation Detector

This detector was recently described [470] for measurements in a wide range of wavelengths. The detector is fabricated of slant-angle-deposited, 100 to 6000 Å-thick metal films of Ni, Ti, Bi, Te, and chromelcopel alloys on a dielectric grating with a ripple period of 0.3–10μm. Rough-abraded surfaces also were used as substrates for the deposition of metal. It was found that a fairly high electrical signal proportional-to-the-incident radiant flux can be picked up from the film in the direction perpendicular to the ripples.

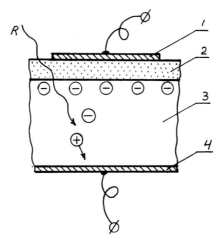

Figure 10.16 Structure of MOS capacitor sensitive to radiation. R = radiation, 1 = gate's metal transparent layer, 2 = SiO_2, 3 = p-Si, 4 = metallization.

Pyrometer Sensor

A photodetector intended for the measurement of radiation is usually part of an optical system including such elements as protection windows, lenses, prisms, optical fibers, and so on. As the light from the source of radiation experiences transformation, the magnitude of the primer signal and the spectrum of wavelengths can be significantly changed. In this paragraph, we will consider a basic calculation for a wide-band pyrometer as an example of a relatively simple sensor optical system including several typical components. A more comprehensive analysis of different optical systems can be found in books on optics and optoelectronics, such as those listed in the beginning of this section.

The pyrometer sensor that we consider (Fig. 10.17) includes a lens directed to the target whose radiant flux or temperature is to be measured, and a photosensor with a wide-band response. A specific spot at the target is focused on by a lens on the sensor's surface, delivering and concentrating the radiant energy to the photosensitive area. The sensor can contain a bundle of optical fibers between the lens and photodetector as shown in Figure 10.17. This addition is sometimes needed for placing the photodetector remotely from the target, whose temperature can be too high. In order to find the sensor's response to temperature, we should consider together the radiation characteristics of the target, transmittance of the optics, and conversion in the detector. The radiant emittance as a function of wavelength for different source temperatures (Fig. 10.18) is given by *Planck's Radiation Formula:*

$$R = \frac{C_1 \cdot \Delta\lambda \cdot \epsilon_\lambda}{\lambda^5 \left(\exp\dfrac{C_2}{\lambda T} - 1\right)} \tag{10.60}$$

where R = radiant emittance (radiant excitance) per unit area of source, at wavelength λ, over a spectral range $\Delta\lambda$, W/m^2;
  λ = wavelength, m;
  ϵ_λ = emissivity of source at wavelength λ (ϵ_λ = 1 for a blackbody radiation), dimensionless;
  C_1 = first radiation constant = $3.7413 \times 10^{-16} W \cdot m^2$;
  C_2 = second radiation constant = $1.4388 \times 10^{-2} m \cdot K$; and
  T = absolute temperature, K.

In some practical calculations, this formula is reduced; it gives a spectrum concentration of radiant emittance r per $\Delta\lambda$ with r expressed in W/m^3 or $W/cm^2 \cdot \mu m$. In this formula $\Delta\lambda$ is deleted from the numerator, ϵ_λ is assumed to be unity, and the unity is neglected in the denominator. For this case, $C_1 = 37413 \times 10^{-2}$ and $C_2 = 14388$:

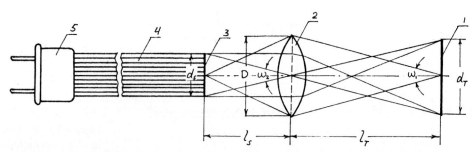

Figure 10.17 Optical system of pyrometer sensor. 1 = target area to be monitored, 2 = lens, 3 = focused image of target, 4 = fiber-optic guide, 5 = photodetector.

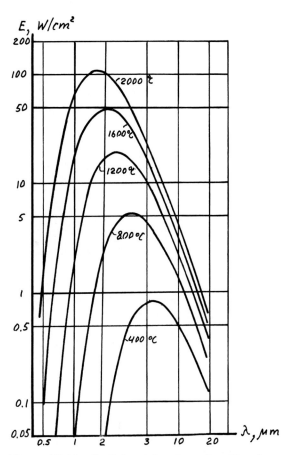

Figure 10.18 Radiation intensity as function of wavelength for different source temperatures when $\epsilon = 1$ (adopted from [1]).

$$r = \frac{C_1 \epsilon_\lambda}{\lambda^5 \cdot \exp (C_2/\lambda T)}. \tag{10.61}$$

The current i [A] produced by the detector is also a function of λ:

$$i = \eta K \int_{\lambda_1}^{\lambda_2} r_T(\lambda) \cdot S(\lambda) \cdot \tau(\lambda) d\lambda \tag{10.62}$$

where
η = detector's efficiency, dimensionless;
K = geometrical factor of the optical system, dimensionless;
$r_T(\lambda)$ = r for a given temperature T, W/m^3;
$S(\lambda)$ = spectral characteristic of the detector's sensitivity, A/W;
$\tau(\lambda)$ = spectral characteristic of the optics' transmission factors, dimensionless; and
λ_1 and λ_2 = limits of integration, m.

In order to find i, $S(\lambda)$ and $\tau(\lambda)$ (Fig. 10.19) are approximated by one of the appropriate expressions (a fifth-power polynomial is usually acceptable); then the integral is solved, preferably by computer simulation.

In the following calculations we will consider a simplified case for which characteristics S and τ are assumed to be known and independent of λ, $\epsilon_\lambda = 1$, and the limits of integration are 0 and ∞. Therefore, the output current will be

$$i = \eta \tau K S \int_0^\infty r_T(\lambda) d\lambda. \tag{10.63}$$

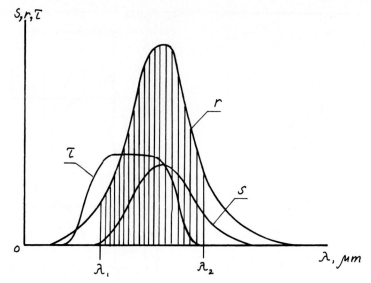

Figure 10.19 Characteristics of components of radiation source and optical system. r = spectral radiance of source, S = detector sensitivity, τ = transmission factor, λ_1 and λ_2 = wavelength operational limits. Because of spectral sensitivity of components, only radiation within hatched area can be measured.

In order to solve this integral for the given temperature T, we will make the following substitutions:

$$x = C_2/\lambda T \quad \text{or} \quad \lambda = C_2/xT \quad \text{and} \quad d\lambda = -C_2 \cdot dx/x^2 T. \quad (10.64)$$

Inserting these terms in equation 10.61 gives the integral radiant emittance at the surface:

$$R = \frac{C_1 T^4}{C_2^4} \int_0^\infty \frac{x^3}{e^x - 1}\, dx = \frac{C_1 T^4}{C_2^4} \cdot \frac{\pi^4}{15} = \sigma T^4 = 5.6697 \times 10^{-8} \cdot T^4 \quad (10.65)$$

where $5.6697 \times 10^{-8} = \sigma$ [W/m$^2 \cdot$ K^4] = Stefan-Boltzmann constant.

Now we will find geometrical factor K for the optical system. This factor can be defined as a ratio of the incident irradiance I_0 [W/m^2] on the sensor surface to the integral radiant emittance given by formula 10.65 without considering the losses for reflection and absorption of the radiant energy.

Radiant flux F_1 [W] coming to the lens from the target is

$$F_1 = J\omega_1 \quad (10.66)$$

where J = radiant intensity, W/steradian (W/sr), and
ω_1 = solid angle as in Fig. 10.17, sr.

The incident irradiance on the lens's front surface, I [W/m^2] is

$$I = \frac{J}{l_T^2} \quad (10.67)$$

where l_T = distance between the target and the lens's front surface, m.

For a circular area of the target having area S [m^2] and diameter d_T [m], the radiant intensity can be calculated using the relationship for radiance B [W/m$^2 \cdot$ sr]:

$$B = \frac{\sigma}{\pi} T^4, \qquad J = BS = \frac{\sigma d_T^2 T^4}{4}. \tag{10.68}$$

The solid angle at the target's side ω_1 depends on l_T and on the diameter of lens D [m]. Assuming that ω_1 is small, its amount is found as

$$\omega_1 = \frac{\pi D^2}{4l_T^2}. \tag{10.69}$$

Now flux F_1 can be expressed by the combination of expressions 10.66–10.68:

$$F_1 = \frac{\pi \sigma T^4 d_T^2 D^2}{16 l_T^2}. \tag{10.70}$$

The part of this flux F_2 will be reflected from the front surface:

$$F_2 = \rho F_1 \tag{10.71}$$

where ρ = reflection coefficient, dimensionless.
Flux F_3 entering inside the lens is

$$F_3 = F_1 - \rho F_1 = F_1(1 - \rho). \tag{10.72}$$

Due to the absorption of light in the lens material, the flux reaching the back surface of the lens F_4 will be reduced and can be calculated using an exponential cofactor:

$$F_4 = F_3 \exp(-wl) = F_1(1 - \rho) \exp(-wl) \tag{10.73}$$

where w = attenuation coefficient for the material of lens, 1/m, and
l = average length of light path in lens, m.

w is defined as a gradient of flux change over the lens thickness referred to the flux ($w = dF/dl \cdot F$). Flux F_5 reflected from the back surface of the lens is calculated similarly to F_2:

$$F_5 = \rho F_4. \tag{10.74}$$

Flux F_6 leaving the lens's back surface and directed toward the detector or to the fiber-optic guide will be

$$F_6 = F_4 - F_5 = F_1(1 - \rho)^2 \exp(-wl). \tag{10.75}$$

Therefore, the transmission of the lens γ, calculated as a ratio F_6/F_1 and the loss of flux β are

$$\gamma = (1 - \rho)^2 \exp(-wl), \qquad \beta = (1 - \gamma) \times 100\%. \tag{10.76}$$

For example, for a sapphire lens 1mm thick, $\gamma = 0.85$ if λ lies between 0.3 and 1μm (sapphire is a high-temperature material whose transparency to light extends up to the wavelength of 6.5μm).

The intensity of light from the back side of the lens as from the source of light J' can be found as follows:

$$J' = \frac{F_6}{\omega_2} \quad \text{where} \quad \omega_2 = \frac{\pi d_S^2}{4l_S^2} \quad \text{and} \quad (10.77)$$

ω_2 (assumed to be small), d_S and l_S are the solid angle, the diameter of the image, and the distance between the lens and the plane of the image, respectively (Fig. 10.17).

If the fiber-optic part is included in the construction, the transmission term should contain a cofactor $(1 - \rho_f)^2 \exp(-w_f l_f)$ allowing for the attenuation in fibers (subscript f at ρ, w, and l refers the parameters to fibers). The total transmission for this optical system τ will be

$$\tau = (1 - \rho)^2 (1 - \rho_f)^2 \exp[-(wl + w_f l_f)]. \quad (10.78)$$

Typically, τ is between 0.06 and 0.16.

Considering all these facts, expression 10.72 can be written as

$$F_6 = F_1 \tau. \quad (10.79)$$

Incident irradiance I_0 at the sensitive area of the detector will be

$$I_0 = \frac{J'}{l_S^2} = \frac{4F_1 \tau}{\pi d_S^2 l_S^2} = \frac{\sigma T^4}{4} \cdot \frac{d_T^2 D^2}{d_S^2 l_T^2} \tau. \quad (10.80)$$

From the geometrical similarity of triangles in Figure 10.17, we have

$$\frac{l_T}{l_S} = \frac{d_T}{d_S}. \quad (10.81)$$

Making the substitution for l_T taken from ratio 10.81 into equation 10.80 we finally obtain

$$I_0 = \frac{\sigma T^4}{4} \cdot \frac{D^2}{l_S^2} \tau. \quad (10.82)$$

As can be seen from this expression, factor $K = D^2/l_S^2$, and the sensitivity of the system can be increased with the increase of the diameter of the lens.

Applications of Photosensors

Several areas of application of photosensors are being developed extremely quickly. Among them are displacement, position, and dimension sensing devices [471–485]; systems for color distinction [486–490]; devices for sensing UV radiation [491–493]; sensing devices for biomedical application [494–501]; light-to-light transducers [502, 503]; in-process inspection of integrated circuits [504]; and so on.

FIBER-OPTIC SENSORS

A fiber-optic sensor is defined as a device by which the properties of light traveling down a fiber are modulated in response to a physical, chemical, or biological quantity in which all or part of the fiber link is in contact [449, 505, 506].

Most physical quantities can be sensed using fibers. Light intensity, displacement (position), amplitude and phase of vibration, rotation, acceleration, pressure, strain, sound pressure, flow, liquid level, temperature, electrostatic and magnetic field intensities, and radiation are just some of the measurands that can be sensed [507].

There are two basic categories of fiber sensors: *intrinsic and extrinsic.* In the intrinsic sensors, the light is modulated by the measurand within the fiber itself, whereas in the extrinsic sensors, light is modulated by a mechanical or optical structure responding to a particular measurand outside the fiber. The main reasons for the intensive development and application of fiber-optic sensors are their compatibility with intrinsic safety requirements, capability of long transmission distances, immunity of transmitted signals to electromagnetic interference, chemical passivity, inherently small size, and high sensitivity. The optical parameters that are subject to modulation are light intensity, phase, polarization, color, optical frequency (that is, Doppler shift, etc.), and optical pulse decay times. A good number of fiber-optic sensors have been designed and have found practical application during the last few years. The development of new sensors is in progress.

Fibers as Light Guides

A light guide is a conduit for the light beam that keeps it from spreading out of the prescribed direction of propagation. An elementary light guide is an individual transparent-material fiber (Fig. 10.20) with a core of refractive index n_1 (medium 1) and a sheath of cladding having refractive index n_2 (medium 2), such that $n_2 < n_1$. The refractive index is the ratio of the phase velocity of light in a vacuum to that in a specified medium. The diameter or thickness of the core lies usually between 50 and $100\mu m$, and the thickness of the cladding is within a few micrometers. In a sensor application, fibers are circular or rectangular cross-sectionally and are surrounded by a relatively thick plastic layer intended for the protection and strengthening of the fiber.

Due to the total internal reflection at the boundary between the core and the cladding, the light beam is guided through the core with small losses. The relationship between n_1, n_2, and angles of incidence θ_1, reflection θ_1', and refraction θ_2 is (Fig. 10.21)

$$\theta_1' = \theta_1, \qquad \sin\theta_1/\sin\theta_2 = n_2/n_1. \qquad (10.83)$$

The fiber provides light guiding due to the effect of *total internal reflection* at the boundary between the two media. The maximum value of θ_2 is 90°. If $n_1 \sin\theta_1 > n_2$, there is no refracted beam in medium 2. The incident beam will be totally reflected at the core-cladding interface and will propagate along the fiber with minimal losses. The critical angle θ_c is defined by the following condition:

$$n_1 \sin\theta_c = n_2 \qquad \text{or} \qquad \sin\theta_c = n_2/n_1 = n \qquad (10.84)$$

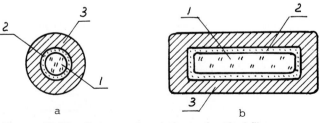

Figure 10.20 Cross-sectional views of optical fibers.
a = circular-section fiber, b = rectangular-section fiber;
1 = core with refractive index n_1, 2 = cladding with
refractive index n_2, 3 = protective jacket.

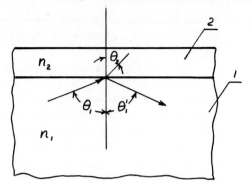

Figure 10.21 Light reflection and refraction in fiber. 1 = core, 2 = cladding.

θ_c is the smallest angle with the normal to the boundary at which total reflection occurs. Since the maximum of $\sin \theta_c$ is unity, n_2 must be equal to or less than n_1 for a total internal reflection; therefore, if the angle of incidence is greater than the critical angle, a total internal reflection takes place. In the reverse conditions, the rays with small angles of incidence will penetrate into and be absorbed in the cladding (Fig. 10.22).

Snell's law establishes a restriction for the maximum value of the angle of incidence φ called the maximum *acceptance angle* (Fig. 10.23). This angle relates to the critical angle as

$$\sin \varphi = n_1 \sin (90° - \theta_c) = n_1 \cos \theta_c = n_1 \sqrt{1 - \sin^2 \theta_c}. \qquad (10.85)$$

By substituting $\sin \theta_c = n_2/n_1$ in this equation, we have

$$\sin \varphi = n_1 \sqrt{1 - (n_2/n_1)^2} = \sqrt{n_1^2 - n_2^2} = NA. \qquad (10.86)$$

Sin φ is called the *numerical aperture* (NA) of the light guide, and φ is the maximum angle for the propagation of rays directed to the fiber's end. If the medium before the fiber's end has an index of refraction n_0, equation 10.86 becomes

$$\sin \varphi = \left(\sqrt{n_1^2 - n_2^2}\right)/n_0. \qquad (10.87)$$

In the event that n_0 is larger than the index of refraction of air, φ becomes smaller, and an approximation for the angle can be

$$\varphi \approx \sin \varphi = \sqrt{n_2^2 - n_1^2} = \sqrt{(n_2 - n_1)(n_2 + n_1)} \approx \sqrt{\Delta n \cdot 2n_1} \qquad (10.88)$$

where $\Delta n = n_2 - n_1$.

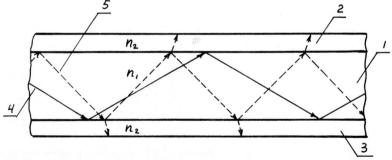

Figure 10.22 Light propagation in fiber. 1 = core, 2 and 3 = cladding, 4 = propagating beam, 5 = attenuating beam.

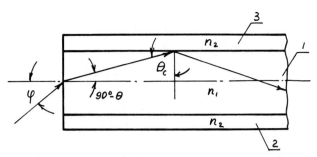

Figure 10.23   Illustration of path of acceptance angle.

By the variation of the difference between the indices of refraction of the core and cladding, the acceptance angle can be changed according to the requirements of a particular design. For example, a large acceptance angle is needed if the light should be gathered from a big area. However, for a small distortion of the optical signal and for a wider bandwidth, a small angle φ is preferable.

It is proved in the theory of light propagation in fibers that light energy can be transmitted along the guide only as discrete modes that have definite values of phase velocity. In the designs of fiber-optic sensors, single-mode and multimode fibers have found application. In order to distinguish the types of fibers, a characteristic, dimensionless parameter V, called *modal volume*, should be calculated:

$$V = \frac{2\pi r n}{\lambda}\sqrt{2\Delta} \tag{10.89}$$

where r = core radius, m;
$\quad n = n_2/n_1$;
$\quad \lambda$ = wavelength of propagating wave, m; and
$\quad \Delta = (n_1^2 - n_2^2)/2n_1^2$.

For a large number of V, the number of principle modes is approximately equal to V. As the core diameter becomes smaller, and the dimension of the core approaches the wavelength of the light beam, the propagation reaches the *cut-off condition* [508], at which $V < 2.405$. Under this condition, only one mode is allowed to propagate down the fiber. With an increase in V, additional modes can be transmitted, and higher-order modes appear in the fiber.

In the foregoing description, we have considered some characteristics of *step-index fibers;* in other words, fibers in which the index of refraction changes abruptly at the core-cladding boundary. There exists another type of fibers, *graded-index fibers,* whose index of refraction symmetrically varies about the axis along which the waves propagate. Due to this arrangement, the fibers act like a series of lenses to a light beam propagating down its length. The number of modes and temporal spreading of modes propagating through a fiber become reduced. The variation of the core refractive index $n(\rho)$ is often expressed as a function of a changing radius ρ [m] in the following form [448]:

$$n(\rho) = n_1\left[1 - 2\Delta\left(\frac{\rho}{r}\right)^\gamma\right]^{1/2} \text{ for } \rho < r \quad \text{ and }$$

$$n(\rho) = n_1(1 - 2\Delta)^{1/2} \text{ for } \rho \gg r \tag{10.90}$$

where $\Delta = (n_1^2 - n_2^2)/2n_1^2 \approx (n_1 - n_2)/n_1$;
$\quad n_1$ = axial refractive index; and
$\quad \gamma$ = profile parameter, dimensionless.

This presentation of the profile is made to simplify the mathematical operations when the propagation of waves is analyzed.

Basic Concepts of Fiber Sensors

A large number of fiber-optic devices have been developed to the stage of engineering prototypes in the laboratory and few are available as commercial products. Using material given in [448, 449, 505, 506], we will briefly review the basic concepts of fiber-optic sensors.

As mentioned above, the fiber sensors are reliable, robust, corrosion-resistant, and intrinsically safe and free from external electromagnetic interference. To a considerable extent, these characteristics are defined by the physical nature of light, by the features of the light guides, and by the methods of the modulation of light. In extrinsic sensors, the light modulation is provided without the alternating of the optical properties of fibers, whose functions are limited by the transmission of the radiation. In the intrinsic sensors, the change of the fiber properties plays an important role in the amplitude, phase, or state of polarization in the modulation of light. In multimode fibers, mode coupling and random situation with phases and the polarization states of the modes make impossible the phase or polarization modulation; therefore, the amplitude modulation is used exclusively. For single-mode fibers, the phase or polarization modulation is acceptable, which makes possible the application of extremely sensitive and accurate interferometric techniques.

In the following list are examples [506] of phenomena that relate changes in the physical domains to changes in frequency ω, intensity I, and polarization P of light:

Mechanical domain: triboluminescence I, ω, piezo-optic effects I, effects of local symmetry on transition I, P, Raman-scattering I, ω.

Electric domain: cathodoluminescence I, ω, electrochronic effects I, electro-optic Kerr effect P, Pockels effect P, Franz-Keldysh effects P.

Magnetic domain: magnetic-absorptive effect P, magnetic birefringence I, magnetic dispersion I, Faraday rotation P, chemoluminescence I, ω, Franz-Keldysh effect I, electrochemical effects I, specific absorption I.

Thermal domain: thermoluminescence I, temperature dependence of effects (e.g., quenching of luminescence I).

Radiative domain: photoluminescence I, ω, cathodoluminescence I, ω, X-ray luminescence I, ω, photochromic effects I.

Based on various effects, a good number of sensors have been designed and a few of them have found industrial application. The material of the following paragraphs contains descriptions of several representative sensors.

Sensors with Movable Shutters, Gratings, or Light-carrying Elements

Almost invariably, multimode fibers are used in movement or position measurement transducers that modulate the amount of light passing through the gap between the fibers (Fig. 10.24). Shutters moving between two fibers (a), (c), (d), and (e) or deflecting fibers (f) are the most typical constructions. The sensitivity of the system can be increased by using a grating (b). For the increase of the range of measurement, a beam expansion system (c) is inserted in the gap between the light guides. The moving element of these constructions can be driven by a stream of fluid, as shown in (d), (e), and (f), or by a force due to acceleration (construction with a proof mass fixed on a spring-suspended shutter).

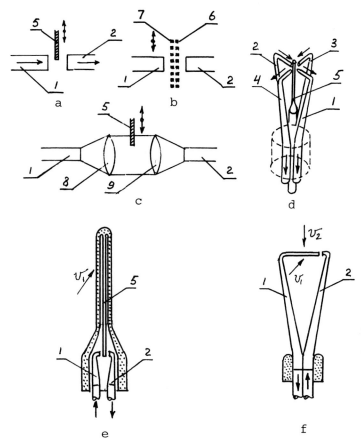

Figure 10.24 Fiber sensors with movable shutter, grating, or light-carrying element (sources: [448] and [509]). a = device with movable shutter, b = device with movable grating, c = device with beam expansion system, d = device with two-dimensional deflecting shutter, e = device with isolation of optical system from medium, f = device with deflecting light-carrying fiber; v_1 and v_2 = fluid speeds, 1, 2, 3, and 4 = optical fibers, 5 = shutter, 6 = fixed grating, 7 = moving grating, 8 and 9 = lenses.

A transfer characteristic of a shutter-displacement sensor can be derived [509] if we consider a simple construction (Fig. 10.25) composed of typical parts: source of light 1, transmitting fiber 2, shutter 3, receiving fiber 4, and photodetector 5. In cases where the shutter's height d [m] is greater than the receiver fiber's diameter $2r$ [m], and the radiation flux is uniformly distributed along the fiber cross section, the flux reaching the detector will be proportional to the area that is not shaded by a shutter (free-of-dashes area in Fig. 10.25). Assuming that the detector's output is proportional to the radiant flux reaching its sensitive area, we can express the detector output as

$$\overline{U} = U/U_m = \frac{1}{\pi} [\arccos (1 - \overline{\Delta r}) - (1 - \overline{\Delta r})(2\overline{\Delta r} - \overline{\Delta r}^2)^{1/2}] \qquad (10.91)$$

where $\overline{U}$, U and U_m = relative, absolute, and maximum values of output signals. (U and U_m are expressed in V, A, Ω, or in other relevant units.); $\overline{\Delta r} = (\Delta r_0 + \Delta r)/r$; and Δr = shutter's displacement from the offset Δr_0, m.

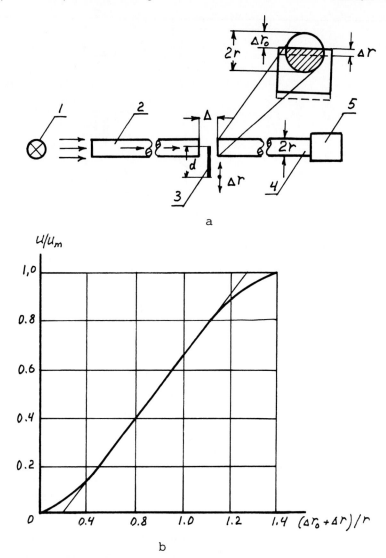

Figure 10.25 Schematic diagram of shutter fiber sensor (a) and its normalized transfer characteristic (b). All notations are given in text.

Function $\bar{U}(\overline{\Delta r})$ is shown in Figure 10.25. This function is close to linear for $\overline{\Delta r}$ between 0.5 and 1.3. Since Δr_0 is defined by the initial adjustment, it is convenient to express the transfer characteristic as

$$\bar{U} = (U - U_0)/U_0 \tag{10.92}$$

for the argument $\overline{\Delta r} = \Delta r/r$, assuming that $\overline{\Delta r}_0 = \Delta r_0/r$, and U_0 corresponds to U at Δr_0.

By a simple modification of formula 10.91, $\bar{U}$ is obtained as follows:

$$\bar{U} = \frac{\arccos(1 - \overline{\Delta r}_0 - \overline{\Delta r}) - (1 - \overline{\Delta r}_0 - \overline{\Delta r})[2(\overline{\Delta r}_0 + \overline{\Delta r}) - (\overline{\Delta r}_0 + \overline{\Delta r})^2]^{1/2}}{\arccos(1 - \overline{\Delta r}_0) - (1 - \overline{\Delta r}_0)(2\overline{\Delta r}_0 - \overline{\Delta r}_0^2)^{1/2}} - 1.$$

$$\tag{10.93}$$

Reflection Sensors

A Fotonic sensor [510] employs the measurement of the light reflected from a target surface (Fig. 10.26) to find the distance from the end of a fiber-optic bundle to the target. One branch of the bundle (transmitting fibers) is exposed to a light source and carries light to the tip of the bundle, where it is emitted and directed to the target surface. The light reflected from the surface is picked up by the receiving branch and directed to the photodetector, whose reading is a function of the distance.

The output-input characteristic of the sensor is given in formula 10.92 where $U_0 = U$ for $h = 0$ (Fig. 10.26).

A typical $\bar{U}(h)$ characteristic shown in Figure 10.26 is obtained [509] for a 0.025-fiber bundle. As is seen from the characteristic, the maximum of the $U(h)$ characteristic is at $h \approx 2r$. This condition is observed for light guides made of fibers having different diameters. A front-slope and back-slope operating region can be used for measurements in a small and in an extended range of gaps.

If light is transmitted from one end of a fiber to another through a separation having an index of refraction different from that of the fiber core, a useful formula for the calculation of losses R_F due to reflection (Fresnel losses) is

$$R_F = \left(\frac{n_1 - n_0}{n_1 + n_0} \right)^2 \tag{10.94}$$

where n_0 and n_1 = refractive indices of the medium between the fibers and fiber core, respectively.

Some of the reflection sensors utilize a special mirror attached to the moving object [511]. For example, a fixed-diameter laser beam is projected on an isosceles-triangle-shaped dual reflecting mirror attached to the object. The laser beam direction is perpendicular to the direction of the motion of the object. The reflected beams strike two photodiodes. The amounts of light received by the diodes is defined by the motion.

Fiber-optic Pressure Sensors

Using principles of the measurement of light reflection from a displaced target, we can build a pressure sensor (Fig. 10.27). Light directed from the light-transmitting

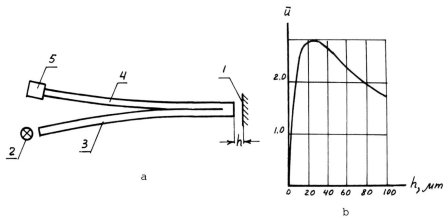

Figure 10.26 Fotonic sensor (a) and its transfer characteristic (b) (source: [509]). 1 = target, 2 = light source, 3 = transmitting branch of lightguide, 4 = receiving branch of lightguide, 5 = photosensor.

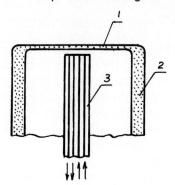

Figure 10.27 Fiber-optic pressure sensor. 1 = diaphragm, 2 = hollow cylinder, 3 = fiber bundle.

part of the fiber bundle is reflected from the inner side of a diaphragm. Under applied pressure the diaphragm changes its curvature, varying the portion of the light reflected toward the light-receiving part of the bundle. A photosensor on the remote part of the receiving fibers will respond to the change of the amount of light, allowing the reading of pressure. Using a glass-blowing technique or the processes described in Other Technologies in Chapter 8 the diaphragm can be fabricated as a thin film that is quite sensitive to the variation in pressure. The calculation of the mechanical behavior of the diaphragm can be performed using formulas in Chapter 5, Elastic Elements and Acoustical Elements. The sensor can be fabricated of high-temperature materials, such as SiO_2. Because of the small diameter of the diaphragm, its natural mechanical frequency can be high. For example, a diaphragm that is $1\mu m$ thick, 1.2mm in diameter, made of SiO_2, and intended for measuring pressure up to $1 \times 10^3 Pa$ (100mm H_2O), has a natural frequency of about 30kHz.

If a diaphragm is made of a transparent material and has air or another gas at both of its sides, R_F is calculated [509] as

$$R_F = 2\left(\frac{n-1}{n+1}\right)^2 \tag{10.95}$$

where n = refractive index for the diaphragm material.

Intrinsic Multimode Sensors

An example of an *intrinsic multimode fiber sensor* for the measurement of displacement, force, strain, pressure, or other mechanical quantities is a microbend sensor [509, 512]. In this sensor (Fig. 10.28), the modulation of light passing through the fiber is caused by the leakage of some of the light out of the fiber. The application of pressure to a fiber laid between a pair of grooved plates creates short-radius bends that cause some light energy to leak out of the fiber core. The loss

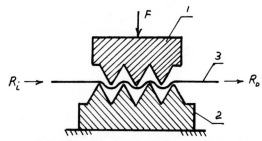

Figure 10.28 Microbend sensor. F = applied force, R_i and R_o = incoming and outgoing radiations, respectively, 1 and 2 = upper and lower grooved plates, 3 = fiber.

of energy occurs because of the change, with bending, of the incident angles of the modes guided along the fiber. Some of these angles become less than the critical angle, causing total internal reflection. A fraction of the modes propagating along the fiber will be radiated to the cladding, and the amount of energy reaching the end of the fiber is reduced. The maximum sensitivity in the conversion of the displacement into the change of light intensity is reached when the spacing Λ [m] between two successive bends is specified. For a graded-index fiber with the profile parameter $\gamma = 2$, the value of Λ is

$$\Lambda = \frac{2\pi r}{\sqrt{2\Delta}}. \tag{10.96}$$

The magnitude of Λ calculated for the real values of r, n_1, and n_2 is close to 1mm. By its nature, this sensor has a nonlinear output-input characteristic. This disadvantage is not essential in some instances; such as when an alarm must be triggered as the signal exceeds some predetermined level.

Temperature Sensors

One simple version of a temperature sensor (Fig. 10.29a) is made of a piece of semiconductor plate, such as GaAs or CdTe, separating the ends of two fibers. The position of the semiconductor band gap edge is temperature-dependent. The band edge moves about 0.35 and 0.31nm/°C for GaAs and CdTe, respectively [513]. If the light with a wavelength corresponding to the band gap is sent from one end of the fiber toward the semiconductor, the amount of light passed through the plate and reaching the other end of the fiber can be a measure of the semiconductor temperature (band-edge temperature sensor) [514].

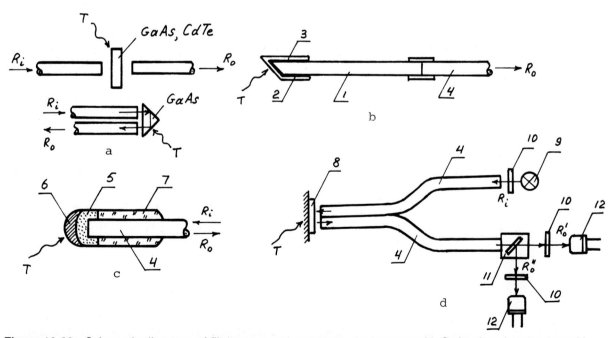

Figure 10.29 Schematic diagrams of fiber temperature sensors. a = sensor with GaAs plate, b = sensor with metallized sapphire rod, c = sensor with phosphor-coated fiber tip end, d = sensor with phosphor attached to object; R_i and R_o = incoming and outgoing radiation, T = temperature, 1 = sapphire rod, 2 = metallization, 3 = Al$_2$O$_3$-protection, 4 = fiber, 5 = clear elastomer, 6 = phosphor elastomer, 7 = cladding, 8 = phosphor at object, 9 = light source, 10 = optical filter, 11 = beam splitter, 12 = photosensor.

Instead of the plate, a prism made of a semiconductor can be used (Fig. 10.29a) to provide a different configuration of sensing and light-carrying parts [515]. A construction that is similar, but based on a different physical concept, is designed [516] using a plate made of lithium niobate ($LiNbO_3$). This is one of the most widely used materials for electro-optic devices [517]. It reveals the dependency of its birefringent characteristics on temperature. Using a relevant analyzing technique allows quite distant ($\sim$ 1km) measurements with 1°C-accuracy.

Another example of a fiber thermometer [518] is a device (Fig. 10.29b) consisting of a single-crystal sapphire rod that is coated at the measurement end with high-melting-point metal and alumina protective film. This end of the rod is placed in a high-temperature measurement zone and glows, radiating light energy. The opposite end of the rod is coupled to a low-temperature fiber outside the high-temperature zone. The metallized tip forms a blackbody cavity that transmits a temperature-dependent radiation via rod and fiber to a photodetector. As we showed above (Fig. 10.18), the radiation spectrum follows Planck's radiation law, which establishes the dependence of the spectral radiant emittance of a blackbody on its temperature. With the temperature increase, the emittance at a given wavelength also increases and the maximum of the emittance shifts toward the shorter waves. This sensor can operate at temperatures up to 1900°C. The rod has a diameter of about 0.25mm and its length is between several centimeters and 30cm. The deposited metal is platinum about $10\mu m$ thick. As the rod is heated in the range of temperatures between 600 and 300°C, the radiant flux changes by 20% per 1% of temperature change.

The transduction of temperatures below 400°C can be performed [449] using the effects of temperature-induced changes in fluorescence or absorption spectra of specific materials. In one of the designs utilizing this principle (Fig. 10.29c), a luminescing phosphor is placed at the tip of one end of an optical fiber in the measuring zone. A light excitation pulse is sent from another end of the fiber and causes luminescence of the phosphor. The time required for the luminescence to decay away depends on the measuring temperature.

In another light-radiating sensor [448], a small amount of europium-doped lanthanum oxysulfide (Eu: La_2O_2S) is placed on the object whose temperature is to be measured (Fig. 10.29d). The fluorescence of the material is excited by illuminating it with ultraviolet light transmitted via a silicon fiber. Another fiber picks up the emitted fluorescence and directs it to a detector system. The phosphor emits at more than one wavelength, and the ratio of the two lines is a function of the temperature. The measuring system of the instrument includes a beam splitter, optical band-pass filters, and photosensors that are coupled with the remote end of the fiber. These elements separate the received fluorescent radiation into two parts corresponding to two wavelengths and produce adequate electrical signals.

Various pyrometers contain fibers for transmitting radiation energy directly to the photosensor. The concept of one of these instruments is considered earlier in this chapter in Photodetectors.

A single-mode fiber that changes its length under temperature can serve as a temperature sensor [519]. The change of the optical pathlength is translated into a change of phase and then into an electrical signal.

A multimode fiber with a removed-cladding tip immersed into mineral oil [520] has found biomedical application. The oil has a strongly pronounced temperature-dependent index of refraction whose change modulates the light flux. Another similar design [521] uses two contacting, parallel-oriented fibers with partly removed cladding. The contact between the fibers is wet with a fluid (e.g., motor oil) having a temperature-dependent index of refraction. With the change in temperature, the transmission of light from one fiber to another is changed as well.

Temperature determination of a high-temperature (greater than 1000°C) gas by fiber-optic infrared spectroscopy [522] is one of the developing techniques. Using a

Fourier transform infrared spectrometer, the shape of the spectrum is examined as an index for the temperature.

The fiber itself can operate as the sensing part of a temperature sensor [513]. As the fiber is heated up, it radiates infrared rays whose intensities are proportional to the temperature.

A fiber sensor made of Eu^{3+} doped fluoride glass [523] is applicable for low-temperature measurements. The core of the fiber is made of glass composed of 61.3% ZrF_4, 33.7% BaF_2, and 5.0% GdF_3. The glass is doped by Eu^{3+} (0.87%) ions. The material changes its transparency to the light of 1.8 to 2.3μm wavelength when the temperature changes. It is reported that in the range 77 to 150K, the accuracy of measurement is about 0.5K.

Principles of Phase-modulated Sensors

The operation of a number of single-mode-fiber sensors is based on the effects of modulation of the light phase (or mode velocity) in the fiber by the measurand. The classical system for this kind of measurement is the Mach-Zehnder interferometer (Fig. 10.30a). In this instrument, a light beam from a source (typically a laser) is split into two branches. One of them is phase-modulated in response to the change of the refractive index of the medium or the length of the path; the other one remains at a stable phase. Two beams combined at the output of the instrument produce, via photodetectors, electrical signals whose difference gives the reading of the phase. The measurement technique using interferometers is quite accurate and allows detection of such small displacements as 0.1μm [524]. The design of a fiber-optic force cell with sensing cavities operating as interferometers has been reported by [525]. The resolution of this device is 2.5N in the range of 0 to 100N.

A fiber-optic version of the interferometer (Fig. 10.30b) has a similar composition and principle of operation. Here light is split by the fiber coupler and directed

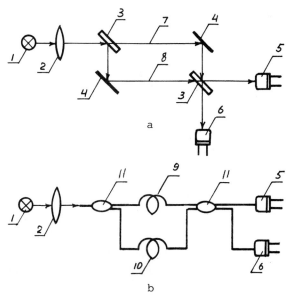

Figure 10.30 Bulk-optic (a) and fiber-optic (b) Mach-Zehnder interferometer. 1 = light source, 2 = lens, 3 = beam splitter, 4 = mirror, 5 and 6 = photosensors, 7 = signal path, 8 = reference path, 9 = sensing fiber, 10 = reference fiber, 11 = fiber coupler.

to the sensing fiber responding to the measurand and to the reference fiber whose physical conditions are kept stable. Another coupler combines two beams and transmits them to the photosensors. The variation in the transmittance characteristics of the sensing fiber is translated into the electrical signal at the outputs of the photodetectors.

In order to find the phase response to different physical quantities [448], we will simplify the conditions by assuming that we are dealing with a single-mode fiber, and that the mode behaves like a ray that travels down the center of the fiber with velocity c/n_1, (c is the speed of light). If the fiber has length L [m], and λ_0 [m] is the wavelength of the radiation in a vacuum, then the phase φ [rad] associated with the fiber's length is

$$\varphi = \frac{2\pi L n_1}{\lambda_0}. \tag{10.97}$$

An increment ΔX of physical quantity X affecting the index of refraction and fiber's length causes the change in the phase $\Delta\varphi$:

$$\Delta\varphi = \frac{2\pi L}{\lambda_0} \cdot \frac{dn_1}{dX} \Delta X + \frac{2\pi n_1}{\lambda_0} \cdot \frac{dL}{dX} \Delta X. \tag{10.98}$$

It is easy to prove that φ is practically insensitive to the change in the fiber diameter. Now we can find the relations between $\Delta\varphi$ and force F [N] stretching the fiber, radial uniform pressure P [Pa] affecting the fiber, and temperature T [K] of the fiber body (the corresponding $\Delta\varphi$'s are denoted by $\Delta\varphi_F$, $\Delta\varphi_P$, and $\Delta\varphi_T$).

According to Hooke's law,

$$\frac{\Delta L}{L} = \frac{F}{A} \cdot \frac{1}{E} \tag{10.99}$$

where ΔL = change in L, m;
      A = cross-sectional area of fiber, m^2; and
      E = Young's modulus, N/m^2.

Assuming that strain does not cause any change in the refractive index n_1 and $\Delta L \approx dL$, we will substitute ΔL from formula 10.99 into formula 10.98, and obtain

$$\Delta\varphi_F = \frac{2\pi n_1 L F}{\lambda_0 A E}. \tag{10.100}$$

The relative change in the fiber length exposed to pressure is expressed as

$$\frac{\Delta L}{L} = \frac{P}{E} 2(1 - \nu) \tag{10.101}$$

where ν = Poisson's ratio.

Neglecting the influence of the radius change on magnitude of $\Delta\varphi$, we will find that

$$\Delta\varphi_p = \frac{4\pi n_1 L P}{\lambda_0 E} (1 - \nu). \tag{10.102}$$

It should be taken into account that both n_1 and L are affected by temperature; therefore,

$$\Delta\varphi_T = \frac{2\pi L}{\lambda_0} \cdot \frac{dn_1}{dT}\Delta T + \frac{2\pi n_1}{\lambda_0} \cdot \frac{dL}{dT}\Delta T = \frac{2\pi L n_1}{\lambda_0}(\alpha_n + \alpha_L)\Delta T \quad (10.103)$$

where $\alpha_n = (1/n_1) \cdot dn_1/dT$ = temperature coefficient of refractive index, 1/K, and
$\alpha_L = (1/L) \cdot dL/dT$ = temperature coefficient of fiber expansion, 1/K.

In calculations for fibers made of fused silica, we can accept the following: $E = 73 \times 10^9$, $\nu = 0.17$, $n_1 = 1.45$, $\alpha_L = 0.55 \times 10^{-6}$, and $\alpha_n = 7.0 \times 10^{-6}$.

The electrical output of one of the photodetectors in the interferometer U_1 is proportional to the cosine function of $\Delta\varphi$:

$$U_1 \propto C_1[1 + \cos(\Delta\varphi)] \quad (10.104)$$

where C_1 = coefficient defined by the parameters of the construction (it is proportional to the light intensity).

The response of the second photodetector U_2 is similar to U_1 but has a phase shift of π radians:

$$U_2 \propto C_2[1 + \cos(\Delta\varphi + \pi)] = C_2[1 - \cos(\Delta\varphi)]. \quad (10.105)$$

If the output of the instrument makes the difference between U_1 and U_2, and $C_1 = C_2$, the sensitivity of the interferometer is doubled.

It should be taken into account in practical design that the resolution of an interferometer of 1×10^{-6} radians is achievable; however, the high sensitivity of the system (signal and reference branches) to temperature variation can be a serious restriction in obtaining the high sensitivity. The other limits are defined by noise at detectors, and fluctuations in the intensity and frequency of light radiation created by the source (usually laser).

Fiber-optic Gyroscopes

Two light beams traveling in opposite directions in a round fiber coil rotating about an axis perpendicular to the plane of the coil will induce a phase shift between each other (Sagnac effect). This effect is exploited in the fiber gyroscope [448, 526, 527], whose operation can be simply illustrated if we consider a model containing a single circular turn of fiber having radius R (Fig. 10.31), and assume that light travels in the fiber as in a vacuum. Besides the turn, the basic circuit of the gyroscope includes the light source, beam splitter, and photodetector. As is clear from the diagram, the gyroscope's system represents an interferometer in which two counter-propagating beams formed by the source and splitter modulate the phase of light translated into the intensity of illumination of the photodetector. Therefore, the output of the photodetector is a function of the angular speed.

If the turn is at still, both beams pass the circumference at the same time τ_0 [s]:

$$\tau_0 = \frac{2\pi R}{c}. \quad (10.106)$$

With rotation of the turn at angular speed Ω [rad/s], the speeds for the clockwise and counterclockwise beams will differ: $c + \Omega R$ and $c - \Omega R$, respectively. The times τ_1 and τ_2 needed for the beams to travel along the same path $2\pi R$ will be

$$\tau_1 = \frac{2\pi R}{c + \Omega R} \quad \text{and} \quad \tau_2 = \frac{2\pi R}{c - \Omega R}. \quad (10.107)$$

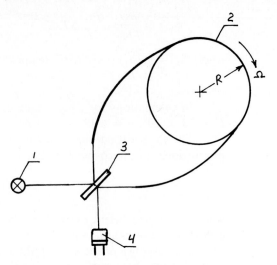

Figure 10.31 Schematic diagram of Sagnac-effect fiber gyroscope. R = radius of turn, Ω = angular speed, 1 = light source, 2 = fiber turn, 3 = beam splitter, 4 = photodetector.

The difference between these transit times $\Delta\tau$ is

$$\Delta\tau = \tau_2 - \tau_1 = 2\pi R\left(\frac{1}{c - \Omega R} - \frac{1}{c + \Omega R}\right) = \frac{4\pi\Omega R^2}{c^2 - \Omega^2 R^2} \approx \frac{4\Omega S}{c^2} \qquad (10.108)$$

where $S = \pi R^2$ = ring area, m^2.

It is assumed in the approximation for $\Delta\tau$ that $c^2 \gg \Omega^2 R^2$. The difference in phases φ [rad] of the two counter-propagating waves coming to the splitter is

$$\varphi = \omega\Delta\tau = 2\pi\nu\Delta\tau = \frac{2\pi c}{\lambda_0}\Delta\tau = \frac{8\pi S}{\lambda_0 c}\Omega \qquad (10.109)$$

where $\omega = 2\pi\nu$, rad/s;
$\quad \nu$ = light's frequency, Hz; and
$\quad \lambda_0 = c/\nu$ = light's wavelength, m.

In practical designs, the gyroscope contains many turns wrapped on a spool, and the overall length of the beam path will be longer, which gives an increase in sensitivity. The phase—angular speed relationship for the spool of N turns will be

$$\varphi = \frac{8\pi SN}{\lambda_0 c}\Omega = \frac{4\pi RL}{\lambda_0 c}\Omega \qquad (10.110)$$

where L = total length of fiber, m.
The sum of the fields E of the two beams producing irradiance affecting the photosensor is given as

$$E = A_1 \exp\left[j(\omega t + \varphi_1)\right] + A_2 \exp\left[j(\omega t + \varphi_2)\right] \qquad (10.111)$$

where A_1 and A_2 = amplitudes of fields,
$\quad \omega$ = angular frequency of light beam, rad/s; and
$\quad \varphi_1$ and φ_2 = phase shifts of fields, rad.

According to the gyroscope principle of operation we have

$$\varphi_1 = \varphi_0 + \frac{1}{2}\varphi \quad \text{and} \quad \varphi_2 = \varphi_0 - \frac{1}{2}\varphi. \quad (10.112)$$

The incident irradiance and the photodetector output U [V or A] are proportional to the sum of the fields times its complex conjugate:

$$U \propto \{A_1 \exp[j(\omega t + \varphi_1)] + A_2 \exp[j(\omega t + \varphi_2)]\} \cdot \{A_1[-j(\omega t + \varphi_1)] \\ + A_2[-j(\omega t + \varphi_2)]\} \quad \text{or} \quad (10.113)$$

$$U \propto A_1^2 + A_2^2 + 2A_1A_2 \cos(\varphi_1 - \varphi_2) = A_1^2 + A_2^2 + 2A_1A_2 \cos\varphi. \quad (10.114)$$

Assuming as before $A_1 = A_2$, we obtain an expression similar to that given in formula 10.104:

$$U \propto C(1 + \cos\varphi). \quad (10.115)$$

This function is periodic (Fig. 10.32). The sensitivity of the device is $dU/d\varphi \propto \sin\varphi$. It is low for small signals (small values of φ). The highest sensitivity can be obtained if the system is tuned to operate at the "quadrature" point (Fig. 10.32), which is halfway between the maximum and minimum values of U. The required phase shift is realized by the insertion into the loop of a special crystalline plate that provides the $\pi/2$ phase shift between the two counter-propagating beams.

Calculations and experiments show [527, 528] that for a typically constructed gyroscope ($R = 0.1$m, $L = 500$m, $\lambda_0 = 0.6328\mu$m), the reliable measurement of an angular speed of 1°/hour requires an ability to measure the difference in phases of 10×10^{-6}rad. For measurements in the range 0.1 to 0.01°/hour, a reliable reading of 1×10^{-6} to 1×10^{-7}rad is necessary. Graphs in Figure 10.33 illustrate the function $\varphi(\Omega)$ for several values of RL.

Due to the advantages of its solid-state nature, the fiber-optic gyroscope has become a serious competitor of some of the traditional spinning-rotor gyroscopes. With the progressing improvement of this optic device, it will find a wider application, especially in those cases where a low-cost system is required, and a moderate resolution is acceptable.

Single-mode-fiber Polarization Sensors

The effect of polarization within a fiber has found its application in several sensors.

In a single-mode fiber, two orthogonally polarized modes can freely propagate and be birefringent to some extent because of geometrical imperfections in circularly

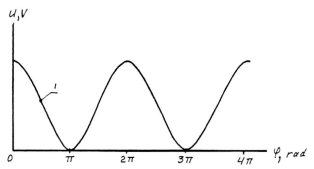

Figure 10.32 Output of Sagnac gyroscope.
1 = "quadrature" point.

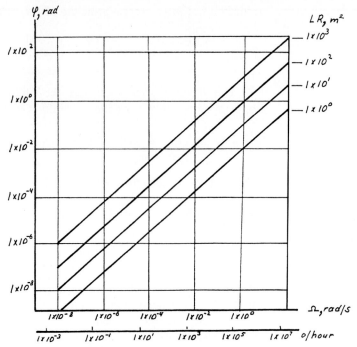

Figure 10.33 Phase of optical signal versus angular speed to be measured for different values of fiber length times turns radius (source: [527, 528]).

symmetrical fibers and possible variations in the refractive index. The extrinsic bire-fringence can be generated by external factors, such as deformation of the fiber. Specially designed, so called Hi-Bi (high-birefringent) fibers are capable of maintaining the initial polarization state over a large distance [529]. If two polarized modes are launched at one end of the fiber, the disturbances in the physical state of the fiber lead to the exchange of energies from one mode to another ("mode coupling"), and, as a result, to the change between the modes' phases. It can be proved that

$$\varphi = \frac{2\pi L(n_x - n_y)}{\lambda_0} \tag{10.116}$$

where n_x and n_y = effective refractive indices for two modes.

Therefore, if physical quantity X affects the fiber, the sensitivity of the system measuring the phase shift will be

$$\frac{d\varphi}{dX} = \frac{2\pi}{\lambda_0}\left[\frac{dL}{dX}(n_x - n_y) + L\left(\frac{dn_x}{dX} - \frac{dn_y}{dX}\right)\right]. \tag{10.117}$$

The angle of polarization of a mode can be altered by a strong magnetic field (Faraday rotation). This phenomenon is realized in sensors for the monitoring of large currents in electricity generating stations.

Other Fiber Sensors

A number of approaches have been demonstrated to measure speed, acceleration, fluid level, flow rate of fluids, amplitude and phase of vibration, chemical composition, and so forth. One interesting approach uses integrated solid-state devices [530] in which light-guiding layers are formed by means of standard semiconductor technologies. In the following short review, we will give descriptions of several designs that have been reported recently.

In well-known fiber-optic *magnetostrictive magnetic field sensors,* magnetostrictive cladding changes its length in response to an applied magnetic field. The length of the optical fiber changes, as does the phase of the optical signal propagating through the fiber. The technology of this fiber is more economical when a fiber has a circular cross-sectional area including the central core made of magnetostrictive material (iron, cobalt, nickel, or various alloys), the inner glass cladding next to the core, the middle annulus of glass for the transmission of light, and the outer glass cladding [531].

Dynamic *holographic interferometry* has been used *for detection* of acoustically induced *phase* modulation of coherent light propagating within a multimode optical fiber [532]. As this fiber is mechanically strained by the action of an acoustic pressure wave, the phase shift, proportional to the pressure, is read out using a holographic phase-detection technique. This method allows a performance comparable with that of a monomode interferometric sensor to be obtained, but with considerably relaxed alignment tolerances.

A *moisture-sensitive sensor* [533] is made of 600-μm fiber coated by a film of cobalt chloride whose optical absorption depends on the presence of moisture that forms hydrates with molecules of bound water. Humidity can be measured in the range of 40 to 80% RH.

A *versatile twisted optical fiber sensor* [534] consists of two multimode fibers twisted together. Some of the guided modes of one of the fibers convert into radiating modes interacting with the medium surrounding the fibers. These modes enter the second fiber and are detected in a photosensor. The medium influences the interaction between the two fibers in several ways: through changes in the index of refraction, absorption characteristics (in dyes, pigments, suspended in liquid particles), and physical interaction with the fiber. The sensor is sensitive to the presence of water or some solvents in oil, and to the impurities in various liquids. Another useful function would be reaction rate monitoring.

An in situ *fiber-optic* technique *for determining* the endpoint of *cure* of a thermoset polymer uses a conventional multimode silicon fiber that is in contact with the resin [535]. The fully cured resin-sensing fiber has a larger refractive index than does uncured or partly cured resin, and, therefore, can modulate light that is transmitted through the fiber arrangement. A near-infrared light from LED has been used for the detection.

A silicon *micromechanical mirror combined with* input and output *fibers* directing light to and receiving it reflected from the mirror are the essential parts of a fiber-optic switching and multiplexing device [536]. A simultaneous torsional moment driving the mirror and providing a 1.2° angular deflection is obtained electrostatically. The mechanical resonance is at 40kHz. The quality factor is approximately 8 (mainly due to the viscous damping in air).

Replacing the passive cladding material over a small length of fiber by an active material such as electro-optic, magnetooptic, or piezoelectric material allows one to obtain the responses of a fiber to electrical or magnetic influences [537, 538]. If an external field applied to the material can modify its refractive index, we then know that the boundary conditions between the core and cladding are altered along with the optical transmission loss and modal power redistribution. The typical arrangement of a sensor is a fiber with a striped cladding surrounded by a nematic liquid crystal that is provided by two electrodes. The voltage applied to the electrodes causes the change in the optical transmission.

The number of experimental designs of fiber-optic sensors for chemical and biomedical instruments is being increased rapidly (see, for example, [505, 535, 539–543]). This growth reflects the general trend toward the intensive development of chemical and biomedical solid-state measuring devices.

chapter 11

Miscellaneous Miniature Sensors

In this chapter, we review a number of sensors that have not been discussed in the foregoing topics. We complete our description of microsensors and devote the final section to the summary of the general principles of microsensor design. The first section of this chapter deals with magnetic microsensors, whose technology has been well established over the last several decades. The other section is related to solid-state chemical sensors, whose technology is still progressing and is very promising.

MAGNETIC SENSORS

The structures, principles of operation, and some of the aspects of design of magnetic sensors will be considered in this section. The sensors are presented in related groups such as Hall-effect devices, magnetoresistors, and magnetotransistors.

Hall-effect Sensors

A model of a Hall element (Fig. 11.1) consists of a rectangular plate made of a conducting or semiconducting material. As the plate carrying current is placed in a magnetic field, the voltage will be induced at the side faces of the plate in a direction perpendicular to the direction of the feeding current in the plate and to the direction of the field. The effect arises from the Lorentz force F_1 [N] developed on the charged particles moving in a magnetic field [544] as shown in Figure 11.1:

$$\vec{F_1} = q(\vec{v} \times \vec{B}) \quad \text{and} \quad F_1 = qvB \tag{11.1}$$

where q = particle's charge, C;
$\quad v$ = particle's velocity, m/s; and
$\quad B$ = magnetic induction, T.

Under the Lorentz forces, charges are separated and deflected along the y-axis, creating an electric field of intensity ϵ_y [V/m]. The charge experiences an additional force F_2 [N] due to this field:

$$F_2 = q\epsilon_y. \tag{11.2}$$

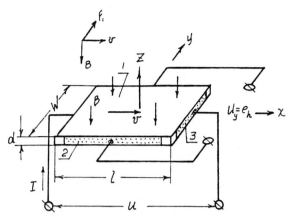

Figure 11.1 Schematic diagram of Hall element.
1 = element, 2 and 3 = electrodes.

Forces F_1 and F_2 are balanced; therefore,

$$\epsilon_y = vB. \tag{11.3}$$

Assuming that the plate is made of n-material and has cross-sectional area $w \times d$ [m^2], the carriers' velocity is expressed as

$$v = \frac{I}{qnwd} \tag{11.4}$$

where
$\quad I =$ current through the plate, A;
$\quad n =$ concentration of carriers, 1/m^3; and
w and $d =$ plate's width and thickness, respectively, m.

Substitution of v into formula 11.4 gives

$$\epsilon_y = \frac{IB}{qnwd}. \tag{11.5}$$

Hall's emf e_h [V] is simply derived from the following equation:

$$e_h = \epsilon_y w = \frac{IB}{qnd} = \frac{K}{d} IB \tag{11.6}$$

where $K = 1/qn =$ Hall's coefficient, m^3/C.

This coefficient is defined by the nature of the material. For a p-type semiconductor, $K = 1/qp$. Using formula (7.7) for mobility and expressing current through the element in terms of external field ϵ, resistivity ρ, or conductivity of material σ, we can obtain another relation for e_h:

$$I = \frac{\epsilon d \cdot w}{\rho} = \epsilon \sigma d \cdot w \tag{11.7}$$

$$e_h = \frac{\mu\rho}{d} BI = \frac{\mu}{d\sigma} BI \tag{11.8}$$

where $K = \mu\rho = \mu/\sigma$.

e_h can be expressed in terms of bias voltage U applied to the element rather than current I:

$$e_h = \frac{w}{l} \mu U B \qquad (11.9)$$

where l = plate's length, m.

As is seen in equation 11.9, the ratio e_h/U is directly proportional to the magnetic field induction.

For a p-n semiconductor, Hall's coefficient depends on the n and p carriers' mobilities:

$$K = \frac{G}{q} \cdot \frac{n\mu_n^2 - p\mu_p^2}{(n\mu_n^2 + p\mu_p^2)^2} \qquad (11.10)$$

where G = geometric factor [545], $m^4/V^2 \cdot s^2$.

III-V compound semiconductors such as InAs ($\mu_n = 33\,000cm^2/V \cdot s$) and InSb ($\mu_n = 80\,000cm^2/V \cdot s$) are widely used materials for Hall-effect sensors, because of their high mobility of carriers ($e_h \propto \mu$). The mobility of lightly doped silicon is relatively low, about $1\,400cm^2/V \cdot s$ (see Table 7.1), and this material is not the best one for the measurements of a weak magnetic field.

Usually, a Hall plate is part of a monolithic integrated circuit containing signal conditioners. Epitaxial growth (Fig. 11.2) is a typical technology for the sensor. The order of magnitude of the Hall voltage can be evaluated if we take into account the following real numbers [546]: a small, a large, and a very large permanent magnet can produce inductions of around 0.1, 1, and $2T$, respectively; dimensions w and l can be of the order $200\mu m$ each; and the epitaxial layer can be $10\mu m$ thick; the resistivity of the epitaxial layer for $N_D = 1 \times 10^{16} cm^{-3}$ can be $0.6\Omega \cdot cm$ (the sheet resistance of the epitaxial layer is $0.6/1 \times 10^{-3} = 600\Omega/\square$), a power of several hundredths of a watt is acceptable for dissipation in the element due to the bias voltage (it corresponds to U of 5 to 10V).

The ability of the sensor to provide an electrical output proportional to the magnetic induction is widely used in the practice of measurement of magnetic values.

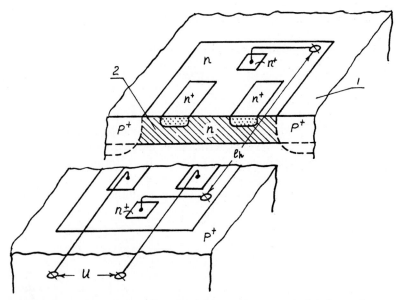

Figure 11.2 Hall plate with epitaxial layer. 1 = p-material, 2 = n-type epitaxial layer.

Different devices based on the analog multiplication of two electrical values also utilize the Hall sensor, since $e_h \propto IB$ or $e_h \propto \mu B$. Measuring e_h is a standard technique to determine the mobility of carriers in the semiconductor material.

Magnetoresistors

A magnetoresistor changes its resistance proportionally to the magnetic field intensity to which it is exposed. If a conductor or semiconductor carrying current is not influenced by the magnetic field, the charge carriers pass a straight-lined mean free path l_0 [m] along the direction of current (Fig. 11.3). After the field is applied, the path of the carrier is deflected and becomes an arc of the same length l_0 [547]. The path of carriers l' along the direction of current is smaller than l_0:

$$l' = r \sin \varphi \tag{11.11}$$

where r = radius of curvature of the carriers' path, m,
$\quad \varphi$ = carriers' deflection angle, degree.

The centripetal Lorentz force is perpendicular to the speed v [m/s] of the charged particle and proportional to this speed and to the induction of the field B [T]. The equation of force balance is

$$qvB = mv^2/r \tag{11.12}$$

where m = effective mass of the particle, kg.
From the last equation,

$$r = mv/qB. \tag{11.13}$$

Angle φ expressed in radians is

$$\varphi = l_0/r = l_0 qB/mv. \tag{11.14}$$

The conductivities of the semiconductors σ and σ' [S/m] are proportional to the amount of the mean free path:

$$\sigma = nq^2 l_0/2mv \quad \text{and} \quad \sigma' = nq^2 l'/2mv \tag{11.15}$$

where n = electron concentration, $1/m^3$.

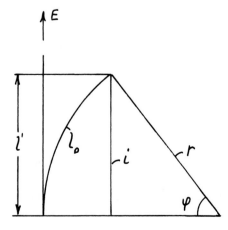

Figure 11.3 Path of charged particles in magnetic field. E = electric field. Magnetic field is perpendicular to plane of drawing.

The absolute ($\Delta\sigma$) and relative ($\Delta\sigma/\sigma$) changes in conductivity due to the influence of the magnetic field are written as

$$\Delta\sigma = \sigma - \sigma' = \frac{nq^2}{2mv}(l_0 - l'), \qquad \frac{\Delta\sigma}{\sigma} = \frac{l_0 - l'}{l_0} \qquad (11.16)$$

where

$$l_0 - l' = l_0 - r\sin\varphi. \qquad (11.17)$$

Expansion of $\sin\varphi$ in the Maclaurin series gives the first four terms:

$$\sin\varphi = \sin 0 + \frac{1}{1!}\left(\frac{d\sin\varphi}{d\varphi}\right)_0 \varphi + \frac{1}{2!}\left(\frac{d^2\sin\varphi}{d\varphi^2}\right)_0 \varphi^2 + \frac{1}{3!}\left(\frac{d^3\sin\varphi}{d\varphi^3}\right)_0 \varphi^3$$

$$= \varphi - \frac{\varphi^3}{6}. \qquad (11.18)$$

Substitutions of φ, taken from formula 11.14, and $\sin\varphi$, taken from formula 11.18, into formula 11.17 and then into formula 11.16 yield

$$l_0 - l' = \frac{l_0\varphi^2}{6} \quad \text{and} \quad \frac{\Delta\sigma}{\sigma} = \frac{\varphi^2}{6} = \frac{1}{6}\left(\frac{l_0qB}{mv}\right)^2. \qquad (11.19)$$

Allowing for the statistical distribution of carriers' speeds by using dimensionless coefficient C, we finally get

$$\frac{\Delta\sigma}{\sigma} = C\left(\frac{l_0qB}{mv}\right)^2. \qquad (11.20)$$

As is seen from this expression, the semiconductor's resistance change in the magnetic field is directly proportional to the square of the magnetic induction.

We have considered the case for the intrinsic semiconductor. For the material with the pronounced n and p carriers having mobilities μ_n and μ_p, the change in the material resistivity $\Delta\rho/\rho_0$ will be [547].

$$\frac{\Delta\rho}{\rho_0} = \frac{3\pi}{8}B^2\left[\frac{4}{\pi}\cdot\frac{\mu_n^3 n + \mu_p^3 p}{\mu_n n + \mu_p p} - \left(\frac{\mu_n^2 n - \mu_p^2 p}{\mu_n n - \mu_p p}\right)^2\right]. \qquad (11.21)$$

A simplified expression for $\Delta\rho/\rho_0$ is used for engineering calculations:

$$\Delta\rho/\rho_0 = (D\mu B)^\chi \qquad (11.22)$$

where D = geometric factor (for μ expressed in $\text{m}^2/\text{V}\cdot\text{s}$ and B in T, D is between 0.2 and 1, and

χ = power for $D\mu B$ ($1 \leq \chi \leq 2$; for $\mu B \ll 1$, $\chi \approx 2$; for $\mu B \gg 1$, $\chi \approx 1$).

The standard constructions of magnetoresistors are dies or thin-film strips of semiconductor with metallization at the ends for the attachment of leads. The highest sensitivity has a Carbino disk in which one electrode is at the center of the disk and another one is formed on its circumference. The carriers move from the center toward the edge of the disk. The materials with a high mobility of carriers, such as InSb, InSb-NiSb, InAs, and HgTe are most appropriate for use in magnetoresistors.

Self-heating of the element must be limited. The maximum current I_m [A] can be calculated as

$$I_m = \sqrt{\frac{2\eta(T - T_m)(w + d)wd}{\rho_0(1 + \Delta\rho/\rho_0)}} \qquad (11.23)$$

where
η = heat transfer coefficient, W/m$^2 \cdot$ K;
T and T_m = temperatures of the element and medium, respectively, K;
w and d = width and thickness of the element, respectively, m;
ρ_0 = resistivity of material, $\Omega \cdot$ m; and
$\Delta\rho/\rho_0$ = relative change of ρ_0 for maximum of B.

It can be easily proved that the temperature coefficient of a magnetoresistor β_R is

$$\beta_R = \beta_\rho + \beta_\mu \frac{2\mu^2(T_0)B^2}{1 + \mu^2(T_0)B^2} \qquad (11.24)$$

where β_ρ and β_μ = temperature coefficients of ρ and μ, respectively, 1/K, and
$\mu(T_0)$ = μ for the reference temperature T_0, m^2/V $\cdot$ s.

Other Sensors

The contemporary trend in the development of the theory and design of magnetic sensors is threefold: (1) deeper study of charges transportation in the material exposed to the magnetic field [548] and analysis of transfer characteristics [549]; (2) improvement of traditional field-sensitive structures; and (3) development of new structures.

Highly sensitive Hall devices based on AlGaAs/GaAs can be fabricated as heterostructures [550, 551]. Experiments show that for a variety of device geometries and operating conditions, the devices exhibit a high output signal as well as a relatively low output noise power when operated in a differential mode at low device currents. The reached sensitivity is 1200V/AT with a small temperature coefficient of -0.1%/°C at room temperature. The heterostructure semiconductors realize a magnetic sensitivity as large as that of InSb devices with a temperature coefficient as small as that of GaAs devices. It is believed that this structure is one of the most promising constructions for the intelligent magnetic sensor IC.

An offset-reduction method for Hall plates has been developed [552]. This method minimizes the influence of stress and temperature on the offset. Generally, the Hall plate exhibits unpredictable output voltage at zero magnetic field. The voltage is called the offset, when the Hall contacts are not on the same equipotential line. Induction of the voltage occurs because of imperfections in the fabrication process, residual stress, temperature gradients, and so forth. The new method uses only one single symmetrical Hall plate in which the direction of current is made to spin by contact commutation with steps $\pi/6$ or smaller. The consecutive Hall voltages are read out and averaged over time, and the offset cancels out. The residual offset is about 10 times less than that for commercially available Hall plates. In this design, the increased number of contacts (e.g., 16) is arranged along a circle, on the surface of the symmetrical Hall plate. The current direction is made to spin by a contact commutation with steps given by the radial distance between the contacts, that is, $\pi/8$. Simultaneously, the Hall voltage perpendicular to each current direction is measured.

Magnetoresistive, thin-film, ferromagnetic elements [553–555] developed during the last several years can measure fields of less than 1×10^{-3}A/m up to several 1×10^4A/m—much lower fields than normal Hall sensors can measure. Magnetoresistive sensors make use of the fact that the electrical resistance of certain ferromagnetic alloys is influenced by external fields. Solid-state processing is typical in the sensor fabrication. This process offers uniformity and repeatability for producing the ferromagnetic thin film. The specific resistance $\rho(\theta)$ [$\Omega \cdot$ m] of the film depends on the angle θ between the internal magnetization M and the current:

$$\rho(\theta) = \rho_\perp + (\rho_\parallel + \rho_\perp) \cos^2 \theta \tag{11.25}$$

where $\rho_\parallel$ and $\rho_\perp$ = resistivities parallel and perpendicular to M. The quotient $\Delta\rho/\rho = (\rho_\parallel - \rho_\perp)/\rho_\perp$ is the measure of magnetoresistive effect. Appropriate materials for the films are binary and ternary alloys of Ni, Fe, and Co. Compositions like NiFe 81/19 (81% of Ni and 19% of Fe) are preferable for sensors since they are free of magnetostriction. This ensures a low influence of temperature-induced or mechanically induced stresses in the film. The comparison of some materials used in magnetoresistive sensors is given in the following list where the first figures in the parentheses denote ρ ($1 \times 10^{-8}\Omega \cdot$ m), and the second figures denote $\Delta\rho/\rho$ (%): NiFe 81/19 (22, 2.2), NiFe 86/14 (15, 3.0), NiCo 50/50 (24, 2.2), NiCo 70/30 (26, 3.7), and CoFeB 72/8/20 (86, 0.07).* The films are deposited by thermal evaporation or cathode sputtering. During the deposition, a low oxygen pressure ($< 1 \times 10^{-4}$Pa) is required, and low deposition rates ($\leq$ 5 nm/min) are preferred. Deposition temperatures in the range between 25 and 300°C are reported. Proven substrate materials are glass, alumina, and oxidized silicon. Passivation of the sensitive layer by inorganic or organic insulators improves stability. An annealing process under vacuum conditions in a homogenous field of 1kA/m is recommended after the deposition.

The highest sensitivity obtained using sensors is of the order of magnitude of 2 000V/T. The typical presentations of the sensitivity are volts at output per volts across the resistor, per A/m of field intensity—for example, 100mV/V/(kA/m); or volts at output per ampere and tesla affecting the cell—that is V/AT. The metal thin film gives high stability and a wide temperature range of operation. Readout techniques for metal-film sensors [556] have many aspects in common with other devices that change their resistance in response to a measurand.

The principle of conversion of the magnetic field intensity into a proportional ac voltage of even harmonic excitation (see Chapter 4, Weak-field Sensors) has been recently realized [557] in a *fluxgate sensor* based on planar technology. In this sensor, a permalloy film is sputtered on a silicon substrate and covered by silicon dioxide. The cores of 0.5μm thickness have an area of 2×4mm^2 that is covered by an evaporated aluminum pick-up coil, whereas an exciting coil is not integrated in the silicon chip. This coil is made of thick film and placed on the ceramic substrate of the hybrid structure.

Another type of Hall cell sensitive to the magnetic field perpendicular to the chip surface is the MOS *Hall cell* [558] in which the thin MOS channel layer is regarded as the active region of the Hall cell. This layer is about 100Å-thick and the sensitivity of the device is around 1000V/AT.

The development of the *"vertical" Hall cell* made possible the creation of a number of two- and three-dimensional magnetic field sensors [559–565] fabricated on one chip. The vertical Hall cell is sensitive to the field oriented parallel to the chip surface; for example, based on the CMOS technology, the current in the active Hall region flows vertically through a substrate ("long" Hall cell). A sensitivity of 450V/AT has been obtained in experiments. In the case of bipolar transistors, the length of the current path is limited by the thickness of the epitaxial layer ($\sim 10\mu$m), and the sensitivity of this device is lower. With the reduction in the width of the vertical current path, the sensitivity is increased. New approaches in the structure's

---

*At fields of 250, 200, 2500, 2500, and 2000 A/m, respectively.

architecture (diffused vertical Hall cell and trenched vertical Hall cell) make the current path narrower. This is achieved by the restrictions using p^+ diffusion in an n-type silicon epitaxial film. In spite of these measures, the sensitivity of these devices is lower in comparison with the CMOS cell. The active area of the trench element is the structure of the field-effect transistor (JFET), and its sensitivity depends on the channel width that, in turn, is defined by the current. This novel device shows a sensitivity higher than 1 200V/AT.

The sensitivity of *magnetodiodes* [566, 567] to a magnetic field is explained by the so-called Suhl or magnetoconcentration effect, which is based on the difference in recombination velocities between two surfaces or between one surface and the bulk of the device. The magnetic field, concentrating the flow of the injected carriers in a diode structure towards or away from the region with different recombination velocities, changes the current-voltage characteristic of the diode. Silicon-on-sapphire is an appropriate structure for the diode. This structure has two interfaces: Si-SiO$_2$, with low recombination velocities; and Si-Al$_2$O$_3$, with relatively high recombination velocities. The in-plane magnetic field deflects the carriers to one of the interfacing surfaces, causing a change in the current-to-voltage ratio.

Experimental magnetotransistors are fabricated in a variety of forms [544]. Three basic effects are present in the magnetotransistor [568]: the Hall effect, carrier deflection, and magnetoconcentration. These effects interact with the transistor's regular bipolar action. The sensitivity to the magnetic field depends on the device's geometry, process of fabrication, and operating conditions. An example of such devices can be a cell having the structure of the standard vertical transistor in bipolar IC's, except for two collector contacts that are located symmetrically with respect to the emitter. The output signal is read as a difference between two collector currents. A two-dimensional magnetic field detection can be simply realized by adding one more collector pair symmetrically with respect to the emitter and perpendicularly to the basic collector pair. When the structure is provided with more contacts, the three-dimensional sensor can be easily realized.

A structure based on the bipolar lateral transistor can also be used as a magnetotransistor. The main difference in the operation of lateral and vertical transistors is in the direction of the detectable magnetic field defined by the direction of the main carrier path from the emitter to the collectors. The lateral magnetotransistor detects the field perpendicular to the chip surface.

In a carrier domain magnetometer structure [544, 558], there exists a region with a nonequilibrium carrier density in which the electrons and holes form a plasma. The magnetic field to be measured develops forces on the charge carriers moving in this domain, changing the operation currents. A high sensitivity in this device is obtained by using the spatial positive feedback for domain enhancement. The structure of this device is somewhat similar to that of a vertical transistor; however, the mode of operation is quite different.

The high sensitivity of transistor-type magnetic sensors allows measurement of weak magnetic fields. However, the main limitation is electrical noise [569], and the limit in the detection of the magnetic field is defined [570] by the minimum field B_{lim} that can be observed when working in a narrow frequency bandwidth, or by the noise equivalent magnetic field B_{eq} in a large frequency band. For different designs $0.3 \mu T < B_{lim} < 20 \mu T$, and $20 \mu T < B_{eq} < 1.3 mT$.

By their nature, all solid-state magnetic transduction elements match conveniently to the structure of an *integrated sensor* [558, 571]. Typical examples of the application of such sensors are contactless keyboards, brushless motors, approximate switches, and so on. [535, 572–74]. For example, the IC keyboard element normally includes the Hall cell, amplifier, the Schmitt trigger circuit, and two emitter followers. A magnetic IC-detector combines a lateral magnetotransistor and differential amplifier only.

Highly functional magnetic sensors [559] are capable of performing some intelligent operations. An example can be an IC sensor that combines in one chip one lateral and two vertical Hall cells, along with a bipolar translinear circuit performing an algebraic operation. The Hall cells develop electrical signals at their outputs that are equivalents of three orthogonal components of the induction: B_x, B_y, and B_z. The magnitude of the induction B [T] is calculated in the circuit as

$$B = \sqrt{B_x^2 + B_y^2 + B_z^2}. \tag{11.26}$$

One translinear circuit consists only of five transistors; it allows the operation $\sqrt{a^2 + b^2}$. The required operation $\sqrt{a^2 + b^2 + c^2}$ can be carried out by means of a cascade combination of three circuits, and is performed as

$$B = \sqrt{\left[\sqrt{(\sqrt{0^2 + B_x^2})^2 + B_y^2}\right]^2 + B_z^2} \tag{11.27}$$

where each combination of terms denoted by T_1, T_2, and T_3 corresponds to one translinear circuit.

The reading of B allows one to perform an omni-directional magnetic measurement in real time. A similar device [575], comprised of two vertical Hall elements and an analog circuit calculating the characteristic angle $\theta = \arctan(B_x/B_y)$, makes it possible to measure the direction of the magnetic field.

There is a pronounced trend toward application of a *magneto-operational amplifier* [576] and making this amplifier a standard component similar to the conventional operational amplifier, but intended for magnetic measurements. The experimental models of such an amplifier [571] contain a suppressed side-wall injection magnetoresistor (SSIMT) [577] as a detector, and typical amplifying stages (emitter follower, voltage amplifier, power amplifier, temperature compensation circuit, etc.). It is assumed that this OP amp can offer many functions for the manipulation of magnetic signals; for example, the OP amp can be a linear magnetic sensor with conveniently controlled sensitivity, a magnetic filter, a magnetic integrator, and so on. By simple external additions of passive and active components, the characteristics of the amplifier can be predetermined.

SOLID-STATE CHEMICAL SENSORS

Intensive research and development activity on solid-state chemical sensors is stimulated by a high demand in contemporary tools for fast, precise, and reliable analysis on the one hand, and by still-existing imperfections in the sensors' performances on the other hand. A great increase in the number of publications on the subject in periodical literature and recent issues of fundamental monographs [578, 579] provide valuable material for understanding the sensors' principles and the specific character of their designs. A number of issues related to regular electrochemical elements (see Chapter 2, Electrode Elements) can be instructive for understanding the performances of solid-state elements. Considering the specificity of the subject, high diversity of material, and limited room in this book, we will briefly review the basic principles of the sensors. The mentioned monographs are recommended for deeper studies.

The operation of the majority of solid-state chemical sensors (SSCS) is based on their primary electrical response to the chemical environment. There are a number of devices whose primary conversion is not electrical, but acoustical, thermal, and optical. However, almost inevitably the consecutive conversion in the sensor is electrical. The application of solid-state structures for sensing the chemical values is attractive because of the small size, simplicity of operation, and potentially low cost (batch, planar processes) of solid-state sensors. Unfortunately, many existing SSCS's suffer from a lack of stability, reproducibility, and selectivity; and from insufficient sensitivity to certain measurands [578]. Current advances in design alleviate these problems. Special features and additions, such as filters, membranes, catalysts, and arrays of sensors, make the measurements more accurate. In some cases, using disposable sensors can be the solution to the problem. The term *solid-state* used in reference to the SSCS's does not always explicitly reflect the essence of the structure. For example, a sensor can contain not only solid semiconducting fractions, but also organic liquid membranes with a gelling agent.

Silicon-based Sensors

The most typical silicon-based chemical sensor is the FET in which the charge at the gate region is developed by the carriers brought from the substance directly contacting the gate dielectric. In this arrangement, the FET's current will respond to the ions and molecules in the solution or gas. The selective sensitivity of the FET can be obtained if the transistor is provided with a membrane that selectively concentrates the ions at the sensitive area. Instead of a broad term like *ChemFET* describing a chemically sensitive FET, the sensor has more specific names: ISFET, ion-sensitive field-effect transistor containing ion-selective membranes; EnFET, enzyme-sensitive FET containing enzymes (an enzyme is any of catalytic protein groups that are produced by living cells and that mediate and promote the chemical processes of life without themselves being altered or destroyed); ImmFET and immuno-FET, containing antibodies or antigens; BioFET, containing whole tissue layers; and BLMFET, containing bilayer lipid membranes (BLM). These sensors belong to a new class of electronic devices that are being developed at the present time. It is expected that the main attributes of the sensors will be small size, rugged construction, availability of batch fabrication, low cost, and the possibility of having a disposable version. It is feasible to use these sensors in measurement system arrays of sensors for a simultaneous measurement of several parameters, the duplication of a failed sensor, and averaging the reading. On-chip integration can also be beneficial. Such functions as filtering, multiplexing, and high-input to low-output impedance conversion can be desirable in many applications. Like any microelectronic sensor, the chemical sensor can include the elements of "smart" functions, such as read-only memory (ROM) and random-access memory (RAM) for the correction and readjustment of output-input characteristics. It should be mentioned that similar functions can be obtained on hybrid constructions, but integration in one chip is attractive for economical reasons. Some technological problems must be solved before the FET-based sensors find their mass application. Among these problems are incompatibility between silicon and chemically sensitive gate materials, the light sensitivity of the FET gate, electrolyte leakages affecting the electrical operation of the amplifier, and other problems. When sensors are used in a living body, their biocompatibility is also a problem. As a result of the advances in sensor technology, some promising alternatives to the conventional ChemFET have been developed, such as the ion-controlled diode (ICD), the extended-gate field-effect transistor (EGFET), electrostatically protected field-effect transistors, and others [578].

One of the fundamental problems associated with the design of FET-based devices is the inability of researchers to develop a structure with a stable reference potential on either side of the chemically-sensitive membrane placed directly on top of the insulating gate. This problem causes drifts and poor reproducibility of sensor characteristics. A similar problem exists in the coated-wire electrodes (CWE's). It is difficult to arrange an external reference electrode for grounding at microlevel. The fabrication of the second kind of electrode, such as Ag/AgCl (see Chapter 2, Electrode Elements), on a microscale is a real problem. It is important to find a material for ImmFET devices that can provide a perfectly polarizable interface. The deposition of thin lipid membranes based on Langmir-Blodgett technology may give a positive result. This technology permits layer-by-layer application of organic film onto a solid substrate. The molecule layers are first formed on a water surface and then transferred onto a substrate.

At present, FET chemical sensors cannot be regarded as long-time operating components. The application of sensors as disposable devices for biomedical measurements is more realistic.

Potentiometric-based microsensors for measurement of the potential difference between the sensing and reference electrodes correlated with the concentration of chemicals have been researched and developed for a long time. It was found recently that amperometric-based devices in which the response curve is current versus concentration offer some advantages over the potentiometric ones.

Semiconducting Metal Oxide Sensors

This type of sensors is based on pressed powder. It is used for gas sensing. In spite of some problems with reproducibility, stability, sensitivity, and selectivity, these sensors are produced commercially because they are acceptable for many measurements and are relatively inexpensive. SnO_2 is a widely exploited semiconductor for gas sensors. This material is chemically quite inert, which is important for the stability and reproducibility of sensor characteristics. The use of a compressed powder has substantially improved the sensor's performance. The progress of investigation of this material over recent years is quite remarkable. The contemporary trend is toward wider application of thin films rather than pressed powder; for example, H_2S sensors fabricated by sputtering are commercially available. The electrical resistance of the sensor changes in response to the reaction between the semiconductor and gases in the atmosphere. The organic vapor contacting the semiconductor lowers the cation/oxygen ratio in the oxide, causing the change in the electrical conductivity. The reaction is stoichiometrical by nature (there is a change in the numerical relationship of elements and compounds as reactants and products in the reaction); and the change in the stoichiometry is accompanied by a significant change in the conductivity. Another explanation for the change in conductivity is related to the mechanism of gas absorption. As oxygen comes in contact with the surface of a semiconductor made of n-material, it extracts and absorbs electrons from the surface, reducing the conductivity. Organic vapors react with oxygen, forming H_2O- and CO_2-liberating electrons that return in the material and increase its conductivity. In many instances, it is difficult to tell which process prevails—stoichiometrical reaction or absorption. If the change in the resistance is fast, it is an indication that the surface rather the bulk process is taking place, because the diffusion rate of the oxygen vacancies is much lower than the rate of interactions at the surface. An intermediate process between those just described is the ion exchange. A sulfide ion can replace an oxide ion at the semiconductor surface. Sulfides are more conductive than oxides; therefore, the replacement gives a reduction in the surface resistance. As mentioned above, one of the characteristics that are being improved is reproducibility. Among the factors affecting this characteristic are intergranular resistance that is inclined to variation, irreversible reactions of the material with the impurities in gases leading to stoichiometry changes, and other factors. Intensive investigations on the subject will very likely lead to solving these problems in the near future.

Catalysts

Catalytic reactions are typical for the operation of oxide-based solid-state sensors. Semiconductors can be catalysts or they can be combined with catalysts for the improvement of sensor performance (sensitivity, selectivity, response time, etc.). Catalysts intensify the oxidation process (sensors for CO, hydrogen, organic reducing agents, etc.), enhancing the change in resistance, as described above. The oxidation process can be selective if the chosen catalyst is selectively active with respect to the particular substance.

Solid Electrolyte Sensors

In solid electrolytes, electroconductivity is defined by the dominating ions rather than electrons. These electrolytes are used in gas and ion sensors as nonporous membranes separating two chemicals of different concentration. One of the substances is used for reference. The potential measured across the membrane determines the concentration of the substance in question. Examples of practical applications of the solid electrolytes can be yttria (Y_2O_3); stabilized zirconia (ZrO_2), which is an O^{2-} conductor at an elevated temperature (see Chapter 2, Oxygen-sensitive Elements); and LaF_3, which is an F^- conductor. These electrolytes are used for the determination of oxygen and fluorine content, respectively. The role of solid electrolytes in sensor application has been expanding quite rapidly. The following are several examples of recently developed electrolytes [578]: Ag^+ conductors, such as α-AgI, Ag_3SI, $Ag_6I_4WO_4$, and $RbAgI_4$; good N^+ conductors, such as sodium β-alumina ($NaAl_{11}O_{17}$) and NASICON ($Na_3Zr_2PSi_2O_{12}$); and new Li^+ conductors, such as $Li_{14}Zr(GeO_4)_4$ and Li_3N. Another membrane material of interest is solid polymer electrolyte (SPE). After the penetration of water through the solid it becomes an ionic conductor. One of the popular SPE's is Nafion (a trademark of Dupont). This material is a perfluorinated hydrophonic ionomer with an ionic cluster. It is widely used in a variety of room-temperature electrochemical sensors.

A high protonic conductivity is present in solid-state proton conductors, such as hydrogen uranyl phosphate (HUP), zirconium phosphate, and dodecamolybdo-phosphoric acid. To remain conductive, these materials need a source of water (similarly to the SPE's). One of the recently developed room-temperature proton conductors, polyvinyl alcohol/H_3PO_4, is reported not to need water to be conductive.

The possibility of fabricating microionic structures has had a strong impact on the development of solid electrolyte sensors. Besides the economical advantages, the reduction of the sizes, especially of the thickness of sensitive layers, leads to the amelioration of the functional characteristics: reducing the electrolyte resistance, response time, level of the operational temperature, and so on. However, the solid ionic conductor can be found for a limited number of atomic species. The circle of the species can be increased due to the possibility of having a difference in the ionic carriers in the electrolyte and in the species under investigation. For example, the Cl^- conductor $SrCl_2$ and the F^- conductor $PbSnF_4$ can be used for oxygen sensing in air [580]. There are indications [581] that the partial pressure of gaseous species can also be measured by solid-state electrochemical techniques. Unfortunately, many solid electrolytes are unstable. The materials that are used in commercially produced instruments are ZrO_2, LaF_3, and Nafion.

Membranes

In solid-state sensors, a membrane is used as a sensing element, a filter, or an intermediate agent making a sensor biocompatible, or as an inert matrix for active sensor components (e.g., enzymes and antibodies) [578]. In addition to the solid electrolyte type, many more membranes such as liquid- and polymer-based ion exchange, neutral carrier and charged carrier membranes, heterogeneous membranes, and others are available. The polymer-based membranes play an important role in biosensors. Along with the search for a new membrane composition for low-temperature application, intensive work is being performed on the investigation of membranes for better anion detection, elimination of solvents within polymer-based materials, and simplification (ultimately, elimination) of the internal liquid reference system.

OTHER SENSORS

Short descriptions of microsensors that have not been discussed in the preceding sections are introduced in the following text. Even a condensed presentation of the material can only begin to cover the rapidly growing body of information on the subject. This review, then, covers the types of sensors that are being developed most intensively.

A *capacitive sensor coupled to an* FET *gate* [582] works as a displacement-measuring device. The FET gate electrodes, which are sensitive to external fields, can be positioned at emitters in order to sense rotary or translational displacements. The onboard circuitry produces signals that can be transmitted via a two-wire bus. The standard VLSI processes are applicable for the fabrication of the sensor.

A *semiconductor strain-gage mass-measuring structure* [583, 584] makes possible the measurement of small masses; for example, in the range of 0.1 to 50mg, with an accuracy of about ± 0.1mg. The structure contains a rigid central part and suspension arms carrying piezoresistors. When a load is applied through a force concentrator, such as a stylus, to the central part, the produced and measured strain in the arm precisely reflects the magnitude of the load.

A relatively thin ribbon (5 × 25μm) made of $Fe_{70}Ni_8Si_{10}B_{12}$, $Ni_{39}Fe_{39}Mo_4Si_6B_{12}$, or $Co_{75}Si_{15}B_{10}$ being strained can serve as a spring and transduction element of a *magnetoelastic force sensor* [585]. Similarly to the traditional transducer, this sensor contains coils in which an induced electromotive force is modulated by a strained element; in this case, the ribbon.

A film made of a highly magnetostrictive amorphous alloy ($Fe_{67}Co_{18}B_{14}Si_1$), attached to the surface of a flywheel that interacts with a magnetic-field-inducing coil, forms a torque sensor operating in the range between 20 and 120N · m [586]. The output of the sensor is quite high, reaching 2V.

A *floating-membrane thermal vacuum sensor* [587] consists of a suspended silicon plate (the floating membrane) surrounded by a wafer-thick rim. The beams supporting the plate contain thermopiles for measuring the temperature elevation of the floating membrane with respect to the rim and the ambient. The heating resistors located at the junction between the beams and the membrane elevate the plate's temperature. The pressure is measured by detecting the heat loss of the hot plate.

Several constructions of *subminiature microphones* designed recently show the possibility of realizing highly miniaturized devices at a low cost with IC-compatible technology.

A micromachined microphone with an optical interface readout [588] contains a square silicon diaphragm (1 000 × 1 000 × 5μm^3) whose displacement is measured interferometrically. The sensor's optical system has a stationary optical flat as a reference plane. Laser beams reflected from the flat and diaphragm are combined to create interference fringes whose shift, read by a photodetector, is proportional to the diaphragm's deflection.

A diaphragm (2 000 × 2 000 × 5μm^3) etched in lightly doped silicon is the principle component of a miniature capacitive microphone [589] that has a traditional construction and can operate with an external bias voltage. The microphone can be highly miniaturized due to the application of IC-technology. The significant points of this design are low stray capacitance (< 0.5pF), and a large number of ventilation holes in the back plate, allowing a small air gap (4μm).

A similar device is one in which a teflon layer used as an electret material for the bias is combined with a microphone amplifier that is an ordinary MOS transistor. [590].

In the past decades, *polyvinylidene fluoride* (PVDF) *films* possessing piezo-electric and pyroelectric properties have been extensively studied. They are already utilized in various types of sensors, such as conical-cavity laser detectors, light pressure transducers, multifunctional alarm sensors [591], and finger pulse and breathing wave sensors [592]. By the application of PVDF for the measurement of heat flow from a heat source to an unknown object, the thermal properties of the object can be determined; thus, the material of the object is identified. The sensor for this purpose [593] is composed of three layers: a thermally conductive rubber, a PVDF film, and a film of an active heat source. When the rubber layer contacts a surface, the voltage drop across the PVDF film gives a reading of the heat flow.

A flowing fluid sets up a gradient of velocity near the wall of a manifold carrying the fluid. The resulting wall shear stress is one of the key parameters characterizing the structure of the flow above the wall. A floating-element *shear sensor* [594, 595] is a suspended polysilicon plate supported by four tethers. The plate experiences lateral deflection in response to the friction forces produced by the flow. A deflection proportional to the shear stress is sensed by a capacitive transduction structure. The operation of another micromachined, shear-stress-sensitive sensor is based on the relation between the flow rate and the total heat loss measured by a sensor chip at the wall [596]. The chip measures 4×3mm^2, is realized in a standard bipolar IC process, and contains a transistor for measuring the chip temperature, resistors for heating the chip, and a thermopile Seebeck-effect sensor for the measurement of the temperature difference across the chip.

As we know, classic thermal anemometers, such as the hot-wire type, and most *semiconductor flow sensors* rely on the dependence of the total amount of heat transferred from the sensor to the fluid on flow velocity. Semiconductor flow sensors with high sensitivity and fast response are obtained by the fabrication of small, micromachined thermal structures [597–607]. In spite of their diversity, the micromachined thermal flow-rate sensitive devices include two common components: heaters and temperature-sensitive elements. The heaters can be metal films, heavily doped diffused layers in silicon, or transistors that radiate heat at their junctions. Metal films, diffused layers, and diodes are typical detectors of temperature in these sensors. As in other miniature silicon sensors, there is a trend in some of the constructions to integrate on one chip temperature detectors, heaters, circuits treating signals from the detectors, and circuits providing the temperature stabilization of the heated substrate. In some of the designs (e.g., [601]) circuits modify the output signal and provide a pulse-modulated output. A self-test circuit also can be included into the on-chip electronics.

The recent discovery of high-temperature *superconducting* materials has stimulated their employment in *sensors* [608, 609]. The first experiments with magnetic, X-ray, and infrared radiation superconductor sensors indicate the high potential of these devices in future instruments.

Some *humidity sensors* can be regarded as microsensors; for example, the humidity-sensitive MOS capacitor [610]. However, most of these sensors are miniature devices containing a substrate with moisture-absorbent substances placed between two electrodes. In order to measure humidity, the electrodes are connected to an electrical circuit and one of three readings is usually taken: resistance [611–13], capacitance [614–16], or impedance [617–19]. In [610], one can find a systematic description of different substances used for sensing humidity, along with the presentation of various mechanisms of changes in electroconductivity as a function of humidity. The basic principles of these sensors are as follows:

1. A lithium chloride solution immersed in a porous binder changes its ionic conductivity depending on the relative humidity of the surrounding atmospheric air.

2. A group of organic polymers having constituent ionic monomers, such as sodium stylene sulphonate, exhibit ionic conductivity and are called polymer electrolytes. Their ionic conductivity increases with an increase in water absorption, due to intensification of the ionic mobility and/or charge carrier concentration.

3. The relative permittivities of polymers such as polyimides and cellulose acetates are between 3 and 6; water's relative permittivity is 78.54 at 25°C. When polymers absorb water, their capacitance changes, and this property can be utilized for sensitivity to humidity.

4. In terms of mechanical strength, temperature capability, and resistance to chemicals, ceramic materials suit chemical sensors quite well. The impedance of the material correlates with humidity and can be used for humidity detection.

Resuming the review of sensors, we will refer readers to periodicals containing reports on the new design of such devices as detectors of nuclear radiation [620–29], microelectrodes [630–32], micromachined silicon electron tunneling sensors [633–35], micro-telemetering, multi-sensor capsule systems [636], power sensors [637], and smart sensors [638]. Detailed information on the application of microsensors for the measurement of mechanical properties of materials can be found in [639, 640]. In [641, 642], a description of liquid-to-solid-phase-transition detectors is given. Issues related to the transmission of power to microsensors and wireless communication with the sensors are presented in [643–46].

Much attention in current research programs is focused on the study of different aspects of the elastic state of micromachined silicon spring elements: plastic deformation of highly doped silicon [647], residual stress and mechanical properties of boron-doped p^+-silicon films [648], modeling of thermal and mechanical stresses in silicon microstructures [649], creep of a sensor's elastic elements (metals versus nonmetals) [650], fracture toughness characterization of brittle thin film [651], liberation of integrated sensors from encapsulation stress [652], buckling of silicon diaphragms [653], stress compensation in ultrasensitive microstructures [654], and mechanical properties, including strain gradients, in thin films [655, 656].

Progress in the design of microstructures is supported by modeling of the structures' architecture, functions, and fabrication processes [657–68] with wide application of computer simulation.

BASIC FACTORS OF MICROSENSOR DESIGN

The design of a microsensor is very specific for a particular type. However, there are several common principles [669, 670] that can be summarized as follows:

1. The driving forces in the development and design of microsensors are lithography and planar technology, batch fabrication (low unit cost), microelectronic infrastructure (capital equipment), and micromachining technologies (mechanical structures).

Improvements in IC fabrication are usually translated directly into microsensors' yield improvement.

Bulk and surface micromachining along with sculptural fabrication techniques make it possible to create a variety of mechanical devices with cavities, diaphragms, suspended structures, and freely moving parts.

An increase in the wafer diameter and thickness does not always benefit the fabrication of microsensors. For example, when a silicon diaphragm is created by a through-wafer etching, the thicker wafer requires greater area to form the same-sized diaphragm.

Usually, in IC design, smaller sizes of device are associated with advantages (faster response, higher integration, fewer interconnections), but these qualities are not always true for microsensors. Some mechanical responses get worse with a decrease in physical size (e.g., bending sensing elements get stiffer, mechanical precision is more critical, etc.).

During the design, the mechanical and microelectronic problems of design are solved simultaneously.

2. The typical combination of the design steps is partitioning, specification of interfaces, design specifications, and detailed design.

When a microsensor measurement system is designed, a decision must be made regarding how much of the system is to be merged into the batch-fabricated microsensor part (the "chip"), and how much is to be "off-chip." The general rule is that the "on-chip" part of the system has to be minimized. The microsensor should contain only the parts that are essential for its major functions.

When a microsensor is fabricated by combination with elements of a standard IC chip, the compatibility between IC steps and microsensor fabrication steps is always a problem.

Special processes in the fabrication of microsensors (e.g., deposition of piezoelectric zinc oxide or gold) impose a specific order to the process steps.

System *partitioning* for microsensors has two aspects: (1) the partitioning of the system into a microfabricated device (the microsensor) and the rest of the system, which can be a hybrid or conventional electronic system; and (2) the partitioning of the calibration of the system among its components.

The separation of the analog electronics from the sensing chip allows the operation of the sensor at a temperature much higher than that possible if the electronics were combined with the sensor.

So-called smart sensors incorporate sophisticated electronics into the microfabricated device. The improved complexity is justified only if it is accompanied by improved performance and a reasonable cost of the device. As mentioned above, it is expedient to minimize the complexity, size, and functionality of the microsensor. In this case, attention is focused on the required microsensor functions and the device can thus be produced using a simpler process. However, the integration of the transduction element with the elements of a signal conditioner can be beneficial. For example, the addition of field-effect transistors to a high-impedance electrode for the measurement of properties of liquids expands the desirable impedance of the electrodes by as much as several orders of magnitude.

Before the prototype of the microsensor is built, the *interfaces* of the system incorporating the sensor must be defined. They include an electronic interface to the measurement system; a physical interface with the environment (with attendant chemical, thermal, and pressure/stress characteristics); a cabling or interconnect requirement with the measurement system; and material requirements for chemical stability, thermal stability, mechanical properties, and, in some cases, biocompatibility. Along with these definitions, it is useful to make some predictions of possible parasitic effects (mechanical, electrical, thermal, chemical, etc.).

Design specifications for a microsensor are formulated as the system and interfacing become clear. The ideas of construction and packaging become more explicit, including the following aspects of design: the physical concept of the device, circuit requirements, and expected nominal system performance; layout-related issues such as chip size; special structures such as a diaphragm, cantilever, and so on; locations of pinouts; principles of passivation; packaging technology and materials; assembly; provisions for connection of the device to the outside system; methods of tests and calibration; and criteria of acceptance.

In this phase of the design, special attention should be paid to packaging.

Microsensors are intended to measure physical variables. They can be calibrated and tested if the devices are packaged (at least partially). Many microelectronic devices are sensitive to temperature, light, magnetic field, and strain. The package must provide environmental access for the measurand along with satisfactory mechanical support and environmental protection against adverse influences from the media [671].

It is necessary to design the microsensor itself and its package at the same time. Packaging and the need to optimize process technology impose constraints on partitioning and design decisions. Packages of microsensors can easily cost as much as an order of magnitude more than the microsensor structure. Several constraints should be considered when materials and processes are selected for a particular design. The microsensor geometry and structure mainly determine the materials and process requirements. For example, the elements of structures with sizes smaller than 1μm require advanced pattering techniques (X-ray or electron-beam lithography); for insulating films with thickness greater than 1μm, sputtering must be used. Deposited films of about 100Å thick are not integral and need a special treatment to have continuity. Interaction among process steps must be carefully considered. Treatment at temperatures higher than approximately 900°C sometimes leads to redistribution in dopants. Chemical changes and shifts in the composition and morphology can occur at temperatures of about 600°C. Temperatures of about 300°C, or ultraviolet light and soft X-rays radiated during plasma etching or metallization, can affect the interfacial properties of different materials. The semiconductor part of a microsensor passing different treatments can be examined by using the semiconductor process-modeling program SUPREM. The combination of structural and electronic elements in one construction (e.g., diaphragms, beams, and p-n junctions) can make optimization of both elements difficult.

A standard integrated-circuit established process is always desirable for microsensor fabrication. However, if mechanical and structural elements, novel materials, nonstandard processes, or unusual packaging is used, overall process sequences with partitioning of the specific steps must be designed.

Detailed design is a final phase of the design work when preparing a process flow, a set of masks along with drawings for various parts, and tools. Specific operations for assembly, intermediate acceptance tests, calibration, trim, and final test are designed and described in detail.

Compensation for device-to-device variation is accomplished by (1) adjusting or trimming each microsensor, or (2) treating the overall system calibration. A significant cost advantage in the correction of characteristics can be achieved if a software trim is available.

3. CAD *tools* support the design process for microsensors. These tools are useful for mask layout, process and structure modeling (geometry and doping, mechanical properties, and residual stress), and finite element modeling (FEM) analysis (stress distribution, risk of fracture, resonance frequencies, response to loads, and thermomechanical effects).

One important factor should be taken into account when predictions of mechanical properties by CAD are developed. Many mechanical properties of the microfabricated materials such as residual stress, elastic moduli, yield and fracture point, density, and thermal expansion coefficient are process and specific-equipment dependent. This means that the process simulator for microsensors can be effective if it is possible not only to model geometric quantities such as layer thickness and dopant distribution, but also to obtain from a database and consider process-dependent properties of the constituent materials.

Structure simulation, such as the creation of a three-dimensional solid model from a description of mask layout and process sequence, is a formidable task. However, progress in this business is remarkable and can be illustrated by one of the first programs called OYSTER [672, 673].

appendix 1:

Transducer Schemes

In this appendix, we will introduce schemes of transducers that are primarily composed of the elements presented in the foregoing chapters. However, some schemes containing elements that have not been considered before are also described. The order of presentation of the material corresponds to that for the measurands in Chapter 1, Measurands. Each scheme is provided with a list of its essential components, along with a brief description of the transduction mechanism. A given collection of transducers includes traditional and novel devices. Because of the huge number of transducer designs, we must limit the amount of material presented in this appendix.

Displacement, Position, Motion, Length, and Thickness Transducers	
1 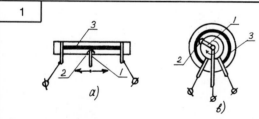 Translational-displacement (a) and angular-displacement (b) resistive transducers. 1 = shaft, 2 = wiper, 3 = resistive layer.	A *resistive transducer* contains a translational- or angular-displacement sensing shaft. This shaft drives a wiper in the transduction element that slides along the resistive layer. The resistance measured between the wiper and one of the ends of the layer is a function of the position, motion, or displacement.
2 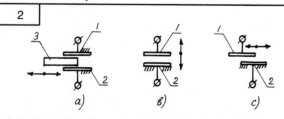 Capacitive displacement transducers with variation in dielectric constant (a), gap between plates (b), and area of capacitor's plates (c). 1 and 2 = capacitor's plates, 3 = dielectric.	The sensing shaft in a *capacitive transducer* changes the position of the dielectric between the capacitor's plates in the transduction element, or it changes the distance and area between the plates. A change of these three parameters leads to a change in capacitance, which is a measure of the quantity to be measured.

3 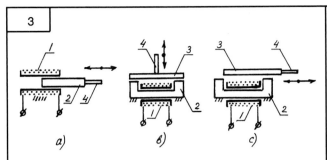 Inductive displacement transducer with variation in core's position (a), gap (b), or area (c) in magnetic system. 1 = coil, 2 = core, 3 = armature, 4 = sensing shaft.	A sensing shaft in an *inductive transducer* is attached to the core or to the armature. Due to the change in position of these parts, the reluctance of the magnetic path is changed. This causes a change in the coil's self-inductance.
4 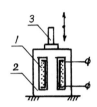 Magnetoelastic or magnetostrictive transducers. 1 = coil, 2 = magnetoelastic core, 3 = sensing shaft.	In a *magnetoelastic or magnetostrictive transducer,* the change in the position of the sensing shaft creates stress in the stress-sensitive core. The permeability of the core material alters with stress, effecting the inductance of the winding wound around the core. The inductance is a function of the shaft's position.
5 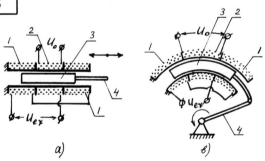 Transformer transducer. a = LVDT, b = RVDT. U_{ex} = excitation voltage, U_o = output voltage, 1 = excitation coil, 2 = output coil, 3 = moving core or armature, 4 = sensing shaft.	Similar to the inductive transducer, a *transformer transducer* contains coils. The change in the magnetic circuit of the transduction element due to the motion of the sensing shaft provides a change in the mutual inductances between the coils. As a result, a voltage proportional to the change in the circuit is induced in the output coil. Two types of transducers are popular in this measuring technique: LVDT and RVDT. LVDT denotes a linear variable differential transformer, and RVDT denotes a rotary variable differential transformer.
6 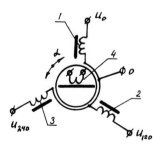 Synchro. α = angular displacement, 1, 2, and 3 = stator, 120°-disposed coils, 4 = rotor single coil.	A *synchro* or selsin is a rotating-transformer type of transducer. Its stator has three 120°-angle disposed coils with voltages induced from a single rotor coil. The ratios of the voltages in the stator are proportional to the angular displacement of the rotor.

7	
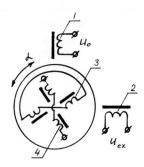 Resolver. α = angular displacement, 1 and 2 = stator, 90°-disposed coils, 3 and 4 = rotor, 90°-disposed coils.	A *resolver* is similar to the synchro, but it has two two-phase windings at the stator and at the rotor spaced 90° apart. The output voltage is proportional to the sine or cosine function of the rotor's angle.
8	
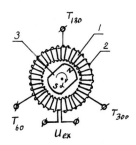 Magnesyn. α = angular displacement, T_{60}, T_{180}, T_{300}, = taps, U_{ex} = excitation voltage, 1 = core, 2 = coil, 3 = permanent magnet.	A *magnesyn* is also an inductive transducer. Its stator is a toroidal coil wound on a high-permeability material core. The coil is tapped at three points (60°, 180°, and 300° from the beginning of the coil). The rotor is a permanent magnet. It rotates inside the stator and saturates different sections of the stator core. The interaction of an ac magnetic field developed by the coil (it is fed by an ac current) and a permanent field from the magnet induces in the coil a voltage of twice the frequency of the feeding current. This voltage picked up from the taps at 60° and 300° is proportional to the cosine of the magnet's axis angle of deviation from the axis of the 180° tap.
9	
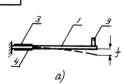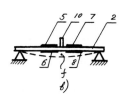 Strain-gage displacement transducers with bending cantilever (a) and beam (b). f = deflection, 1 = cantilever, 2 = bending beam, 3 through 8 = strain gages, 9 and 10 = sensing shafts.	In a *strain-gage displacement transducer*, metal foil- or semiconductor gages are attached to a bending beam. The tip of the beam or a sensing plunger pushing the bar acts as a sensing shaft. The displacement of the sensing elements causes bending of the bar and tension or contraction of the strain-gages. These strains are proportional to the measuring deflection. The strain-gages form two or four arms of an electrical bridge.
10	
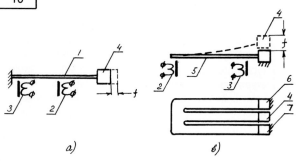 Vibrating-element displacement transducers with wire (a) and E-shaped cantilever (b). f = displacement, 1 = vibrating wire, 2 = excitation coil, 3 = pickup coil, 4 = sensing element, 5 = E-shaped, flexible cantilever, 6 and 7 = stationary parts.	The *vibrating-element transducer* relies on the change of the natural mechanical frequency of vibration of a spring element that experiences stress proportional to the displacement of the sensing element. A prestressed wire is usually used as a spring element. Excitation and pickup coils are also parts of the transduction element. They are connected to the feedback loop with an amplifier maintaining oscillations in the system. The sensing shaft, being deflected from the original position, develops longitudinal stress in the wire, and this stress affects the frequency. Another version of the vibratory transducer is a device with a vibrating element whose stiffness is controlled by a change in the oscillating body shape. In this case, the motion of the sensing shaft does not directly create the stresses governing the frequency. Instead, the change in the shape modifies the moment of inertia of the vibrator's most sensitive area, and this change gives the shift of frequency.

11	The *electrooptical displacement transducer* is based on the measurement of the intensity of the reflected light from the target. It has a light source sending light to the moving target and a light sensor receiving the light. The output signal from the sensor decreases exponentially with the increase of the distance to the measured object. Typically, infrared light-emitting diodes (LED's) and photosensitive diodes are used in this transducer.

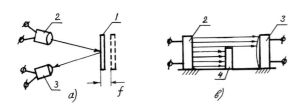

Electrooptical displacement transducers. a = transducer with moving target, b = dimension sensor; f = displacement, 1 = moving target with light reflector, 2 = light source, 3 = light sensor, 4 = object to be gaged.

Electrooptical dimensional gaging is provided by placing the object to be gaged between the light source and the light sensor. The amount of occultation is a measure of the specific dimension to be measured. The sensor output is inversely proportional to the amount of occultation. A multiline array of light-sensitive elements and a light-beam scanning technique determines and qualifies the shape of the measured object by processing data from the elements.

12	A *laser triangulation measuring system* includes a laser light source and two light sensors. The triangulation method of sensing the displacement of the surface consists of viewing a spot of laser light by two light sensors at the same angle, but from opposite directions. Using the laser allows the system to obtain a well-collimated and coherent light beam. There exist different variations of this method, but the general approach remains the same: the displacement is calculated from known reference parameters and from parameters determined by the geometrical relations.

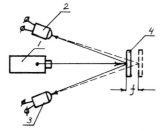

Laser triangulation measuring system. f = displacement, 1 = laser light source, 2 and 3 = light sensors, 4 = surface.

13	In a *laser interferometer* the laser beam is split into two beams by a beam splitter. One of the beams is directed to a mirror located at a certain distance to provide a reference in measurements. The other beam reaches a mirror or reflector fixed on the moving object. This beam, being reflected, recombines and optically interferes with the reference beam (or beams). The superimposed light sensed by a photodetector has increased (constructive interference) and decreased (destructive interference) illumination phases, which are correlated with the measuring displacement of the object.

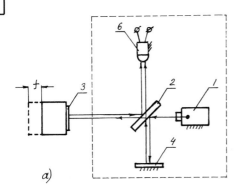

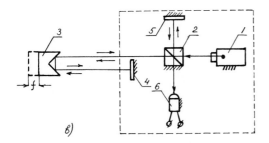

Laser interferometer with one (a) and two (b) mirrors for reference. f = displacement, 1 = laser light source, 2 = beam splitter, 3 = reflector on moving object, 4 and 5 = mirrors for reference, 6 = photodetector.

14 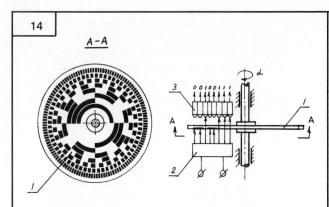 Optical angular encoder. α = angular displacement, 1 = code disk, 2 = light source, 3 = photodetector system.	An *optical linear or angular encoder* is made of a transparent disk or a strip with digital code marks on it. The linear position of the strip or angular position of the disk is defined by a binary system of notation as a train of digital pulses. Two symbols in the binary system, "0" and "1," are provided by the alternation of the transparent and opaque areas that are sensed by a light source-photodetector system.
15 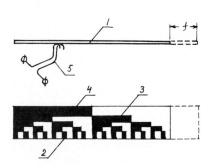 Brush-type encoder. f = displacement, 1 = strip, 2, 3, and 4 contacting segments, 5 = pick-off brushes.	A *brush-type encoder* contains a disk or strip with the digital code marks made of contacting and noncontacting segments. A pick-off brush in contact with the segments closes or opens an electrical circuit, providing a digital signal in response to the displacement of moving parts.
16 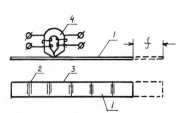 Magnetic encoder. f = displacement, 1 = magnetic strip, 2 = magnetized area, 3 = nonmagnetized area, 4 = magnetic head.	The principle of digitizing in a *magnetic encoder* is similar to that used in optical and in contact devices. The carriers of the digital code marks are ferromagnetic disks or strips with a pattern of magnetized and nonmagnetized areas. A magnetic head responding to the magnetization is in close proximity to the moving parts and produces "0" or "1" pulses when magnetized or nonmagnetized areas pass the head. A contemporary technique allows the inscription of the magnetic pattern very precisely, providing a high resolution for the transducer.
17 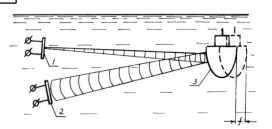 Ultrasonic transducer. f = displacement, 1 and 2 = piezoelectric elements emitting and receiving acoustic pulses, 3 = target.	An *ultrasonic transducer,* used for the measurement of distance (primarily underwater), is provided with a piezoelectric element emitting pulses of acoustic energy directed to the target, which is a small area on the object. The signal reflected from the target travels back to the transducer, generating electrical pulses in the element. The time between transmitting and receiving the pulses is a measure of the distance between the transducer and the target. In this sonic radar, a separate or the same element can be used for generating and receiving the signals.

| 18 | A *microwave radar* system is also used in the remote measurement of distance (radar trackers, radar altimeters). The concept of measurement is typical: generating and receiving the reflected signals, and counting the time of the pulse travel. The transducing elements are separate antennas for transmitting and receiving the signals. |

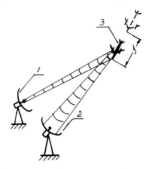

Microwave radar. f = displacement, 1 and 2 = microwave antennas for sending and receiving pulses, 3 = target.

| 19 | A part of the *lidar* is a ruby laser that generates intense infrared pulses in a beam width as small as 30 seconds of an arc. A radar technique is used for measuring the distance to the object. The principle of the operation is based on sensing a beam reflection from the object, whose reflectory properties differ from those of the media. |

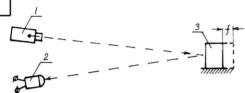

Lidar. f = displacement, 1 = ruby laser, 2 = infrared pulses receiver, 3 = object.

| 20 | A *thermoionic diode valve* with a movable anode has been used for the measurement of displacement. The current in such a diode is controlled by the change in the distance between the cathode and anode. The anode is driven by a sensing pin, whose one end is attached to the anode. Its other end is extended outside the valve's shell. This end is used for contacting the moving object. |

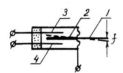

Thermoionic diode valve. f = displacement, 1 = sensing pin, 2 = anode, 3 and 4 = cathodes.

| 21 | An *ionization sensor* is introduced by a sealed gas-filled tube with two flat electrodes inside the tube. The tube is placed between two plates that are connected to a source of radio-frequency voltage. This voltage is high enough to develop a glow discharge inside the tube. The difference in potentials between the two electrodes is developed due to the space potential inside the tube and also the potential produced by the capacitive coupling between the outside plates and electrodes. When the tube is placed in the middle between the plates, the potential between the electrodes is zero. A displacement of the tube from the central position creates a potential proportional to the displacement. |

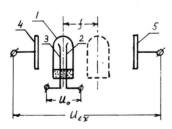

Ionization sensor. f = displacement, U_{ex} = radio-frequency voltage, 1 = sealed gas-filled tube, 2 and 3 = flat electrodes, 4 and 5 = plates.

| 22 | A *Hall-effect sensor* is composed of a semiconductor plate that can be displaced in a nonuniform magnetic field formed by two poles of a permanent magnet. A constant electric current flows through the plate and the voltage developed across the plate in the direction perpendicular to the current's direction (Hall voltage) is proportional to the flux density or, for the nonuniform field, to the plate's displacement. |

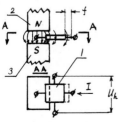

Hall-effect sensor. f = displacement, I = current, U_h = Hall voltage, 1 = semiconductor plate, 2 and 3 = two poles of a permanent magnet.

23	
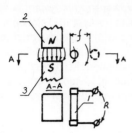 Magnetoresistive transducer. f = displacement, R = probe's resistance, 1 = probe, 2 and 3 = poles of permanent magnet.	A *magnetoresistive transducer* uses a metal or semiconductor probe, whose resistance is sensitive to the magnetic field intensity across the probe. If the probe moves along a field having a linear variation of intensity in the direction of motion, the probe's resistance can be used to indicate the displacement. By special shaping of the permanent magnet's poles and by the arranging of the appropriate distance between the poles, a linear output from the transducer can be obtained.
24	
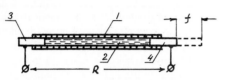 Electrode transducer. f = displacement, R = transducer's resistance, 1 = elastic tube, 2 = electrolyte, 3 and 4 = electrodes.	An *electrode transducer* is made of an elastic tube filled with an electrolyte and two electrodes at the ends of the tube. It is used for a rough measurement of displacement. The change of the distance between two electrodes and the cross-sectional area of the current path in the electrolyte causes a change in the resistance between the electrodes. The resistance is proportional to the elongation of the tube.
25	
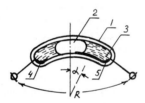 Electrolytic potentiometer. α = angular displacement, R = transducer's resistance, 1 = curved tube, 2 = trapped air bubble, 3 = electrolyte, 4 and 5 = electrodes.	An *electrolytic potentiometer* is a transducer consisting of a curved tube, an electrolyte, and electrodes in the tube with a trapped air bubble. This construction is similar to that of a spirit level, which is used to find a horizontal line or plane. Displacement of the tube moves the bubble, varying the impedance between the electrodes by more or less immersion of them in the electrolyte.
26	
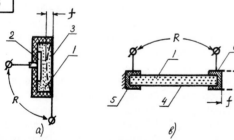 Junction-resistance transducer with particles (a) or elastic member (b). f = displacement, R = transducer's resistance, 1 = electroconductive particles, 2 and 3 = electrodes, 4 = elastic member, 5 and 6 = leads.	A *junction-resistance transducer* has multiple-contact small electroconductive particles that are placed between conducting electrodes or constitute a filler in a rubber or plastic elastic member. Under the displacement of the plates or with the deformation of the member, the resistance of the contacts between the particles is changed proportionally to the measurand. This change in the resistance is read from the electrodes or from the leads attached to the ends of the members.
27	
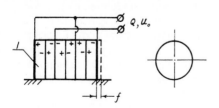 Piezoelectric transducer. f = deformation, Q or U_o = charge or voltage at output, 1 = stack of piezocrystals.	A *piezoelectric transducer* contains disk or bar elements with electrodes on their surface. The elements are made of a material that develops electrical charges on the surface proportional to the deformation of the element's body.

28	In *position-sensing switches,*

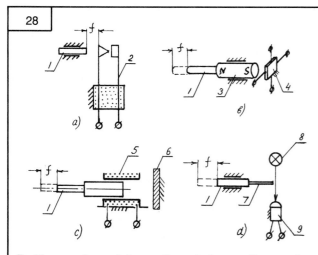

a)

б)

c)

d)

In *position-sensing switches,*

    a) when the moving part approaches a certain position, an *electrical contact* is closed or opened.

    b) when the moving part, carrying a permanent magnet or a ferromagnetic armature, approaches the *Hall element,* it generates a voltage.

    c) when a metal or a ferromagnetic body approaches the coil of an *eddy-current* proximity switch, the coil's inductance changes abruptly.

    d) when the moving object approaches a certain limit, it blocks or unblocks a light path, originating at a light source and directed toward the photodetector.

Position-sensing switches. a through d = position-sensing switches with different transduction elements; f = displacement, 1 = sensing shaft, 2 = contacts, 3 = permanent magnet, 4 = Hall element, 5 = coil, 6 = metal object, 7 = shutter, 8 = light source, 9 = photosensor.

29	Many *displacement transducers* can be used for *measuring thickness* by noting the two levels forming the thickness.

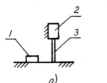

a)

б)

Displacement transducer for measuring thickness.
f = thickness, 1 = object, 2 = displacement transducer, 3 = sensing shaft, (a) and (b) = positions of transducer for noting areas forming thickness.

30	A *resistive thickness or length transducer* employs electrodes that are in contact with an electroconductive body (often film) of known electroconductivity. The resistance between the electrodes is proportional to the measurand.

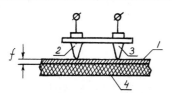

Resistive thickness or length transducer. f = thickness, 1 = electroconductive film, 2 and 3 = electrodes, 4 = nonelectroconductive substrate.

31	An *inductive thickness transducer* contains a coil whose magnetic circuit's reluctance is varied by:

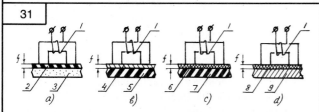

a)

б)

c)

d)

Inductive thickness transducer. a through d = inductive thickness transducer at different combinations of test pieces and bases; f = thickness, 1 = coil, 2 = ferromagnetic test piece, 3 = nonmagnetic base, 4 = nonmagnetic test piece, 5 = magnetic base, 6 = nonmagnetic and nonconductive test piece, 7 = magnetic base, 8 = nonmagnetic and nonconductive test piece, 9 = nonmagnetic but electroconductive base.

An *inductive thickness transducer* contains a coil whose magnetic circuit's reluctance is varied by:

    a) the thickness of the ferromagnetic test piece attached to the nonmagnetic base;

    b) the thickness of the nonmagnetic but conductive test piece attached to the magnetic base;

    c) the thickness of the nonmagnetic and nonconductive test piece attached to the magnetic base; and

    d) the thickness of the nonmagnetic and nonconductive test piece attached to the nonmagnetic but electroconductive base.

32	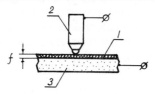	A *capacitive thickness transducer* allows measurement of the thickness of a thin insulating layer on the electroconductive base. The capacitance is a measure of the thickness that is formed by two flat electrodes. One of them is applied to the surface of the layer and the other one is the base.
	Capacitive thickness transducer. f = thickness, 1 = thin insulating layer, 2 = surface electrode, 3 = electroconductive base.	
33		An electrical *breakdown transducer,* used to *measure the thickness* of a nonconductive layer on the conductive base, has a concentrator of the electrical field (small metal ball) attached to the layer. The breakdown voltage measured between the ball and the base is proportional to the square of the layer's thickness.
	Breakdown transducer for measuring thickness. f = thickness, 1 = nonconductive layer, 2 = concentrator of electric field, 3 = conductive base.	
34	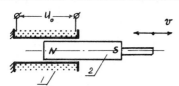	In a *standing-wave transducer* a source of ultrasonic energy excites vibrations at the surface of a test piece whose thickness is to be measured. In the presence of standing waves, the power absorption from the source is increased. The lowest noted frequency of the excitation, corresponding to the maximum absorption, is used for calculating the thickness, which is inversely proportional to this frequency.
	Standing-wave transducer. f = thickness, 1 = test piece, 2 = source of ultrasonic energy, 3 = substrate.	

Linear and Angular Velocity Transducers

35	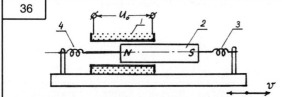	An *electromagnetic linear-velocity transducer* is composed of a stationary coil with a permanent-magnet core moving within the coil. The core is attached to the object whose velocity is to be measured. When the core moves, magnetic lines of the field created by the core cross the turns. An electromotive force induced in the turns is proportional to the speed of the core.
	Electromagnetic linear-velocity transducer. v = velocity, U_o = output voltage, 1 = coil, 2 = permanent magnet.	
36		An *electromagnetic vibratory, linear-velocity transducer* is built from a coil attached to a moving object and a spring-suspended permanent magnet within the coil. When the vibratory speed of the coil is such that the frequency of vibration exceeds the resonance frequency of the mechanical system, the magnet remains almost immovable. Therefore, the voltage induced in the coil due to the motion of the turns across the field is proportional to the speed of the object.
	Electromagnetic vibratory-velocity transducer. v = velocity, U_o = output voltage, 1 = coil, 2 = permanent magnet, 3 and 4 = springs.	
37	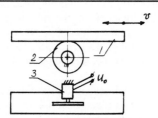	A *tachometer generator with* a *wheel* or a *disk* fastened on the shaft develops a voltage proportional to the linear velocity of the object that the wheel touches. A translational speed at the wheel's or disk's circumference is equal to the linear speed of the object. At the same time, the angular speed of the shaft and the output voltage are linear functions of the translational speed.
	Tachometer generator with wheel or disk. v = velocity, U_o = output voltage from tachometer, 1 = moving object, 2 = wheel, 3 = tachometer.	

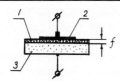

38 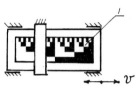 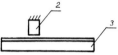Doppler-effect devices. v = velocity, 1 = signal transmitter, 2 = signal receiver, 3 = moving object.	*Doppler-effect devices* are intended for the remote sensing of linear velocity. If a narrow-beam light, radio, or ultrasonic signal is sent, and received reflected from the moving object, the change in the frequency of the received signal is proportional to the velocity of the object. The frequency becomes higher if the object travels toward the receiver and lower if the object travels away.
39 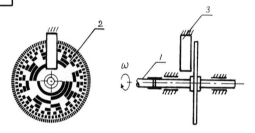 Encoder-type linear-speed transducer. v = velocity, 1 = strip with marks, 2 = electronic pickup, 3 = moving object.	An *encoder-type linear-speed transducer* contains a strip with optical, magnetic, or electroconductive marks on it. The strip is attached to the moving object whose velocity is to be measured. A stationary electronic pickup receives pulses from the marks that pass through its sensitive element. The frequency of the pulses or their width is proportional to the measuring velocity.
40 Encoder-type angular-speed transducer. ω = angular speed, 1 = driving shaft, 2 = disk, 3 = electronic pickup.	An *encoder-type angular-speed transducer* is similar to the encoder transducer for measuring linear velocity with one difference: instead of a strip, the moving part is a disk with marks. This disk is attached to the shaft whose speed is to be measured.

41
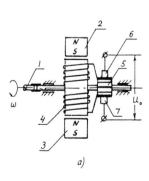
	In an *electromagnetic tachometer generator*, a dc or ac voltage at the output is proportional to the angular speed of its shaft. This shaft is driven by a rotating member whose speed is measured. Three types of tachometers are in use; all of them contain an electrical-machine-like stator and rotor.
	    a) The stator of a dc-*output tachometer* contains several permanent-magnet poles or dc-excited electromagnetic poles. A wound rotor is provided with a commutator and brushes. The output voltage is picked up from the brushes.
	    The frequency of this voltage is also proportional to the speed.
	    b) The stator of an ac-*induction tachometer* contains two windings around its poles. One of
	
	(continued)

41	(continued)
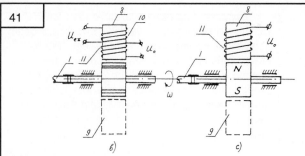 Electromagnetic tachometer generators. a = dc-output tachometer, b = ac-induction tachometer, c = permanent-magnet rotor tachometer; ω = angular speed, U_{ex} = excitation voltage, U_o = output voltage, 1 = driving shaft, 2 and 3 = permanent-magnet poles, 4 = wound rotor, 5 = commutator, 6 and 7 = brushes, 8 and 9 = poles, 10 = excitation coil, 11 = output coil.	them is excited by an ac signal and induces a voltage in the other one as in the secondary of a regular transformer. The rotor has a short-circuited path for the current similar to that of an induction motor. When the rotor rotates, the flux distribution in the stator is changed, providing the change in the coupling coefficient between the coils. The output voltage across the secondary follows this change indicating the speed. c) The rotor of a *permanent-magnet rotor tachometer* carries a rotating magnet whose field induces, in the stationary stator coils, ac voltage having an amplitude and frequency proportional to the speed.
42 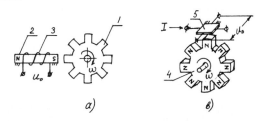 Toothed-rotor magnetic tachometers. a and b = toothed-rotor magnetic tachometer with coil (a) and with Hall element (b); ω = angular speed, U_o = output voltage, I = excitation current, 1 = ferromagnetic rotor with protrusions, 2 = permanent magnet, 3 = coil, 4 = permanent-magnet rotor with protrusions, 5 = Hall element.	A *toothed-rotor magnetic tachometer* composed of a soft ferromagnetic or magnetic-material rotor with protrusions that modulate the magnetic flux in the magnetic circuit when the rotor rotates. The magnetic circuit includes a ferromagnetic rotor, permanent magnet, conductors of the magnetic flux, and a coil, or it is composed of a magnetic rotor and Hall element. The voltage developed in the coil or Hall element is proportional to the speed. The change in the magnetic flux of the system occurs due to alternating the reluctance of the magnetic path, redistributing the flux, or inducing the eddy currents.
43 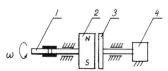 Drag-torque tachometer. ω = angular speed, 1 = shaft, 2 = revolving permanent magnet, 3 = electroconductive disk, 4 = torque-sensitive cell.	In a *drag-torque tachometer* a revolving permanent magnet mounted on the shaft induces eddy currents in an electroconductive disk or cup. These currents produce a flux that interacts with the flux from the magnet. As a result, a drag force on the disk or cup is generated. This force is proportional to the speed of the shaft and is converted into an electrical signal by a torque- or force-sensitive cell.
44 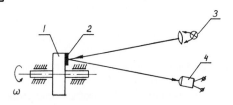 Photoelectric tachometer. ω = angular speed, 1 = rotating object, 2 = reflecting mark, 3 = light source, 4 = photosensor.	A *photoelectric tachometer's* sensing system is composed of a source of light directing a light beam toward the rotating object, reflecting marks affixed to the object, and a photosensor focused on the area with the marks. When the object rotates, it modulates light by reflecting marks, producing a tray of pulses whose frequency is proportional to the speed. Instead of marks, some inherently reflective spots on the object can be used to change the intensity of light reflected toward the photosensor.

45 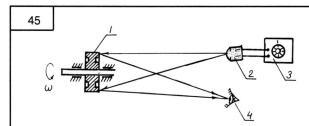 Stroboscopic system. ω = angular speed, 1 = rotating object, 2 = source of flashing light, 3 = manually adjusted source of electrical pulses, 4 = human eye.	In a *stroboscopic system* a high-intensity source of flashing light is used to illuminate the rotating object. The frequency of flashes is adjusted manually until the rotating object appears immovable. For this adjustment, the number of flashes per minute corresponds to the number of revolutions per minute.
46 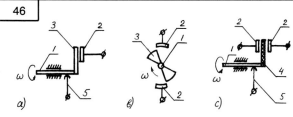 Capacitance tachometers with variation of area between plates (a and b) or dielectric constant of gap (c). ω = angular speed, 1 = rotating shaft, 2 = stationary plate, 3 = moving plate, 4 = moving spacer, 5 = sliding contact.	In a *capacitance tachometer,* one of the plates or spacers forming the capacitor is a part of a rotating object and moves with it; another plate (or plates) is stationary. When the object rotates, it periodically changes the relative position of the plates or the dielectric constant of the gap, producing variations in the capacitance. The frequency of alternations is proportional to the speed of the object.
47 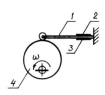 Strain-gage tachometer. ω = angular speed, 1 = beam, 2 and 3 = bonded strain gages, 4 = eccentric disk.	A *strain-gage tachometer* consists of a beam with bonded strain gages. The beam is in contact with a cam or with an eccentric disk rotating with the shaft whose speed is measured. The number of deflections of the beam per unit of time is proportional to the speed. Every deflection of the beam causes a change in the resistance of the gages, which can generate an electrical signal.
48 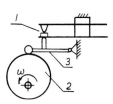 Switch tachometer. ω = angular speed, 1 = electrical contacts, 2 = eccentric disk, 3 = mechanical link.	In a *switch tachometer,* a periodical breaking and closure of electrical contacts are provided by a cam or an eccentric disk fixed on the rotating shaft and controlling the operation of the contacts through a mechanical link. The number of closures of the electrical contacts is proportional to the speed.
49 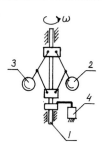 Centrifugal-force-tachometer transducer. ω = angular speed, 1 = shaft, 2 and 3 = masses, 4 = displacement-sensitive element.	A *centrifugal-force-tachometer transducer* contains masses that rotate with a shaft. Under a centrifugal force, masses are displaced proportionally to the speed. This displacement is sensed by one of the displacement-type elements (e.g., inductive, resistive, etc.) that produce an output electrical signal.

50

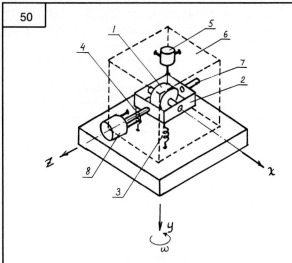

A *gyroscopic tachometer* is used when a reference point for the measurement of the angular speed is not available; for instance, in space. If the gyroscopic system is rotated around the axis (*y*-axis) perpendicular to the axis of the spinning flywheel (*x*-axis), a gyroscopic moment proportional to the measuring angular speed is developed around the *z*-axis, which is perpendicular to the *x*- and *y*-axes. This moment, being sensed by torque- or force-sensing elements, is converted to an electrical output signal.

Gyroscopic tachometer. ω = angular speed, x = spin axis, y = input (measurand) axis, z = output axis, 1 = flywheel, 2 = gimbal, 3 and 4 = restraining springs, 5 = damper, 6 = case, 7 = gimbal axis, 8 = torque sensitive element.

Accelerometers

51

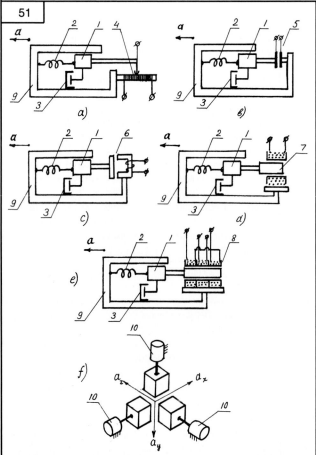

Potentiometric, capacitive, and inductive accelerometers contain a spring-supported damped mass, which is connected to one of the components: the sliding wiper of the potentiometer, the moving plate of the capacitive element, the moving armature, or the core of an inductive element. The displacement of the mass due to the acceleration provides variation of the resistance, capacitance, or inductance. An LVDT (linear variable differential transformer) is also used to convert the displacement of the mass into an ac output voltage. Triaxial accelerometers employ a system with triple elements. Each element in the system is intended for the measurement of a one-direction acceleration.

Potentiometric, capacitive and inductive accelerometers. a through e = one-direction accelerometers with various transduction mechanisms, *f* = triaxial accelerometer;

(continued)

51	*(continued)*	
a = acceleration, a_x, a_y, and a_z = triaxial accelerations, 1 = seismic mass, 2 = spring, 3 = damper, 4 = potentiometric element, 5 = capacitive element, 6 = inductive element with moving armature, 7 = inductive element with moving core, 8 = differential-transformer element (LVDT), 9 = case, 10 = one-directional accelerometer.		

52	
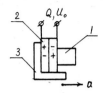 Piezoelectric transducer. a = acceleration, Q = electrical charge at output, U_o = output voltage, 1 = seismic mass, 2 = piezoelectric crystals, 3 = case.	In a *piezoelectric transducer,* the acceleration acts on the seismic mass that develops a force on piezoelectric quartz, or ceramic crystal, or on several crystals. The force causes charges on the crystals proportional to the acceleration.

53	
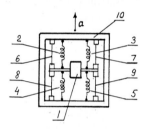 Unbonded strain-gage accelerometer. a = acceleration, 1 = seismic mass, 2, 3, 4, and 5 = springs, 6, 7, 8, and 9 = strain gages, 10 = case.	A *unbonded strain-gage accelerometer* contains pretensioned strain-gage wires supporting a seismic mass. Together with special springs, the wires provide an elastic suspension of the mass and, being deformed by the acceleration, they form an electrical signal proportional to the acceleration. The wires can be connected in a half-bridge or full-bridge circuit.

54	
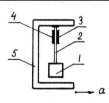 Bonded strain-gage accelerometer. a = acceleration, 1 = seismic mass, 2 = spring element, 3 and 4 = strain gages, 5 = case.	A *bonded strain-gage accelerometer* has wire or foil gages bonded directly to the spring elements. The acceleration causes deformation of the spring element, stretching one group of gages and contracting the other. The gages connected in a bridge circuit produce an electrical signal as a response to the acceleration.

55	
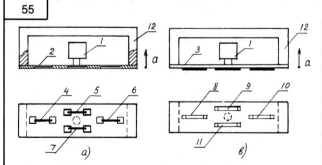 Piezoresistive accelerometers with diffused (a) and cemented (b) semiconductor strain gages. a = acceleration, 1 = seismic mass, 2 = semiconductor beam, 3 = ceramic or metal beam, 4, 5, 6, and 7 = diffused semiconductors, 8, 9, 10, and 11 = cemented semiconductor strips, 12 = case.	A *piezoresistive accelerometer* incorporates a crystal semiconductor beam that works as a spring element. The crystal carries a seismic mass. Several strain-sensitive gages are placed in the crystal's body, and they are physically an integral part of the beam. The gages are usually connected in a Wheatstone bridge. In a similar version, a spring element is made of metal or ceramic material, and semiconductor strips are bonded on the deforming surfaces of the element, similar to the foil or wire gages. The deflection of the spring element under the acceleration causes deformation of the gages, producing an electrical output.

56 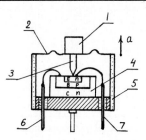 Piezotransistor accelerometer. a = acceleration, B = base, C = collector, E = emitter, n = n-silicon, p = p-silicon, 1 = seismic mass, 2 = spring element, 3 = stylus, 4 = n-p-n planar transistor, 5 = case, 6 and 7 = leads.	A *piezotransistor accelerometer* contains a stylus attached to a seismic mass. The sharp end of the stylus is in contact with the sensitive-to-stress area of the p-n junction of a transistor. The force developed due to the acceleration is transmitted to the stylus, whose sharp edge develops a concentrated stress in the semiconductor. The electrical characteristics of the transistor are changed, causing a change of the current in the electrical circuit.
57 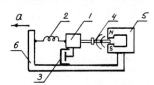 Hall-effect accelerometer. a = acceleration, 1 = seismic mass, 2 = spring, 3 = damper, 4 = Hall element, 5 = source of nonuniform magnetic field, 6 = case.	In a *Hall-effect accelerometer,* the Hall element is attached to a spring with a seismic mass deflecting because of the forces due to acceleration. The element moves in a nonuniform, linear-gradient-intensity magnetic field. Under these conditions, the generated transverse Hall voltage is proportional to the measured acceleration.
58 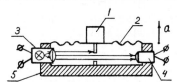 Photoelectric accelerometer. a = acceleration, 1 = seismic mass, 2 = spring element, 3 = light source, 4 = photosensor, 5 = case.	In a *photoelectric accelerometer,* a seismic mass deflected by forces due to acceleration drives a shutter placed between a photosensor and a light source. The illumination of the photosensor is changed because of variation of the area of the orifices in the shutter through which the light beam passes. As a result, a photocurrent in the sensor becomes proportional to the amount of acceleration.
59 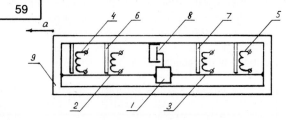 Vibrating-element accelerometer. a = acceleration, 1 = seismic mass, 2 and 3 = vibrating wire, 4 and 5 = pickup coils, 6 and 7 = driving electrical magnet, 8 = damper, 9 = case.	A *vibrating-element accelerometer* contains a seismic mass supported by prestressed wires made to vibrate at their natural frequency by an electromagnetic driving system. When the acceleration develops force along the wires, their frequency of vibration is changed, producing the signal correlated with the acceleration.
60 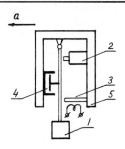 Servo accelerometer. a = acceleration, 1 = seismic mass, 2 = position-sensing device, 3 = servomechanism, 4 = damper, 5 = case.	A *servo accelerometer* contains a seismic mass and servo mechanism that controls the position of the mass. The acceleration causes the mass to move. The position-sensing device detects the motion and produces the error signal in the servo loop, developing a current flowing through the force-generating element, which balances the force due to the acceleration. The current in the feedback loop is the measure of the acceleration.

61

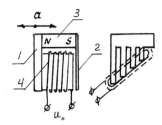

Frequency-of-vibration relay. a = vibratory acceleration, U_o = output voltage, 1 = armature, 2 = vibrating reeds, 3 = permanent magnet, 4 = coil.

A *frequency-of-vibration relay* operates on a resonance principle. An array of metal reeds, which differ in length or in thickness, are fixed on a support. Each reed has a specific resonant frequency of mechanical resonance. The support is attached to a vibrating object. When the frequency of vibration coincides with the resonance frequency of one of the reeds, its amplitude of vibration is appreciably increased and is detected by an electromagnetic or another type of pickup element. As a result, the relay responds only to the predetermined frequencies.

62

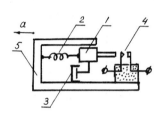

Acceleration switch. a = acceleration, 1 = seismic mass, 2 = spring, 3 = damper, 4 = contacts, 5 = case.

An *acceleration switch* has a construction similar to that of the accelerometer. It contains a seismic mass and element that responds to the force developed under acceleration. When the force reaches a predetermined level, the element responds by closing electrical contacts, or by triggering an electronic gate.

Attitude Transducers and Sensors

63

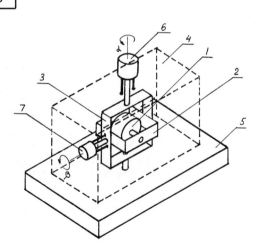

Attitude gyro. α and β = angular displacement in space, 1 = spinning rotor, 2 = inner gimbal, 3 = outer gimbal, 4 = case, 5 = object, 6 and 7 = angular displacement elements.

Several types of *attitude gyros* are in use. Among them are a single-degree-of-freedom gyro, a two-degree-of-freedom gyro (free gyro), a floated gyro, a gas-bearing gyro, an electrostatic gyro, a cryogenic gyro, a tuned-rotor gyro, a vertical gyro, and a directional gyro. The differences among them are in the shape of the rotating wheel, method of suspension of the wheel and types of bearings, some features related to the adjustment of the reference positions, and in the application for a specific measurement.

The most practical model of gyro is a heavy, rapidly spinning rotor. Its axle is equipped with bearings mounted in a rotatable inner frame (gimbal). This frame can rotate within an outer gimbal, providing freedom for the rotor to orient itself in the space. The second frame can also freely rotate within a case that is attached to the moving object. The displacement of the inner and outer gimbals from the reference positions is detected by potentiometric, reluctive, capacitive, and optical transduction elements. The signals from the element are proportional to the angular displacements of the object in space. The rotor is driven by direct-current and alternating-current motors. Some constructions (in rockets) contain a hot-gas turbine spinning the rotor.

| 64 | An *attitude-rate gyro* with a spinning rotor and one gimbal suspended in bearings within its case is similar to the attitude gyro. The difference lies in restraining springs and in dampers that are in contact with the gimbal and create an elastic and viscous-friction load on the gimbal. When the case experiences a change in attitude, the gimbal deflection angle (output angle) is proportional to the attitude rate that can be sensed by the potentiometric, reluctant, or other elements. |

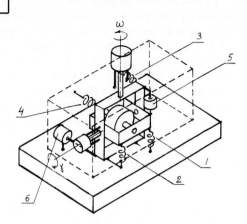

Attitude-rate gyro. γ and ω = attitude rates, 1 and 2 = inner-gimbal restraining springs (ends of springs are connected to inner and outer gimbals), 3 and 4 = outer-gimbal restraining spring (ends of springs are connected to outer gimbal and case), 5 = damper for inner gimbal, 6 = damper for outer gimbal. All other parts are similar to those in Scheme 63.

| 65 | A *ring-laser gyro* contains a ring-shaped cavity (the original design) or triangular-shaped cavity with circulating light beams generated by a laser. The frequency difference between the two circulating beams is a function of the attitude rate that acts perpendicularly to the plane of the moving beams. This difference in frequencies is sensed by a photodetector using an interferometric technique. |

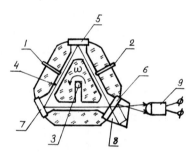

Ring-laser gyro. ω = attitude rate, 1 and 2 = anodes, 3 = cathode, 4 = laser gas-discharge cavity, 5, 6, and 7 = mirrors, 8 = beam mixer, 9 = photodetector.

| 66 | A *fiber-optic gyro* contains a coil with a long (up to 5km) wound optical fiber. Two light beams travel along the fiber in opposite directions. An optical system with a beam splitter directs the beams on a photodetector. When the attitude rate is zero, the phase shift between the two beams is 180°; they cancel each other and the output photocurrent is minimized. With the attitude rate oriented along the fiber (around the coil's axis), the original phase shift is changed. This change occurs because of the increase in the light path for one beam and decrease in the path for another beam. As a result, the photodetector's current responds to the increased illumination and becomes larger. |

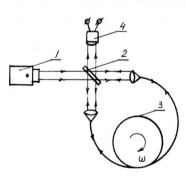

Fiber-optic gyro. ω = attitude rate, 1 = laser light source, 2 = beamsplitter, 3 = wound optical fiber, 4 = photosensor.

| 67 | In *special designs,* the output signals can be proportional to the *integral* or to the *double-integral* of the *attitude rate.* These designs constitute a single-degree-of-freedom gyro having a viscous restrain of the spin axis about the output axis (for the integrating device), and massive inertial gimbals (for the double-integrating system). The integrating effects are achieved due to the viscous friction or the inertia in the elements of construction. |

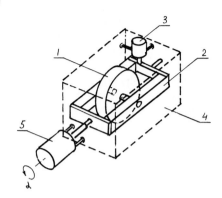

Integral attitude-rate gyro. α = angular displacement in space, 1 = flywheel, 2 = heavy gimbal, 3 = damper, 4 = case, 5 = angular-displacement sensitive element.

| 68 | A *pendulum-type transducer* contains a suspended mass, which remains vertical within a case attached to the displacing object. The mass is linked with the potentiometric transduction element that responds to the deviation of the case from the reference position. In this design, other angular-displacement transduction elements can be used instead of the potentiometric one. |

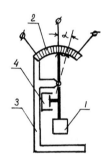

Inclinometer. α = inclination, 1 = suspended mass, 2 = potentiometric transduction element, 3 = case, 4 = damper.

| 69 | In an *electrolytic-potentiometer transducer,* a glass curved container is partially filled with an electroconductive liquid. Several electrodes are placed inside the container and are in contact with the liquid and gas (air) filling a part of the tube. As the case of the transducer is angularly deflected from the originally adjusted position, the areas of the contacts between the electrodes and liquid are redistributed. The resistance measured between the electrodes is a function of the deflection. |

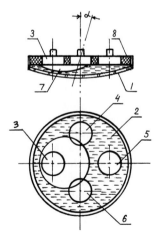

Electrolytic potentiometer. α = inclination, 1 = curved container, 2 = electroconductive liquid, 3, 4, 5, and 6 = electrodes, 7 = gas bubble, 8 = case.

| 70 | A *capacitive-mercury transducer* contains flat electrodes close to the surface of a pool of mercury sealed in a case that is electrically insulated from the electrodes. A deviation of the case causes a change in the capacitances between the electrodes and the mercury's surface since the surface tends to stay horizontal. The electrodes and moving surface form a differential system, where the difference between the capacitances forms the output signal proportional to the attitude. |

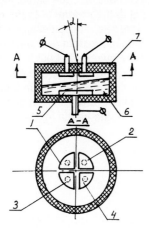

Capacitive-mercury transducer. α = inclination, 1, 2, 3, and 4 = flat electrodes, 5 = common electrode, 6 = mercury pool, 7 = case.

| 71 | The operation of a *servoinclinometer* is typical for any servo-type transducer. A deviation of a case carrying a pendulous mass is detected by a displacement-sensitive element. An electrical signal of error caused by the deviation is fed into a feedback-loop amplifier and torque-restoring mechanism, which tends to return the mass to the original position. The current required for restoring the mass is a measure of the inclination. |

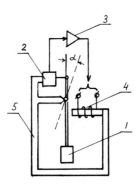

Servo inclinometer. α = inclination, 1 = pendulous mass, 2 = displacement-sensitive element, 3 = feedback-loop amplifier, 4 = torque-restoring mechanism, 5 = case.

| 72 | *Compass-type sensors* are used to indicate the attitude of the moving object with respect to the earth's magnetic field. |

The *simplest version* of the *compass* is a magnetic bar that orients itself along the lines of the field and deflects the sensing element of the displacement-sensing device in order to produce an electrical output signal. The device can be a variable potentiometer, inductance, etc.

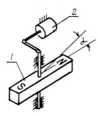

Magnetic-bar compass. α = angular deviation from earth's north-south reference line, 1 = magnetic bar, 2 = displacement-sensitive element.

73	
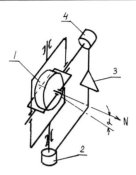	A *gyrocompass* is a sensing device in which the spinning rotor's axis is forced to be aligned by a control system along the lines of the earth's magnetic field.
Gyrocompass. α = angular deviation from earth's north-south reference line, 1 = gyro, 2 = angular-displacement element, 3 = amplifier, 4 = control motor.	

74	
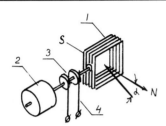	An *induction compass* contains a rapidly rotating coil. The voltage induced in the coil, due to crossing the lines of the earth's field, is at a maximum when the axis of the coil is perpendicular to the lines of the field.
Induction compass. α = angular deviation from earth's north-south reference line, 1 = coil, 2 = motor, 3 = slip rings, 4 = brushes.	

75	
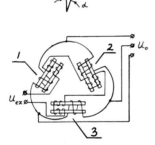	A *flux-gate compass* incorporates an ac electromagnetic system that is disbalanced at the presence of a directed outside magnetic field. This disbalance induces voltages in the coils of the system. The amplitudes and phases of the voltages indicate the relative orientation of the system and field.
Flux-gate compass. α = angular deviation from earth's north-south reference line, U_{ex} = excitation voltage, U_o = output voltage, 1, 2, and 3 = cores with coils of electromagnetic system.	

76	
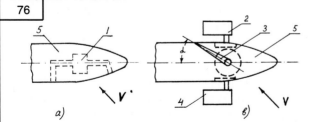	The reference for the *angle-of-attack transducer* is the direction of the fluid stream ambient to the moving object in air or in water. In one of these transducers, the difference between two pressures at the surface of the object is used to evaluate the angle. In another transducer, a freely suspended vane in the stream aligns itself along the airstream. The rotation of the vane around the axis of suspension is transmitted to an angular-displacement transduction element that produces an electrical signal proportional to the angle.
Fluid-stream-angle-of-attack transducers with differential-pressure transducer (a) and vanes (b). α = angle of attack, V = direction of fluid stream, 1 = differential-pressure transducer, 2, 3, and 4 = vanes, 5 = streamlined case.	

77	*Celestial-reference attitude sensors* contain an optical system and photodetectors that sense the attitude with reference to the sun, planets, stars, light-to-dark boundary, and other targets. The sensors are directed toward the targets by closed-loop control systems that produce electrical signals correlated with the attitude.

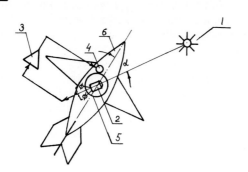

Celestial-reference attitude sensors. α = angular deviation from celestial reference line, 1 = light-emitting body, 2 = photosensor, 3, 4, and 5 = elements of closed-loop control system, 6 = vehicle.

78	In a *radio compass* system, a highly directional antenna is rotating, and the receiving part of the system is searching for the maximum signal from the transmitter. The position of the antenna, corresponding to the maximum signal, indicates the orientation of the moving object with respect to the transmitter, which is a target for determining the attitude.

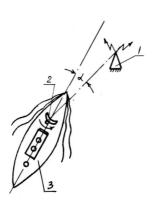

Radio compass. α = angle between direction of vehicle and line connecting vehicle and radio transmitter, 1 = radio transmitter, 2 = highly directional rotatable antenna, 3 = vehicle.

Strain Gages

79	*Metal-wire bonded strain gages* have been extensively used in recent years. They are made of a thin wire formed in a zigzag pattern that is cemented to a nonelectroconductive substrate. The ends of the pattern are soldered or welded to leads for connection to an electrical circuit. The gage is glued to the measured surface and senses deformation of this surface. Nowadays the wire gages are replaced by strain-sensitive metal-foil gages, which are simpler in production and have superior characteristics.

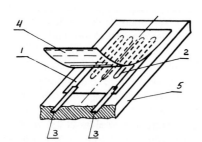

Metal-wire bonded strain gage. 1 = substrate, 2 = strain-sensitive wire, 3 = leads, 4 = protective film, 5 = measured surface.

80 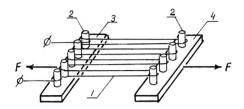 Unbonded metal-wire strain gage. F = force, 1 = strain-sensitive stretched wire, 2 = post, 3 and 4 = moving parts.	*Unbonded metal-wire strain gages* have become obsolete; however, they still have some applications, e.g., in accelerometers. The typical construction of a gage consists of a thin wire stretched between posts. When the distance between the posts is changed, the wire resistance is changed as well, producing the output signal. In the practical design, several loops of wire usually form arms of the Wheatstone bridge. The bridge circuit is arranged in such a manner that elongation of the two arms is accompanied by shortening of the other two arms.
81 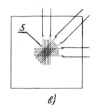  Metal-foil strain gage. a through d = different configurations of metal-foil strain gages; 1 = strain-sensitive foil, 2 and 3 = leads, 4 = substrate, 5 = three-element, 45° stacked rosette, 6 = 90° two-element planar rosette, 7 = strain-sensitive foil pattern for attachment to diaphragm.	A *metal-foil strain gage* is produced in several configurations. Its simplest form is a strain-sensitive, metal-foil thin grid attached to a substrate made of an electrical insulator. The grid is terminated with tabs for connecting wires. When the substrate is cemented to the surface, its deformation is transferred to the grid through the substrate, providing a change in the grid's resistance. Several gages placed on the same substrate for measuring strains simultaneously in more than one direction are called rosettes. The gages' axes, along which the stresses are measured, can be oriented 45°, 60°, or 90° to each other. This allows determination of the magnitudes and directions of principal stresses, shear stress, etc. Foil gages are commonly used for building different transducers of mechanical quantities. In these devices, the gage is cemented to the surface of a spring element whose deformation is proportional to the measurand (force, torque, acceleration, and others). For instance, special patterns are designed for attachment to a flat, round, pressure-sensitive diaphragm. A deformation of the diaphragm under pressure provides elongation and contraction of different parts of the strain-sensitive pattern, resulting in a change in the gage resistance proportional to the pressure.
82 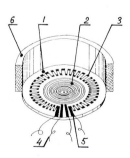 Deposited-metal strain gage. 1 = pressure-sensitive diaphragm made of insulating material, 2 and 3 = strain-gage deposited patterns, 4 and 5 leads, 6 = case.	Several technologies are in use to fabricate *deposited-metal strain gages.* These gages have patterns similar to those of the foil-type elements. They are fabricated by evaporating thin films on insulating substrates, which are also made by the vaporization of insulators on metal surfaces. Flame-spray technology is also used to build strain-sensitive films, ceramic substrates, and protective coatings. They are primarily used for measurements at high temperatures.

83 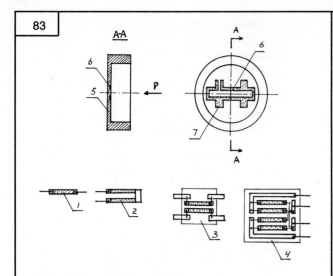	*Semiconductor strain gages* are characterized by a much larger gage factor than that of the metal gages. They are produced in the form of very fine and short strips with leads attached to the ends of the strips. Quite often, one or several strips are affixed to a substrate and form rosettes, a full-bridge, or a dual-element arrangement. Another version of the gage is a spring element made of a semiconductor (silicon diaphragm, cantilever, etc.). The element incorporates strain-sensitive areas formed primarily by diffusion. These areas are located at the places with maximum strains. The gages form full-bridge or half-bridge structures, which are usually combined with elements of compensation for temperature.

Semiconductor strain gages. 1 = single strip, 2 = *U* gage, 3 = dual element on substrate, 4 = full bridge, 5 = pressure-sensitive silicon diaphragm, 6 = diffused strain gages, 7 = lead, *P* = pressure.

84 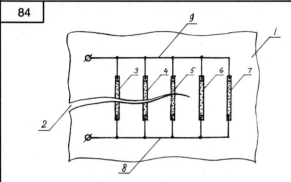	*Special-application gages* are made of materials and have a design that makes possible measurements of post-yield stress, evaluation of crack propagation in constructions, determination of the fatigue life during cyclic loading tests, etc.

Crack-propagation strain gage. 1 = construction, 2 = propagating crack, 3, 4, 5, 6, and 7 = elements of foil grid (sequentially broken conductors), 8 and 9 = common buses.

Force Transducers

85 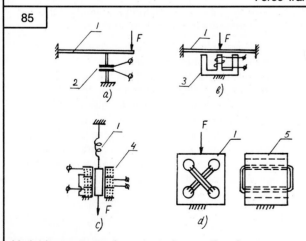	In a *variable-reactance transducer,* force applied to a spring element causes its deformation, which is transferred to moving parts of the variable-capacitance or the variable-inductance elements. The linear variable differential transformer (LVDT), combined with the sensitive element, produces an output voltage proportional to the force. A magnetostrictive element with windings wound around a core made of stress-sensitive material is also used to convert force into an electrical signal. This signal occurs as a result of the change in the mutual inductance between two coils. One of them is fed by an ac voltage that induces a voltage across another coil when the stress affects the permeability of the core.

Variable-reactance force transducers. *F* = force, 1 = spring element, 2 = variable capacitor, 3 = variable

(continued)

85	*(continued)*	
reluctance, 4 = linear variable differential transformer (LVDT), 5 = magnetoelastic element.		

86		

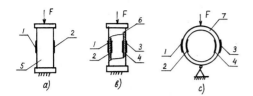

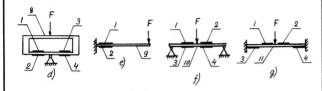

Spring-element force transducers with strain gages affixed to surfaces of different spring elements (a through g). F = force, 1, 2, 3, and 4 = strain gages, 5 = solid cylinder, 6 = hollow cylinder, 7 = round proving ring, 8 = flat proving ring, 9 = cantilever, 10 = simply supported beam, 11 = restrained beam.

A *strain-gage force transducer* contains a spring element with strain gages affixed to its surface. The force causes deformation of the element and a change in the resistances of the strain gages, which are usually made of a metal foil or of a semiconductor material. The most typical spring elements are solid and hollow cylinders, round and flat proving rings, cantilevers, simply supported and restrained beams, and washers. In these designs, the gages are attached to the sensing elements at places of concentrated stresses. The most common arrangement of a cell is the combination of two gages sensing a change in length and two others responding to contraction. These gages, almost inevitably connected in the Wheatstone bridge, form the output-voltage signal.

87		

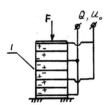

Piezoelectric force transducer. F = force, Q or U_o = charge or voltage at output, 1 = stack of piezocrystals.

In a *piezoelectric force transducer* the transduction element is composed of one or a stack of piezoelectric crystals experiencing the force. The output quantity is an electrical charge developed on the faces of the crystals; this charge is proportional to the force. By a special shaping and orientation of the crystals, a three-directional measurement of forces can be performed. Inherently, the crystal is intended to convert the force contracting the crystal. However, some of the transducers contain a preloading member that permits measurement of forces in the directions of contraction and tensioning of the crystals. The input impedance of the signal conditioner connected to the crystals must be high (ideally infinitely large) in order to measure the static force. Generally speaking, a piezoelectric transducer is more suitable for measuring variable rather than static forces.

88		

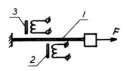

Vibrating-element force transducer. F = force, 1 = vibrating wire, 2 = excitation coil, 3 = pick-up coil.

A *vibrating-element force transducer* is similar to the displacement element because it utilizes the same principle of operation. It contains a vibrating wire converting the force applied along its length into the change in the vibration frequency. The wire is one of the elements in the closed-loop feedback system maintaining the vibrations.

Torque Transducers

| 89 | *Reluctance-type torque transducers* contain a torsion bar, whose angular deflection is measured by a separate variable-reluctance transduction element or by a built-in system with a variable-gap magnetic circuit, for instance, TVDT (torsional variable differential transformer). A variation in the reluctance can also be achieved when a magnetostrictive core that constitutes a part of the torsion bar is deformed under the applied torque. The core is provided with windings. The inductances of the windings are functions of the core's permeability that is controlled by the stress in the bar. Stress, in turn, is a function of the applied torque. |

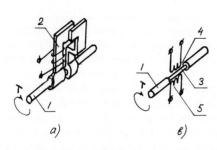

Torque transducer with variable-reluctance transduction element (a) or magnetostrictive core (b). *T* = torque, 1 = shaft, 2 = magnetic system with variable reluctance, 3 = magnetostrictive core, 4 and 5 = coils.

| 90 | In a *photoelectric torque transducer,* a displacement of the ends of the torsion bar is sensed by a photodetector whose illumination is controlled by disks with transparent segments. The disks are attached to the ends of the bar, whose angular displacement varies the intensity of the collimated light beam. The beam is directed from a light source through the segments toward the photodetector. |

Phase-displacement between two trains of pulses, developed in two photodetectors, is also used for measuring the torque. In this case, two code patterns are placed near the ends of the torsion bar. When the bar is twisted, the relative position of the patterns changes, so the light reflected from the moving patterns generates pulses in the photodetector. The phase shift between the pulses, modulated by two patterns, is proportional to the torque.

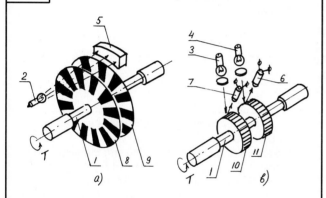

Photoelectric torque transducers with displaced segments (a) or code patterns (b) on torsion bar. *T* = torque, 1 = torsion bar, 2, 3, and 4 = light sources, 5, 6, and 7 = photodetectors, 8 and 9 = disks with transparent segments, 10 and 11 = code patterns.

| 91 | The *phase-displacement* of pulse trains can be induced by *magnetic-code marks* placed at the ends of the torsion bar. The principle of operation is similar to that of the photoelectric code-mark system. The pulses are induced in the coil when the moving magnetic marks pass close to the coil. In another design, gearlike ferromagnetic teeth work as marks. When the teeth pass close to the coil, its inductance is periodically changed. |

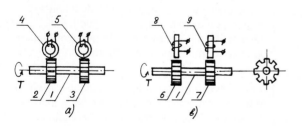

Magnetic-marks (a) and ferromagnetic-teeth (b) torque transducers. *T* = torque, 1 = torsion bar, 2 and 3 = disks with magnetic-code marks, 4 and 5 = sensing coils, 6 and 7 = gears with ferromagnetic teeth, 8 and 9 = variable-inductance coils.

| 92 | In a *strain-gage torque transducer,* strain gages are affixed to a torsion bar. The gages respond to the bar's shear stress, which is proportional to the torque. Several configurations of the bars are used in the constructions of transducers. They can be circular, square, and cruciform solid shafts, or hollow and cruciform tubes. The signals from the gages mounted on the rotating shafts are picked up by brushes contacting slip rings attached to the rotating shaft. A rotating transformer is used to eliminate the sliding contacts. The transformer contains concentrically wound primary and secondary coils placed on rotating and stationary members. Due to the electromagnetic coupling between the coils, signals from the gages are transmitted from the rotating to the stationary coil that is coupled to a signal conditioner. |

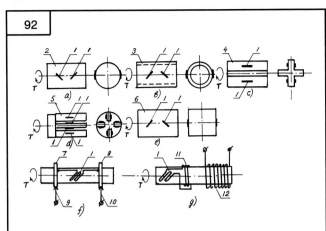

Strain-gage torque transducer. a through e = different-shape torsion bars with affixed strain gages for measurements of torque, *f* and *g* = systems for picking up signals: slip rings (*f*) and rotating transformer (*g*); *T* = torque, 1 = strain gage, 2 = circular shaft, 3 = hollow shaft, 4 = cruciform shaft, 5 = hollow cruciform shaft, 6 = square shaft, 7 and 8 = slip rings, 9 and 10 = brushes, 11 and 12 = concentrically wound primary and secondary.

Pressure Transducers

| 93 | A *variable impedance pressure transducer* contains an elastic element and potentiometric, capacitive, or inductive element. Displacement of the elastic element is transferred to the moving part of the transduction element and provides the change in resistance or reactance.

The LVDT and bridge-type circuits are quite commonly used to obtain the voltage output. |

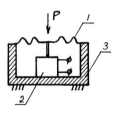

Variable-impedance pressure transducer. *P* = pressure, 1 = pressure-sensitive elastic element (diaphragm), 2 = potentiometric, capacitive, inductive element, or LVDT, 3 = case.

| 94 | *Resistive pressure transducers* contain materials sensitive to pressure, or particles whose contact resistance undergoes changes when pressure is applied to them. Manganin wire, monocrystalline tellurium, and indium antimonide reveal the change in resistance when they are exposed to pressure. |

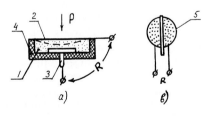

Resistive pressure transducers with electroconductive particles (a) or manganin-wire wound-coil (b). *P* = pressure, *R* = transducer's resistance, 1 = electroconductive particles, 2 = electroconductive diaphragm, 3 = electrode, 4 = case, 5 = manganin-wire wound coil.

95

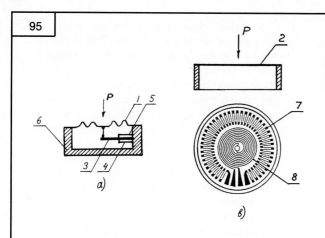

Strain-gage pressure transducers having separate diaphragm and sensing element (a) or deposited gages (b). P = pressure, 1 and 2 = pressure-sensitive diaphragm, 3 = cantilever, 4 and 5 = strain gages, 6 = case, 7 and 8 = deposited strain-gage patterns.

Two basic types of *strain-gage pressure transducers* are used in practical measurements. In one of them, the deflection of the pressure-sensitive spring element bends the beam with affixed strain gages. In the other type, the strain gages are affixed directly to the spring element and respond to the stress developed in the element's material. The gages are made of metal foil, vacuum-deposited or sputtered films and semiconductor material.

96

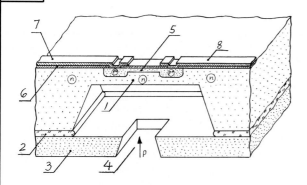

Semiconductor pressure-sensitive structure with diffused strain gages. P = pressure, 1 = diaphragm, 2 = glass seal, 3 = silicon support wafer, 4 = pressure port, 5 = p-type diffused piezoresistive element, 6 = silicon dioxide, 7 and 8 = metallization.

Integrally *diffused strain-gage diaphragms* or force-sensing beams are very popular. It is typical to diffuse a four-arm strain gage in the diaphragm and integrate the sensing part of the transducer with signal conditioner members.

97

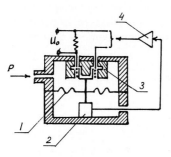

Servo-type pressure transducer. P = pressure, U_o = output voltage, 1 = pressure-sensitive element, 2 = displacement transduction element, 3 = force-restoring mechanism, 4 = amplifier.

Servo-type pressure transducers have several modifications. The basic model contains a pressure-sensitive element, whose deflection under pressure provides an error signal that is fed into an amplifier. The amplified signal excites a force-restoring mechanism that balances the force developed in the spring element due to the applied pressure. The current or voltage feeding the restoring system at the state of balance is a measure of the pressure. Force-balancing is performed by a servomotor, the force coil of an electromagnetic system, or by a capacitive actuator. The displacement of the sensing element is detected by a transduction element (differential transformer, photoelectric cell, variable capacitor, etc.).

| 98 | In a *piezoelectric pressure transducer,* a pressure-sensing diaphragm transduces the force to a stack of disks made of piezoelectric ceramics or crystalline quartz. The electrical charges, picked up from the faces of the stack, are proportional to the pressure. |

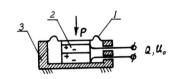

Piezoelectric pressure transducer. P = pressure, Q or U_o = charge or voltage at output, 1 = diaphragm, 2 = stack of piezocrystals, 3 = case.

| 99 | Two basic designs are used in the construction of *vibrating-element pressure transducers.* |

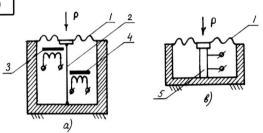

A pressure-sensitive element (diaphragm or bellows) develops a force against a *tensioned wire* or *quartz crystal* whose natural mechanical frequency of vibration depends on the stress due to the force. As a result, the frequency is a function of pressure. The vibrations are maintained by a feedback loop with an amplifier in the loop. Electromagnetic and piezoelectric effects are utilized to pick up signals proportional to the deflection of the elements and to exert the driving forces on the vibrating member.

Vibrating-element pressure transducers with pressure-sensitive element changing strain in vibrating wire (a) or quartz crystal (b). P = pressure, 1 = pressure-sensitive element, 2 = tensioned wire, 3 = pickup coil, 4 = driving magnetic system, 5 = quartz crystal.

| 100 | Several transducer designs contain pressure-sensitive spring elements (*cylinder, metal or silicon diaphragm, capsule, Bourdon tube,* etc.) vibrating at their natural frequencies. With the change in pressure, the effective stiffness of the element is changed, providing a deviation in the resonant frequency. Typically, electromagnetic pickup and driving elements are used to maintain vibrations. |

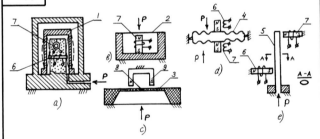

Pressure transducers with different pressure-sensitive vibrating elements (a through e). P = pressure, 1 = vibrating cylinder, 2 = vibrating metal diaphragm, 3 = vibrating silicon diaphragm, 4 = vibrating capsule, 5 = vibrating Bourdon tube, 6 and 7 = elements of electromagnetic pickup and drive, 8 = diffused electroconductive area, 9 = permanent magnet.

| 101 | *Pressure switches* close or open electrical contacts when the pressure to be monitored reaches a predetermined level. The representative construction of the switch incorporates a pressure-sensitive elastic element, a counterbalancing spring working against the element, and a lever linkage carrying the contacts. When the pressure reaches the set-up point, the levers, carrying the contacts, trigger. |

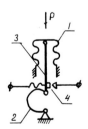

Pressure switch. P = pressure, 1 = pressure-sensitive element, 2 = counterbalancing spring, 3 = lever linkage, 4 = contacts.

Microphones	
102 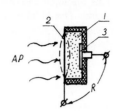 Carbon-button microphone. *AP* = acoustic pressure, *R* = variable resistance, 1 = electroconductive particles, 2 = diaphragm, 3 = electrode.	In a *carbon-button microphone,* the sound field acts upon an electroconductive diaphragm that develops pressure on a packet of carbon granules. The contact resistance between the granules depends on the pressure. When a dc voltage is applied across the packet, the alternating resistance produces an ac voltage drop, which is proportional to the sound intensity.
103 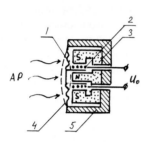 Moving-coil microphone. *AP* = acoustic pressure, U_o = output voltage, 1 = diaphragm, 2 = coil, 3 = permanent magnet, 4 = protection grid, 5 = case.	A *moving-coil microphone* contains a diaphragm exposed to sound waves. The diaphragm carries a coil placed in the magnetic field. The voltage induced in the coil is proportional to its amplitude of vibration, which, in turn, depends on the sound pressure.
104 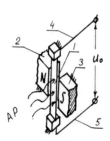 Ribbon microphone. *AP* = acoustic pressure, U_o = output voltage, 1 = light metal ribbon, 2 and 3 = poles of magnet, 4 and 5 = electrical leads.	In a *ribbon microphone,* the sound field acts on both sides of a light metal ribbon placed between the poles of a magnet. A vibratory motion of the ribbon induces a voltage along the ribbon. The voltage is proportional to the difference between the acoustic pressures at both sides of the ribbon.
105 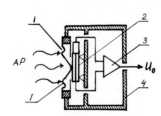 Piezoelectric microphone. *AP* = acoustic pressure, U_o = output voltage, 1 = diaphragm, 2 = ceramic or quartz crystals, 3 = built-in preamplifier, 4 = case.	*Piezoelectric microphones* contain ceramic or quartz crystals linked with a diaphragm or directly exposed to acoustic waves. Stresses in the crystals, resulting from a sound field, generate an output proportional to the acoustic pressure. Many designs incorporate a built-in preamplifier next to the crystal. This arrangement reduces the electrical noise and output impedance.

106	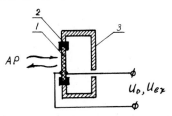	*Piezoelectric hydrophones* are similar to the microphone construction but adapted for underwater operation. The elements of the construction must be well sealed. The hydrophone should operate at a wide range of frequencies. Quite often, it also acts as a generator of acoustic signals.

Piezoelectric hydrophone. AP = acoustic pressure, generated or received, U_o = output voltage, U_{ex} = excitation voltage, 1 = piezoelectric crystal, 2 = seal, 3 = case.

107	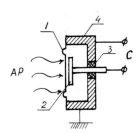	A *condenser microphone* incorporates a stretched metal diaphragm that forms one plate of a capacitor. A metal disk placed close to the diaphragm acts as a back-plate. When a sound field excites the diaphragm, the capacitance between the two plates varies according to the variation in the sound pressure. A stable dc voltage is applied to the plates through a high resistance to keep electrical charges on the plate. The change in the capacitance generates an ac output proportional to the sound pressure. In order to convert ultralow-frequency pressure variations, a high-frequency voltage (carrier) is applied across the plates. The output signal is the modulated carrier.

Condenser microphone. AP = acoustic pressure, C = variable capacitance, 1 = metal diaphragm, 2 = metal disk, 3 = insulator, 4 = case.

108	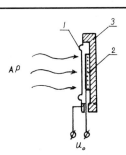	An *electret-type microphone* is a condenser microphone in which the electrical charges are created by a thin layer of polarized ceramic or plastic films (electrets). The ability of the electrets to keep the charge obviates using the source for a high-voltage polarization.

Electret-type microphone. AP = acoustic pressure, U_o = output voltage, 1 = diaphragm, 2 = electret, 3 = case.

Flowmeters

109	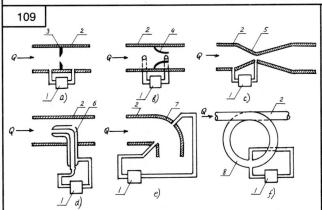	A *differential-pressure flowmeter* includes a flow-restricting element. It can be an orifice or nozzle narrowing the pipe's diameter, Venturi or Pitot tubes, a centrifugal elbow, or loop. A transduction element is introduced by a differential-pressure transducer with pressure ports connected to the pipe at points where the pressure drop is developed (e.g., before and after the restriction). The flow rate to be measured is proportional to the square root of the differential pressure.

Differential-pressure flowmeters with variety of flow restricting elements (a through f). Q = flow, 1 = differential-pressure transduction element, 2 = pipe, 3 = orifice, 4 = nozzle, 5 = Venturi tube, 6 = Pitot tube, 7 = centrifugal elbow, 8 = centrifugal loop.

110	

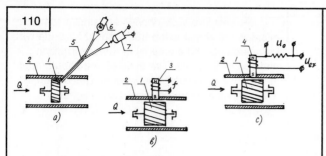

Turbine flowmeter with optical pickup (a), electromagnetic pickup (b), or RF excitation (c). Q = flow, f = frequency of pulses, U_{ex} = excitation voltage, U_o = output voltage, 1 = turbine-type rotor, 2 = pipe, 3 = coil with magnet, 4 = coil with soft-ferromagnetic-material core, 5 = optical fiber, 6 = light source, 7 = photodetector.

In a *turbine flowmeter*, a turbine-type rotor is mounted in the pipe with the moving fluid. The turbine is rotated by the passage of the fluid. The rate of rotation is proportional to the flow velocity and is converted into electrical pulses by a coil with a magnet inserted in the side of the pipe. The blades of the turbine pass close to the coil and induce pulses in the coil. The number of pulses per unit of time is proportional to the flow rate. In order to reduce the drag imposed on the turbine by the magnet, RF excitation with an eddy-current sensing coil is also used. Electro-optical pickup with the chopping of a light beam by a turbine's blades is another arrangement for counting the pulses.

111	

Rotating-cup anemometers with generator (a) or optical system (b). Q = flow, f = frequency of pulses, U_o = output voltage, 1 = vane, 2 = ac generator, 3 = light interrupter, 4 = photodetector, 5 = light source.

A *rotating-cup anemometer* is one of the most popular transducers for measuring wind-speed velocity. An assembly with rotating vanes drives a small ac generator, whose output voltage and/or frequency is proportional to the velocity. In another construction, the angular speed of the rotating member is converted into pulses by an electrooptical system including a light source, light interrupter, and photodetector.

112	

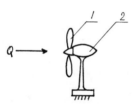

Propeller-type flowmeter. Q = flow, 1 = propeller, 2 = pulse generator.

Sensing elements of *propeller-type flowmeters* are contoured propellers similar to those utilized in aircraft or in ships. The angular speed of the propeller is proportional to the velocity of the flow passing the propeller. Conversion of the angular speed into an electrical analog output (frequency) is provided by an element generating pulses similar to those in the rotating-cup anemometer.

113	

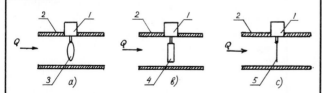

Flowmeters with different targets: disk (a), cylinder (b), or bristle (c). Q = flow, 1 = force-sensitive transduction element, 2 = pipe, 3 = disk, 4 = cylinder, 5 = bristle.

A *target flowmeter* contains a target (disk, cylinder, or bristle) immersed in the flow stream and oriented perpendicularly to the direction of the flow. The drag force on the target is a function of the flow velocity. The target links a force-sensitive transduction element, producing an electrical input proportional to the velocity.

114

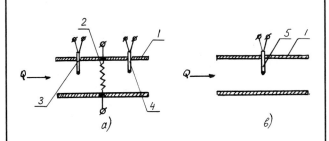

Thermal flowmeter with heating element (a) or anemometer with self-heated thermistor (b). Q = flow, 1 = pipe, 2 = heating element, 3 and 4 = temperature-sensitive probes, 5 = temperature-sensitive, self-heating resistive element.

Thermal flowmeters are divided into two basic classes. In one, a heating element raising the temperature of the fluid in a pipe is mounted between two temperature-sensitive probes immersed in the upstream and downstream of the heater. When the fluid is still, heat from the heating element propagates uniformly along the pipe, and the probes' temperatures are balanced. The flow of the fluid causes a temperature disbalance. The temperature difference between the two probes (thermistors or thermocouples) is a measure of the flow rate.

The other type of thermal flowmeter is the anemometer. It contains a current heated wire, thin-film probe, or thermistor immersed in a stream of fluid. The cooling effect of the fluid on the transduction elements leads to a change in their resistances, which are functions of the flow rate.

115

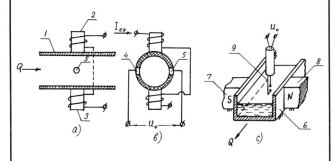

Electromagnetic flowmeter (a), side view (b), and flowmeter with immersed electrodes (c). Q = flow, I_{ex} = electromagnet excitation current, U_o = output voltage, 1 = pipe, 2 and 3 = poles of electromagnet, 4 and 5 = electrodes, 6 = duct, 7 and 8 = poles of magnet, 9 = immersed electrodes.

In an *electromagnetic or magnetic flowmeter,* voltage induced in a moving electroconductive liquid, crossing lines of the magnetic field, is directly proportional to the flow rate. The flowmeter consists of a nonconductive pipe (or a conductive pipe with an insulating inner surface liner), two electrodes usually flush with the inside surface of the pipe, and an electromagnet (sometimes a permanent magnet). The electrodes are in contact with the liquid and are oriented perpendicularly to both the direction of flow and the lines of the magnetic field. The pipe is mounted in the gap of the magnet. The liquid must be electroconductive; however, the electroconductivity can be very low.

There are a number of modifications that can be made to the flowmeter. For instance, the local velocities of the liquid in a duct can be measured by immersing two electrodes into a stream and mounting a magnet outside the duct.

116

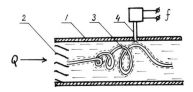

Swirl flowmeter. Q = flow, f = frequency of pulses, 1 = pipe, 2 = swirl blades, 3 = swirl, 4 = detector of oscillation.

A *swirl flowmeter* is designed so that it converts the effects caused by vortices in a flow into the frequency of the output signal. The transducer incorporates a set of fixed swirl blades in a straight pipe. These blades impart a swirling motion to the fluid, producing an unstable oscillation and precession of fluid around the axis of the tube. The frequency of oscillation is proportional to the volumetric flow rate. A variety of transduction elements are used to detect such fluid oscillation (piezoceramic disk, hot-wire anemometer, etc.).

| 117 | An *ultrasonic flowmeter* employing the Doppler principle contains a piezoelectric transmitter that generates ultrasonic waves in the liquid traveling in a pipe. Two sonic receivers (also piezoelectric) are located at equal distances upstream and downstream from the transmitter. Because of the additive and subtractive effects of the liquid's velocities, the frequency of signals obtained from the receivers are functions of the sum and difference of the velocities. By taking the difference between the two frequencies, the output that is proportional to the liquid velocity can be used for signal conditioning. |

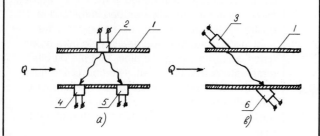

Flowmeter with one ultrasonic transmitter and two receivers (a) or flowmeter with one sonic transmitter and one receiver (b). Q = flow, 1 = pipe, 2 and 3 = ultrasonic-wave transmitters, 4, 5 and 6 = sonic receivers.

In other designs of ultrasonic transducers, measurement of the pulse propagation time or the pulse's phase shift are employed to detect the liquid's velocity. Physical configurations of transducers differ from each other by the number of transmitting and receiving elements, their location along the liquid-carrying pipe, and the wave-propagation path.

Level Sensors

| 118 | A *pressure-type level sensing system* contains a pressure transducer mounted at the bottom of a liquid-filled tank. The transducer responds to the pressure developed by the weight of the liquid's column. This pressure is directly proportional to the measured height. |

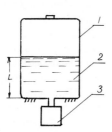

Pressure-type level sensing system. L = level, 1 = tank, 2 = liquid, 3 = pressure transducer.

| 119 | A *weighing sensing system for measuring level* determines the level with load cells placed underneath the bottom of the tank or connected to the tank by a mechanical link. If the tank's weight and liquid's density are known, the level is readily calculated using data obtained with the cells. |

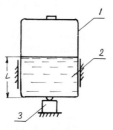

Weighing sensing system for measuring level. L = level, 1 = tank, 2 = liquid, 3 = load cell.

| 120 | In a *float-type level sensor* the buoyancy force holds the float on the surface of the liquid. The float carries a member having a magnetic coupling with a transduction element (coil, magnetic reed, or Hall-effect switch), that is mounted on the outside wall of the tank and can be actuated by the proximity of the float. |

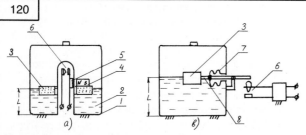

Float-type sensors with magnetic coupling (a) or mechanical link (b). L = level, 1 = tank, 2 = liquid, 3 = float, 4 = magnet, 5 = magnetic armature, 6 = contacts, 7 = bellows, 8 = lever.

In some designs, the float mechanically links the switching mechanism through the sealing in the wall (e.g., bellows). The switching system can respond to the restraining force developed by a spring element connected to the float or by an actuator of a force-balance servo system.

121

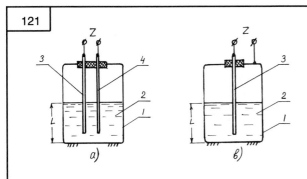

Capacitive level sensors with two (a) or one (b) electrodes. L = level, Z = impedance, 1 = tank, 2 = liquid, 3 and 4 = electrodes.

Conductivity and capacitive level sensors serve as a continuous and point-level sensors by measuring the impedance between two electrodes immersed in the liquid or between one electrode and the electroconductive tank's wall.

122

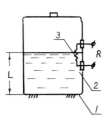

Variable-inductance level transducer (a), and transformer-type level transducer (b). L = level, Z = impedance, 1 = tank, 2 = liquid, 3 = coil, 4 = core.

An *inductive level transducer* finds its application in the measurement of the level of liquid metals and other electroconductive liquids. In one of the designs, a coil is wound around a tube containing the liquid. The inductance of this coil changes rapidly as the liquid moves and approaches the coil. In another design, the transducer is introduced by a transformer with a primary coil wound on one limb of a twin-limbed iron core. The other limb is enclosed by a tube containing the liquid and forming one turn of the secondary winding. The effective resistance of this turn is inversely proportional to the height of the liquid column in the tube. The change in the height can be sensed by measuring the power consumption at the primary coil.

123

Heat-transfer level sensor. L = level, R = resistance, 1 = tank, 2 = liquid, 3 = resistive heated element.

Heat-transfer level sensors are built from a heated (usually self-heated) wire, thermistor, or thermocouple, whose heat transfer undergoes a step change when the transition from gas to liquid takes place. This change causes the change in the element's resistance or electromotive force.

124

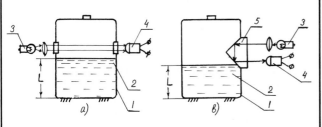

Transmittance-mode (a) and reflectance-mode (b) photoelectric level sensors. L = level, 1 = tank, 2 = liquid, 3 = light source, 4 = photodetector, 5 = prism.

Photoelectric level sensors operate in transmittance or reflection modes. In the transmittance mode, a sensing system, including a light beam source and a photodetector, responds to the interruption or the attenuation of the light beam when the liquid breaks the beam path from the source to the detector. In the reflection mode, an optical prism mounted inside a tank changes the reflectance of the light when it is immersed in the liquid. The construction of the transducer is arranged so that a light source and photodetector for sensing the change in the light's intensity are mounted on the outside wall of the tank. The light beam passes through and is reflected from the faces of the prism.

125	In a *vibrating-element level sensor,* the oscillations of a member (paddle) are damped when it is immersed in the liquid. The attenuation of oscillations indicates that the liquid has reached the measured level. The oscillations are stimulated and sensed by electronic means.

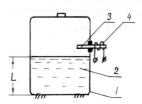

Vibrating-element level sensor. *L* = level, 1 = tank, 2 = liquid, 3 = vibrating paddle, 4 = excitation coil.

126	Several sensing techniques are used in *ultrasound level sensors.*

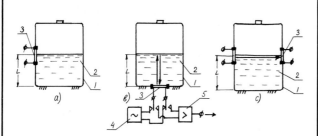

Ultrasound level sensors. a, b, and c = level-sensing systems with one crystal at side (a), bottom (b), and two crystals at side (c) of tank; *L* = level, 1 = tank, 2 = liquid, 3 = piezoelectric crystal, 4 = pulse generator, 5 = pulse receiver.

a) *Oscillations of quartz,* ceramic or magneto-strictive elements at an ultrasound frequency have a greater amplitude in gas than in liquid. Wetting the elements causes a decrease in the amplitude, providing the detection of the liquid level.

b) Point-level or continuous-level sensing is provided by *measuring* the *time lapse between the transmission and reception* of the *ultrasound pulses* generated by ceramic crystals at the bottom of the tank. Usually one crystal acts, alternately transmitting and receiving pulses that pass along the liquid height and are reflected from the surface back to the tank bottom. Some constructions contain separate elements for generating and receiving the pulses.

c) A *point-level detection* is also performed by two piezoceramic crystals oriented toward each other across the inside of a tank. One of the crystals transmits ultrasonic waves and the other one receives them. The transmission is intensified when the liquid wets the crystals. The increase in the output voltage of the receiving crystal indicates that the level has reached the specific point.

Humidity and Moisture Sensors

127	*Resistive and/or capacitive hygrometers* have sensing elements that absorb or give up moisture in a hygroscopic layer until equilibrium is reached with the ambient water vapor pressure. Two electrodes are in contact with the layers, whose resistance and/or capacitance (often leaky capacitance) varies with the humidity. A variety of materials are used to form the layer; among them are polymer films, lithium chloride, an aqueous solution of hygroscopic salt, a carbon-powder suspension in a gelatinous cellulose, aluminum oxide, some experimental materials like lead iodide, polyelectrolyte combinations with ion exchange resins, and other materials.

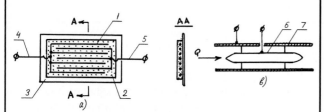

Hygrometer with hygroscopic layer (a) and capacitive-type hygrometer (b). *Q* = gas flow, 1 = electrode, 2 = hygroscopic layer, 3 = substrate, 4 and 5 = leads, 6 and 7 = internal and outer concentric electrodes.

One of the modifications of the capacitive hygrometer is a capacitor formed by two concentric electrodes. The measured gas flows between them. The dielectric constant of the gas depends on the quantity of water vapor in the gas.

128	

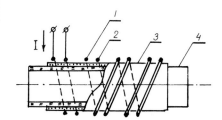

Electrolytic hydrometer. I = current, 1 and 2 = spiral electrodes, 3 = P_2O_5 film, 4 = pipe.

An *electrolytic hygrometer* contains a thin pipe coated with a thin layer of phosphorus pentoxide (P_2O_5) that fills the space between two platinum electrodes. When the humid gas flows through the pipe, the moisture in the gas is absorbed by the layer. A direct current flowing through the wires decomposes water into gaseous hydrogen and oxygen. The current required for electrolysis is directly proportional to the concentration by volume of the water in the gas.

129	

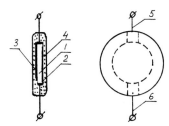

Oscillating-crystal humidity sensor. 1 = quartz crystal, 2 and 3 = electrodes, 4 = hygroscopic coating, 5 and 6 = leads.

In an *oscillating-crystal humidity sensor,* the frequency of oscillation of a quartz crystal exposed to the humid gas depends on the water content in the gas. The crystal is coated with a thin hygroscopic layer that absorbs water from the gas until equilibrium is reached with the ambient water vapor pressure. The crystal's effective vibrating mass depends on the amount of water absorbed by the layer. The change in this mass alters the frequency.

130	

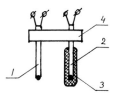

Psychrometer. 1 and 2 = temperature-sensitive elements (thermistors), 3 = wet wick, 4 = holder.

Humidity measurements by *psychrometers* are based on the electronic sensing of temperatures of dry and wet elements that are in contact with the humid gas. Thermistors and thin-wire platinum or nickel resistors are usually used for reading the temperatures.

131	

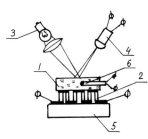

Dew-point sensor. 1 = mirror, 2 = cooler, 3 = light source, 4 = photosensor, 5 = heat sink, 6 = temperature sensor.

The dew point, measured by *dew-point sensors,* is the temperature at which the liquid and vapor phases of a fluid are in equilibrium. The most commonly used sensor contains a mirror cooled by a Peltier-effect semiconductor cooler. The temperature of the mirror is sensed by a resistive thermometer embedded in the mirror. Light from a light emitting diode is directed toward the mirror surface. The reflected light from the surface is picked up by a photosensor. A condensate is formed on the mirror surface when its temperature is decreased. The condensate causes a scattering of light that is detected by the photodetector. The reading of the temperature at this condition gives the dew point.

132	

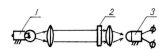

Infrared-type humidity sensor. 1 = light source, 2 = optical filter, 3 = photodetector.

The operation of *infrared-type humidity sensors* relies on the effects of absorption of infrared radiation by humid gases. It is noteworthy that the absorption is selective and is pronounced at specific wavelengths. Sensors contain light sources, filters, and photodetectors, and operate similarly to the spectroscopic devices. The moisture content is a function of photocurrent.

The absorption of a microwave is also indicative for the measurement of humidity.

Vacuum Sensors

<table>
<tr>
<td>

133

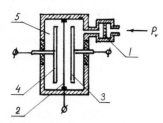

Diaphragm-type manometer. P_v = vacuum, 1 = gas filter, 2 = diaphragm, 3 and 4 = electrodes, 5 = evacuated and sealed chamber.

</td>
<td>

The most typical *diaphragm-type manometer* used for measuring a vacuum is a variable-capacitance transducer with a metal diaphragm deflected by the measured pressure. The diaphragm is positioned close to stationary plates. The pressure is measured by sensing the capacitances formed by the diaphragm and plates. There is no significant difference between this transducer and the conventional capacitive pressure gage, which is used for measuring absolute or differential pressure. The specific conditions for measuring a vacuum require using filters to keep the sensing chamber free of contamination. Incorporating a getter stabilizes the operation of the transducer by absorbing gas molecules in the reference cavities. Some transducer designs require a high-temperature bake-out of the elements to drive gas molecules out of the surfaces that are in contact with the fluid. This outgassing is performed prior to measurements.

</td>
</tr>
<tr>
<td>

134

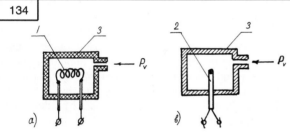

Filament (a) and thermistor (b) Pirani gages. P_v = vacuum, 1 = self-heated filament, 2 = self-heated thermistor, 3 = case.

</td>
<td>

The *Pirani gage* operation is based on the effect of heat dissipation from a heated wire or thermistor exposed to the vacuum under measurement. The loss of heat leads to a change in the temperature and resistance of the heated elements. Read-out of the change in the resistance is provided by means of a Wheatstone bridge, one arm of which can be used for compensating for temperature.

</td>
</tr>
<tr>
<td>

135

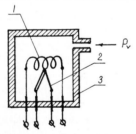

Thermocouple gage. P_v = vacuum, 1 = filament, 2 = thermocouple, 3 = case.

</td>
<td>

Thermocouple gages employ the Pirani gage's physical principle with one difference: the temperature change is sensed by a thermocouple generating an electromotive force proportional to the temperature or to the decrease in the measured pressure. Heating is provided by an alternating current. Another thermocouple incorporated in the sensor's circuit is used to compensate for ambient temperature variation.

</td>
</tr>
<tr>
<td>

136

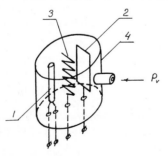

Hot-cathode ionization gage. P_v = vacuum, 1 = hot cathode, 2 = negatively charged plate, 3 = grid, 4 = glass envelope.

</td>
<td>

Hot-cathode ionization gages make use of the effects of positive ions produced because of the collisions of electrons with gas molecules. The number of positive ions is proportional to the gas concentration. The hot-cathode ionization is created by electrons emitted from a heated cathode. A negative-charged plate placed inside the pressure chamber attracts the positive ions. The current through the gage is a measure of the vacuum.

</td>
</tr>
</table>

137 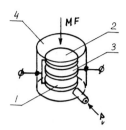 Cold cathode gage. P_v = vacuum, MF = magnetic field, 1 and 2 = cathodes, 3 = anode, 4 = glass envelope.	A source of ionization in a *cold-cathode gage* is a high voltage impressed between two plates in a pressure chamber. The voltage creates ionization of molecules and their flow between the electrodes. The flow path is increased by the addition of a transverse magnetic field that stimulates the helical-path motion of the electrons. With the increase of the path, the number of ionizing collisions is also increased, producing a greater current, which allows easier measurement of the vacuum.
138 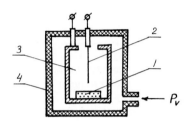 Alpha-particle ionizing sensor. P_v = vacuum, 1 = alpha-particle source, 2 = ion collector, 3 = vacuum chamber, 4 = case.	An *alpha-particle ionizing sensor* makes use of the radioactive ionization effect due to the bombardment of gas molecules by alpha particles. A negatively-charged electrode in the pressure chamber attracts positive ions of the gas. The flow of ions is measured and indicates the vacuum degree.
139 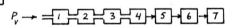 Gas-analysis vacuum sensor. P_v = vacuum input of mass spectrometer, 1 = vacuum chamber, 2 = ion source, 3 = mass analyzer, 4 = electron multiplier, 5 = preamplifier, 6 = amplifier and signal conditioner, 7 = display.	*Gas-analysis vacuum sensors* are virtually quadrupole mass spectrometers, with which the molecular density of a gas components and the equivalent partial pressure can be determined. The density of the gas composition can be then easily translated into total pressure.

Densitometers	
140 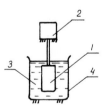 Tubular-float displacer densitometer. 1 = float, 2 = force-sensing element, 3 = liquid, 4 = container.	The operation of the *tubular-float displacer densitometer* is based on the effect explained by Archimedes' Law. A displacer, fully submerged in the measured liquid, is buoyed up by a force equal to the weight of the displaced liquid. The buoying force is proportional to the density of the liquid. The float is mechanically connected to the force-sensing element that provides an electrical output proportional to the density.
141 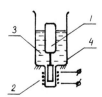 Hydrometer-type densitometer. 1 = float, 2 = displacement-sensing element, 3 = liquid, 4 = container.	A *hydrometer-type densitometer* contains a weighted float immersed in the measured liquid. The buoyancy of the liquid causes the float to rise or sink depending on the liquid density. The displacement of the float is converted into an electrical signal by an inductive, photoelectric, or other transduction element.

142

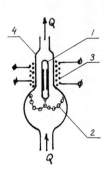

Chain-balanced-float density transducer. Q = flow, 1 = float, 2 = chain, 3 = displacement-sensing element, 4 = container.

A *chain-balanced-float density transducer* also uses the weighted float. This float is totally submerged and connected by a chain to the chamber with the liquid to be measured. The force sinking the float depends on the effective chain weight acting on the float. The vertical displacement of the float is a function of the density. A linear variable differential transformer, responding to the float position, provides an output voltage proportional to the density.

143

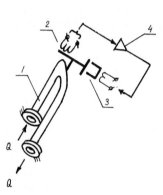

Vibrating U-shaped-tube densitometer. Q = flow, 1 = U-shaped tube, 2 = pickup coil, 3 = driving mechanism, 4 = feedback-loop amplifier.

In a *vibrating U-shaped-tube densitometer,* the natural mechanical frequency of vibration of the fluid-filled tube changes when the density of the fluid varies.

The transducer includes electromechanical elements and a feed-back loop amplifier maintaining the vibrations and providing a frequency output determined by the density.

144

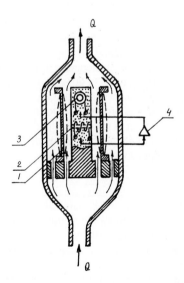

Vibrating-cylinder densitometer. Q = flow, 1 = vibrating hollow cylinder, 2 and 3 = elements of electromagnetic pickup and drive, 4 = feedback-loop amplifier.

A sensitive element of the *distributed-spring-mass-vibrating-element-densitometer* transducer is an elastic member (diaphragm, hollow cylinder, end-fixed beam, vane, and quartz crystal), which is constrained to vibrate at its resonance frequency. The member is in contact with the measured fluid. With a change in the fluid density, the effective mass of the vibrating member is also changed, providing the deviation of the resonance frequency.

145

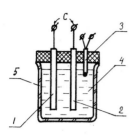

Capacitive density sensor. C = capacitance, 1 and 2 = electrodes, 3 = temperature sensor, 4 = liquid, 5 = container.

In *capacitive sensors,* the density of liquid is measured using a known relationship between the density and the dielectric constant of the liquid. Two electrodes are immersed in the measured liquid whose dielectric constant variation is sensed by measuring the capacitance between the electrodes. For adequate readings, the liquid's temperature is measured simultaneously.

146

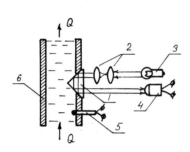

Photoelectric density sensors. Q = flow, 1 = prism, 2 = lenses, 3 = light source, 4 = photosensor, 5 = temperature sensor, 6 = pipe.

Photoelectric density sensors make use of the correlation between the refractive index of the fluid and its mass density. The measurement of the index of refraction of a substance is provided with a refractometer, which allows observation of an interference pattern produced by passing light through the substance.

147

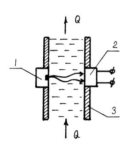

Nucleonic density measuring system. Q = flow, 1 = source of gamma rays, 2 = radiation detector, 3 = pipe.

A *nucleonic density measuring system* includes a source of gamma rays that directs radiation through a measured liquid. A radiation detector, receiving the passed rays, produces an electrical output depending on the density; the higher the density, the greater the attenuation, the lower the output.

148

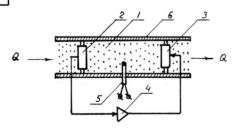

Acoustic density sensor. Q = flow, 1 = fluid, 2 = microphone, 3 = sound generator, 4 = feedback-loop amplifier, 5 = temperature sensor, 6 = pipe.

The *sonic sensing of density* relies on measuring the sound velocity in fluids. The changes in velocity are proportional to the density at constant pressure and temperature. The transducer includes a sound source and microphone. The wave's propagation time is measured by sensing the phase difference between the transmitted and received signals. Indirectly, the sound velocity can be determined by measuring the resonant frequency of a cavity that is excited by a microphone-sound source-feedback-amplifier circuit.

Viscometers	
149 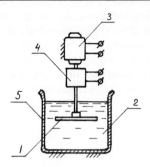 Rotating-element viscometer. 1 = rotating disk, 2 = liquid, 3 = constant-speed motor, 4 = torque-sensing element, 5 = container.	*Rotating-element viscometers* contain a member (disk, cone, or cylinder), which is forced to rotate in the measured fluid. A torque-sensing element detects the torque developed due to viscous drag. The torque is a measure of the viscosity.
150 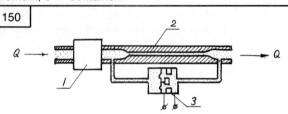 Differential-pressure viscometer. Q = flow, 1 = pump, 2 = capillary tube, 3 = differential pressure transducer.	In a *differential-pressure viscometer,* the measured fluid is made to flow with a constant rate through a capillary tube. The pressure drop across the tube, measured by a differential pressure transducer, is directly proportional to the viscosity if the flow is laminar.
151 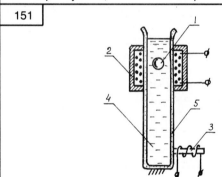 Falling-piston or falling-ball viscometer. 1 = ferromagnetic ball, 2 = electrical magnet, 3 = magnetic switch, 4 = liquid, 5 = container.	A *falling-piston or falling-ball viscometer* is comprised of a tube with a measured liquid, piston or ball inside the tube, an electrical magnet, and a magnetic switch. The piston or ball is made of a ferromagnetic material. It is first lifted to the top of the tube by the magnet and is then permitted to drop to the bottom of the tube under the force of gravity. The moment when the piston or ball reaches the bottom is sensed by a magnetic switch. The time required to pass the length of the tube is proportional to the viscosity.
152 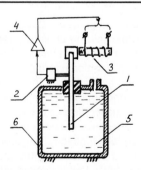 Vibrating-element viscometer. 1 = vibrating beam, 2 = electronic sensor of vibrations, 3 = electromagnetic actuator, 4 = feedback-loop amplifier, 5 = liquid, 6 = container.	The operating principle of the *vibrating-element viscometer* is based on the measurement of the damping effect of a viscous fluid on the vibration of a solid member immersed in a fluid. The member can be shaped as a disk, beam, paddle, or hollow cylinder. The vibrations are maintained by an electromechanical system. The amplitude of vibration of the member is increased with the decrease of viscosity. This amplitude is electronically sensed and is indicative of the viscosity.

| 153 | In a *float-type viscometer,* a viscosity-sensitive float is displaced by a drag force developed by the constant-rate flow of the fluid to be measured. The float is placed in a tapered tube similar to that in a rotameter. The displacement of the float along the tube axis is proportional to the viscosity. The displacement is electronically measured and calibrated in terms of viscosity. |

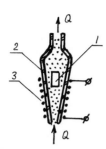

Float-type viscometer. Q = constant-rate flow, 1 = float, 2 = tapered tube, 3 = variable-inductance coil.

Thermometers

| 154 | A *thermocouple thermometer* contains a junction of two dissimilar metals. The contact potential at the junction varies with temperature. Usually a thin wire is used for the thermocouple's fabrication. However, foil-made design versions are also used. |

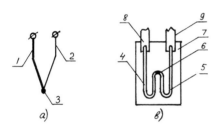

Wire (a) and foil (b) thermocouples. 1 and 2 = conductors made of different metals, 3 = sensing junction, 4 and 5 = foil conductors, 6 = butt-bonded junction, 7 = matrix, 8 and 9 = lead ribbons.

| 155 | *Metal-resistance thermometers* are made, as a rule, of a thin platinum or platinum-alloy wire wound on a ceramic mandrel, or of a wire held by an insulated mount. Nickel and nickel-alloy wires are also used. In some designs, the temperature-sensitive layer is formed by a deposition of platinum on a ceramic substrate. The resistance of the thermometers is a close-to-linear function of temperature. |

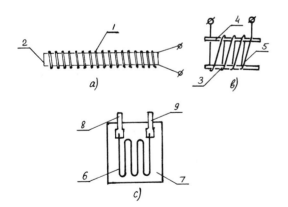

Wound-wire (a), held-wire (b), and deposited resistive (c) thermometers. 1 = coil of platinum wire, 2 = mandrel, 3 and 4 = insulating holders, 5 = platinum-wire grid, 6 = deposited platinum grid, 7 = ceramic substrate, 8 and 9 = lead ribbons.

| 156 | A *thermistor* is made of a mixture of semiconductor powder compounds (e.g., NiO, Mn_2O_3, and Co_2O_3), which are shaped as cylinders, disks, or beads. The material is compressed, sintered at a high temperature, provided with two leads, and insulated. The change in temperature is read as a change in resistance. The majority of thermistors have a negative temperature coefficient; however, devices with a positive coefficient (posistor) are also in use. |

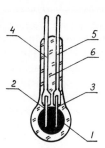

Thermistor. 1 = semiconductor powder compound, 2 and 3 = leads contacting compound, 4 and 5 = leads for outside connection, 6 = glass hermetic seal.

| 157 | A *germanium resistance thermometer* is used primarily for the cryogenic range of temperatures. The transduction element is fabricated of an arsenic-, gallium-, or antimony-doped germanium crystal that is provided with leads assuring a good contact with the crystal. The element is sealed in a case that has good thermal conduction. The resistance measured between the leads is a nonlinear function of temperature. Another version of the element is a voltage-output, four-lead structure fed by a stable current. |

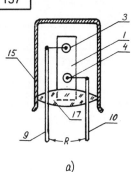 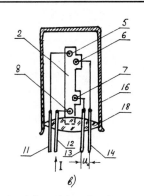

a)   b)

Two-lead germanium thermometer (a) and four-lead germanium thermometer (b). I = feeding current, R = resistance, U_o = output voltage, 1 and 2 = germanium elements, 3 through 8 = gold wire bonds, 9 through 14 = leads, 15 and 16 = heat-conductive cases, 17 and 18 = glass hermetic seals.

| 158 | *Gallium-arsenide or silicon diode or transistor thermometers* provide a voltage output proportional to the measured temperature. The voltage is measured across a forward-conducting p-n junction when the diode or transistor is fed with a stable current. |

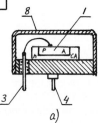 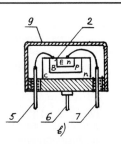

a)   b)

Temperature-sensitive diode (a) and temperature-sensitive transistor (b). A = anode, B = base, C = collector, CA = cathode, E = emitter, n = n-silicon, p = p-silicon, 1 = temperature-sensitive *p-n* junction, 2 = temperature-sensitive n-p-n structure, 3 through 7 = leads, 8 and 9 = cases.

| 159 | A *capacitance thermometer* contains a multilayer glass/ceramic metallized structure whose dielectric constant varies with the temperature change. The dielectric constant change is sensed by using the capacitance measuring technique. |

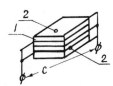

Capacitive thermometer. C = variable capacitance, 1 = glass or ceramic layer, 2 = metallized area.

160	The operating principle of the *inductance thermometer* is based on the effect of change in the magnetic permeability of certain materials with temperature. A variation in the permeability is sensed by measuring the inductance of a coil containing the temperature-sensitive core.
Inductance thermometer. 1 = ferromagnetic temperature-sensitive core, 2 = coil.	
161 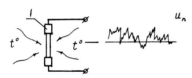 Acoustic thermometer with flowing gas (a) or enveloped gas (b). Q = flow of gas whose temperature is to be measured, 1 = flowing gas, 2 and 3 = microphones, 4 and 5 = sound generators, 6 = pipe, 7 = object whose temperature is to be measured, 8 = enveloped gas, 9 = gas container, 10 and 11 = feedback-loop amplifiers.	The transduction in an *acoustic thermometer* relies on the principle that the sound velocity in substances depends on temperature. In gases, the velocity is proportional to the square root of the absolute temperature. In solids and liquids, the velocity decreases as the temperature increases. Two categories of acoustic sensing systems are used in temperature measurements. a) The *system measures the acoustic characteristics of the medium* whose temperature is measured. b) The *system measures the acoustic characteristics of the object* that is in thermal equilibrium with the measured medium. In both cases, the most common arrangement for detecting the temperature is to measure the resonance or pulse-propagation characteristics of the materials.
162 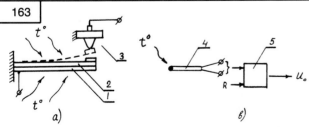	A *noise thermometer* is based on the effect of electron random motion in a conductor if its temperature is above absolute zero. This motion of electrons produces an alternating difference in potentials at the conductor terminals. The magnitude of the potentials increases with temperature. These potentials, after amplification, are a measure of the temperature.
Noise thermometer. $t°$ = temperature, U_n = noise voltage, 1 = resistor.	
163	Two types of *temperature switches* are used in control and measurement. a) *Directly actuating temperature switches* contain a bimetallic element that changes its shape with temperature. The change of shape occurs due to the difference in the thermal coefficients of expansion of the two mechanically jointed elastic parts. When the temperature reaches the prescribed level, the deformed parts provide closure of an electrical contact. b) An *analog thermometric sensor* can operate as a *switch* if its output is fed into an electronic circuit triggering at the prescribed level of the thermometer output signal.
Directly actuating temperature switch (a) and switch with analog thermometric sensor (b). R = reference signal, $t°$ = temperature, U_o = output signal, 1 and 2 = bimetal layers having different coefficients of expansion, 3 = electrical contact, 4 = analog thermometric sensor, 5 = comparator.	

Radiation Pyrometers and Heat Flux Sensors	
164 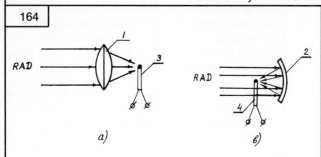 Total radiation pyrometer. a = calorimeter with lens, b = calorimeter with mirror; RAD = radiation, 1 = lens, 2 = concave mirror, 3 and 4 = temperature-sensing elements.	A *total radiation pyrometer* consists of a lens or mirror focusing radiation on thermocouples, thermopiles, thermistors, or bolometers. The transduction elements produce an electrical signal output, which gives a temperature reading directly.
165 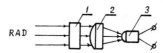 Narrow-band pyrometer. RAD = radiation, 1 = filter, 2 = lens, 3 = photodiode.	In a *narrow-band pyrometer,* the incoming radiation passes through optical filters and is directed to an IR photon detector that responds to a narrow band of wavelengths. The output of the detector is proportional to the measured temperature.
166 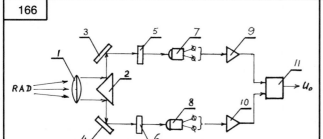 Ratio pyrometer. RAD = radiation, U_o = output signal, 1 = lens, 2 = prism, 3 and 4 = mirrors, 5 and 6 = selective filters, 7 and 8 = photosensors, 9 and 10 = amplifier, 11 = signal divider.	A *ratio pyrometer* contains optics with filters, which transmit the radiation energy from the heat-radiating body in two narrow wavelength bands. The split energy is focused onto separate detectors. The ratio of signals from the two detectors is indicative of temperature.
167 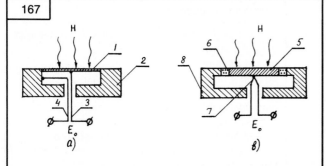 Foil-type calorimeter (a) and sensing-mass calorimeter (b). E_o = electromotive force, H = heat flux, 1 = foil (constantan), 2 = heat sink (copper), 3 and 4 = copper leads, 5 = sensing mass, 6 = thermal insulation, 7 = thermocouple, 8 = case.	In a *calorimeter,* a sensitive area can be introduced by a foil, sensing mass, or wafer that is opened to the measured flux. The temperature of the sensitive elements is directly proportional to the flux intensity. Thermocouples are primarily used for the measurement of elements' temperature. They can be integrated with the sensitive area (foil and thermopile calorimeters), or be attached to the sensing mass ("slug calorimeter"). The voltage developed by the thermocouples is a measure of the heat flux.

168

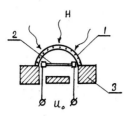

Radiometer. H = radiant flux, U_o = output signal,
1 = transparent window, 2 = temperature-sensitive element, 3 = housing.

Radiometers have a construction that is similar to that of radiation pyrometers. They respond to an incident radiant flux and have an output that is calibrated in units of heat flux. The sensing surface of the transduction element is isolated by a transparent window from convective heat flux. Unlike a pyrometer, the radiometer has a much wider viewing angle. Thermopiles, resistive elements, pyroelectric detectors, and bolometers are used for the conversion of the radiant flux into an electrical output.

169

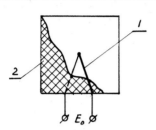

Surface heat-flow sensor. E_o = output electromotive force,
1 = thermocouple, 2 = blackened sensing surface.

A surface heat-flow sensor contains a differential thermocouple or thermopile, which is mounted on a thin surface (round or square) convenient for measuring heat conduction through the sensor. The sensing surface is blackened outside to provide good emittance or absorptance of the heat flow. Depending on the application, the output signal can be proportional to one of the components of the heat flow (due to conduction, radiation, or convection) or to the combination of the components.

Optical Detectors

170

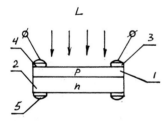

Photovoltaic detector. L = light, 1 = p-type silicon,
2 = n-type silicon, 3 = plated electrode, 4 = positive terminal contact, 5 = negative terminal contact.

Photovoltaic detectors generate an electrical output proportional to the illumination of the junction of two different materials or of the junction of similar materials containing different impurities. Typical materials widely used in the past were combinations of iron and selenium or copper and copper oxide. Contemporary detectors almost invariably contain materials such as silicon, germanium, indium arsenide, and indium antimonide. Doping of different parts of the material forms a sensitive-to-light p-n junction. Usually the p layer is exposed to the incident light and produces electrical charges. These elements belong to homojunction detectors. A heterojunction detector employs a combination of dissimilar materials, for instance, glass and crystal silicon, nGe-nSi, pCu_2S-nCdS, pSi-nCdS, and other compounds.

A typical structure of the photovoltaic detector contains a thin layer of a p-type material exposed to light. This layer forms a junction with the more massive, n-type portion of the semiconductor. P and n surfaces are provided with plated metal electrodes and solder terminals for connecting leads.

171

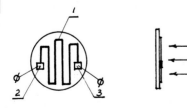

Photoconductive detector. L = light, 1 = photoconductive material, 2 and 3 = conducting electrodes.

Photoconductive detectors are based on the effects of change in the semiconductor's resistance when the material is exposed to light. Almost all semiconductors decrease in resistance with the increase of incident illumination. This decrease takes place because the number of charge carriers is proportional to the energy absorbed from the photon's energy. The basic photoconductive cell contains a semiconducting material

(continued)

171	*(continued)*

between two conducting electrodes, which are used to connect the leads. The most common materials used for the cells are lead sulfide and lead selenide polycrystalline films and also doped germanium and silicon bulk single-crystals.

Mercury cadmium telluride detectors can operate both in photoconductive and photovoltaic modes.

172

Photodiode. a = photodiode symbol, b = photodiode package, c = photodiode structure; *L* = light flux, 1 = photodiode chip, 2 = transparent window, 3 = case, 4 and 5 = leads, 6 = photodiode anode, 7 = n-type substrate, 8 = n⁺-diffused layer, 9 = silicon dioxide, 10 = metallization for anode contact.

173

Phototransistor. a = phototransistor symbol, b = phototransistor structure; *L* = light flux, 1 = silicon dioxide, 2 = p-type diffused layer, 3 = n-type epitaxial layer, 4 = n⁺ substrate, 5 = emitter contact.

In a *photoconductive/photovoltaic detector,* the p-n junction resistance and junction photocurrent change in response to the incident light.

Reverse-biased diodes and transistors are typical elements for this category of photodetectors. They can operate in both the photoconductive and photovoltaic modes. The difference between the two constructions is defined primarily by semiconductor structures, which are introduced by p-n junction *photodiodes,* p-i-n (*i* = intrinsic) photodiodes, avalanche photodiodes, *photofets,* pnp or npn *junction transistors,* and *photodarlingtons.*

174

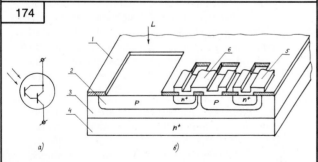

Photodarlington. a = photodarlington transistor symbol, b = photodarlington structure; *L* = light flux, 1 = silicon dioxide, 2 = p-type diffused layer, 3 = n-type epitaxial layer, 4 = n⁺ substrate, 5 = emitter contact, 6 = metallization.

175	*Photoemissive detectors* are either evacuated or gas-filled tubes containing a cathode and one or more anodes. When photons impinge on the cathode, the electrons are ejected from the cathode surface and are accelerated toward the anode that is at a positive potential with respect to the cathode.

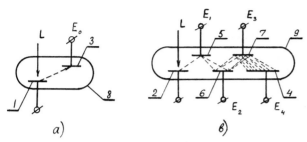

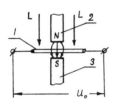

Photoemissive cell (a) and photomultiplier (b). L = light flux, E_0 through E_4 = differences in potentials between cathode and one of electrodes, 1 and 2 = cathodes, 3 and 4 = anodes, 5, 6, and 7 = dynodes, 8 and 9 = glass envelopes.

The photoelectric current increases proportionally to the intensity of the illumination. This design is the simplest diode version of the detector. In order to increase the sensitivity, several more electrodes (dynodes) are added to the construction of the detector. In this device, called a photomultiplier, the electrons that are ejected from the cathode are focused on one of the dynodes. When the surface of the dynode is struck, an increased number of electrons are liberated. The electrons flow to the next dynode, where the process is repeated with a progressively increasing number of electrons. The dynodes have sequentially higher positive potentials with respect to each other. As a result, the output current has a significantly increased magnitude, which defines the high sensitivity of the detector.

176	Indium antimonide and mercury cadmium telluride semiconductors are used in *photoelectromagnetic* optical *detectors.* The magnetic field applied to the semiconductor separates the electrical charges created by the illumination of the semiconductor surface. The positive and negative carriers (holes and electrons) migrate to different parts of the semiconductor, developing an electromotive force proportional to the light intensity. The emf is picked up at the end terminals mounted at the areas with concentrated charges.

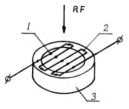

Photoelectromagnetic optical detector. L = light flux, U_o = output voltage, 1 = semiconductor crystal, 2 and 3 = poles of magnet.

177	*Thermoelectric detectors* contain thermopiles (multijunction thermocouple devices) whose sensing junctions become warmer than the reference junctions when the incident radiant flux strikes the sensitive area. The reference junctions are mounted on a heat sink and remain at the temperature of the sink. This temperature can be assumed constant if the time of the measurement is not too long. The voltage developed by the thermopile is a measure of the radiant flux. To preserve the sensitive area, most detectors are sealed and back-filled with an inert gas.

Thermoelectric detectors. RF = radiant flux, 1 = sensing junction, 2 = reference junction, 3 = heat sink.

178	A *radiation bolometer* consists of a fine temperature-sensitive resistive detector (a thermistor, or a germanium, silicon, or indium antimonide die), which absorbs radiation. The radiation causes a change in the detector's temperature and consequent variation in its resistance. The radiation-sensing detector is usually combined with a similar element that is shielded from the radiation. This reference element is used for thermal compensation. The two elements are connected as arms of a Wheatstone bridge circuit that is balanced when the sensitive element is not subject to the radiation. The output voltage of the bridge circuit is proportional to the incident radiant flux.

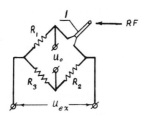

Radiant bolometers. RF = radiant flux, R_1, R_2, and R_3 = resistors of electrical bridge, U_{ex} = excitation voltage, U_o = output voltage, 1 = temperature-sensitive resistive element.

179	*Pyroelectric detectors* contain parts made of materials that exhibit electrical polarization as a response to the change in the material's temperature. Changes in incident radiant flux cause the change in the temperature. As a result, charges proportional to the flux are developed in the material. These charges produce a difference in potentials between the two electrodes contacting the material. Since the internal impedance of the element is capacitive in nature, a high input-impedance voltmeter should be used to measure the output signals. Chopping the flux and forming an ac output signal are a common technique used for measurements. Among the materials that reveal the pyroelectric effect are triglycine sulfate (TGS), lithium tantalate ($LiTaO_3$) and polyvinyl fluoride (PVF_2).

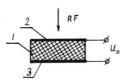

Pyroelectric detectors. RF = radiant flux, U_o = output voltage, 1 = pyroelectric material, 2 and 3 = electrodes.

180	A *pressure-actuated detector* contains a capsule with a diaphragm at each end. One of the diaphragms is blackened and exposed to light. Another diaphragm is provided with a mirror. The capsule is sealed and filled with an inert gas. Radiation causes an increase in the diaphragm's temperature. Heat is transferred to the gas, which expands and causes a deflection of the diaphragm with the mirror. The change of the mirror's position is sensed by an optical system. It consists of a light source directing the light beam to the mirror, and of a photodetector responding to the deflection of the beam. The electrical output from the photodetector is proportional to the measured radiant flux intensity.

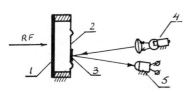

Pressure-actuated detectors. RF = radiant flux, 1 = blackened diaphragm, 2 = pressure-sensitive diaphragm, 3 = mirror, 4 = light source, 5 = photodetector.

Transducers of Electric and Magnetic Quantities

181

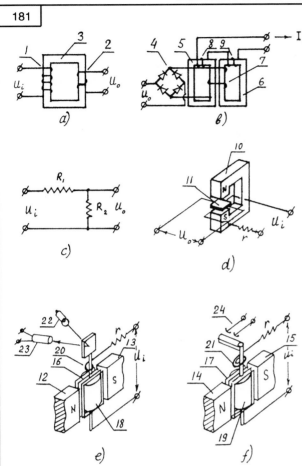

Voltage transducers. a = transformer, b = magnetic amplifier, c = voltage divider, d = Hall element, e = D'Arsonval meter's mechanism with electrooptical sensor, f = D'Arsonval meter's mechanism with contacts; I_o = output current, r = series resistor, R_1 and R_2 = voltage-divider resistors, U_i = input voltage, U_o = output voltage, 1 = primary coil, 2 = secondary coil, 3 = core, 4 = rectifier, 5 and 6 = saturable reactors, 7 = control coil, 8 and 9 = gate coils, 10 = permanent magnet, 11 = Hall element, 12 through 15 = poles of permanent magnets, 16 and 17 = coils, 18 and 19 = armatures, 20 and 21 = restraining springs, 22 = light source, 23 = photosensor, 24 = electrical contacts.

One of the typical *voltage transducers* is a voltage transformer that converts a high voltage into a low-voltage signal. Usually, a rectifier is connected to the secondary coil of the transformer to provide a dc output signal.

Another version of the voltage transducer is a magnetic amplifier with a saturable reactor. If the measured voltage is alternating, it is first rectified and then applied to the control coil of the amplifier. The current in the output (gate) coil is proportional to the measured voltage.

A simple voltage divider is also used for voltage transduction. The divider's end terminals are connected to the measured voltage, and the voltage drop across one of the resistances of the divider is regarded as the output signal.

In a Hall-effect voltage transducer, the control current is developed by the measured voltage, which is applied to the semiconductor crystal through a series-connected resistor. The crystal is placed in the gap of a permanent magnet. The Hall voltage, produced across the crystal in the direction perpendicular to the current flow and magnetic field, is directly proportional to the voltage that is measured.

A transducer with a D'Arsonval meter's mechanism permits the conversion of the measured voltage into a different electrical quantity (e.g., capacitance, photocurrent, etc.) that is convenient for signal conditioning. The meter contains a permanent magnet and a coil carrying current. The coil is pivoted and can rotate between the magnet's pole pieces. The coil's angular displacement is proportional to the current passing through the winding. This displacement is detected by a transduction element (capacitive, electrooptical, etc.) and is a function of the voltage to be measured. Although a D'Arsonval galvanometer is a current-measuring device, it can also be used to measure a dc voltage by limiting the current through the moving coil (inserting a resistance in series with the coil). The coil can be provided with a lever carrying an electrical contact that reaches a stationary contact when the coil moves. The latter contact is adjusted at a set-point voltage. Closure of the contacts provides a discrete output.

182

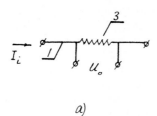

a)

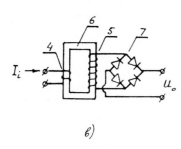

b)

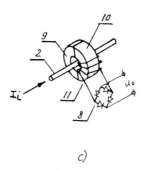

c)

Current transducers. a = shunt, b = current transformer, c = clamp-on sensor; I_i = input current, U_o = output voltage proportional to I_i, 1 and 2 = conductors, 3 = shunt resistor, 4 = primary coil, 5 = secondary coil, 6 = core, 7 and 8 = rectifiers, 9 and 10 = two halves of clamp-on core, 11 = secondary coil.

The most common *current transducers* are shunts for direct currents and current transformers for alternating currents. The shunt is a low-resistance insertion connected in series with the conductor carrying the current to be measured. The voltage drop across the shunt is the measure of the current. The current transformer contains two windings with known turns ratios. One of them (the primary) is wound with a thick wire, consists of a small number of turns, and is connected in series with the conductor in which the current is to be measured. The other winding (the secondary) is wound with a thin wire and consists of a large number of turns. It is used to form a current in the secondary that is proportional to the current in the primary. This current is usually rectified and introduced at the transducer output.

A clamp-on current sensor allows measurement of a large alternating current without direct electrical contact with the conductor. The sensor is formed by a conductor carrying current that passes through a circular laminated-iron or ferrite core. The conductor consists of a one-turn primary winding, while the secondary winding consists of a fine wire wound a number of times around the core. The core can be continuous or in two halves that are clamped together around the conductor. The current in the winding, loaded with a low resistance, provides an accurate sample of the conductor's current.

Current-controlled magnetic amplifiers, Hall-effect current sensors, and D'Arsonval galvanometers are used for measurements of small and large currents. The difference between the voltage and current sensing elements, operating on the same principle, lies in the composition of the control circuits. For current measurements, these circuits typically have low resistances. The Hall-element current meter contains an electromagnet whose field is controlled by the measured current, whereas the current through the crystal is held constant.

A number of semiconductor elements are used for transduction and amplification of small measured currents (MOSFET's, FET's, etc.) The conversion of small direct currents into alternating currents is provided by a vibrating-capacitor electromechanical system. Accurate signal conditioning of low-level ac signals is frequently easier than that of dc signals, and dc-to-ac conversion is a common technique. A high-vacuum electron tube is still in use when a very high impedance at the input of a circuit is required.

183

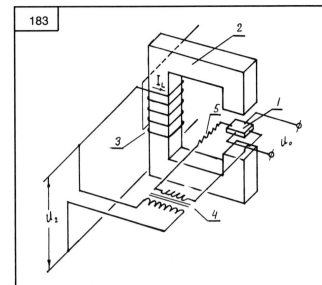

Power transducer. I_L = load current, U_L = load voltage, U_o = dc output voltage proportional to product I_L, U_L, and power factor, 1 = semiconductor crystal, 2 = core, 3 = current coil, 4 = voltage transformer, 5 = series resistance.

Power transducers are based on sensing the voltage and current and providing a product of them in order to obtain the measured power. In ac systems, the power factor and the sine of the power factor angle are taken into account for measuring the real and reactive powers, respectively. The multiplication is performed in analog or digital circuits. Convenient devices for this purpose are the Hall-effect elements. A multiplier can be built by putting a semiconductor crystal in the gap of an electromagnet such that an alternating current passing through the electromagnet's winding generates the flux, and ac voltage produces the control current. In this case, the dc Hall voltage is proportional to the true power measured in watts (voltage times current, times power factor). Besides this, a double-frequency ac voltage component is developed at the output, and it is proportional to the volt-amperes in the circuit. Measuring the power in a three-phase system requires two Hall-effect elements.

184

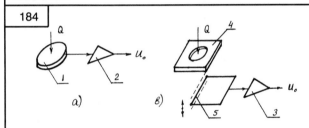

Charge transducers. a = charge-collecting type, b = vibrating-member type; Q = electrical charge, U_o = output voltage, 1 = surface accumulating charges, 2 and 3 = high-input-resistance amplifiers, 4 = metallic plate, 5 = vibrating electroconductive member.

There are several techniques for *measuring electrostatic charges*. They are primarily based on collecting the charges at an electroconductive surface and consequently measuring the overall potential of the surface. Another approach is to move (vibrate) an electroconductive member in the field and to sense the variation in the charges of this member.

185

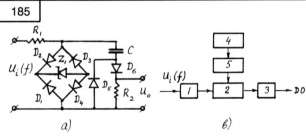

Frequency converters. a = frequency-to-dc converter, b = block diagram of frequency-to-digital converter; DO = digital output, $U_i(f)$ = input voltage of frequency f, U_o = output voltage proportional to f, R_1 = series resistor, D_1 through D_4 and Z_1 = rectifying diodes and Zener diodes providing near-square-wave, constant-amplitude voltage, C = charging capacitor, D_5, D_6, and R_2 = rectifying network, 1 = pulse shaper, 2 = gate, 3 = output counter, 4 = crystal oscillator, 5 = preset time base counter.

Unlike most transducers, *frequency- and time-interval converting devices* are solely electrical circuits converting periodic signals into their digital or analog equivalents. They are mainly frequency-to-dc converters and frequency-to-digital converters.

186

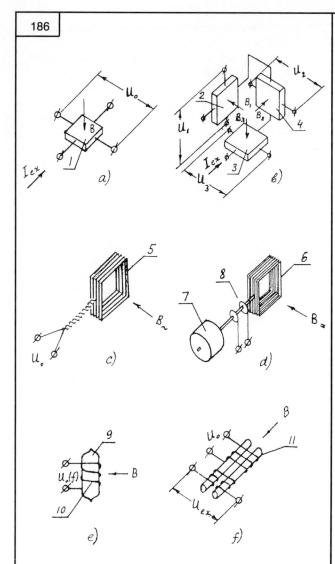

Magnetic-flux-density sensors. a = one-directional-Hall-element sensor, b = three-directional-Hall-element sensor, c = inductor probe for alternating field, d = spinning-coil probe for steady-state magnetic field, e = nuclear-resonance-magnetometer sensor, f = flux-gate magnetometer sensor; B = flux density, B_1, B_2, and B_3 = flux densities along three mutually orthogonal axes, $B_=$ = flux density of a steady-state field, $B_\sim$ = flux density of an alternating field, I_{ex} = excitation current, U_{ex} = excitation voltage, U_o, U_1, U_2, and U_3 = output voltages, U_o (f) = output voltage with frequency depending on B, 1, 2, 3, and 4 = semiconductor crystals, 5 and 6 = coils, 7 = motor, 8 = sliding rings and brushes, 9 = capsule with substance containing lithium or hydrogen, 10 = coil around capsule, 11 = saturable reactors.

Magnetic-flux-density sensors provide an electrical output proportional to the flux density. The most popular device for measuring the flux density is the Hall-effect sensor, in which the excitation current is held constant and the semiconductor crystal placed in the field generates voltage along the axis perpendicular to the direction of the field and current flow. This voltage is proportional to the flux density. The devices are designed as small-size probes, which can contain in their tip one, two, or three crystals for measuring the field in one, two, or three mutually orthogonal transverse directions.

An inductor probe is made of an air-core inductor. The voltage induced in a coil placed in an alternating magnetic field is proportional to the measured flux density. Steady-state magnetic fields can be measured by spinning the coil in the field at a constant speed. The ac voltage developed in the coil due to crossing the lines of the field is a function of the flux density.

A nuclear magnetic resonance magnetometer is based on the effect of the sensitivity of the nuclear resonance frequency of certain substances placed in the magnetic field to the strength of the field. The transduction element consists of a small container with a substance in it such as deuterium or lithium. A coil, excited by a radio-frequency current, is wound around the container. The resonant frequency is identified by a sharp increase in the power consumed by the substance because of the characteristic energy absorption at resonance. Counting the frequency allows measurement of the flux density.

A flux-gate magnetometer, used for measuring extremely small flux densities, is composed of several saturable reactors that are excited by an ac voltage. This voltage keeps the induction in the cores of the reactors close to saturation. When the outside steady field acts against the core, the component of a second-harmonic current is produced in the reactor's circuit. It is noteworthy that this component does not manifest itself at all in the absence of the outside field. With the increase of the field intensity, the second-harmonic voltage can be detected, giving a measure of the flux density. Three mutually orthogonal reactors allow measurement of the flux density along the three special axes.

Nuclear Radiation Detectors

187

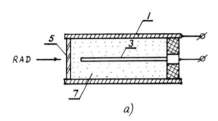

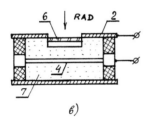

a)

b)

Ionization detector, proportional counter, and Geiger counter. a = chamber with end window, b = chamber with side window; RAD = radiation, 1 and 2 = cathodes, 3 and 4 = anodes, 5 and 6 = windows, 7 = gas.

An *ionization detector* contains a sealed ionization chamber ("ion chamber"), filled with a gas that can produce a substantial ionization potential. Argon, krypton, neon, xenon, helium, and other gases are used for this purpose. Typically, the chamber is made of metal. It carries a negative potential (cathode) and is grounded. The anode, such as a stretched wire, rod, or disk, is mounted inside the chamber. The window, sufficiently transparent to the measured radiation (alpha, beta, gamma, and X-rays), is placed at one of the side- or end-surfaces of the chamber, and allows the radiation to penetrate inside the chamber. When the electrodes are connected to a power supply, and the chamber is exposed to radiation, the average current caused by the ionization or a number of pulses and/or their amplitudes can be measures of the amount of radiation.

A *proportional counter* has a construction similar to that of the ionization detector, but it operates at a different mode. When a low voltage is applied to the electrodes (ion-chamber operation), no additional ions are created by the collision of particles, and the initial ionizing event produces the only electron flow. With the voltage increase (proportional counter operation), the primary electrons produce secondary electrons because of collisions with gas molecules. The characteristic of the counter has a proportional region where the total number of produced electrons is proportional to the voltage across the chamber. In this region, each discharge is a single avalanche, originated by a primary ion pair. The counter's output is proportional to the number of ions formed in the initial ionizing event. This makes it possible to differentiate between alpha and gamma or alpha and beta radiations, since they have different energies of incident radiation and produce different output pulses with amplitudes proportional to the primary ionization.

Geiger counters usually contain a sealed, gas-filled cylinder and a fine axial wire. The cylinder is operated at ground potential and the wire at a positive potential that is higher than that in the proportional counter. In this region, the discharge due to ionization spreads along the whole length of the wire. The total number of the produced electrons does not depend on the amount of primary ionization, and all of the discharge pulses become equal in amplitude. In other words, each particle crossing the cylinder produces a gas ionization that is roughly independent of the particle's nature and energy. In order to have discrete rather than continuous discharges, a quenching agent (e.g., halogen or alcohol vapor) is added to the mixture of gases. The overall ionization is measured by counting the pulses.

188

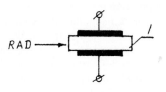

a)

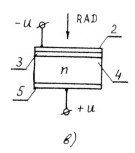

δ)

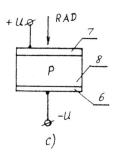

c)

Semiconductor nuclear radiation detectors. a = intrinsic-semiconductor radiation detector, b = surface-barrier type extrinsic-semiconductor radiation detector, c = diffused-junction type extrinsic-semiconductor radiation detector; RAD = radiation, 1 = intrinsic semiconductor (Ge or Si), 2 = evaporated gold contact, 3 = p-type oxide layer, 4 = n-type single-crystal silicon, 5 and 6 = ohmic contacts, 7 = phosphorus-doped n-type layer, 8 = p-type single-crystal silicon.

Semiconductor detectors have several modifications.

a) *Intrinsic-semiconductor detectors* contain pure crystals with electrodes on them. Incident radiation creates electron-hole pairs when a difference in potentials exists between the electrodes. The electrons and holes flow to the electrodes, creating the output current. The typical materials for the detector are germanium and silicon. For this application, the carriers' concentration rather than impurity or imperfections of the material is the determining factor for performance. Electron-hole pairs are produced by gamma- and X-rays because of

1. the elastic scattering of photons by electrons,
2. the ejection of an electron from an atom, and
3. the conversion of a photon into an electron and a positron when the photon moves within a nuclear electric field.

High-purity germanium detectors are intrinsic semiconductors used for the detection of gamma radiation. The detectors require a cryogenic temperature and a vacuum for normal operation.

b) A surface-barrier *extrinsic-semiconductor detector* is made of n-type single-crystal silicon with a p-type layer of silicon dioxide formed on the surface facing the radiation. A thin gold electrode film is evaporated above the silicon dioxide. The back side of the crystal is also covered by a thin golden or aluminum film providing an ohmic contact.

c) A diffused-junction detector contains p-type single-crystal silicon with diffused phosphorus forming an n-type layer on the sensitive surface. The electrode films are deposited on both surfaces of the crystal. Under incident radiation, electron-hole pairs are produced in the depletion region, close to the junction. The depletion depth is defined by the resistivity of the silicon wafer and the voltage applied across the crystal. A p-i-n junction structure (n- and p-regions are separated by an intrinsic region) is used in order to increase the depletion depth, which is adjusted according to the type of particles and their energy: the higher the energy, the greater the depletion depth. A few other materials are used in radiation detectors; for instance, GaAs, CdTe, and HgI_2.

189 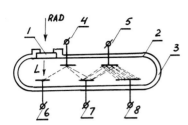 Scintillation detector. RAD = radiation, *L* = light flux, 1 = scintillation phosphor, 2 = photomultiplier, 3 = light-tight enclosure, 4 through 8 = photomultiplier's terminals.	A *scintillation detector* or radiation, commonly called a scintillation counter, consists of a scintillation phosphor optically coupled to a photomultiplier. The particles to be detected excite and ionize phosphor molecules, which radiate light. Some of this light falls on the photocathode of the photomultiplier tube and develops an electrical output proportional to the measured radiation.
190 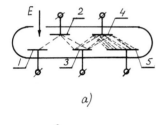 *a)* 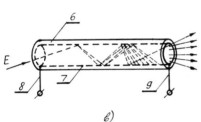 *b)* Electron multipliers. a = multiple-dynode electron multiplier, b = continuous-dynode electron multiplier; *E* = electron flux, 1 = cathode, 2, 3, and 4 = dynodes, 5 = anode, 6 = glass pipe, 7 = resistive-material inner coating, 8 and 9 = terminals.	*Electron multipliers* are utilized in instruments for the detection of electrons. They exist in two major modifications: multiple-dynode and continuous-dynode types. The construction and operation of the multiple-dynode detector is similar to that of the photomultiplier that acts in a vacuum. Each incoming electron that reaches the first dynode ejects several other electrons by secondary emission. This process of multiplication is repeated at each succeeding dynode by applying a higher potential to it than to the preceding dynode. When it arrives at the anode, the electron flow is significantly amplified and can be measured. The continuous-dynode multiplier contains a glass pipe with a coating on its inner surface. This coating is made of a resistive material that provides a secondary emission for incoming electrons. The electron flow moves along the pipe, reflecting from the inner wall and progressively gaining electrons. The electrical field accelerating the flow is formed by the high voltage applied across the two ends of the pipe. The output signals are interpreted similarly to those in a photomultiplier.
191 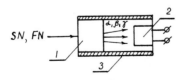 Slow- and fast-neutron detectors. *SN* and *FN* = slow- and fast-neutron radiation, 1 = conversion material, 2 = transduction element for α-, β-, or γ-radiation, 3 = case.	*Slow- and fast-neutron detectors* contain conversion materials that react to incident neutrons by generating charged particles. The materials used for this purpose (a fill gas, coating, foil, etc.) are stable isotopes having a high efficiency of conversion for the given type of radiation. The produced charges are detected by transduction elements having structures similar to those used for α-, β-, γ-radiation. The representative conversion materials for slow-neutron detection are isotopes Li^6 and B^{10}, and, for fast-neutron detection, H^3 and Li^6F.

appendix 2:

Tensors

Anisotropic properties and behavior of matter are conveniently described by tensors [674, 675].

If x_1, x_2, and x_3 are orthogonal coordinates of a certain point M in one system of coordinates, and $x_{1'}$, $x_{2'}$, and $x_{3'}$ are its coordinates in another orthogonal system, the conversion of the coordinates from one system into another is provided with the following formulas:

$$x_{i'} = \alpha_{i1}x_1 + \alpha_{i2}x_2 + \alpha_{i3}x_3 = \sum_{k=1}^{3} \alpha_{ik} \cdot x_k \qquad (i = 1, 2, 3) \qquad (A2.1)$$

where minuses at indices indicate that the quantity relates to a new system of coordinates.

$$\alpha_{ik} = \cos(x_{i'}, x_k) \qquad (A2.2)$$

The reverse conversion is provided with similar formulas.

$$x_i = \sum_{i=1}^{3} \alpha_{il} \cdot x_{i'} \qquad (l = 1, 2, 3) \qquad (A2.3)$$

A scalar is a tensor of rank zero. A tensor of rank one is a set of three quantities, a_1, a_2, and a_3, which is converted into another set $a_{1'}$, $a_{2'}$, and $a_{3'}$, when the system of coordinates is changed from x_k to $x_{k'}$:

$$a_{i'} = \sum_{k=1}^{3} \alpha_{ik} \cdot a_k \qquad (i = 1, 2, 3). \qquad (A2.4)$$

Therefore, any spacial vector is a tensor of the first rank.

A set of $3^2 = 9$ quantities a_{kl} is called a tensor of rank two when the given quantities are converted into new set of quantities $a_{i'j'}$ as the coordinate systems are changed, and the conversion is provided using the formulas

$$a_{i'j'} = \sum_{k=1}^{3} \sum_{i=1}^{3} a_{kl} \cdot \alpha_{ik} \cdot \alpha_{jl} \qquad (i, j = 1, 2, 3). \qquad (A2.5)$$

The higher-ranking tensors can be found similarly. A tensor of rank one or vector $\bar{a}\ \{a_1,\ a_2,\ a_3\}$ can be introduced by a general term a_i. A tensor T of rank two defined by components a_{kl} is written as a matrix:

$$T = \begin{bmatrix} a_{11} & a_{12} & a_{13} \\ a_{21} & a_{22} & a_{23} \\ a_{31} & a_{32} & a_{33} \end{bmatrix}. \tag{A2.6}$$

Similarly, higher-ranking tensors are introduced by higher-order matrices.

For purposes of brevity, the summation symbol (Σ) in the expressions for tensors is usually omitted, and an index that occurs twice is considered as a summation index (Einstein convention). For example, the transformation given in equation A2.4 is written as $a_{i'} = \alpha_{ik}a_k$, and k is the summation index (the term with the repeated index is to be summed for all values of the index).

Quite often, tensors are used for the description of stresses and strains. For example, let us consider a certain volume V inside of a solid that is confined by surface S. Each infinitesimal element of this surface dS experiences an outside force $\bar{p}_n dS$. The force $\bar{p}_n$, referring to the unity area (stress), depends on the orientation with respect to the normal $\bar{n}$ to the surfaces. In general, $\bar{p}_n$ and $\bar{n}$ do not coincide.

Relating $\bar{p}_n$ to the rectangular Cartesian coordinate system $(x_1,\ x_2,\ x_3)$, we can define the corresponding stresses as vectors:

$$\bar{p}_1\{p_{11},\ p_{12},\ p_{13}\}, \qquad \bar{p}_2\{p_{21},\ p_{22},\ p_{23}\}, \qquad \bar{p}_3\{p_{31},\ p_{32},\ p_{33}\}. \tag{A2.7}$$

A stress tensor Π is determined by these three vectors or a matrix composed of the elements p_{ik}.

D'Alembert's principle states that the resultant of the external forces and the kinetic reaction acting on a body equals zero, and the following relationship is valid for any direction of $\bar{n}$:

$$\bar{p}_n = \bar{p}_1 \cos{(\bar{n},\ x_1)} + \bar{p}_2 \cos{(\bar{n},\ x_2)} + \bar{p}_3 \cos{(\bar{n},\ x_3)}. \tag{A2.8}$$

If we want to find a vector of stresses $\bar{p}_{i'}$ related to new axes $x_{1'},\ x_{2'}$, and $x_{3'}$, it can be written as a function of the old vector.

$$\bar{p}_{i'} = \sum_{k=1}^{3} \alpha_{ik}\bar{p}_k \qquad (i = 1,\ 2,\ 3) \tag{A2.9}$$

Assuming that $\{p_{i'1'},\ p_{i'2'},\ p_{i'3'}\}$ are the coordinates of vector $\bar{p}_{i'}$ in the new system of coordinates, $p_{i'j'}$ can be found as the following:

$$p_{i'j'} = \Pi p_{xj'}\bar{p}_{i'} = \sum_{k=1}^{3} \alpha_{ik}\Pi p_{xj'}\bar{p}_k. \tag{A2.10}$$

Since

$$\Pi p_{xj'}\bar{p}_k = \sum_{i=1}^{3} \alpha_{jl}p_{kl}, \qquad \text{therefore,} \tag{A2.11}$$

$$p_{i'j'} = \sum_{k=1}^{3}\sum_{l=1}^{3} \alpha_{ik}\alpha_{jl}p_{kl} \qquad (l,\ j = 1,\ 2,\ 3). \tag{A2.12}$$

The last expressions can be regarded as formulas for the conversion of p_{kl} when the coordinate systems are changed. Since these formulas are identical to formula A2.5, p_{kl} constitutes a tensor.

Another example is a unity tensor J, which is written

$$J = \begin{vmatrix} 1 & 0 & 0 \\ 0 & 1 & 0 \\ 0 & 0 & 1 \end{vmatrix}. \tag{A2.13}$$

Assuming for one coordinate system $a_{ij} = 1$, $a_{ik} = 0$ ($l \neq k$), in any other system we will have

$$a_{i'j'} = \sum_{k=1}^{3} \alpha_{ik}^2 = 1, \qquad a_{i'j'} = \sum_{k=1}^{3} \alpha_{ik}\alpha_{jk} = 0 \qquad (i \neq j), \tag{A2.14}$$

which can be obtained using formulas (A2.5).

A dyad, denoted as $\bar{a}\bar{b}$, is a tensor that is defined by the following matrix, written for two given vectors $\bar{a}$ $\{a_1, a_2, a_3\}$ and $\bar{b}$ $\{b_1, b_2, b_3\}$:

$$\begin{vmatrix} a_1b_1 & a_1b_2 & a_1b_3 \\ a_2b_1 & a_2b_2 & a_2b_3 \\ a_3b_1 & a_3b_2 & a_3b_3 \end{vmatrix}. \tag{A2.15}$$

The values $a_{kl} = a_k b_i$ are modified with formula A2.5 because the expression in A2.4 gives

$$a_{i'} = \sum_{k=1}^{3} \alpha_{ik}a_k \qquad \text{and} \qquad b_{j'} = \sum_{l=1}^{3} \alpha_{jl}b_l. \tag{A2.16}$$

After taking the product $a_{i'}b_{j'}$ and using formula A2.5 we obtain

$$a_{i'}b_{j'} = \sum_{k=1}^{3} \sum_{l=1}^{3} \alpha_{ik}\alpha_{jl}a_k b_i \qquad (i, j = 1, 2, 3), \tag{A2.17}$$

that is with assignment $a_{i'j'} = a_{i'}b_{j'}$, the elements of the matrix are converted with formula A2.5. A dyad $\bar{b}\bar{a}$ differs from the dyad $\bar{a}\bar{b}$:

$$\bar{b}\bar{a} = \begin{bmatrix} b_1a_1 & b_1a_2 & b_1a_3 \\ b_2a_1 & b_2a_2 & b_2a_3 \\ b_3a_1 & b_3a_2 & b_3a_3 \end{bmatrix}. \tag{A2.18}$$

A sum (or difference) of two tensors a_{ij} and b_{ij} of the same rank is a new tensor c_{ij}, which is determined as

$$c_{ij} = a_{ij} + b_{ij}. \tag{A2.19}$$

Tensor a_{ij} is called symmetric if $a_{ij} = a_{ji}$ and skew-symmetric if $a_{ij} = -a_{ji}$. A tensor of rank two can be decomposed into symmetrical and skew-symmetrical parts (the terms in the first and second parentheses of formula A2.20, respectively):

$$a_{ij} = \frac{1}{2}(a_{ij} + a_{ji}) + \frac{1}{2}(a_{ij} - a_{ji}). \tag{A2.20}$$

When a deformation of solids is determined, these parts define separately strain and angular displacements in elements of an object.

A product of two tensors is also a tensor having a rank equal to the sum of the vector's ranks. The component of the product tensor is formed by multiplying each component of one tensor by each component of another tensor. For example, a dyad is a product of two rank-one tensors. A product of a tensor of rank two a_{ij} by a tensor of rank one b_k is a new tensor of rank three, $c_{ijk} = a_{ij}b_k$ ($i, j, k = 1, 2, 3$). The product of two tensors is called *an outer product*.

The process of obtaining one tensor from another is known as *contraction,* which leads to a reduction in the rank of the tensor. For example, sums can be written for a certain tensor, say a_{ijk}:

$$a_{11k} + a_{22k} + a_{33k} = c_k \quad (k = 1, 2, 3). \tag{A2.21}$$

In this case, the contraction is provided with respect to the two indices i and j, and the rank of the new tensor c_k is reduced by two units. A transformation of a tensor of rank two a_{ij} gives a scalar quantity $a_{11} + a_{22} + a_{33}$ called the invariant of the tensor. When the process of contraction is applied to the outer product of two tensors in such a way that one of the two indices involved in the contraction belongs to the first factor and the other to the second, the resulting tensor is termed *the inner product* of the two tensors with respect to the given set of indices. For example, two inner products c_i and c_j' can be obtained from tensor $a_{ij}b_k$:

$$c_i = \sum_{k=1}^{3} a_{ik}b_k \quad \text{and} \quad c_j' = \sum_{k=1}^{3} a_{kj}b_k. \tag{A2.22}$$

These are tensors of rank one; they are vectors. Vector $\bar{c}$ (c_1, c_2, c_3) (as well as $\bar{c}'$) is called a linear vector function of vector $\bar{b}$ (b_1, b_2, b_3). The components of $\bar{c}$ are linear functions of components of $\bar{b}$ and the factors are the components of tensor a_{ik}. Denoting this tensor as T, we can write

$$\bar{c} = T\bar{b}. \tag{A2.23}$$

Tensor T plays the role of an operator when vector $\bar{b}$ is transformed into $\bar{c}$.

In general, tensor calculus, employing different operations (summing, multiplying, contracting, etc.), helps to obtain data related to physical transformations and makes the process independent from the changes in the coordinate systems. The set of a tensor's coefficients is a physical quantity whose presentation (format and numerical values) depends on the coordinate system, but the quantity remains the same with the change of the systems. The following examples illustrate some calculations with tensors:

1. According to formulas A2.1 and A2.2, a vector $\bar{p}$ defined by the components p_1, p_2, and p_3 in one system of coordinates (axes 1, 2, and 3 in Fig. A2.1) can be introduced as $\bar{p}'$ in a new system of coordinates (1', 2', and 3') with the components:

$$\begin{aligned} p_1' &= p_1 a_{11} + p_2 a_{12} + p_3 a_{13} \\ p_2' &= p_1 a_{21} + p_2 a_{22} + p_3 a_{23} \\ p_3' &= p_1 a_{31} + p_2 a_{32} + p_3 a_{33} \end{aligned} \tag{A2.24}$$

    or briefly

$$p_i' = p_j a_{ij}$$

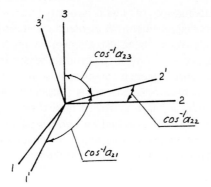

Figure A2.1 Two systems of coordinates and direction cosines.

where a_{ij} are the direction cosines whose meaning is clear from Figure A2.1 and the notations presented in the following table:

$$
\begin{array}{c}
\begin{array}{ccc} 1 & 2 & 3 \end{array} \\
\begin{array}{c} 1' \\ 2' \\ 3' \end{array}
\begin{bmatrix}
a_{11} & a_{12} & a_{13} \\
a_{21} & a_{22} & a_{23} \\
a_{31} & a_{32} & a_{33}
\end{bmatrix}
\end{array}
\qquad (A2.25)
$$

2. The second-rank tensor (see formula A2.5) T_{kl}, having nine components in one system of coordinates, is presented in a new system as

$$T'_{ij} = a_{ik} a_{jl} T_{kl}. \qquad (A2.26)$$

Expanding T'_{ij} with respect to the summation index l first, we obtain [16]:

$$T'_{ij} = a_{ik} a_{j1} T_{k1} + a_{ik} a_{j2} T_{k2} + a_{ik} a_{j3} T_{k3}. \qquad (A2.27)$$

The further expansion with respect to k gives

$$
\begin{aligned}
T'_{ij} = {} & a_{i1} a_{j1} T_{11} + a_{i1} a_{j2} T_{12} + a_{i1} a_{j3} T_{13} + a_{i2} a_{j1} T_{21} + a_{i2} a_{j2} T_{22} \\
& + a_{i2} a_{j3} T_{23} + a_{i3} a_{j1} T_{31} + a_{i3} a_{j2} T_{32} + a_{i3} a_{j3} T_{33}. \qquad (A2.28)
\end{aligned}
$$

As an example, we will calculate the value of component T'_{12}:

$$
\begin{aligned}
T'_{12} = {} & a_{11} a_{21} T_{11} + a_{11} a_{22} T_{12} + a_{11} a_{23} T_{13} + a_{12} a_{21} T_{21} + a_{12} a_{22} T_{22} \\
& + a_{12} a_{23} T_{23} + a_{13} a_{21} T_{31} + a_{13} a_{22} T_{32} + a_{13} a_{23} T_{33}. \qquad (A2.29)
\end{aligned}
$$

3. We have already shown the general application of tensors for the calculation of three-dimensional stresses in solids. The following discussion gives a more detailed picture of the stress state for representative cases.

    We assume that an incremental cube in the vicinity of a point O (Fig. A2.2) of a stressed object has a small volume for which the state of stress

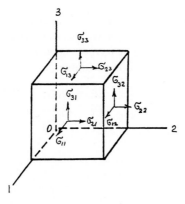

Figure A2.2 Incremental cube under stresses. σ_{ij} = normal stresses when $i = j$, σ_{ij} = shear stresses when $i \neq j$.

might be considered homogeneous. The total stress acting on a cube's face can be resolved into three components: one along the normal to the face (σ_{11}, σ_{22}, σ_{33}), called normal stress, and two in the plane of the face, termed shear stresses (σ_{21}, σ_{31}, σ_{12}, σ_{32}, σ_{13}, and σ_{23}). Normal tensile stresses are considered positive and compressive stresses negative. Shear stresses are said to be positive if they cause the elongation of the face's diagonal that is located between the axes having the same sign. If the diagonal is located between the axes of different signs, the stresses are negative.

Stresses that are the same, but opposite in sense, occur on the hidden faces of the element. The system of forces applied to the cube must satisfy the conditions of equilibrium; therefore $\sigma_{ij} = \sigma_{ji}$ ($\sigma_{12} = \sigma_{21}$, $\sigma_{23} = \sigma_{32}$, and $\sigma_{31} = \sigma_{13}$). These conditions reduce the number of the independent components of the stress tensor from nine to six. In general, the stress tensor is a tensor of rank two having nine components.

The indices of the tensor are simplified as is shown below:

$$\begin{bmatrix} \sigma_{11} & \sigma_{12} & \sigma_{13} \\ \sigma_{21} & \sigma_{22} & \sigma_{23} \\ \sigma_{31} & \sigma_{32} & \sigma_{33} \end{bmatrix} \rightarrow \begin{bmatrix} \sigma_{11} & \sigma_{12} & \sigma_{13} \\ \sigma_{12} & \sigma_{22} & \sigma_{23} \\ \sigma_{13} & \sigma_{23} & \sigma_{33} \end{bmatrix} \rightarrow \begin{bmatrix} \sigma_1 & \sigma_6 & \sigma_5 \\ \sigma_6 & \sigma_2 & \sigma_4 \\ \sigma_5 & \sigma_4 & \sigma_3 \end{bmatrix} \quad \text{(A2.30)}$$

$11 \rightarrow 1$, $22 \rightarrow 2$, $33 \rightarrow 3$, 23 and $32 \rightarrow 4$, 31 and $13 \rightarrow 5$, 12 and $21 \rightarrow 6$.

Following the scheme given in Formula A2.30, we can compose the tensors for three typical cases illustrated in Figure A2.3: one-directional tension (Formula A2.31a), two-directional tension (Formula A2.31b), and uniform compression (Formula A2.31c).

$$\begin{bmatrix} 0 & 0 & 0 \\ 0 & \sigma & 0 \\ 0 & 0 & 0 \end{bmatrix} \quad \begin{bmatrix} \sigma & 0 & 0 \\ 0 & \sigma & 0 \\ 0 & 0 & 0 \end{bmatrix} \quad \begin{bmatrix} -\sigma & 0 & 0 \\ 0 & -\sigma & 0 \\ 0 & 0 & -\sigma \end{bmatrix} \quad \text{(A2.31)}$$
$$\qquad\qquad \text{(a)} \qquad\qquad\qquad \text{(b)} \qquad\qquad\qquad \text{(c)}$$

4. In an isotropic conductor, the vector of current density $\bar{J}$ $\{J_1, J_2, J_3\}$ coincides with the vector of electrical intensity $\bar{E}$ $\{E_1, E_2, E_3\}$ (see Fig. A2.4). If γ is an electric conductivity, the components of the current density are found as follows:

$$J_1 = \gamma E_1, \qquad J_2 = \gamma E_2, \qquad \text{and} \qquad J_3 = \gamma E_3. \quad \text{(A2.32)}$$

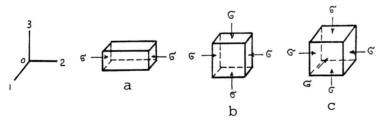

Figure A2.3 Three stress states of elementary parallelepiped. a and b = one- and two-directional tension, c = uniform compression.

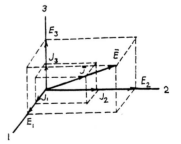

Figure A2.4 Electric field intensity coinciding with current density in isotropic conductor.

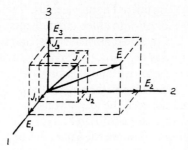

Figure A2.5 Electric field intensity and current density having different directions.

In an anisotropic conductor, Figure A2.5, the conductivity is described by a second-rank tensor containing nine components:

$$\begin{bmatrix} \gamma_{11} & \gamma_{12} & \gamma_{13} \\ \gamma_{21} & \gamma_{22} & \gamma_{23} \\ \gamma_{31} & \gamma_{32} & \gamma_{33} \end{bmatrix} . \tag{A2.33}$$

The components of the current densities are calculated from the equations written in matrix form:

$$\begin{bmatrix} J_1 \\ J_2 \\ J_3 \end{bmatrix} = \begin{bmatrix} \gamma_{11} & \gamma_{12} & \gamma_{13} \\ \gamma_{21} & \gamma_{22} & \gamma_{23} \\ \gamma_{31} & \gamma_{32} & \gamma_{33} \end{bmatrix} \begin{bmatrix} E_1 \\ E_2 \\ E_3 \end{bmatrix} . \tag{A2.34}$$

It can be assumed for many anisotropic materials that $\gamma_{12} = \gamma_{21}$, $\gamma_{23} = \gamma_{32}$, and $\gamma_{13} = \gamma_{31}$.

REFERENCES

Abbreviations

A = Proceedings of the 5th International Conference on Solid-State Sensors and Actuators and Eurosensors III, Part I 1990. Montreux, Switzerland, 1989. In *Sensors and Actuators* B1 (nos. 1–6).

B = Proceedings of the 5th International Conference on Solid-State Sensors and Actuators and Eurosensors III, Part II 1990. Montreux, Switzerland, 1989. In *Sensors and Actuators* A21 (nos. 1–3).

C = Proceedings of the 5th International Conference on Solid-State Sensors and Actuators and Eurosensors III, Part III 1990. Montreux, Switzerland, 1989. In *Sensors and Actuators* A22 (nos. 1–3).

D = Proceedings of the 5th International Conference on Solid-State Sensors and Actuators and Eurosensors III, Part IV 1990. Montreux, Switzerland, 1989. In *Sensors and Actuators* A23 (nos. 1–3).

E = *Proceedings*. 1990. 1990 IEEE Micro Electro Mechanical Systems Workshop. Napa Valley, Calif.

F = *Technical Digest*. 1990. 1990 IEEE Solid-State Sensor and Actuator Workshop. Hilton Head Island, S.C.

G = *IEEE Transactions on Electron Devices* 35 (no. 6). 1988.

H = *Sensor Review* 8(4). 1988. London, U.K.: IFS Publications.

I = Digest of Technical Papers. Transducers' 87. The 4th International Conference on Solid-State Sensors and Actuators. The Institute of Electrical Engineers of Japan, Tokyo, 1987.

IP = *IEEE Proceedings*.

IT = *IEEE Transactions*.

J = *Sensors and Actuators* 10. 1986.

K = *Technical Digest*. 1986. 1986 IEEE Solid-State Sensors Workshop. Hilton Head Island, S.C.

P = Published in St. Petersburg, Russia.

M, MP = Published in Moscow, or Moscow and St. Petersburg, Russia, respectively.

MC = *Measurements & Control* (Journal).

SA = *Sensors and Actuators* (Journal).

1. Norton, H. N. 1982. *Sensor and analyzer handbook*. Englewood Cliffs, N.J.: Prentice-Hall, Inc.

2. Seippel, R. G. 1983. *Transducers, sensors, & detectors*. Reston, Va.: Reston Publishing Co., Inc. (a Prentice-Hall Co.).

3. Middelhoek, S., and Hoogerwerf, A. C. **J**, 1–8.

4. Shannon, C. E. 1948. *Bell System Tech. J.* 27:379–423.

5. Luk'ianov, E. K., and Khazan, A. D. 1967. *Medical Instrument Industry News* 1:110–16. **M**.

6. Luk'ianov, E. K., and Khazan, A. D. 1977. *Medical Research Engineering* 12 (no. 6): 10–19.

7. Ageikin, D. I., et al. 1975. *Transducers for measurements and control*. **M**.

8. Turichin, A. M. 1975. *Electrical measurements of non-electrical quantities*. **MP**.

9. Kvartin, M. I. 1979. *Electromechanical and magnetic devices of automation*. **M**.

10. Feodosyev, V. I. 1973. *Strength of materials*. **M**.

11. Khazan, E. A., and Silvestrovich, S. I. 1968. *Semiconductor Strain Gaging* 2:306–18. Transactions of III Conference, Novosibirsk, Russia.

12. Khazan, E. A., and Silvestrovich, S. I. 1971. *Silicates, Transactions of MKHTI* 68: 67–72. **M**.

13. Davidenkov, N. N., and Shevandin, E. M. 1939. *Journal of Theoretical Physics* (no. 9): 1116. **M.**

14. Osipovich, L. A. 1979. *Transducers of physical quantities.* **M.**

15. Handbuch der Industriellen Messtechnik. Herausgegeben von P. Profos. Essen, Germany: Vulkan-Verlag, 1980.

16. Levshina, E. S., and Novitski, P. V. 1983. *Electrical measurement of physical quantities.* **P.**

17. Davies, C. W., and James, A. M. 1979. *A dictionary of electrochemistry.* London, U.K.: The MacMillan Press Ltd.

18. Shramkov, E. G. 1972. *Electrical measurements.* **M.**

19. Halliday, D., and Resnick, R. 1988. *Fundamentals of physics.* N.Y.: John Wiley & Sons, Inc.

20. Neubert, H. K. P. 1963. *Instrument transducers.* London: Oxford University Press.

21. Ioffe, V. K., et al. 1979. *Handbook on acoustics.* **M.**

22. Kharinskiy, A. L. 1971. *Basic design of radio equipment.* **M.**

23. Plonus, M. A. 1978. *Applied electromagnetics.* N.Y.: McGraw-Hill Book Co.

24. Fedotov, A. V. 1979. *Calculation and design of inductive measuring devices.* **M.**

25. Guru, B. S., and Hiziroglu, H. R. 1988. *Electric machinery and transformers.* San Diego, Calif.: Harcourt Brace Jovanovich, Inc.

26. Kaznelson, O. G., and Edelshtein, A. S. 1970. *Automatic measuring instruments with magnetic suspension.* **M.**

27. *Radiotron designer's handbook.* 1952. Sydney, Australia: Wireless Press.

28. *Radio engineering handbook.* 1959. N.Y.: McGraw-Hill Book Co.

29. Vigoda, U. A. 1962. *Instruments* (no. 10). **M.**

30. Kaman Instrumentation Corp. 1979. *Appl. note 108.* Colorado Springs, Colo.

31. Mikhlin, B. Z. 1960. *High-frequency capacitive and inductive transducers.* **MP.**

32. Bradley, F. N. 1971. *Materials for magnetic functions.* N.Y.: Hayden Book Co., Inc.

33. Bozorth, R. M. 1951. *Ferromagnetism.* N.Y.: D. Van Nostrand Co. Copyright ©, 1951, American Telephone & Telegraph Company, reprinted by permission.

34. Ribalchenko, U. J. 1981. *Magnetostrictive torsion-moment transducers.* **M.**

35. Golubzov, M. G. 1957. *RF electromechanical filters.* **M.**

36. Bell, D. A. 1988. *Fundamental of electric circuits.* Englewood Cliffs, N.J.: Prentice-Hall, Inc.

37. Oliver, F. J. 1972. *Practical instrumentation transducers.* N.Y.: Pitman Publishing Corp.

38. Fedotov, A. V. 1974. *Measuring Technique* (no. 1). **M.**

39. Bessonov, L. A. 1977. *Nonlinear electrical circuits.* **M.**

40. Cushing, V. 1958. *The Review of Scientific Instruments* 29 (no. 2).

41. Khazan, A. D. 1960. *Biomedical Instruments* (no. 3): 16–28. **M.**

42. Voland, G. 1986. *Control system modeling and analysis.* Englewood Cliffs, N.J.: Prentice-Hall, Inc.

43. Stephanopoulus, G. 1984. *Chemical process control.* Englewood Cliffs, N.J.: Prentice-Hall, Inc.

44. Timoshenko, S., et al. 1975. Vibration problems in engineering. N.Y.: John Wiley & Sons.

45. Gusev, E. D. 1965. *Instrument making* (no. 4). **M.**

46. Cussi, G. R., and Tushine, D. R. 1982. **IP,** NAECON.

47. Schuster, G. M. 1975. Resonant pressure instrumentation. Dissertation, N.Y.: Polytechnic Institute of New York.

48. Intraub, J., and Kahn, E. H. 1970. Solid-state, digital type pressure transducer. Papers of 6th International Aerospace Instrumentation Symposium, England: P. Peregrinus, Ltd.

49. Ioffe, V. K., et al. 1954. *Graphs and tables on electroacoustics.* **MP.**

50. Fung, Y. C., et al. 1957. *Journal of the Aeronautical Sciences* 24 (September).

51. Arnold, R. N., and Warburton, G. B. 1953. *Journal and Proceedings of the Institute of Mechanical Engineers* 167. London.

52. Busse, D. W. 1981. **MC** (October): 122–28.

53. Paros, J. M. 1976. *Measurements & Data* (now **MC**, March–April): 74–79.

54. Malov, V. V. 1978. *Piezoelectric resonance transducers*. **M.**

55. Busse, D. W. 1978. Air data symposium 7:1–15. Colorado Springs, Colo.: U.S. Air Force Academy.

56. Weisbord, L. 1969. *Single beam force transducer with integral mounting isolation*. U.S. Patent 3,470,400 (Sept. 30).

57. Albert, W. C. 1982. Proc. 28th Int. Instrum. Symp. (3–6 May): 33–44. Las Vegas, Nev.

58. Ingard, U. 1953. *The Journal of the Acoustical Society of America* 25 (no. 6).

59. Butler, G. A. 1987. *Engineering Design News* (January 8).

60. Eckert, E. R. G., and Drake, R. M. 1988. *Analysis of heat and mass transfer*. Bristol, Penn.: Hemisphere Publishing Corporation.

61. Kreith, F. 1973. *Principles of heat transfer*. N.Y. Reprinted by permission of Harper-Collins Publishers.

62. Martincek, G. 1965. *J. Sound Vib.* 2(2).

63. Chizhevski, K. G. 1977. *Calculation of round and circular plates*. **P.**

64. Ageikin, D. I., and Kuznetsova, N. N. 1961. *Frequency-output transducer of small displacement*. Russia Patent 150,040.

65. Etkin, L. G. 1961. *Measuring Technique* (no. 12). **M.**

66. Roark, R. J. 1965. *Formulas for stress and strain*. N.Y.: McGraw-Hill Book Co.

67. Andreeva, L. E. 1966. *Elastic elements of instruments*. Jerusalem: IPST.

68. Pjatin, U. M. 1977. *Design of element of measuring instruments*. **M.**

69. Osadchij, E. P. 1979. *Design of transducer for measuring mechanical values*. **M.**

70. Jerman, J. H. **D**, 988–92.

71. Bäcklund, Y., et al. **B**, 58–61.

72. Ko, W. H. **J**, 303–20.

73. Miloserdin, U. B., and Lakin, U. G. 1978. *Calculation and design of mechanisms of instruments and devices*. **M.**

74. Feodosyev, V. I. 1949. *Elastic elements of precision instruments*. **M.**

75. Khazan, A. D. 1971. *Medical Instrument Industry News* (no. 2): 10–15. **M.**

76. Khazan, A. D. 1984. **MC** (no. 108, Dec.): 150–53.

77. Khazan, A. D. 1984. Measurement. *Journal of the International Measurement Confederation* 2 (no. 2): 92–97. London, UK.

78. Dynasil. Fused silica. *Dynasil Corporation of America Catalogue* 702-A.

79. Preobrazhenski, V. P. 1953. *Heat-engineering measurements and instrumentation*. **MP.**

80. McDonald, D. A. 1952. *J. Physiol.* 118:328.

81. Womersley, J. R. 1955. *J. Physiol.* 127:553.

82. Taylor, M. G. 1958. *Physics Med. & Biol.* 2:324.

83. McLachlan, N. W. 1955. *Bessel functions for engineers*. Oxford, England: University Press.

84. Baumoel, J. 1984. **MC** (June).

85. Mylvaganam, K. S. 1989. **MC** (Dec.).

86. Lynnworth, L. C. 1975. *Inst. Tech.* (Sept.).

87. Smalling, J. W., et al. 1984. Proc. 39th Texas A & M Instrumentation Symposium for the Process Industries (Jan. 18–20): 27–38.

88. Iberall, A. S. (Trans.). 1950. *Amer. Soc. Mech. Engrs.* 72:689.

89. Humphreys, J. D. 1953. *Tele-Tech.* (April).

90. Cook, N. H., and Rabinowicz, E. 1963. *Physical measurement and analysis*. Reading, Mass.: Addison-Wesley Publishing Co., Inc.

91. Lieneweg, F. 1980. *Handbuch der technischen temperaturmessung.* Braunschweig, Germany: Vieweg.

92. Smith, L. S. 1954. *Physical Review* 94 (no. 1, April).

93. Shockley, W. 1963. *Electrons and holes in semiconductors.* Princeton, N.J.: D. Van Nostrand Co.

94. Morin, F. J., and Maita, J. P. 1954. *Phys. Rev.* 96 (Oct.): 28–35.

95. Morin, F. J., and Maita, J. P. 1954. *Phys. Rev.* 94 (June): 1525–29.

96. Petersen, K. E. 1982. **IP** 70 (no. 5, May).

97. Knott, J. F. 1973. *Fundamentals of fracture mechanics.* London, England: Butterworth.

98. Kontsevoi, U. A., et al. 1982. *Plasticity and strength of semiconductor material and structures.* **M.**

99. Burenkov, U. A., and Nikanorov, K. P. 1974. *FTT* 16 (no. 5): 1496–98. **M.**

100. Burenkov, U. A., et al. 1975. *FTT* 17 (no. 7): 2183–86. **M.**

101. Wortman, J. J., and Evans, R. A. 1965. *J. Appl. Phys.* 36 (no. 1): 153–56.

102. Colclaser, R. A. 1980. *Microelectronics: Processing and device design.* N.Y.: John Wiley & Sons, Inc.

103. Sze, S. M. (Ed.). 1988. *VLSI technology.* N.Y.: McGraw-Hill Book Co.

104. Soclof, S. 1985. *Analog integrated circuits.* Englewood Cliffs, N.J.: Prentice-Hall, Inc.

105. Ten Kate, W. R. T., and Audet, S. A. **I**:103–6.

106. Yen, A., et al. 1990. *Research on microsystem technology.* Annual Report, MIT (May).

107. Bean, K. E. 1978. *IT on Electron Devices* ED-25 (no. 10, Oct.).

108. Seidel, H. **I**:120–25.

109. Seidel, H. **F**:86–91.

110. Clark, L. D., Jr., et al. 1988. IEEE Solid-State Sensor and Actuator Workshop (June): 5–8. Hilton Head Island, S.C.

111. Schnakenberg, U., et al. **D**:1031–35.

112. Hoff, A. M. **F**:52–54.

113. Wu., X., and Ko, W. H. **I**:126–29.

114. Nakamura, M., et al. **I**:112–15.

115. McNeil, V. M., et al. **F**:92–97.

116. Brown, R. B., et al. **E**:77–81.

117. Ikuta, K., et al. **E**:38, 39.

118. Cline, H. E., and Anthony, T. R. 1978. *J. Appl. Phys.* 49:2777.

119. Parfenov, O. D. 1977. *Microcircuit technology.* **M.**

120. Khazan, A. D., et al. 1973. *Biomedical Instrumentation News* (no. 2): 20, 21. **M.**

121. Wallis, G., and Pomerantz, D. I. 1969. *J. Appl. Phys.* 40:3946.

122. Van den Vlekkert, H. H., et al. **I**:730–33.

123. Knecht, T. A. **I**:95–98.

124. Harendt, C., et al. **D**:927–30.

125. Kanda, Y., et al. **D**:939–43.

126. Esashi, M., et al. **D**:931–34.

127. Field, L. A., and Muller, R. S. **D**:935–38.

128. Barth, P. W. **D**:919–26.

129. Huang, Q., et al. **B**:40–42.

130. Lüder, E. **J**:9–23.

131. Obermeier, E., et al. **K.**

132. Takahashi, K. **I**:235–40.

133. Kamins, T. I. **D**:817–24.

134. Walker, J. A., et al. **E**:56–60.

135. Prudenziati, M., and Morten, B. **J**:65–82.

136. Prudenziati, M. **I**:85–90.

137. Doolittle, H. D., and Ettre, K. S. 1966. Ceramic-metallizing tape, U.S. Patent 3,293,072 (Dec. 20).

138. Vita Corporation. *Bulletin* (no. M02). Wilton, Conn.

139. Fu, S., et al. 1984. Proc. 1984 Int. Symp. Microel., 115–20. Dallas.

140. Fu, S., et al. 1985. *Solar Energy Mater.* 12:309–17.

141. Matsumoto, H., et al. 1982. *J. Appl. Phys.* 21 (Suppl. 21-2): 103–107.

142. Matsumoto, H., et al. 1984. *Solar Cells.* 11:367–73.

143. Lachowicz, H. K., and Szymczak, H. J. 1984. *J. Mag. Magnetic Mater.* 41:327.

144. Ikegami, A., et al. 1983. 4th Europ. Hybrid. Microel. Conf., 211–18. Copenhagen.

145. Tien, T.-Y. 1977. Sensor device and method of manufacturing same. U.S. Patent 4,007,435 (Feb. 8).

146. Neuman, M. R. 1982. *Med. Prog. Technol.* 9:95–104.

147. Cobbold, R. S. C. 1970. *Transducers for biomedical measurements* (ch. 9). N.Y.: Wiley Interscience.

148. Kanayama, T., et al. **I**:532–35.

149. Smith, R. L., and Collins, S. D. **D**:830–34.

150. Guckel, H., et al. **K**.

151. Mastrangelo, C. H., et al. **D**:856–60.

152. Smith, R. L., et al. **D**:825–29.

153. Anderson, R. C., et al. **D**:835–39.

154. Shoji, S., et al. **I**:91–94.

155. Tavrow, L. S., et al. **D**:893–98.

156. Lindberg, U., et al. **D**:978–81.

157. Yoshida, A., et al. **I**:249–51.

158. Brown, C. 1990. *Electronic Engineering Times* (Dec. 17).

159. Lenkkeri, J., and Leppävuori, S. **D**:1011–14.

160. Miyoshi, S., et al. **I**:309–11.

161. Suzuki, K., et al. **D**:915–18.

162. Hjort, K., et al. **E**:73–76.

163. LePore, J. J. 1980. *J. Appl. Phys.* 51:6441.

164. Hök, B., et al. 1983. **SA** 4:341.

165. Herb, J. A., et al. **D**:982–87.

166. Bower, R. W. **D**:993–98.

167. Polla, D. L., and Muller, R. S. **J**.

168. Muller, R. S. **I**:107–11.

169. Vellekoop, M. J., et al. **D**:1027–30.

170. Vellekoop, M. J., and Visser, C. C. G. 1988. **IP**, Ultrasonics Symp.: 575–78. Chicago, Ill.

171. Johnsen, C. A., et al. 1988. **IP**, Ultrasonics Symp.: 279–84. Chicago, Ill.

172. Akamine, S., et al. **D**:964–70.

173. Cross, L. E. **E**:72.

174. Chang, S.-C., and Hicks, D. B. **K**.

175. Chang., S.-C. 1979. *IT on Electron Devices* ED-26:1875.

176. Kreider, K., and Tarlov, M. **F**:42, 43.

177. Kreider, K. **K**.

178. Frye, G. C., et al. **F**:61–64.

179. Lazorina, E. I., and Soroka, V. V. 1974. *Sov. Phys. Crystallogr.* 18:651–53. **M**.

180. Ueda, T., et al. 1985. Proc. 3rd Int. Conf. Solid-State Sensors and Actuators— Transducers '85 (June 11–14): 113–16. Philadelphia, Penn.

181. Judge, J. S. 1971. *J. Electrochem. Soc.* 118:1772–75.

182. Vondelig, J. K. 1983. *J. Mater. Sci.* 18:304–14. London: Chapman & Hall.

183. Vig, J. R., et al. 1977. Proc. 31st Ann. Symp. Frequency Control, Monmouth, N.J.: 131. Washington, D.C.: Electronic Industries Association.

184. Tellier, C. R. 1982. *J. Mater. Sci.* 17:1348–54.

185. Bartlett, P. N., et al. **D**:911–14.

186. Buncick, M. C., and Denton, D. D. **F**:102–106.

187. Maseeh, F., and Senturia, S. D. **F**:55–60.

188. Mehregany, M., et al. **K**.

189. Pan, J. Y., et al. **F**:70–73.

190. Knapp, J., et al. **D**:1080–83.

191. Masaki, T., et al. **E**:21–26.

192. Walker, J. A., et al. **B**:243–46.

193. Brown, C. 1990. *Electronic Engineering Times* (Sept. 10).

194. Howe, R. T., and Muller, R. S. 1986. *IT on Electron Devices* ED-33:499–506.

195. Howe, R. T., and Muller, R. S. 1983. *J. Electrochem. Soc.* 130.

196. Moser, D., et al. **D**:1019–22.

197. Mehregany, M., et al. 1987. **SA** 12:341–48.

198. Ikeda, K., et al. **D**:1007–10.

199. Pitcher, R. J., et al. **B**:387–90.

200. Chang, S.-C., et al. **B**:342–45.

201. Mehregany, M., et al. **G**:719–23.

202. Fan, L.-S., et al. **I**:849–52.

203. Behi, F., et al. 1990. **IP** (cat. no. 90CH2832-4).

204. Mehregany, M., et al. **B**:173–79.

205. Huster, R., and Stoffel, A. **D**:899–903.

206. Guckel, H., et al. **I**:277–82.

207. Sandmaier, H., and Kühl, K. **B**:142–45.

208. Shegliang, Z., et al. **I**:130–33.

209. Petersen, K., et al. 1988. Advanced sensor designs. In *Silicon sensors and microstructures*. Fremont, Calif.: NovaSensor Publication.

210. Diem, B., et al. **D**:1003–6.

211. Keller, H. W., and Anagnostopoulos, K. **I**:316–19.

212. Ishida, M., et al. **B**:267–70.

213. Ishida, M., et al. 1988. *Appl. Phys. Lett.* 52:1326–28.

214. Lu, S.-J., et al. **D**:961–63.

215. Kamimura, K., et al. **D**:958–60.

216. Zhao, G., et al. **D**:840–43.

217. Kloeck, B., and de Rooij, N. F. **I**:116–19.

218. Petersen, K., and Barth, P. 1987. Wescon Proceedings: 24/1. San Francisco, Calif.

219. Petersen, K., and Barth, P. 1989. *Conference Record: Wescon/89* (Nov. 14–15): 220. San Francisco, Calif.

220. Bryzek, J., and Mallon, J. R., Jr. 1988. Silicon sensor & microstructure markets. In *Silicon sensors and microstructures*. Fremont Calif.: NovaSensor Publication.

221. Maszara, W. P., et al. 1988. *J. Appl. Phys.* 64:4943–50.

222. Barth, P. W., et al. 1989. Sensors Expo West Proceedings (May 23–25: Paper 206B-1).

223. Mitani, K., et al. **F**:74–77.

224. Horenstein, M. 1990. *Microelectronic circuits and devices*. Englewood Cliffs, N.J.: Prentice-Hall, Inc.

225. Millman, J. 1979. *Microelectronics: digital and analog circuits and systems*. N.Y.: McGraw-Hill Book Co.

226. Lorenz, H., et al. **D**:1023–26.

227. Bousse, L., et al. **I**:99–102.

228. Lambrechts, M., et al. **K**.

229. Wong, A. S., and Cheung, P. W. **K**.

230. Sarro, P. M., et al. **I**:227–30.

231. Sarro, P. M., and van Herwaarden, A. W. 1986. *J. Electrochem. Soc.* 133 (no. 8): 1724–29.

232. Wohltjen, H. **I**:471–77.

233. Huang, P. H. **G**:744–49.

234. Martin, S. J., et al. **I**:478–81.

235. Wenzel, S. W., and White, R. M. **G**:735–43.

236. Martin, B. A., et al. **C**:704–708.

237. Haartsen, J. C. and Venema, A. **C**:675–78.

238. Venema, A., et al. **I**:482–86.

239. Zellers, E. T., et al. **K**.

240. Hauden, D., et al. **IP** 1981 Ultrasonics Symp.: 148–51.

241. Neumeister, J., et al. **C**:670–72.

242. Cullen, D. E., and Reeder, T. M. 1978. Differential surface acoustic wave transducer. U.S. Patent 4,100,811 (July 18).

243. Cullen, D. E., and Montress, G. K. **IP** 1980 Ultrasonics Symp.:696–701.

244. Dias, J. F. 1981. *Hewlett-Packard J.* (Dec.): 18–21.

245. Tiersten, H. F., et al. **PI** 1981 Ultrasonics Symp.:163–66.

246. Bonbrake, T. B., et al. **PI** 1985 Ultrasonics Symp.:591–94.

247. Joshi, S. G. 1983. *Rev. Sci. Instrum.* 54(8):1012–16.

248. Toda, K., and Mizutani, K. 1983. *J. Acoust. Soc. Am.* 74:667–79.

249. Ishido, M., et al. 1987. *IT Instrum. and Meas.* IM-36(1):83–86.

250. Ahmad, N. **IP** 1985 Ultrasonics Symp.:483–85.

251. Kovnovich, S., and Harnak, E. 1977. *Rev. Sci. Instrum.* 48(7):920–22.

252. Clarke, P., et al. **F**:177–80.

253. Adler, R., et al. **IP** 1980 Ultrasonics Symp.:139–41.

254. Wohltjen, H., and Dessy, R. E. 1979. *Anal. Chem.* 51(9):1458–75.

255. Ricco, A. J., and Martin, S. J. **F**:5–8.

256. Ricco, A. J., et al. 1985. **SA** 8:319.

257. Zellers, E. T., et al. **I**:459–61.

258. Huang, P. H. **I**:462–66.

259. Brace, J. G., et al. **I**:467–70.

260. Martin, S. J., et al. **K**.

261. Venema, A., et al. **J**:47–64.

262. Brace, J. G., et al. **K**.

263. Ballantine, D. S., et al. 1986. *Anal. Chem.* 58:3058–66.

264. Roederer, J. E., and Bastiaans, G. J. 1983. *Anal. Chem.* 55:2333–36.

265. Muramatsu, H., et al. **B**:362–68.

266. Howe, R. T. **I**:843–48.

267. Schmidt, M. A., and Howe, R. T. **K**.

268. Sniegowski, J. J., et al. **F**:9–12.

269. Buser, R. A., and de Rooij, N. F. **B**:323–27.

270. Buser, R. A., and de Rooij, N. F. **E**:132–35.

271. Roszhart, T. V. **F**:13–16.

272. Linder, C., and de Rooij, N. F. **D**:1053–59.

273. Guckel, H., et al. **B**:346–51.

274. Elwenspoek, M., et al. Proceedings, 1989 IEEE Micro Electro Mechanical Systems Workshop: 126–32. Salt Lake City, Utah.

275. Lammerink, T. S. J., et al. **B**:352–56.

276. Benecke, W., et al. **I**:838–42.

277. Grosch, G. **D**:1128–31.

278. Stokes, N. A. D., et al. **B**:369–72.

279. Brennen, R. A., et al. **E**:9–14.

280. Tang, W. C., et al. **B**:328–31.

281. Pisano, A. P., and Cho, Y.-H. **D**:1060–64.

282. Behi, F., et al. **E**:159–65.

283. Fan, L.-S., et al. **G**:724–30.

284. MacDonald, N. C., et al. 1989. **SA** 20:123–33.

285. Söderkvist, J. **B**:293–96.

286. Meldrum, M. A. **B**:377–80.

287. Andres, M. V., et al. 1988. **SA** 15:417–26.

288. Ikeda, K., et al. **B**:146–50.

289. Stemme, E., and Stemme, G. **B**:336–41.

290. Kawamura, Y., et al. **I**:283–86.

291. Bouwstra, S., et al. **B**:332–35.

292. Hetrick, R. E. **B**:373–76.

293. Wenzel, S. W., and White, R. M. **C**:700–3.

294. Näbauer, A., et al. **A**:508, 509.

295. Lasky, S. J., et al. **F**:1–4.

296. Robinson, C., et al. **I**:410–13.

297. Yamada, K., et al. **B**:308–11.

298. Tsugai, M., and Bessho, M. **I**:403–5.

299. Kanda, Y., and Yamamura, K. **I**:406–9.

300. Sandmaier, H., et al. **I**:399–402.

301. Dell'Acqua, R. 1989. SAE Conf., (Feb. 27–Mar. 2, 1982). Detroit, Mich.

302. Puers, B., et al. **G**:764–70.

303. Puers, B., et al. **I**:324–27.

304. Leuthold, H., and Rudolf, F. **B**:278–81.

305. Seidel, H., et al. **B**:312–15.

306. Danel, J.S., et al. **D**:971–77.

307. Bill, B., and Wicks, A. L. **B**:282–84.

308. Boxenhorn, B., and Greiff, P. **B**:273–77.

309. Henrion, W., et al. **F**:153–57.

310. Suzuki, S., et al. **B**:316–19.

311. Rudolf, F., et al. **I**:395–98.

312. MacDonald, G. A. **B**:303–7.

313. Satchell, D. W., and Dreenwood, J. C. 1989. **SA** 17:241–45.

314. Starr, J. B. **F**:44–47.

315. McArthur, S. P., and Holm-Kennedy, J. W. **K**.

316. Rudolf, F., et al. **B**:297–302.

317. Allen, H. V., et al. **B**:381–86.

318. De Bruin, D. W., et al. **F**:149–52.

319. Timoshenko, S., and Goodier, J. N. 1970. *Theory of elasticity*. 3rd ed. N.Y.: McGraw-Hill Book Co.

320. Szabo, I. 1984. *Höhere technische mechanik*. Berlin, Germany: Springer-Verlag.

321. Christel, L., et al. **B**:84–88.

322. Reimann, H., and Fathi, Y. **K.**

323. Terry, S. C., et al. **I**:76–78.

324. De Bruin, D. W., et al. **B**:54–57.

325. Mallon, J. R., Jr., et al. **B**:89–95.

326. Bao, M.-H., et al. **B**:137–41.

327. Germer, W., and Kowalski, G. **D**:1065–69.

328. Wu, X.-P. **B**:65–69.

329. Chau, K. H.-L., et al. **F**:181–83.

330. Guckel, H., et al. 1985. *J. Appl. Phys.* 57 (no. 5, March): 1671–75.

331. Guckel, H., Burns, D. W., and Rutigliano, C. R. **K.**

332. Petersen, K., et al. **B**:96–101.

333. Shuwen, G., et al. **B**:133–36.

334. Qinggui, C., et al. **I**:320–23.

335. Kowalski, G. 1987. **SA** 11:367–76.

336. Ansermet, S., et al. **B**:79–83.

337. Shoji, S., et al. **I**:305–7.

338. Hanneborg, A., and Øhlckers, P. **B**:151–54.

339. Suminto, J. T., et al. **I**:336–39.

340. Puers, B., et al. **B**:108–14.

341. Cho, S. T., et al. **F**:184–87.

342. Kjensmo, A., et al. **B**:102–7.

343. Schiller, P., et al. **F**:188–91.

344. Suminto, J. T., and Ko, W. H. **B**:126–32.

345. Schörner, R., et al. **B**:73–78.

346. Neumeister, J., et al. 1985. **SA** 7:167–76.

347. Voorthuyzen, J. A., and Bergveld, P. **I**:328–31.

348. Barth, P. W., et al. **K.**

349. Oki, A. K., and Muller, R. S. **K.**

350. Sugiyama, S., et al. **I**:444–47.

351. Allen, H. V., et al. **I**:448–50.

352. Regtien, P. P. L. **I**:451–55.

353. Bryzek, J. 1987. Sensors Expo Proceedings:279–84.

354. Terry, S., et al. **D**:1070–79.

355. Øhlckers, P., et al. **B**:49–53.

356. Chau, H.-L., and Wise, K. D. **I**:344–47.

357. Puers, B., et al. **D**:944–47.

358. Tenerz, L., et al. **I**:312–15.

359. Trimmer, W. S. N., et al. **I**:857–60.

360. Trimmer, W. S. N., and Gabriel, K. J. **H**:189–206.

361. Mahadevan, R., et al. **B**:219–25.

362. Sakata, M., et al. **B**:168–72.

363. Mehregany, M., et al. **F**:17–22.

364. Tai, Y.-C., and Muller, R. S. **B**:180–83.

365. Muller, R. S. **B**:1–8.

366. Mehregany, M., et al. **E**:1–8.

367. Egawa, S., and Higuchi, T. **E**:166–71.

368. Fujita, H., and Omodaka, A. **G**:731–34.

369. Fujita, H., and Omodaka, A. **I**:861–64.

370. Allen, M. G., et al. **B**:211–14.

371. Brennen, R. A., et al. **F**:135–39.

372. Bart, S. F., et al. **H**:269–92.

373. Fukuda, T., and Tanaka, T. **E**:153–58.

374. Ohnstein, T., et al. **E**:95–98.

375. Huff, M. A., et al. **F**:123–27.

376. Kim, C.-J., et al. **F**:48–51.

377. Pister, K. S. J., et al. **E**:67–71.

378. Tang, W. C., et al. **F**:23–27.

379. Fleischer, M., et al. **B**:357–61.

380. Nakagawa, S., et al. **E**:89–94.

381. Higuchi, T., et al. **E**:222–26.

382. Shoji, S., et al. **B**:189–92.

383. Esashi, M. **B**:161–67.

384. Esashi, M., et al. **I**:830–33.

385. Wakamiya, M. **I**:865–68.

386. Blom, F. R., et al. **B**:226–28.

387. Ono, T. **C**:726–28.

388. Van Lintel, H. T. G., et al. **H**:153–67.

389. Smits, J. G. **B**:203–6.

390. Pourahmadi, F., et al. **F**:78–81.

391. Yang, S. J. E., et al. **E**:136–41.

392. Busch-Vishniac, I. J., et al. **E**:142–46.

393. Riethmüller, W., et al. **I**:834–37.

394. Riethmüller, W., and Benecke, W. **G**:758–63.

395. Parameswaran, M., et al. **E**:128–31.

396. Jerman, H. **F**:65–69.

397. Tabib-Azar, M., and Leane, J. S. **B**:229–35.

398. Zdeblick, M., and Angell, J. B. **I**:827–29.

399. Van de Pol, F. C. M., et al. **B**:198–202.

400. Hale, K. F., et al. **B**:207–10.

401. Neukomm, P. A., et al. **B**:247–52.

402. Bergamasco, M., et al. **B**:253–57.

403. Busch, J. D., and Johnson, A. D. **E**:40, 41.

404. Kuribayashi, K., et al. **E**:217–21.

405. Richter, A., and Sandmaier, H. **E**:99–104.

406. Bart, S. F., et al. **B**:193–97.

407. Moroney, R. M., et al. **E**:182–87.

408. Matsumoto, H., and Colgate, J. E. **E**:105–10.

409. Kim, Y.-K., et al. **E**:61–66.

410. Pelrine, R. E. **E**:34–37.

411. Zdeblick, M. J., et al. **K**.

412. Bassous, E. 1978. *IT on electron devices* ED-25 (no. 10): 1178–85.

413. Gabriel, K. J., et al. **I**:853–56.

414. Linder, C., and de Rooij, N. F. 1989. *Abstract Digest, Transducers '89 Montreux, Switzerland* (289–90). COMST, Lausanne.

415. Pisano, A. P. 1989. **SA** 20:83–89.

416. Gragg, J. E., et al. 1984. Technical Digest, IEEE Solid-State Sensor Workshop (June). Hilton Head Island, S.C.

417. Bao, M., et al. **I**:299–304.

418. Tack, P. C., and Busta, H. H. **K**.

419. Matsuda, K., et al. **B**:45–48.

420. Schubert, D., et al. **H**:145–55.

421. French, P. J., and Evans, A. G. R. **I**:379–82.

422. Mason, W. P., and Thurston, R. N. 1957. *J. Acoust. Soc. Am.* 29 (no. 10, Oct.): 1096–101.

423. Dean, M. 1962. *Semiconductor and conventional strain gages.* N.Y.: Academic Press, Inc.

424. Weodarski, W. 1970. *Electronic Letters* (no. 6).

425. Kadlec, C. 1970. *Electro-technology* (Jan.).

426. Pomerantz, D. I. 1967. *Mechanical and Thermoelectric Transducers* (Dec. 5). U.S. Patent 3,356,915.

427. Diamond, H. 1969. *Electromechanical transducer* (Feb. 11). U.S. Patent 3,427,410.

428. Dorey, A. P., and Maddern, T. S. 1969. *Solid-State Electronics* 12 (no. 3): 185–89.

429. Ricketts, D. 1982. *Electronic Progress, Raytheon Co.* 24 (no. 1).

430. Berlincourt, D. A. 1970. *Electro-Technology* (Jan.):33–38.

431. Tareev, B. M., et al. 1988. *Electroradio materials.* **M.**

432. Morosov, A. I., et al. 1981. *Piezoelectric transducers for the radioelectronic devices.* **M.**

433. *Thermistors.* 1988. Catalog no. 183. N.J.: Thermometrics Co.

434. Anwar, A., et al. 1984. *IT on Instrumentation & Measurement* IM-33 (no. 1, March).

435. Zucker, O., et al. **D**:1015–18.

436. Meijer, G. C. M. **J**:103–25.

437. Slotboom, J. W., and de Graaff, H. C. 1976. *Solid-State Electron.* 19:857–62.

438. Van Herwaarden, A. W., and Sarro, P. M. **J**:321–46.

439. Van Herwaarden, A. W., et al. **C**:621–30.

440. Baciocchi, M., et al. **C**:631–35.

441. Lang, W., et al. **C**:473–77.

442. Kölling, A., et al. **C**:645–49.

443. Wolffenbuttel, R. F. **C**:639–44.

444. Urban, G., et al. **C**:650–54.

445. Aoyama, T., et al. **C**:812–14.

446. Kulczyk, W. K., et al. **C**:663–69.

447. Krummenacher, P., and Oguey, H. **C**:636–38.

448. Wilson, J., and Hawkes, J. F. B. 1989. *Optoelectronics.* Englewood Cliffs, N.J.: Prentice Hall, Inc.

449. Udd, E., et al. 1991. *Fiber-optic sensors.* N.Y.: John Wiley & Sons, Inc.

450. Cheo, P. K. 1990. *Fiber optics & optoelectronics.* Englewood Cliffs, N.J.: Prentice-Hall, Inc.

451. Culshaw, B. **J**:263–85.

452. Smith, R. A., et al. 1968. *The detection and measurement of infrared radiation.* Oxford, U.K.: Oxford University Press.

453. Liangzao, F. and Shuduo, W. **I**:509–12.

454. Mader, G., and Meixner, H. **C**:503–7.

455. Blakemore, J. S. 1974. *Solid-state physics.* Philadelphia: Saunders.

456. Pierce, J. 1956. *Proceedings IRE* 44.

457. Masek, J., et al. **C**:461–64.

458. Stillman, G. E., and Wolfe, C. M. 1977. *Avalache photodiodes.* In *Semiconductors and Semimetals* eds. Willard R. K., and Beer, A. C., 12. N.Y.: Academic Press, Inc.

459. Boslau, O., and von Münch, W. **C**:570–73.

460. Kimata, M., et al. **C**:451–55.

461. Pohjonen, H., and Andersson, M. **D**:1124–27.

462. Okuyama, M., et al. **C**:465–68.

463. Debusschere, I., et al. **C**:456–60.

464. Changchun, Z., and Junhua, L. **C**:469–72.

465. Mincai, H., et al. **C**:494–97.

466. Voorthuyzen, J. A., and Bergveld, P. **A**:350–53.

467. Changchun, Z., et al. **I**:231–34.

468. Collet, M. G. **J**:287–302.

469. Okuyama, M., et al. **I**:500–3.

470. Konov, V. I., et al. **C**:498–502.

471. Erb, K. J. **I**:159–62.

472. Audet, S. A., and ten Kate, W. R. T. **I**:267–70.

473. Clark, S. K., and Wise, K. D. 1979. *IT Elec. Dev.* ED-26:1887–96.

474. Jerman, J. H., et al. **F**:140–44.

475. French, P. J., et al. **C**:414–19.

476. Muro, H., and French, P. J. **C**:544–52.

477. Kato, H., et al. **B**:289–92.

478. Meijer, G. C. M., and Schrier, R. **C**:538–43.

479. Ko, W. H., and Miao, C.-L. **K**.

480. Kawasaki, A., and Goto, M. **C**:534–37.

481. Nishida, K., and Kondo, M. **C**:445–48.

482. Yashiro, H., et al. **I**:423–26.

483. Suzuki, K., and Kuwahara, E. **I**:69–72.

484. Takeuchi, M., et al. **C**:522–24.

485. Xing, Y. Z., and Lian, W. J. **I**:427–30.

486. Kato, H., et al. **C**:581–85.

487. Wolffenbuttel, R. F., and de Graaf, G. **C**:574–80.

488. Wolffenbuttel, R. F. **I**:219–22.

489. Kato, H., et al. **I**:79–82.

490. Glass, A. S., and Morf, R. **C**:564–69.

491. Popović, R. S., et al. **C**:553–58.

492. Wolffenbuttel, R. F. **C**:559–63.

493. Nakano, S., et al. **I**:245–48.

494. Sloper, A. N., et al. **A**:589–91.

495. Lukosz, W., et al. **A**:585–88.

496. Nellen, P. M., and Lukosz, W. **A**:592–96.

497. Schelter, W., et al. **A**:495–98.

498. Mårtensson, J., et al. **A**:134–37.

499. Maeda, M., et al. **A**:215–17.

500. Niessner, R., et al. **A**:261–66.

501. Yokoo, T., et al. **I**:565–68.

502. Sasaki, A., et al. **G**:780–86.

503. Sasaki A. **I**:3–10.

504. Kimura, I. **C**:525–28.

505. Culshaw, B. **I**:185–94.

506. Fluitman, J, and Popma, T. **J**:25–46.

507. Krohn, D. A. **K**.

508. Pask, C., et al. 1975. *J. Opt. Soc. Am.* 65:356.

509. Zhilin, V. G. 1987. *Fiber-optic velocity and pressure transducers.* **M**.

510. Kissinger, C. 1988. **MC** (April).

511. Takamatsu, Y., et al. **C**:435–37.

512. Wahl, J. F. 1988. *Photonic Spectra* (Dec.).

513. Krasjuk, B. A., et al. 1990. *Fiber-optic sensors.* **M**.

514. Christensen, D. A., and Vaguine, V. A. **I**:493–95.

515. Christensen, D. A., and Ives, J. T. 1987. Fiber-optic temperature probe using a semi-conductor sensor. Proc. NATO Advanced Studies Institute: 361–67. Dordrecht, the Netherlands.

516. Knox, J. M. 1983. Birefringent filter temperature sensor. In Proc. First Intern. Conf. Opt. Fiber Sensors: 1–3. London.

517. Nishizawa, K. **I**:585–90.

518. Dils, R. R. 1983. *J. Appl. Phys.* 84:1198.

519. Eickhoff, W. 1981. *Opt. Lett.* 6(4):204–6.

520. Scheggi, A. M., et al. 1983. Proc. 1st Int. Conf. on Opt. Fibre Sensors, (April): 13–16. London.

521. Ramakrishnan, S., and Kersten, R. T. 1984. Proc. 2nd Int. Conf. on Opt. Fibre Sensors (September): 105–10. Stuttgart.

522. Maeda, M., and Katsuyama, M. **D**:1132–36.

523. Ohishi, Y., and Takahashi, S. 1986. *Appl. Opt.* 25:720–23.

524. Kitajima, H., et al. **C**:442–44.

525. Watanabe, M., et al. **I**:208–11.

526. Post, E. I. 1967. *Review of Modern Physics* 39 (no. 2).

527. Culshaw, B., and Giles, J. P. 1983. *J. Phys. Sci. Instr.* 16.

528. Bergh, R. A., et al. 1984. *J. of Lightwave Technology* 2 (no. 2).

529. Oka, K., et al. **C**:438–41.

530. Valette, S., et al. **D**:1087–91.

531. Jaeger, R. E., and Aslami, M. 1987. *Lasers and Optronics* (Oct.).

532. Hall, T. J., and Fiddy, M. A. 1980. *Optics Letters* 5 (no. 11).

533. Russel, A. P., and Fletcher, K. S. 1985. *Anal. Chimica Acta* 170:209–16.

534. Smela, E., and Santiago-Aviles, J. J. 1988. **SA** 13:117–29.

535. Bossche, A., and Mollinger, J. R. **C**:754–57.

536. Gustafsson, K., and Hök, B. **I**:212–15.

537. El-Sherif, M. A., et al. **I**:200–3.

538. Imai, M., et al. **I**:204–7.

539. Poscio, P., et al. **D**:1092–96.

540. Carr, R. J. G. **D**:1111–17.

541. Villuendas, F., and Pelayo, J. **D**:1142–45.

542. Merlo, S., et al. **D**:1150–54.

543. Giuliani, J. F., and Bey, P. P., Jr. **I**:195–99.

544. Kordić, S. **J**:347–78.

545. Nathan, A., et al. **I**:536–38.

546. Soclof, S. 1985. *Applications of analog integrated circuits.* Englewood Cliffs, N.J.: Prentice-Hall, Inc.

547. Kotenko, G. I. 1972. *Magnetoresistors.* **M**.

548. Baltes, H. P., et al. **K**.

549. Popović, R. S. **I**:539–42.

550. Nathan, A., et al. **C**:776–79.

551. Sugiyama, Y., et al. **I**:547–50.

552. Munter, P. J. A. **C**:743–46.

553. Lenz, J. E., et al. **F**:114–17.

554. Dibbern, U. **J**:127–40.

555. Eijkel, K. J. M., et al. **C**:795–98.

556. Lenz, J., et al. **K**.

557. Seitz, T. **C**:799–802.

558. Nakamura, T., and Maenaka, K. **C**:762–69.

559. Maenaka, K., et al. **I**:523–26.

560. Ristić, L., et al. **F**:111–13.

561. Kordić, S., and Munter, P. J. A. **G**:771–79.

562. Zongsheng, L., et al. **C**:786–89.

563. Ristić, L., et al. **C**:770–75.

564. Falk, U. **C**:751–53.

565. Kordić, S., et al. **I**:527–31.

566. Kamarinos, G., et al. 1977. *IEDM Tech. Dig.,* Int. Electron Devices Meet.: 114a–114c. Washington, D.C.

567. Lutes, O. S., et al. 1980. *IT on Electron Devices* ED-27:2156–57.

568. Nathan, A., and Baltes, H. P. **C**:758–61.

569. Nathan, A., and Baltes, H. P. **C**:780–85.

570. Chovet, A., et al. **C**:790–94.

571. Ristić, L. J., et al. 1987. *IEEE Electron Device Letter* EDL-8:395–97.

572. Jasberg, H. **C**:737–42.

573. Kohsaka, F., et al. **C**:803–6.

574. Miyashita, K., et al. **I**:431–33.

575. Maenaka, K., et al. **C**:747–50.

576. Maenaka, K., et al. **C**:807–11.

577. Ristić, L., et al. **I**:543–46.

578. Madou, M. J., and Morrison, S. R. 1989. *Chemical sensing and solid-state devices.* N.Y.: Academic Press, Inc.

579. Janata, J., and Huber, R. J., eds. 1985. *Solid-state chemical sensors.* N.Y.: Academic Press, Inc.

580. Pollak, A., et al. 1985. *C & EN* 25 (no. 28).

581. Weppner, W. 1986. Proceedings of the 2nd International Meeting on Chemical Sensors (July). Bordeaux, France.

582. Jacobsen, S. C., et al. **E**:209–16.

583. Tai, Y.-C., and Muller, R. S. **C**:410–13.

584. Øhlckers, P., et al. **I**:332–35.

585. Seekircher, J., and Hoffmann, B. **C**:401–5.

586. Sugiyama, J., et al. **I**:387–90.

587. Van Herwaarden, A. W., and Sarro, P. M. **I**:287–90.

588. Warkentin, D. J., et al. **I**:291–94.

589. Bergqvist, J., and Rudolf, F. **B**:123–25.

590. Sprenkels, A. J., and Bergveld, P. **I**:295–98.

591. Shu-Duo, W. **D**:883–85.

592. Chen, Y., et al. **D**:879–82.

593. Gao, G., et al. **D**:886–89.

594. Schmidt, M. A., et al. **I**:383–86.

595. Schmidt, M. A., et al. **G**:750–57.

596. Van Oudheusden, B. W., and Huijsing, J. H. **I**:368–71.

597. Van Oudheusden, B. W., and van Herwaarden, A. W. **C**:425–30.

598. Tanaka, N., et al. **I**:352–55.

599. Sekimura, M., and Shirouzu, S. **I**:356–59.

600. Tai, Y.-C., and Muller, R. S. **I**:360–63.

601. Stemme, G. **I**:364–67.

602. Ellis, C. D., et al. **F**:132–34.

603. Ohnstein, T. R., et al. **F**:158–160.

604. Yoon, E., and Wise, K. D. **F**:161–64.

605. Van Oudheusden, B. W., and Huijsing, J. H. **C**:420–24.

606. Stemme, G. N. **K**.

607. Huijsing, J. H., et al. **I**:372–75.

608. Decroux, M. **B**:9–14.

609. Kurakado, M., and Matsumura, A. **B**:33–36.

610. Yamazoe, N., and Shimizu, Y. **J**:379–98.

611. Tsurumi, S., et al. **I**:661–64.

612. Yaoquan, X., et al. **I**:669–72.

613. Tsuchitani, S., et al. **H**:375–86.

614. Yamamoto, T., et al. **I**:658–60.

615. Parameswaran, M., et al. **H**:325–35.

616. Denton, D. D., et al. 1990. *IT on Instrumentation and Measurement* 39 (no. 3, June): 508–11.

617. Shimizu, Y., et al. **I**:665–68.

618. Sadaoka, Y., et al. **I**:673–76.

619. Sakai, Y., et al. **I**:677–80.

620. Ten Kate, W. R. T. **J**:83–101.

621. Kemmer, J. A. **I**:252–57.

622. Wouters, S. E., et al. **C**:513–16.

623. Yabe, M., and Sato, N. **C**:487–93.

624. Audet, S. A., et al. **C**:482–86.

625. Racine, G.-A., et al. **C**:478–81.

626. Hughes, R. C., et al. **K**.

627. Naruse, Y., and Hatayama, T. **I**:262–66.

628. Wada, M., et al. **I**:258–61.

629. Otaredian, T., and Audet, S. A. **C**:517–21.

630. Ji, J., and Wise, K. D. **F**:107–10.

631. Kong, L. C., et al. **F**:28–31.

632. Padmadinata, F. Z., et al. **A**:491–94.

633. Kenny, T. W., et al. **E**:192–96.

634. Hollis, R. L., et al. **E**:115–19.

635. Quate, C. F. **E**:188–91.

636. Oshima, Y., et al. **I**:163–66.

637. Kodato, S., and Nakamura, M. **I**:241–44.

638. Glachino, J. M. **J**:239–48.

639. Senturia, S. D. **I**:11–16.

640. Tai, Y.-C., and Muller, R. S. **E**:147–52.

641. Martin, S. J., and Ricco, A. J. **C**:712–18.

642. Hughes, R. C., et al. **C**:693–99.

643. Akin, T., et al. **F**:145–48.

644. Sakamoto, K., et al. **I**:504–8.

645. Neukomm, P. A., and Kündig, H. **B**:258–62.

646. Hetke, J. F., et al. **D**:999–1002.

647. Maseeh, F., and Senturia, S. D. **D**:861–65.

648. Ding, X., et al. **D**:866–71.

649. Pourahmadi, F., et al. **D**:850–55.

650. Bethe, K., et al. **D**:844–49.

651. Fan, L. S., et al. **D**:872–74.

652. Hälg, B., and Popović, R. S. **D**:908–10.

653. Ding, X., et al. **F**:128–31.

654. Cho, S. T., et al. **E**:50–55.

655. Fan, L.-S., et al. **E**:177–81.

656. Rozgonyi, G. A., and Miller, D. C. 1976. *Thin Sol. Films* 31(no. 2): 185–216.

657. DeSalvo, G. J., and Gorman, R. W. 1987. *ANSYS engineering analysis system user's manual, version 4.3 A.* Swanson Analysis System Inc.

658. Gardner, J. W. **A**:166–70.

659. Maseeh, F., et al. **E**:44–49.

660. Buser, R. A., and de Rooij, N. F. **E**:111, 112.

661. Zhang, Y., et al. **F**:32–35.

662. Harris, R. M., et al. **F**:36–41.

663. Sasada, I., et al. **I**:179–82.

664. Kumar, S., and Cho, D. **E**:27–33.

665. Mahadevan, R. **E**:120–27.

666. Crary, S. B. **K**.

667. Nathan, A. N., et al. **I**:519–22.

668. Guvenc, M. G. **I**:515–18.

669. Senturia, S. D. 1990. *Technical Digest:* 3–7. IEEE IEDM.

670. Senturia, S. D., and Smith, R. L. **H**:221–34.

671. Smith, R. L., and Collins, S. D. **G**:787–92.

672. Koppelman, G. M., and Wesley, M. A. 1983. *IBM J. of Res. and Dev.* 27:149–63.

673. Koppelman, G. M. 1989. IP Micro Electro Mechanical Systems Workshop (MEMS '89): 88–93. Salt Lake City, Utah.

674. Wyle, C. R. 1975. *Advanced engineering mathematics.* N.Y.: McGraw-Hill Book Co.

675. Saverin, M. A., ed. 1951. *Handbook of Mechanical Engineer* 1. **M**.

676. Chang, S. C., and Micheli, A. L. 1985. GM Research Publication GMR-5212.

677. Parfenov, O. D. 1977. *Microcircuit technology.* **M**.

INDEX